The Safety Challenges and Strategies of Using Lithium-Ion Batteries

IEEE Press
445 Hoes Lane
Piscataway, NJ 08854

The Safety Challenges and Strategies of Using Lithium-Ion Batteries

Edited by

Michael G. Pecht
University of Maryland, USA

IEEE Press
WILEY

Published by John Wiley & Sons, Inc., Hoboken, New Jersey.
Published simultaneously in Canada.

Library of Congress Cataloging-in-Publication Data applied for

Hardback ISBN: 9781394342907

Cover Design: Wiley
Cover Image: © marian/Getty Images

Set in 9.5/12.5pt STIXTwoText by Straive, Chennai, India

Contents

The Editors, Authors, and Assistants

Michael G. Pecht (60,000+ citations, 105+ H-index) holds a BS in Physics, an MS in Electrical Engineering, and an MS and PhD in Engineering Mechanics from the University of Wisconsin. He is a Professional Engineer, an IEEE Fellow, an ASME Fellow, an ASM Fellow, and an SAE Fellow. He served on three US National Academy of Science studies, two US Congressional investigations in automotive safety, and as an expert to the US FDA. He is a Distinguished Professor and the Director of the Center for Advanced Life Cycle Engineering (CALCE) at the University of Maryland, which is funded by over 150 of the world's leading electronics companies at more than US$6M/year. He has written more than 30 books on product reliability and supply chain management, and a series of books on the electronics industry in China, Korea, Japan, Taiwan, and India. He has written over 700 technical articles and has 12 patents.

Simin Peng received his PhD degree in electrical engineering from Shanghai Jiao Tong University, Shanghai, China, in 2013. From 2016 to 2017, he was a Visiting Scholar at the University of Maryland, College Park, USA. He is an Associate Professor at Yancheng Institute of Technology, China. His research interests include modeling and control of wind power, microgrids, and battery energy storage systems. He is an IEEE Senior Member and the leader of the Advanced Energy Storage Technology and Applications Research Center at Yancheng Institute of Technology. He is the standing director of the IEEE PES Electric Vehicle Satellite Committee-China Battery System Subcommittee and IEEE PES Energy Storage Committee-China Energy Storage System and Equipment Subcommittee. He has authored over 50 academic papers and holds 40 patents.

Rashed A. Islam is a Design for Reliability Lead at Joby Aviation. His group is responsible for the powertrain and electronics reliability for Joby's revolutionary electric vertical takeoff and landing (eVTOL) aircraft. Previously, Rashed was the Head of Product Integrity at Lyft Bikes and Scooters and Technical Lead Manager at Google Devices and Services, where his team managed the Nest home automation products. He also led the Sustainable Materials Reliability efforts for Google's consumer hardware. Rashed has more than 15 years of industry experience in reliability engineering, failure analysis, and materials development (recycled plastics,

Pb-free solder, piezoelectric, and magnetoelectric materials). Before Google, Rashed worked on e-readers and tablets (at Amazon), LEDs (at Philips), and energy harvesters and dielectric antennas (at Eoplex). Rashed has a PhD in Materials Science and Engineering and has published more than 50 journal and conference publications and book chapters. He has also authored three US patents.

Sanjay Tiku has an MS and PhD in Mechanical Engineering from the University of Maryland, College Park. He currently works as Senior Director of Hardware Engineering at Microsoft. He directs the activities of a team of engineers and researchers to ensure the performance and reliability of complex electromechanical products and to drive product qualification. Previously, he worked at the Research Center of Tata Motors and taught mechanical engineering for a while in India. His research interests include the quality and reliability of electronic products and electronic parts selection and management. He has written several papers and book chapters in this area. He is a member of the IEEE and an invited member of the academic honor society Phi Kappa Phi.

Yan Ning serves as the Battery Performance and Reliability Lead at Dell Technologies, dedicated to delivering reliable and enduring batteries through testing and intelligent health management powered by AI. She collaborates closely with tier-one, established battery suppliers and emerging technology providers, fostering partnerships to integrate revolutionary battery technologies. Yan's expertise extends to telemetry data analysis and machine learning applications for batteries and computer systems, enabling her to diagnose field failures and extract valuable product insights for business optimization. Yan is the Vice Chair of the IEEE P1624 Working Group, where she contributes to updating the standard for Organizational Reliability Capability. Yan holds a PhD in Mechanical Engineering from the Center for Advanced Life Cycle Engineering (CALCE) at the University of Maryland. Yan has six US patents and six US patent applications pending.

Hassan Abbas Khan is the Director of the Energy and Power Systems Lab, a Lead for the Horizon EU project (LOCEL-H2), and a founding member of the LUMS Energy Institute. He received his BSc in Electronic Engineering from GIKI, Pakistan, and his MS and PhD in Electrical and Electronic Engineering from the University of Manchester, UK, in 2006 and 2010, respectively. Currently, he is co-Editor-in-Chief for Elsevier's e-Prime, and a Senior Member of IEEE. He is an expert on batteries and their applications in a spectrum of applications from EVs to grid-scale storage. Dr. Khan also evaluates drivers changing the conventional grid, focusing on storage-based systems and EVs to develop evidence-based policymaking for disruptive technologies in the evolving utility grid. He has contributed to over 100 peer-reviewed research publications and has advised the Government of Pakistan on a 100 MWp flagship solar project in Pakistan and has also led various energy efficiency-related projects at LUMS.

Haibo Huo received a PhD degree in Control Theory and Control Engineering from Shanghai Jiao Tong University, Shanghai, China, in 2008. From 2015 to 2016, she was a visiting scholar at the Center for Advanced Life Cycle Engineering, University of Maryland, College Park, USA. She is currently an Associate Professor at Shanghai Engineering Research Center of Marine Renewable Energy, Shanghai Ocean University, China. Her research interests include fuel cell and hydrogen energy storage system integration, modeling, and control and health. She is an IEEE member and the Standing Director of the IEEE PES China Satellite Technical Committee-Electric Vehicle Committee – Battery System Subcommittee. She has authored over 40 academic papers.

Yulun Zhang is a battery reliability engineer at Microsoft Corporation, where he works extensively on battery-related reliability analyses, accelerated testing, and root cause investigations for Microsoft Surface products. He has over 10 years of experience in the areas of electrochemistry and lithium-ion batteries. He has authored and co-authored more than 10 papers in different journals, such as *Nature Energy* and *Proceedings of the IEEE*. Dr. Zhang received a PhD degree in chemistry from the University of Utah, Salt Lake City, UT, USA, in 2018 and a bachelor's degree in chemistry from Xiamen University, Xiamen, China, in 2014. Before joining Microsoft, Dr. Zhang was a postdoctoral fellow at the Idaho National Laboratory (INL), Idaho Falls, ID, USA, from 2019 to 2021.

Kianoush Naeli is a reliability engineer (CRE). He completed his BS, MS, and Ph.D. in Electrical Engineering. His PhD research at the Georgia Institute of Technology was on static and resonant sensors in Micro-Electro-Mechanical Systems (MEMS). After his PhD, he worked as a postdoc in the Department of Mechanical Engineering at Georgia Tech. He joined Hewlett-Packard (now HP Inc.) in 2010 as a process integration engineer, working on inkjet printheads. In 2013, Dr. Naeli shifted his career toward reliability engineering, while still working on inkjet printers. He worked on the reliability of metamaterial antennas at Kymeta Corporation starting in 2016. He joined Microsoft in 2018, where he has been focusing on the reliability of lithium-ion batteries. Since then, he has been leading the battery reliability of multiple generations of Microsoft Surface products.

Virendra Jadhav received his D.Sc. degree from Washington University in St. Louis. He has been a practicing engineer for over 25 years, in the areas of electronics hardware design, mechanical analysis, and reliability. During his tenure at IBM as a Development Engineer, he worked on stress analysis and testing of electronics and packaging, such as low-K dielectric materials in high end ICs, solder fatigue of BGA packages, and thermomechanical design of high-end CPU packages. In his current capacity as a Reliability Engineer at Microsoft, he has been working on the reliability of high-end personal computers. He is currently the Director of Reliability Engineering at Microsoft in Redmond, WA. Since 2017, he and his team have been actively working on the reliability of lithium-ion batteries.

Weiping Diao earned her PhD in Mechanical Engineering from the University of Maryland, College Park, USA, in 2021. She also holds a BE and ME in Electrical Engineering from Beijing Jiaotong University, Beijing, China. Currently, she is an Assistant Professor in the Department of Electrical and Computer Engineering at Binghamton University. Before joining academia, she spent three years at Apple Inc. as a Battery Algorithm Engineer in Cupertino, USA. Her research focuses on lithium-ion battery technologies, including reliability testing, lifetime modeling, degradation mechanisms analysis, battery management strategies for charging, aging, and imbalances, as well as safety/failure analysis.

Niles Perera is a Professor of Supply Chain Management at the Department of Transport Management and Logistics Engineering, University of Moratuwa, and the Founding Director of the Center for Supply Chain, Operations, and Logistics Optimization. He holds a PhD from the University of Sydney, Australia, and a BSc from the University of Moratuwa, Sri Lanka. An award-winning researcher, he focuses on various facets of supply chain management, which have been featured in some of the leading journals in the field.

Aparna Akula (IEEE Senior Member) is a Principal Scientist at the Center of Excellence for Intelligent Sensors and Systems, CSIR-Central Scientific Instruments Organisation (CSIR-CSIO) Chandigarh, India. She obtained her Master's in Advanced Instrumentation Engineering and PhD in Engineering from the Academy of Scientific and Innovative Research (AcSIR), India. She has published over 50 papers, two granted patents, and two filed patents. She has been an Associate Editor of *IEEE Transactions on Artificial Intelligence* since 2023, and *IET Electronics Letters* (2020–2023). She is the founding Advisor of the IEEE WIE affinity group AcSIR-CSIO student branch. She is an Indian National Academy of Engineering Young Associate.

Quanqing Yu, PhD, is an Associate Professor at the Harbin Institute of Technology, China. He mainly engaged in energy storage/power battery system management and development research. He is currently the Secretary General of the Power Battery Sub-Committee of the IEEE PES Electric Vehicle Technical Committee (China). He is an editorial board member of journals including *Chain*, *Batteries*, *Energies*, *Green Energy and Resources*, *Green Energy*, and *Intelligent Transportation*. He has authorized 15 patents, and published more than 70 journal papers, and 11 of which were selected as ESI. He was named in "the World's Top 2% of Scientists" in 2022 and 2023.

Logeeshan Velmanickam received a BSc degree in Electrical and Electronic Engineering from the University of Peradeniya, Sri Lanka, in 2014, and a PhD degree from North Dakota State University, USA, in 2019. He is currently a Senior Lecturer at the Department of Electrical Engineering, University of Moratuwa, Sri Lanka. He has published several research papers in leading international conferences and journals. His research interests include energy storage systems, power systems, AI and IoT applications, instrumentation, smart sensor development, and lab-on-a-chip development.

İlknur Baylakoglu is currently working as a Senior Chief Researcher and Specialty Engineering Department Head at TÜBİTAK BİLGEM. Her focus is on the reliability, maintainability, testability, and safety of high technology electronics for satellite and avionics systems, where she is the Technical Lead of avionic projects. She has served as Vice Chair of the NATO NIAG project for Reliability Prediction of Electronic Equipment during 2022–2024. She received her BS and MS degrees in Chemical Engineering from Middle East Technical University, Ankara, Turkey. She worked as a Research Scholar with CALCE, University of Maryland during 2000–2002 and 2018. Since 2006, she has been working as an expert project proposal evaluator with the EU Research Executive Agency (REA). She has written over 20 technical articles.

Saurabh Saxena received a PhD degree (2020) in mechanical engineering from the University of Maryland, College Park, USA, and a BTech degree (2011) in electrical engineering from the IIT (BHU), Varanasi, India. He is currently an Assistant Professor at the Center for Automotive Research and Tribology (CART) at IIT Delhi, India. Before this position, he worked as an ML Research Engineer at Apple Inc., Cupertino, USA, and as a postdoctoral researcher at the Applied Materials Division at Argonne National Laboratory, USA. He has been involved in various battery and machine learning research projects for the last 10 years. His research interests include electrochemical energy storage, lithium-ion batteries, battery management systems, machine learning for prognostics and health management, accelerated degradation testing, and physics of failure.

Hayder Ali received a BS degree in Electrical Engineering from the University of Engineering and Technology (UET), Lahore, Pakistan, in 2015. He earned his MS and PhD in Electrical Engineering from the Lahore University of Management Sciences (LUMS), Lahore, in 2018 and 2024, respectively. Dr. Ali has engaged in numerous projects centered around lithium-ion batteries, the installation of solar systems, and various sustainability initiatives. His work includes collaborations with prestigious institutes such as NASA, the Center for Advanced Life Cycle Engineering (CALCE) at the University of Maryland, USA, and Texas A&M University, USA. His ongoing research interests encompass a broad range of topics, including the performance evaluation of lithium-ion batteries and solar photovoltaic systems, among others.

Lingxi Kong earned a PhD degree in Mechanical Engineering from the University of Maryland, College Park, MD, USA. He is currently a Postdoc at the Center for Advanced Life Cycle Engineering (CALCE) at the University of Maryland. He is working on research related to the physics-of-failure-based failure analysis, specifically in lithium-ion batteries. His research interests include lithium dendrite growth, battery supply chain, safety and protection devices, and lithium-ion battery testing. Dr. Kong is a committee member of SAE International. He contributes to establishing standards for mitigating battery-related risks.

Mohammed Saquib Khan earned his PhD in Materials Engineering from the Materials Research Centre at Malaviya National Institute of Technology, Jaipur. Currently, he is engaged in postdoctoral research at the Indian Institute of Technology Kanpur, where he is pioneering the development of next-generation energy storage systems, with a particular focus on room-temperature solid-state lithium-sulfur and sodium-sulfur batteries. His research is instrumental in advancing battery technology, with an emphasis on enhancing performance, safety, and sustainability. Dr. Khan has a background in nanomaterial growth for energy storage applications and in in-situ and operando characterization techniques, which are crucial for the development of more robust and safer batteries. Dr. Khan has authored over 16 articles and holds one granted patent with two additional patents filed.

Manoj Kumar Lohana is a forward-thinking business student specializing in Supply Chain Management, blending over five years of impactful writing with expertise in administrative support, market research, and strategic planning. Known for his organizational prowess and ability to thrive under pressure, Manoj brings a fresh perspective to logistics and operational strategies, aiming to revolutionize how organizations approach supply chain complexities.

Sahithi Maddipatla received a B.E. degree in Electrical and Electronics Engineering from the Visvesvaraya National Institute of Technology, Nagpur, India, in 2020. She is currently a PhD candidate in Mechanical Engineering at the University of Maryland, College Park. She is working as a Graduate Research Assistant at the Center for Advanced Life Cycle Engineering (CALCE). Her research interests include failure analysis, testing, prognostics, and health monitoring of lithium-ion batteries. Her current work focuses on the swelling mechanisms in pouch-type lithium-ion cells.

Shalini Dwivedi is currently pursuing a PhD in Engineering Sciences at the Academy of Scientific and Innovative Research, Ghaziabad, India. She has been awarded the CSIR-HRDG GATE-SRF fellowship by the Council of Scientific and Industrial Research, India. She holds a Master's degree (MTech) in Advanced Instrumentation. An active IEEE member since 2021, she chaired the Women in Engineering (WIE) affinity group of the IEEE AcSIR-CSIO student branch from 2022 to 2023. In the following year, from 2023 to 2024, she served as the Secretary for the IEEE Sensors

Council of the student branch. Her research interests include the integration of machine learning techniques for real-time data analysis in battery health monitoring systems.

Thisali S. Rathnayake (Member, IEEE) is a graduate of the University of Moratuwa, Sri Lanka. She received her Bachelor of Science degree in Electrical Engineering in 2023. Her research interests include energy storage systems and automation systems.

Dinithi Senarath is a graduate student at the University of Moratuwa, Sri Lanka. She obtained her Bachelor of Science in Engineering specializing in Electrical Engineering in 2023. Her research interests include power systems, energy storage systems, and optimization.

Gifty Pamela Afun graduated with a BSc degree in Electrical/Electronics from Regional Maritime University, Ghana, in 2018. She has five years of experience as a Production Supervisor in the assembly of electric vehicles. She is currently enrolled in the Mechanical Engineering program at the College of Engineering in Shanghai Ocean University, Shanghai, China. Her research interests revolve around solid-state batteries and hybrid energy management. She has actively contributed to advancing gender equality in STEM, notably by facilitating training for women in engineering roles.

Raveen Sanjaya De Silva (Member, IEEE) is a PhD candidate of University of Technology Sydney, Australia. He is engaged in energy storage systems management and development research for residential microgrids. He obtained his Bachelor of Science in Engineering from the University of Moratuwa, Sri Lanka, specializing in Electrical Engineering. His research interests include modeling and control of energy storage, microgrid, and battery energy storage system.

Prabhashi Amanda Andrahennadi is a scholar in the field of sustainable energy and environmental engineering and currently serves as a PhD student in the MeSSO (Mechanical Energy System Simulation Optimization) group at Munster Technological University, Ireland. Her research is primarily focused on CO_2 reduction in Irish dairy and beef farms through optimizing renewable systems and through the application of life cycle analysis. Amanda received her BSc (Hons) degree in Electrical Engineering from the University of Moratuwa, Sri Lanka, in 2022.

Nipun Iranga Wijesekara is a student in the Department of Transport Management and Logistics Engineering at the University of Moratuwa, Sri Lanka. His research interests lie in the supply chain domain.

Dulja Bamunusinghe (Member, IEEE) is a graduate student at the University of Moratuwa, Sri Lanka. She obtained her Bachelor of Science in Engineering in Electrical Engineering. Her research interests include high voltage, power systems, and battery production.

Jinkang Chen is currently pursuing his Master's degree in Energy and Power Engineering at the School of Electrical Engineering, Yancheng Institute of Technology, China. His research interests include battery system modeling and battery power estimation.

Genkai Xia received his BE degree in Electrical Engineering and Automation from the Qingdao University of Technology, Qingdao, China, in 2020. He is an MS degree candidate in Energy and Power Engineering at the School of Electrical Engineering, Yancheng Institute of Technology. His research interests include failure analysis, testing, prognostics, and capacity monitoring of lithium-ion batteries.

Yue Shen is a graduate student in Energy and Power Engineering at the School of Electrical Engineering, Yancheng Institute of Technology, China. Her research interests include lithium-ion battery modeling and energy state estimation.

Yu Zhang is currently pursuing a Master's degree in Energy and Power Engineering at the School of Electrical Engineering, Yancheng Institute of Technology, China, where his research interests include battery health state estimation, energy storage systems, and optimization.

Jie Wu is a graduate student in the Energy and Power Engineering at the School of Electrical Engineering, Yancheng Institute of Technology, China. He obtained his Bachelor of Science in Engineering specializing in Electrical Engineering in 2022. His research interest lies in the battery energy storage system.

Guanwei Jiang is a graduate student in the Energy and Power Engineering at the School of Electrical Engineering, Yancheng Institute of Technology, China. His research interests include battery modeling and the remaining useful life estimation of lithium-ion batteries.

Yuwei Nie is a master's student in the Electrochemical Energy Storage Safety Laboratory, School of Automotive Engineering, Harbin Institute of Technology, Weihai, Shandong, China. In 2022, he received his bachelor's degree from the Wuhan University of Technology, China. He has been committed to researching data-driven intelligent safety management of battery systems, and his research results have been published in multiple SCI journals such as *Applied Energy* and *Journal of Power Sources*.

Preface

Lithium-Ion Batteries: A History of Fires and Explosion Incidents – Lessons Learned and To Be Learned

Lithium-ion (Li-ion) batteries (LIB) were introduced into the market in the 1980s and have now become widely used in electronic devices, including smartphones, laptops, electric bikes, scooters, electric vehicles, and energy storage systems. Although they have high energy density and environmental benefits, safety is a top concern. Numerous lithium-ion battery fire and explosion incidents have been reported, resulting in property damage and bodily injuries. These incidents were primarily due to thermal runaway, where chemical reactions cause the battery to reach temperatures as high as 900 °C, resulting in fires that can easily spread and explode due to gas generation in the batteries [1].

One of the earliest and most notable fire incidents that brought significant attention to the safety concerns of lithium-ion batteries occurred with laptops in the mid-2000s. In 2006, Dell announced a recall of approximately 4.1 million lithium-ion laptop batteries in cooperation with the US Consumer Product Safety Commission (CPSC) [2]. This recall was initiated after reports that the laptop battery could overheat, causing a fire hazard to consumers. The batteries were manufactured by Sony and were used in laptops sold between April 2004 and July 2006 [3]. Following Dell's recall, Apple and other manufacturers also initiated recalls for similar reasons, with Apple recalling 1.8 million devices with batteries manufactured by Sony [4]. The root cause of these incidents was traced back to manufacturing defects that could lead to internal short circuits and thermal runaway, which at that time was referred to as spontaneous combustion [5].

Starting in the late 2000s, incidents related to lithium-ion battery fires in electronic cigarettes were reported more regularly [6]. Investigations identified 195 incidents of e-cigarette fires and explosions in the United States between 2009 and 2016, and among these incidents, 61 occurred in pockets, 60 occurred when devices were in use, 48 occurred when batteries were charged, and 18 occurred in storage [6]. For example, in 2016, an e-cigarette battery exploded in a man's pocket in New York, causing severe leg burns [7]. In 2018, an e-cigarette battery exploded when in use, causing an instant death [8].

In the 2010s, there were many notable lithium-ion fire incidents. In 2010, a UPS cargo plane transporting batteries crashed in Dubai due to a battery fire, killing two crew members [9]. In 2011, an Asiana Airline cargo plane crashed due to a fire in a battery pack, causing the deaths of two pilots [10]. In 2013, Boeing 787 Dreamliners were grounded worldwide due to two battery fires within nine days of each other, where one required an emergency landing [11]. In 2016, Samsung Galaxy Note 7 smartphones were recalled and production was stopped after the devices burst into flames due to battery faults [12]. In 2018, HP recalled 50,000 laptop and mobile workstation batteries because of battery problems that could result in fires [13]. In 2019, Apple recalled about 458,000 MacBook Pro laptop computers due to fires caused by the batteries [14].

Although there have been numerous improvements in technology and safety standards, incidents have continued to occur, including fires in electric bikes, scooters, and electric vehicles. The Fire Department in New York (FDNY) reported that there were 28 fires with 16 injuries in 2019; 104 fires, 79 injuries, and 4 deaths in 2021; 216 fires, 147 injuries, and 6 deaths in 2022; and 268 fires, 150 injuries, and 18 deaths in 2023 [15], all due to lithium-ion batteries. In February 2024, a battery fire from e-bikes occurred on a residential block, causing 44 injuries and 15 deaths [16]. In March 2024, there were two battery fires on the campus of the University of Maryland due to e-scooters, and students were told not to bring them into buildings anymore.

In 2013, a Tesla Model S involved in high-speed travel resulted in an accident, in which the vehicle caught fire and its front end burst into flames [17]. In 2019, a Tesla Model S caught fire in a garage [18]. In 2020, an electric Porsche Taycan caught fire while parked in a garage [19]. In 2022, Chevrolet recalled 142,000 Bolt electric vehicles due to battery fires caused by defects in LG cells, resulting in a cost of more than $1.8 million [20]. From February 24, 2020, to March 16, 2022, Mercedes Benz, Hyundai, General Motors, Chrysler, and Volkswagen notified the National Highway Traffic Safety Administration (NHTSA) about safety-related battery defects in their EVs and conducted multiple recalls, including the 2019 Smart Fortwo, 2017–2018 Pacifica plug-in hybrid EV, 2017–2022 Chevrolet Bolt EV, 2019–2022 Kona, 2020 Ioniq, 2021 ID4, and 2022 Chevrolet Bolt EUV [21]. In 2023, Stellantis recalled over 90,000 plug-in hybrid electric vehicles (PHEVs) due to battery fires [22].

On July 23, 2024, a truck carrying lithium-ion batteries (container weight exceeding 75,000 pounds) overturned and caught fire near Baker, California, in the northbound lanes of I-15, a major highway to Las Vegas. Due to the fire and hazardous materials and chemicals involved, both northbound and southbound lanes were closed. Drivers were trapped on the highway in 109 °F (43 °C) heat for hours on Friday and Saturday before the lanes were reopened.

On August 1, 2024, an electric Mercedes-Benz sedan caught fire while parked in the underground parking lot of an apartment building in Incheon, South Korea [23]. This incident damaged some 140 cars, led to the evacuation of more than 200 families (more than 700 residents) and injured 23 people. Following the massive fire caused by the Mercedes-Benz vehicle, consumer calls for mandatory disclosure of electric vehicle battery information have been growing in South Korea. In addition, some communities are considering banning electric vehicles from underground parking areas when their charge exceeds 90%, due to the increased risk of fire.

On August 19, 2024, California firefighters had to douse a flaming battery in a Tesla semi-truck with about 50,000 gallons (190,000 L) of water to extinguish flames after a crash on a freeway in California. Firefighters also used an aircraft to drop fire retardant on the "immediate area" of the electric truck as a precautionary measure. Firefighters said that the battery reached temperatures of 1000 °F (540 °C) while it was burning. The freeway was closed for about 15 hours as firefighters made sure the batteries were cool enough to recover the truck (https://www.ntsb.gov/investigations/Pages/HWY24FH015.aspx).

In early September 2024, BMW announced they were recalling over 140,000 Mini Cooper SE electric vehicles due to potential battery defects that could lead to fires, according to documents filed with the NHTSA. The recall affects models produced from 2020 to 2024, when the batteries may have been manufactured outside production specifications (https://www.nj.com/business/2024/09/bmw-recalls-140k-mini-cooper-se-vehicles-worldwide-if-you-have-one-of-these-vehicles-bring-it-to-your-dealer-for-free-repairs.html).

The Federal Aviation Administration (FAA) reported that there were 510 incidents due to lithium-ion batteries, involving smoke, fire, or extreme heat, from 2006 to 2024 [24]. This amounts to an average of two fires in airplanes per month. However, newer FAA statistics show that the annual lithium battery incidents number has surpassed 52, which means there is more than one incident per week on average [25].

Fire incidents have also been a concern in the waste management and recycling of lithium-ion batteries. There have been reported battery fires in waste containers, transportation hubs, intermediate sites, and end-of-life facilities [26]. In 2007, a fire occurred at a waste treatment plant in Preston, United Kingdom, caused by lithium-ion batteries [27]. In 2007, a massive explosion in Canada occurred due to batteries, and an entire waste processing facility was destroyed [28].

Battery fires have not been limited to the Americas. In January 2024, Health Canada started recalling batteries for Dyson vacuums that were sold in Canada from Amazon due to risks of fire and burns. Specifically, the recall involved the FirstPower 4000 mAh V6 Battery Replacement for Dyson V6 Series Vacuums [29]. On June 24, 2024, in Hwaseong, South Korea (28 miles south of Seoul), a lithium battery factory owned by Aricell caught fire after several batteries exploded. The Aricell factory housed an estimated 35,000 battery cells on its second floor, where the batteries were inspected and packaged, with more cells stored elsewhere. The fire killed 22 workers and wounded 8 more.

In China, where the largest percentage of lithium-ion batteries are made, there has clearly been the most incidents of battery fires. In June 2020, NIO recalled 4803 ES8 electric vehicles due to fire concerns in their battery packs. In the first half of 2021, there were about 1.28 million EVs that were recalled due to safety-related battery defects, and from January 2021 to July 2021, seven Chinese EV manufacturers conducted recalls for a total of 56,399 EVs due to defective batteries [30].

In 2022, there were 18,000 reported e-bike fires in China, which increased by 23% compared to 2021. Also in 2022, there were 3242 reported fire incidents caused by rechargeable batteries (mostly for e-bikes) that occurred inside places of residence, which increased by 17% compared to 2021. In the first quarter of 2022, there were 3777 e-bike fires, which resulted in a year-on-year increase of 36% [31]. In 2023, there were 21,000 fires, a 17.4% increase from the previous year.

By 2023, there were 1496 major lithium-ion battery manufacturers in China. According to the vice director of the national testing center in China, 20 out of 100 projects failed to pass basic testing requirements due to quality issues, including 14 that caused fires. This also includes a battery manufacturing facility in Guangdong, which caught fire due to defective batteries on May 11, 2023 [32]. In January 2024, a battery fire caused the deaths of 13 schoolchildren [33]. A few days later, 39 died in another battery fire [34], which led President Xi to state that more needs to be done to prevent such fires [35]. In February 2024, an apartment in Nanjing, China caught fire due to batteries in electric bikes, killing 15 people and injuring more than 40. On May 23, 2024, due to the increasing number of e-bike fires, the Chinese government had a nationwide crackdown and assigned over 74,000 police to seize more than 300,000 e-bikes [36]. On August 23, 2024, a fire broke out at a street shop in Jiangsu, resulting in six deaths and reducing several electric bikes to just their metal frames. On August 28, 2024, a battery fire broke out at an electric bike shop in Foshan, which spread to a nearby grocery store and injured 10 people [37]. This is only a small snapshot of the battery fires in China.

Although lithium-ion batteries have been considered a mature technology, there are consequential safety risks that continue to surface. Part of the problem arises from the widespread proliferation of batteries in everything that we use. However, another part of the problem stems from batteries being a commodity item that are relatively easy to make but with severe safety risks if not made in a high-quality manner. For example, recently, more e-bikes, e-scooters, power banks, and vaping devices have caught fire, and one may ask why this is the case. While this has not been extensively studied yet, there are several likely reasons.

First, EV makers tend to more carefully choose the companies that will make the batteries for their cars. The battery companies typically make many billions of batteries per year and can assemble battery packs as well. These companies most often have facilities that resemble semiconductor clean rooms with inline manufacturing, and they use robotics with many automated quality control methods in place, including optical inspections and measurements of material characteristics.

The companies that make many of the e-bikes, e-scooters, power banks, and vaping devices are typically trying to market a low-cost product, frequently to students and young people, in some cases as a toy. These companies often do not know much about battery reliability and safety, and they typically depend on third-party battery pack manufacturers who buy the batteries (cells) at the lowest possible price. The manufacturers of these battery cells have, in many cases, factories that resemble garages, with a few workers moving materials around, often on many floors of a building. This type of environment can easily result in foreign materials getting into the battery, resulting in opportunities for failure and a resulting fire and explosion.

A second reason, although significantly less of a concern, is that some of the e-bikes, e-scooters, power banks, and vaping devices do not have battery management systems (BMSs). The lack of a BMS, or an improper BMS, may be a result of cost savings by a pack assembler or the final product manufacturer. However, because of the low cost of high-quality BMSs, this has become less of a problem, except in low-cost items.

A third reason pertains to abuse and the blaming of users, battery pack assemblers, and final product manufacturers. For example, overcharging has often been blamed on the users, but charge and discharge profiles and safety are controlled by the BMS, and most e-bikes, e-scooters, and power banks have BMSs with these types of safety-related controls. However, any battery that is abused in the sense of punctures, during pack assembly, storage, transportation, use, and disposal, can cause fires. Unlike in Tesla electric vehicles, where the battery pack is enclosed in a titanium case, most of the e-bikes, e-scooters, power banks, and vaping devices are enclosed in less costly materials like plastics.

This book aims to address the state of practice of lithium-ion batteries with a focus on safety. The reader is presented with a range of companies that are producing batteries, the various lithium-ion chemistries being implemented in batteries by these companies, and which chemistries are being used for various applications. Next, the numerous defects in design and manufacturing that can affect the propensity for fires are presented along with best practices. This is followed by qualification tests, quality assurance methods, and standards needed to ensure a safe design and address the storage and transportation of batteries. Subsequently, an overview of supply chain risks is presented, and the safety risks due to re-wrapped and counterfeit batteries are discussed. Then, there are discussions about how new regulations in the European Union concerning the ability to replace batteries and the right to repair will affect safety risks. Finally, there are discussions on the reuse, repurposing, and disposal of batteries.

After reading this document, the reader should have the information to be able to:

- **Know the industry,** having knowledge of its history and progression, current landscape, and future outlook, including manufacturing capacities, geographic distribution, and demand trends.
- **Analyze batteries with confidence,** learning about the types of batteries, their chemistries, failure modes and mechanisms, and the tradeoff between energy density and safety risks. By understanding the advantages and disadvantages of different cathode materials, anode materials, electrolytes, and separators, readers can make informed decisions when selecting batteries for specific applications.

- **Future-proof your business,** understanding the battery supply chain in the consumer, industrial, electric vehicle, and renewable energy sectors, and the pivotal roles played by key stakeholders. From cell manufacturers to gauge makers, pack houses, and product companies, understand how each entity contributes to safety assurance and risk mitigation throughout the stages from acquiring cells to delivering final products, identifying potential risks and vulnerabilities along the way. Emphasize how audits are crucial in battery manufacturing for quality control and risk management, enhancing safety and user confidence.
- **Ensure product safety and compliance,** gaining insights into the various safety mechanisms (current interrupt device, positive temperature coefficient devices, vents, separators, and flame retardants) that are and are not implemented in batteries, and the value and lack of value of regulations and safety standards governing batteries and their integration into products and systems. Gain insights into best practices for designing products that prioritize consumer safety while maintaining efficiency and performance.
- **Know the safety testing protocols,** exploring the evolution of safety test standards and their impact on design practices within the battery industry. Understand the intricacies of cell, pack, and product safety testing protocols, equipping yourself with the knowledge to assess and ensure compliance with rigorous safety standards. At the same time, understand the limitations of these tests and the fact that they do not guarantee the prevention of fires.
- **Mitigate safety risks,** learning about methods to identify and address potential battery material and production risks, and the storage and transportation of batteries, battery packs, and products incorporating batteries. The reader will also learn about data transparency challenges between manufacturers, end-users, and system designers, pointing out weaknesses in how manufacturers present information in data sheets about battery performance and storage for safe and reliable use.
- **Promote safe sustainability,** understanding the risks of repairing, reusing, repurposing, and recycling batteries.

Michael G. Pecht, Ph.D.

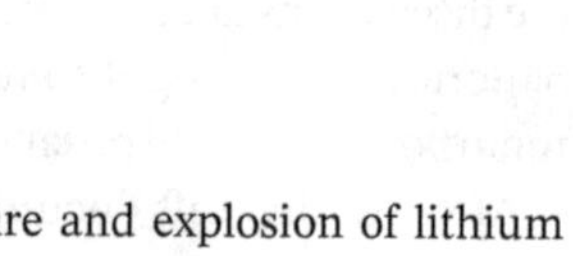

Distinguished Professor and Professional Engineer
University of Maryland, USA

References

1 Wang, Q., Ping, P., Zhao, X. et al. (2012). Thermal runaway caused fire and explosion of lithium ion battery. *Journal of Power Sources* 208: 210–224. https://doi.org/10.1016/j.jpowsour.2012.02.038.

2 U.S. Consumer Product Safety Commission (2024). Dell Announces Recall of Notebook Computer Batteries Due To Fire Hazard. https://www.cpsc.gov/Recalls/2006/dell-announces-recall-of-notebook-computer-batteries-due-to-fire-hazard (accessed 04 March 2024).

3 Darlin, D. (2006). Dell recalls batteries because of fire threat. *The New York Times* (14 August). https://www.nytimes.com/2006/08/14/technology/14cnd-battery.html (accessed 04 March 2024).

4 Darlin, D. (2006). Apple recalls 1.8 million laptop batteries. *The New York Times* (24 August). https://www.nytimes.com/2006/08/24/technology/23cnd-apple.html (accessed 04 March 2024).
5 CNET. Apple recalls 1.8 million batteries. https://www.cnet.com/culture/apple-recalls-1-8-million-batteries/ (accessed 04 March 2024).
6 McKenna, L.A. Jr. (2019). Electronic cigarette fires and explosions in the United States 2009–2016.
7 CNN. Man says e-cigarette battery exploded in his pocket. https://www.cnn.com/2016/02/25/health/e-cigarette-explodes-in-mans-pocket/index.html (accessed 04 March 2024).
8 NBC News. These batteries were never intended for use in e-cigarettes. They can cause deadly explosions. https://www.nbcnews.com/health/health-news/battery-behind-dangerous-deadly-e-cigarette-explosions-n1032901 (accessed 04 March 2024).
9 Signal Integrity Journal (2011). Did lithium batteries cause crash of UPS plane? *Security Magazine* (05 April). https://www.signalintegrityjournal.com/articles/22-did-lithium-batteries-cause-crash-of-ups-plane (accessed 04 March 2024).
10 Sang-Hun, C. (2011). 2 Die as Asiana Cargo Plane Crashes Off South Korea. *The New York Times* (28 July). https://www.nytimes.com/2011/07/29/world/asia/29crash.html (accessed 04 March 2024).
11 Simple Flying. A Look At The Battery Issues That Grounded The Boeing 787 In 2013. https://simpleflying.com/boeing-787-battery-issues/ (accessed 04 March 2024).
12 CNET. Samsung Galaxy Note 7 recall: everything you still need to know about what's coming next. https://www.cnet.com/tech/mobile/samsung-galaxy-note-7-return-exchange-faq/(accessed 04 March 2024).
13 Shilov, A. (2019). HP Expands 2018 Battery Recall Program, Delayed Announcement Due to Govt Shutdown. AnandTech. https://www.anandtech.com/show/14087/hp-expands-2018-battery-recall-program (accessed 04 March 2024).
14 Das, R. (2019). Apple recalls 458,000 laptops due to fire hazard. WLNS-TV 6. https://www.wlns.com/top-stories/apple-recalls-458000-laptops-due-to-fire-hazard/ (accessed 04 March 2024).
15 ny1.com. FDNY chief fire marshal urges Congress to act on lithium-ion batteries. https://ny1.com/nyc/all-boroughs/politics/2024/02/16/fdny-fire-marshal-urges-congress-to-act-on-lithium-ion-batteries (accessed 04 March 2024).
16 The Straits Times. Fire at residential block that killed 15 casts spotlight on e-bikes in China. https://www.straitstimes.com/asia/east-asia/deadly-fire-that-killed-15-sparks-scrutiny-of-e-bike-hazards-in-china (accessed 04 March 2024).
17 Tesla. Model S Fire. https://www.tesla.com/blog/model-s-fire (accessed 04 March 2024).
18 Matousek, M. A Tesla Model S caught fire in a garage in San Francisco. Business Insider. https://www.businessinsider.com/tesla-model-s-catches-fire-in-san-francisco-garage-2019-5 (accessed 04 March 2024).
19 Russo, D. Electric Porsche Taycan catches fire while parked overnight in garage, company confirms. CNBC. https://www.cnbc.com/2020/02/18/electric-porsche-taycan-catches-fire-in-garage-company-confirms.html (accessed 04 March 2024).
20 Gitlin, J.M. Multiple recalls spark Fed investigation of LG's electric car batteries. ARS Technica. https://arstechnica.com/cars/2022/04/multiple-recalls-spark-fed-investigation-of-lgs-electric-car-batteries/ (accessed 04 March 2024).
21 National Highway Traffic Safety Administration (2025). EQ 22-001: LGES High Voltage Battery Failures. U.S. Department of Transportation.
22 electrive.com. Stellantis to recall over 90,000 PHEVs due to fire risk. https://www.electrive.com/2023/05/10/stellantis-to-recall-over-90000-phevs-due-to-fire-risk/ (accessed 04 March 2024).

23 The Korea Herald. Yonhap, Mercedes-Benz's crisis management capacity questioned amid lax stance on S. Korean EV fire. https://m.koreaherald.com/view.php?ud=20240809050608 (accessed 10 August 2024).

24 Federal Aviation Administration. Lithium Battery Incidents. https://www.faa.gov/hazmat/resources/lithium_batteries/incidents (accessed 04 March 2024).

25 Kelleher, S.R. (2023). FAA: Lithium Battery Incidents On Planes Now Happening More Than Once Per Week. Forbes. https://www.forbes.com/sites/suzannerowankelleher/2023/03/03/faa-lithium-battery-fire-smoke-airplane/?sh=7846a2bbb04b (accessed 04 March 2024).

26 O'Connor, P. An Analysis of Lithium-ion Battery Fires in Waste Management and Recycling".

27 letsrecycle.com. 'Major chemical incident' at Veolia's site in Preston. https://www.letsrecycle.com/news/major-chemical-incident-at-veolias-site-in-preston/ (accessed 04 March 2024).

28 Catharines, N.T.W.-S. (2007). Plant completely destroyed by blast: fire chief. NiagaraThisWeek.com https://www.niagarathisweek.com/news/plant-completely-destroyed-by-blast-fire-chief/article_f8a65f09-ff3b-5304-9766-8ce3ab44e3ad.html (accessed 04 March 2024).

29 Daily Hive. Dyson vacuum batteries recall in Canada. https://dailyhive.com/canada/dyson-vacuum-batteries-recall-canada (accessed 04 March 2024)

30 Lu, Q. (2021). Lithium battery safety issues became trending hashtag. *China Energy News* (06 December). http://paper.people.com.cn/zgnyb/html/2021-12/06/content_25892614.htm (accessed 04 March 2024).

31 Guo, J. (2023). Uncovering the safety issue of lithium batteries. *China Economic Weekly* (15 August). http://paper.people.com.cn/zgjjzk/html/2023-08/15/nw.zgjjzk_20230815_3-02.htm (accessed 04 March 2024).

32 Dongguan 5–11 fire incident investigation. https://dgjj.dg.gov.cn/dgjj/wzqk/202312/573fb7acd8e64401a43221df89e995f8.shtml (accessed 04 March 2024).

33 South China Morning Post. At least 11 dead in Shenyang factory fire. https://www.scmp.com/news/china/society/article/3134560/least-11-dead-shenyang-factory-fire (accessed 04 March 2024).

34 Bloomberg (2024). Chinese Public Criticizes Officials Over Fire That Killed 39. Available: https://www.bloomberg.com/news/articles/2024-01-25/chinese-public-lashes-out-at-officials-over-fire-that-killed-39 (accessed 04 March 2024.

35 Xu, W. and Zhao, R. (2024). Xi orders steps to stop repeated accidents after fire kills 39. *China Daily* (25 January). https://www.chinadaily.com.cn/a/202401/25/WS65b1343fa3105f21a507e28c.html. (accessed 04 March 2024).

36 [YouTube video link]. https://youtube.com/watch?v=i_U5fYIVBKI&si=bmQTzMzquFs06rVJ (accessed 31 May 2025).

37 YouTube (2024). Electric Bike Fires Spread Across China, Supermarkets Ablaze. *CCP Suppresses News* (3 September). https://www.youtube.com/watch?v=xSpU6PnKguM (accessed 6 September 2024).

Acknowledgement

To ensure the highest quality, based on the highest expertise, each chapter in this book was written by a team of expert authors and assistants, all with many years of experience in lithium-ion battery and pack design, materials, manufacturing, quality assurance, testing, standards, supply chains, storage, transportation, reuse, and disposal. I had a lot of help in putting this book together, and although I edited the whole book, I need to thank all the chapter authors who worked together to make this happen. Many of the authors are my former students who are well-known for their work at companies and in academia. Many of the other authors are people whom I have gotten to know over the course of my team's study of lithium-ion batteries. In addition, numerous students helped in gathering data and preparing graphs and figures. Finally, a special thanks goes to Amish Vaishnav, who helped with formatting and Gee Abraham for her edits and suggestions to improve both the publishing and technical quality of this content. A very brief summary of each of the co-editors, authors, and assistants is given above in **The Editors, Authors, and Assistants** section.

Acronyms

A123 Systems	A123 Systems LLC
AC	Alternating Current
ACCC	Australian Competition and Consumer Commission
ACF	Anisotropic Conductive Filament
ACIR	Alternating Current Internal Resistance
ADG Code	Australian Dangerous Goods Code
ADR	European Agreement concerning the International Carriage of Dangerous Goods by Road
AHJ	Authority Having Jurisdiction
AI	artificial intelligence
ANOVA	analysis of variance
AQL	acceptance quality limit
AS	Aerospace Standard
ASSLBs	all-solid-state lithium batteries
ATL	Amperex Technology Limited
BASE	B-alumina solid electrolyte
BASF	Badische Anilin und Soda Fabrik
BESS	Battery Energy Storage System
BMS	battery management system
BMW	Bavarian Motor Works
BorgWarner	BorgWarner Inc.
BTR	BTR New Material Group Co., Ltd
BYD	BYD Co., Ltd.
CAAC	Civil Aviation Administration of China
CAGR	Compound Annual Growth Rate
CALB	China Aviation Lithium Battery
CASA	Civil Aviation Safety Authority of Australia
CATL	Contemporary Amperex Technology Co., Ltd.
CCC	China Compulsory Certification
CE	Conformité européenne
CEAP	Circular Economy Action Plan
CFR	Code of Federal Regulations
CID	Circuit Interrupt Devices
Cl_2	chlorine
CMC	carboxymethyl cellulose

CO	carbon monoxide
COTS	commercial off-the-shelf
COVID-19	Coronavirus disease 2019
CPP	casting polypropylene
CPSC	US Consumer Product Safety Commission
CPU	central processing unit
C-rate	charge rate (relative rate at which a battery is charged or discharged)
CT	computed tomography
CTQ	critical to quality
CTS	critical to safety
CT scan	computed tomography scan
DC	direct current
DCIR	direct current internal resistance
DENAN	Electrical Appliance and Material Safety Law
DFMEA	design failure mode and effect analysis
DGCA	Directorate General of Civil Aviation of India
DIY	do it yourself
DMAIC	define, measure, analyze, improve, and control
DMC	dimethyl carbonate
DO-160	Environmental Conditions and Test Procedures for Airborne Equipment (standard)
DOD	depth of discharge
DOE	Department of Energy
DOT	Department of Transportation
DPPB	defects parts per billion
DRC	Democratic Republic of Congo
e-bikes	electric bicycles
EAC	equivalent annual cost
EC	ethylene carbonate
ECGT	Directive on Empowering Consumers for the Green Transition
EICC	Electronic Industry Citizenship Coalition
EIS	electrochemical impedance spectroscopy
EOL	end-of-line
EOS	electrical overstress
EPA	Environmental Protection Agency
EPR	extended producer responsibility
ESA	Energy Storage Association
e-scooters	Electric scooters
ESD	electrostatic discharge
ESPR	Eco-design of Sustainable Products Regulation
ESS	energy storage system
EU	European Union
EVs	electric vehicles
E-waste	electronic waste
FAA	Federal Aviation Administration
FDNY	Fire Department in New York
FEA	finite element analysis

FET	field effect transistor
FG	fuel gauge
FIFO	first in, first out
FMEA	failure mode and effect analysis
FQC	final quality control
FRACAS	failure reporting, analyses, and corrective action system
G	gravitational force
GEM	Green Eco Manufacture
GG	gas gage
GM	General Motors
GPU	graphics processing unit
GWh	gigawatt hour
HALT	highly accelerated life test
HCl	hydrogen chloride
HCN	hydrogen cyanide
HEV	hybrid electric vehicles
HF	hydrogen fluoride
HMR	Hazardous Materials Regulations
HSC	HuaSheng Lithium Battery Co., Ltd.
HSM	hydrodynamic shear mixing
HT	high temperature
HUST	Huazhong University of Science and Technology
HVAC	heating, ventilation, and air conditioning
IATA DGR	International Air Transport Association Dangerous Goods Regulations
IATF	International Automotive Task Force
IC	integrated circuit
ICAO	International Civil Aviation Organization
ICONIQ	Hyundai ICONIQ electric vehicle
IEA	International Energy Agency
IEC	International Electrotechnical Commission
IEEE	Institute of Electrical and Electronics Engineers
IFC	International Fire Code
IGBT	Insulated-gate bi-polar transistors
IMDG	International Maritime Dangerous Goods
IoT	Internet of Things
IP	ingress protection
IPC	Institute for Printed Circuits (now known as IPC Association Connecting Electronics Industries)
IPX	ingress protection rating
IPxx	Ingress Protection Rating (e.g., IP67 for dust and water resistance)
IQC	incoming quality control
ISO	International Organization for Standardization
JISC	Japanese Industrial Standards Committee
KCL	Kirchhoff's current law
KOH	potassium hydroxide
LAM	loss of active material
LCO	lithium cobalt oxide

LFP	lithium iron phosphate
Li_6PS_5X	a type of thiamine ore used in solid electrolytes
LIB	lithium-ion battery
$LiBF_4$	lithium tetrafluoroborate
$LiClO_4$	lithium perchlorate
$LiCoO_2$	lithium cobalt oxide
$LiFePO_4$	lithium iron phosphate
Li-ion	lithium-ion
$LiNiCoMnO_2$	lithium nickel cobalt manganese oxide
LiOH	lithium hydroxide
$LiPF_6$	lithium hexafluorophosphate
Li-S	lithium–sulfur
LMO	lithium manganese oxide
LMT	light means of transport
LNO	lithium nickel oxide
LT	low temperature
LTO	lithium titanate
MAQ	maximum allowable quantities
MCU	micro-controller unit
MEP	Ministry of Environmental Protection
METI	Ministry of Economy, Trade and Industry of Japan
MIIT	Ministry of Industry and Information Technology of China
MIL-STD-810	Environmental Engineering Considerations and Laboratory Tests (standard)
ML	machine learning
MLIT	Ministry of Land, Infrastructure, Transport and Tourism of Japan
MoEF	Ministry of Environment and Forests
MoEFCC	Ministry of Environment, Forest, and Climate Change
MOLIT	Ministry of Land, Infrastructure and Transport of South Korea
MOQ	minimum order quantity
MOSFET	metal-oxide semiconductor field effect transistor
MoU	Memorandum of Understanding
MXJO	Brand of lithium-ion battery company
N/P ratio	negative/positive ratio, anode/cathode ratio
NaCl	sodium chloride (salt)
$Na\text{-}NiCl_2$	sodium-nickel chloride
NaOH	sodium hydroxide
Na-S battery	sodium-sulfur battery
NCA	nickel cobalt aluminum oxide
NCM	lithium nickel manganese cobalt oxide
NMC	lithium nickel manganese cobalt oxide
NFPA	National Fire Protection Association
NHTSA	National Highway Traffic Safety Administration
NiCd	nickel-cadmium
NiMH	nickel-metal hydride
NMC	nickel manganese cobalt oxide
NMP	N-methyl pyrrolidone
NOx	nitrogen oxides

NPI	new product introduction
NTC	negative temperature coefficient
OBA	out of box audit
OBM	original battery manufacturer
OCV	open circuit voltage
OEM	original equipment manufacturer
OQC	outgoing quality control
ORT	ongoing reliability test
OSHA	Occupational Safety and Health Administration
P2D	pseudo-two-dimensional
PC	personal computer
PCB	printed circuit board
PCBA	printed circuit board assembly
PCM	protection circuit module
PCS	power conversion system
PE	polyethylene
PFMEA	Process Failure Mode and Effect Analysis
PHEV	plug-in hybrid electric vehicle
PoF	physics-of-failure
PP	polypropylene
POSCO	Pohang Iron and Steel Company
PP	polypropylene
PRBA	Portable Rechargeable Battery Association
PSA	pressure-sensitive adhesive
PTC	positive temperature coefficient
PV	photovoltaic
PVDF	polyvinylidene fluoride
QHSE	quality, health, safety and environment
QR code	quick response code
R&D	Research and Development
RChevy	Chevrolet (automobile manufacturer)
RCRA	Resource Conservation and Recovery Act
r-CVD	rapid chemical vapor deposition
RH	relative humidity
RID	European Regulation concerning the International Carriage of Dangerous Goods by Rail
RoHS	restriction of hazardous substances
RS	rock salt
RSOC	remaining state of charge
SAE	Society of Automotive Engineers
SBR	styrene-butadiene rubber
SDGs	United Nations' Sustainable Development Goals
SEI	solid–electrolyte interphase
SESS	stationary energy storage system
Shanshan	Shanshan Lithium Industry Co., Ltd
SiC	silicon-carbide
SiOX	silicon oxides

SK Innovation	SK Innovation Co., Ltd.
SOC	state of charge
SOH	state of health
SOx	sulfur oxides
SPC	statistical process control
SPIR	Sinopec Petroleum Industry Research Institute
SPRT	sequential probability ratio test
TCO	thermal cutoff
TVS	transient voltage suppressor
UL	Underwriters Laboratories
UN	United Nations
UNC	Uncontrolled
UNEP	United Nations Environment Programme
UPS	uninterruptible power supply
UPS	united parcel service
USB-C	USB Type-C
USGS	United States Geological Survey
UUTs	units under tests
WEEE	waste electrical and electronic equipment
XFH	Shenzhen Xfh Technology Co., Ltd.
XRR	X-ray reflectivity
XTC	XTC New Energy Materials (Xiamen) Co., Ltd
ZIF	zero insertion force

1

Basics of Lithium-Ion Battery Technology

Simin Peng, Yue Shen, Genkai Xia, Sahithi Maddipatla, Lingxi Kong, and Mohammed Saquib Khan

Batteries are electrochemical devices that convert chemical energy into electrical energy using an electrochemical oxidation–reduction (redox) reaction. Primary batteries are non-rechargeable and rely on irreversible chemical reactions, making them suitable for single-use applications like remote controls and flashlights.

In contrast, secondary batteries are rechargeable, using reversible chemical reactions that allow multiple charge–discharge cycles. They are commonly found in devices like smartphones, laptops, and electric vehicles, and examples include alkaline (primary) and lithium-ion (secondary) batteries. The operation of secondary batteries involves two processes: discharge and charge. During the discharge process, a chemical reaction releases electrons, generating an electric current to power devices. Conversely, during charging, an external electrical current reverses this reaction, restoring the chemical state of the battery and enabling it to store energy for future use.

Batteries can be classified into different categories and types based on their chemical composition, size, form factor, and applications. Lithium-ion batteries (LIBs), a type of rechargeable battery, are widely favored for their lack of memory effect, high energy density, long cycle life, and low self-discharge rate.

1.1 Lithium-Ion Battery Cell Structure and Chemistry

The fundamental structure of an LIB consists of several key components: the anode (negative electrode), cathode (positive electrode), electrolyte, separator, current collectors, and a protective casing. The anode, commonly made of graphite, plays a crucial role during the charging process by storing lithium ions. The cathode, typically composed of lithium metal oxides, such as lithium cobalt oxide (LCO), lithium manganese oxide (LMO), or lithium iron phosphate (LFP), serves as the positive electrode. The electrolyte, which can be a lithium salt in either an organic solvent or solid form, facilitates the movement of ions between the anode and cathode. A porous separator membrane is placed between the electrodes to prevent short-circuiting, allowing ions to pass through while keeping the electrodes separated. Current collectors, made of copper for the anode and aluminum for the cathode, gather and transfer electrons throughout the cell. These components are housed within a protective casing to ensure the battery's integrity.

The Safety Challenges and Strategies of Using Lithium-Ion Batteries, First Edition. Edited by Michael G. Pecht.

The anode and cathode in LIBs typically feature porous structures with carefully tuned porosities, achieved during the calendering process. These electrodes are coated with a powder layer containing active materials, conductive agents (e.g., acetylene black or carbon black), electrode binders (e.g., polyvinylidene fluoride [PVDF], polytetrafluoroethylene [PTFE], or carboxymethyl cellulose [CMC]), and additives. The porous nature of the electrode coatings increases the effective surface area for redox reactions and enhances the wettability of the electrolyte, which helps to reduce electrochemical polarization. Additionally, the porous structure is crucial in mitigating dendrite formation at the solid–liquid interface during charging, thereby decreasing the risk of internal short circuits.

Lithium-ion batteries operate on the principle of lithium-ion intercalation within electrode materials. Intercalation refers to the reversible insertion and removal of lithium ions between the layers of a host material in both the anode and the cathode without causing significant structural changes to these materials. The fundamental operation of an LIB involves the movement of lithium ions between the anode and the cathode. During discharge, lithium ions move from the anode to the cathode, a process known as lithium-ion extraction. Simultaneously, electrons flow from the anode to the cathode through an external electrical circuit, providing power to devices connected to the circuit. Conversely, during charging, lithium ions are inserted back into the anode—a process called lithiation, restoring the battery's energy. This back-and-forth movement of lithium ions, often termed the "rocking chair mechanism," is crucial for the battery's ability to store and release energy efficiently. Materials such as graphite in the anode and metal oxides or metal phosphates in the cathode facilitate this mechanism by supporting the reversible intercalation and deintercalation of lithium ions.

The operation of an LIB involves the movement of lithium ions between the anode and the cathode through the electrolyte. During charging and discharging, the valence states of the anode and cathode materials change. The overall redox reaction equation for an LCO and graphite battery remains unchanged and can be expressed as follows:

$$\mathrm{Li}_{1-x}\mathrm{CoO}_2 + \mathrm{Li}_x\mathrm{C}_6 \rightleftharpoons \mathrm{LiCoO}_2 + 6\mathrm{C}$$

The reaction during discharge can be expressed as follows:
Anode (oxidation reaction):

$$\mathrm{Li}_x\mathrm{C}_6 \rightarrow 6\mathrm{C} + x\mathrm{Li}^+ + x\mathrm{e}^-$$

Cathode (reduction reaction):

$$\mathrm{Li}_{1-x}\mathrm{CoO}_2 + x\mathrm{Li}^+ + x\mathrm{e}^- \rightarrow \mathrm{LiCoO}_2$$

The reaction during charge is reversed:
Anode (oxidation reaction):

$$6\mathrm{C} + x\mathrm{Li}^+ + x\mathrm{e}^- \rightarrow \mathrm{Li}_x\mathrm{C}_6$$

Cathode (reduction reaction):

$$\mathrm{LiCoO}_2 \rightarrow \mathrm{Li}_{1-x}\mathrm{CoO}_2 + x\mathrm{Li}^+ + x\mathrm{e}^-$$

When LCO is used as the cathode material and graphite as the anode material, the redox reaction above demonstrates excellent reversibility, typically resulting in high coulombic efficiency. This significantly enhances the cycling performance observed in LIBs configured with this combination [1]. Figure 1.1 illustrates the schematic of the charge–discharge reaction in an LIB.

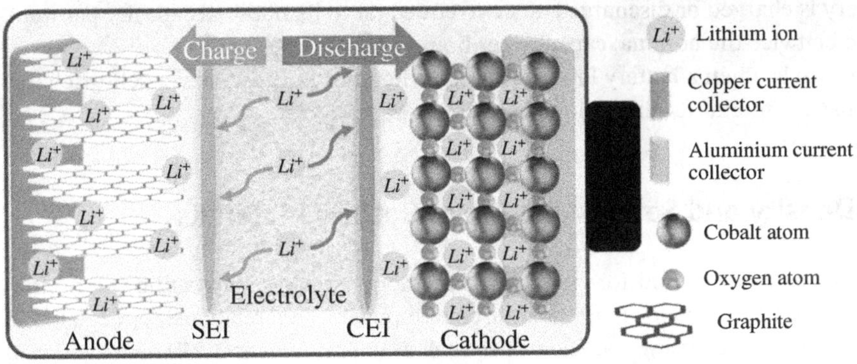

Figure 1.1 Schematic of the charge–discharge reaction in a lithium-ion battery.

1.2 Definitions of Key Battery Performance Metrics

- **Specific capacity** refers to the amount of electrical charge a battery can store per unit mass or volume of the active material. It is usually expressed in milliampere-hours per gram (mAh/g) or milliampere-hours per cubic centimeter (mAh/cm^3). Specific capacity indicates how much charge a battery material can hold relative to its weight or volume, which helps evaluate the material's efficiency and performance.
- **Power density** measures the power a battery can deliver per unit mass or volume. It is typically expressed in watts per kilogram (W/kg) or watts per liter (W/L). Power density is crucial for applications requiring high power output in a compact form, such as electric vehicles or power tools.
- **Energy density** measures the energy a battery can store per unit mass or volume. It is expressed in watt-hours per kilogram (Wh/kg) or watts-hours per liter (Wh/L). Energy density is important for applications with long-lasting energy storage, such as portable electronics or renewable energy storage systems.
- **Coulombic efficiency** refers to the ratio of the total charge extracted from a battery during discharge to the total charge supplied during charging. It is expressed as a percentage. High coulombic efficiency indicates minimal loss of charge during the battery cycle, which is crucial for maintaining battery life and performance.
- **Capacity retention** measures the ability of a battery to maintain its charge capacity over time and usage cycles. It is typically expressed as a percentage of the original capacity. A high-capacity retention signifies that the battery can sustain its performance and energy storage capability even after many charge–discharge cycles, which is essential for long-term reliability and durability.
- **Cycle life** of the battery can be defined as the number of charge–discharge cycles a battery can undergo before its capacity falls to a specified percentage of its original capacity. It reflects the longevity of the battery.
- **Internal resistance** of the battery is the opposition within the battery to the flow of electric current, leading to energy loss in the form of heat. A lower internal resistance generally indicates better performance.
- **C-rate** refers to the rate at which a battery is charged or discharged relative to its maximum capacity. It is expressed as a multiple of the battery's nominal capacity. For example, a 1C rate

means the battery is charged or discharged at a current equal to its nominal capacity per hour. A 2C rate would be twice the nominal capacity per hour.

- **Self-discharge rate** is when a battery loses its charge when not in use. A lower self-discharge rate means the battery retains its charge longer during storage.

1.3 Energy Density and Safety Analysis of Battery Materials

The Ragone plot serves as a method for characterizing energy storage, illustrating the nonlinear relationship between extractable energy and discharge power. This relationship, depicted in a Ragone plot, can be expressed using specific values such as energy/power (ERP) density or normalized metrics. By plotting energy density (Wh/kg) against power density (W/kg), Ragone plots facilitate comparisons across different battery chemistries. Figure 1.2 illustrates the Ragone plot of different energy storage systems [2, 3].

With a slightly lower energy density (approximately 90–120 Wh/kg) than standard LIBs, lead-acid batteries are valued for enhanced safety due to a more stable chemistry. Their reduced susceptibility to thermal runaway makes them ideal for applications prioritizing safety concerns [4]. Nickel-metal hydride (NiMH) batteries, frequently used in hybrid vehicles and consumer electronics, offer an energy density ranging from approximately 60 to 120 Wh/kg. While they exhibit reduced susceptibility to thermal runaway compared to LIBs, they have lower energy density. They are less eco-friendly due to the inclusion of heavy metals like nickel [5]. Lithium–sulfur (Li-S) batteries demonstrate a theoretical energy density of up to 500 Wh/kg, significantly exceeding that of LIBs. However, they are still in the early stages of development, facing challenges related to cycle life and stability [6].

Lithium-ion batteries (LIBs) exhibit a characteristic hook-shaped performance in the ERP domain, where increasing delivered power comes at the expense of reduced retained energy. Nickel-cobalt-aluminum oxide (NCA) and nickel-cobalt-manganese oxide (NMC) cells typically store approximately 140–220 Wh/kg, roughly twice the specific energy of LFP cells (90–160 Wh/kg). However, LFP cells excel in retaining higher specific power across a broader

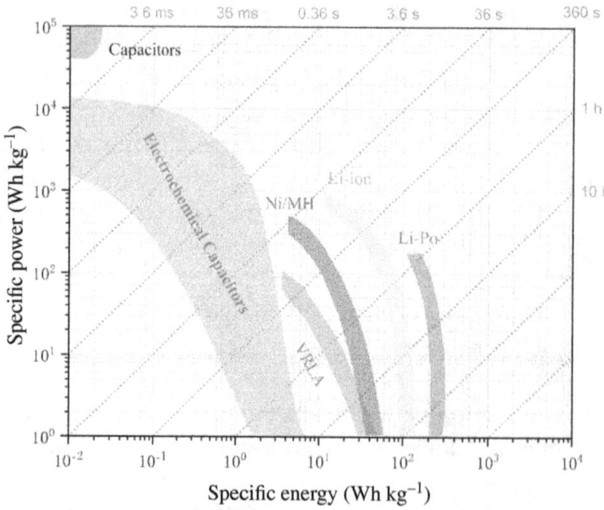

Figure 1.2 The Ragone plot of different types of batteries.

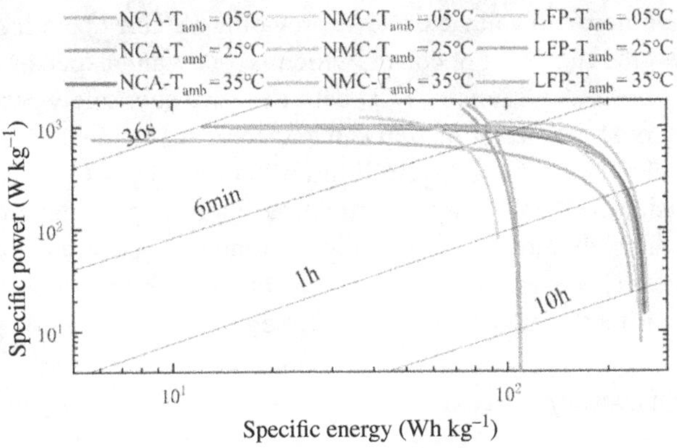

Figure 1.3 The Ragone plot of NCA, NMC, and LFP at different temperatures.

range of C-rates. Specifically, LFP cells can operate effectively at high C-rates (up to 20/C) without disrupting normal cell behavior. In contrast, NCA and NMC cells are limited to a narrower C-rate range to mitigate high Joule losses and prevent rapid voltage drops, limiting their specific power capabilities. Figure 1.3 illustrates the specific power versus specific energy performance of NCA, NMC, and LFP cells at different ambient temperatures [7].

Higher energy density batteries often pose increased safety risks, particularly concerning thermal runaway, whereas safer alternatives may sacrifice some energy density. Achieving the optimal balance between energy density and safety is crucial for advancing and implementing cutting-edge battery technologies.

1.4 Cathode Materials: LCO, LMO, LFP, NMC, NCA, and Li-SPAN

The performance of LIBs is significantly influenced by the choice of cathode materials, which dictate the battery's energy density, cycle life, safety, and cost. Several common cathode materials used in LIBs, including LCO, LMO, LFP, lithium NMC, lithium NCA, and sulfur, are presented in this section. Each material offers unique advantages and limitations, making them suitable for different applications. Understanding these properties is essential for selecting the most appropriate cathode material for a specific battery design.

1.4.1 Lithium Cobalt Oxide (LCO) Battery

Lithium cobalt oxide (LCO) was first synthesized by Johnston's team in 1958 and proposed as a cathode material for LIBs by Goodenough's team in 1980 [8]. Sony introduced the first commercially available LIB utilizing LCO in 1991 [8]. LCO belongs to the hexagonal crystalline system and has an R3m space group, characterized by a typical layered structure.

Several synthesis methods are employed to produce electrochemically active LCO, including the sol–gel method, co-precipitation technique, solid-state reaction, hydrothermal method, freeze-drying, combustion method, microwave synthesis, and mechanochemical synthesis. The choice of synthesis process affects the structure of LCO and the electrochemical performance. Low-temperature synthesis may offer high initial capacity but often leads to rapid capacity loss due to ion disorder and the formation of a spinel-like phase. Thus, optimizing synthesis conditions is crucial for better performance.

LCO batteries are characterized by a high nominal voltage of 3.9 V, providing a stable operating voltage and high energy density. However, the high cost of cobalt, which is a rare and expensive metal, contributes to increased production costs. Additionally, LCO batteries have poor safety performance, with issues such as overheating or short-circuiting that can lead to safety incidents.

Recent advancements include the development of solid-state lithium batteries by a Chinese research team, achieving an energy density of 360 Wh/kg and attracting interest from electric vehicle manufacturers [9]. LCO batteries are well known for their high volumetric energy density and extensive use in consumer electronics. However, interfacial side reactions at high voltages often compromise their cycle life, underscoring the importance of surface optimization.

1.4.2 Lithium Manganese Oxide (LMO) Battery

In 1981, Goodenough's team discovered LMO spinel as a cathode material [10]. Lithium manganate ($LiMn_2O_4$) exists in two forms: spinel and layered, with the spinel form adopting a cubic crystal system and space group Fd3m. Due to abundant manganese resources, $LiMn_2O_4$ is more cost-competitive than LCO and offers economic advantages over other cathode materials such as NMC and LFP. In addition to cost efficiency, LMO offers better safety performance, as it is less prone to overheating or short-circuiting compared to NMC and LCO. This makes it suitable for applications with stringent safety requirements.

LMO also shows minimal performance degradation at low temperatures, ensuring stability in cold regions, and it outperforms LFP materials in this aspect. LMO is devoid of harmful heavy metals, making it more environmentally friendly than NMC and LCO, which contain nickel or cobalt. This sustainability aspect makes LMO a more environmentally conscious choice [11]. However, LMO has drawbacks, particularly its short cycle life, especially at elevated temperatures, leading to quicker performance degradation. Its energy density is intermediate, falling between the high-energy-density LCO and the low-energy-density LFP batteries.

1.4.3 Lithium Iron Phosphate (LFP) Battery

It was first discovered and implemented by Goodenough's team in 1997 [12]. LFP features an olivine structure that contributes to its high stability and safety profile, which reduces the risk of overheating or explosion. While LFP offers lower energy density than other lithium battery types, such as ternary lithium batteries (which use NMC or NCA as cathode materials), it compensates with stability, particularly in high-temperature environments.

Due to their safety attributes and cost-effectiveness, LFP batteries have been widely adopted in electric vehicles and energy storage systems. In 2023, the installed capacity of LFP batteries for power applications increased significantly among Chinese power battery companies, including BYD, CATL, and Hangzhou Zhongxin. In particular, BYD's blade batteries, an advanced LFP technology, have succeeded commercially, especially in the electric vehicle and energy storage sectors. These batteries, named for their distinctive blade-like shape, possess high energy density, safety, and a long lifespan. Their design allows for efficient energy storage and improves thermal management, enhancing charging efficiency and resistance to high temperatures [13]. Additionally, Nanchang Xinda announced the successful development of a 4C LFP battery, indicating its ability to charge fully within 15 minutes, advancing fast charging capabilities [14].

1.4.4 Lithium Nickel-Cobalt-Manganese Oxide (NMC) Battery

These are layered materials with the general formula $Li(Ni_yMn_zCo_{1-y-z})O_2$ and are synthesized by incorporating cobalt into NMO. NMC, initially conceptualized by researchers at the

National University of Singapore [15], shares a structural similarity with LCO, both belonging to the hexagonal crystal system with an R3m space group. The cathode material of NMC is a composite blend composed of oxides of nickel, cobalt, and manganese. The ratios of nickel, cobalt, and manganese can be customized to meet specific requirements, with formulations like $LiNi_{0.6}Mn_{0.2}Co_{0.2}O_2$ (NMC 622), $LiNi_{0.5}Mn_{0.3}Co_{0.2}O_2$ (NMC 532), $LiNi_{1/3}Mn_{1/3}Co_{1/3}O2$ (NMC-111), and $LiNi_{0.8}Mn_{0.1}Co_{0.1}O_2$ (NMC 811), representing varying molar ratios.

NMC integrates the advantages of nickel, cobalt, and manganese, leveraging a synergistic effect among these transition metals to enhance the battery's energy density, cycle life, and charge–discharge performance. Moreover, NMC utilizes cost-effective nickel and manganese to replace cobalt partially, balancing performance and cost considerations. Compared to LCO, NMC offers improved safety features, making it well-suited for extensive adoption in electric and hybrid vehicles.

1.4.5 Lithium Nickel-Cobalt-Aluminum Oxide (NCA) Battery

NCA batteries, or lithium NCA batteries, are high-performance LIBs featuring a cathode material composed of nickel, cobalt, and aluminum. These batteries are widely acclaimed for their high energy density and robust cycling performance. NCA batteries enable long ranges, a crucial factor for electric vehicles and portable electronic devices. Moreover, they maintain excellent performance at high discharge rates, making them suitable for applications requiring rapid response and high power output.

However, NCA batteries face challenges primarily related to thermal stability and safety. Their high nickel content increases the risk of thermal runaway under conditions of overcharging or high temperatures. To mitigate these challenges, researchers often enhance the safety performance of NCA batteries through improvements in electrolyte formulations, incorporation of flame retardants, and optimization of battery structural design. Despite this, NCA batteries are still essential in many high-end applications because of their exceptional performance characteristics.

1.4.6 Lithium-Sulfurized Polyacrylonitrile (Li-SPAN) Battery

The widespread adoption of LIBs has significantly increased demand for their rare constituent materials, particularly cobalt. The dependence on scarce resources poses sustainability challenges that could hinder the growth of the battery industry. To tackle this dependency, researchers are actively investigating alternative materials and developing new battery systems.

Lithium–sulfur (Li-S) batteries hold strong potential as next-generation rechargeable batteries for long-distance transportation and large-scale grid applications thanks to their high theoretical specific capacity (1675 mAh/g) and energy density (2600 Wh/kg). Unlike elemental sulfur cathodes, sulfurized polyacrylonitrile (SPAN) can eliminate the shuttle effect—the inefficiency caused by the dissolution and migration of intermediate polysulfide species during cycling—making it a promising alternative sulfur cathode material for Li-S batteries. SPAN is highly compatible with carbonate-based electrolytes, commonly used in Li-ion batteries, and it can be easily synthesized by heating sulfur powder and polyacrylonitrile (PAN) in an inert atmosphere. Additionally, SPAN can be seamlessly integrated into the manufacturing processes of existing Li-ion batteries.

1.4.7 Summary of Cathode Materials

Different types of LIBs offer distinct performance and cost advantages (Table 1.1). Choosing the right battery type depends on the specific requirements of the application, which include factors

Table 1.1 Comparison of the performance and cost of different types of lithium batteries.

Type	Energy density (Wh/kg)	Average voltage (V)	Cost	Cycle life	Safety
LCO	575	3.6–4.2	Medium	Medium	Medium
LMO	100–120	3.6–3.8	Medium	Medium	Medium
LFP	90–120	3.2–3.4	Low	High	High
NMC	200–260	3.6–4.2	High	Low	Low
NCA	200–270	3.6–4.2	High	Low	Low
SPAN	550	1–3	Low	Medium	Medium

Source: Chen et al. [16]/IEEE.

such as energy density, cycle life, safety, and cost. Comparing cell performance across various Li-ion battery types can provide valuable insights for informed decision-making.

1.5 Anode Materials: Carbon-Based, Silicon-Based, Metal, and Alloying Anodes

Anode materials are crucial in LIBs, as they significantly influence the batteries' performance and efficiency. The diverse range of anode materials is presented here, encompassing carbon-based, silicon-based, and metal/alloying materials, delving into their distinctive properties, challenges, and recent advancements. By comprehending the intricacies of these materials, researchers endeavor to advance battery technology, improve energy storage capacities, and tackle critical challenges in the field.

1.5.1 Carbon-Based Materials

Carbon-based materials can be broadly categorized into graphite materials and various forms of amorphous carbon. Graphite is a mineral with a layered structure of carbon atoms arranged in regular hexagonal lattices. Its layered configuration facilitates the insertion and extraction of lithium ions during charging and discharging, providing optimal channels for lithium-ion movement. Graphite offers excellent electrical conductivity, high cycling stability, cost-effectiveness, and availability. However, challenges such as relatively low energy density and limitations in charge–discharge rates persist.

Amorphous carbon materials, such as amorphous carbon, carbon black, and carbon nanotubes, exhibit a variety of properties and functionalities. Amorphous carbon is noted for its extensive specific surface area and abundant surface-active sites, which form conductive networks during battery operation, aiding in the embedding and release of lithium ions. Materials like carbon black and nanotubes possess exceptional mechanical strength and chemical stability, effectively enduring stresses and fluctuations during cycling, thus ensuring structural integrity and cycle life. Despite their promise, these materials face challenges such as high production costs and optimizing long-term cycling stability, necessitating ongoing exploration of their potential and application prospects [17].

The SEI layer, a protective film formed on the surface of the anode in LIBs, plays a critical role in the stability and lifespan of the battery. A stable SEI layer effectively prevents adverse reactions between the electrode material and the electrolyte, reduces the loss of lithium ions, and thereby prolongs the battery's service life. Recent advancements have highlighted significant breakthroughs in battery carbon-based materials, such as Professor Yunhui Huang's team [18] at Huazhong University of Science and Technology (HUST), developing an artificial inorganic SEI to repair failed graphite batteries. This innovation transforms low-value failed graphite into high-value prelithiated graphite, mitigating SEI regrowth and enhancing battery performance.

1.5.2 Silicon-Based Materials

Silicon, with its low equilibrium potential of approximately 0.3 V versus Li/Li+ and the highest theoretical capacity of 4212 mAh/g among all known materials, is extensively utilized as an anode material in rechargeable batteries. As the second most abundant element, comprising about 27.7% of the Earth's crust, silicon offers significant potential for energy storage applications. Silicon-based materials encompass a variety of anode materials, including silicon oxides (SiO_X), silicon–carbon compounds (SiC), and silicon-based alloys. Silicon oxides (SiO_X), predominantly used as negative electrode materials in batteries, boast high theoretical specific capacity and are considered highly promising for next-generation negative electrode materials following graphite. However, silicon-based negative electrodes face challenges during charge and discharge processes, such as issues with volume expansion and low conductivity. To tackle these challenges, researchers have employed various strategies, including nanostructuring, surface modification, and enhancing mechanical strength by designing porous or composite structures.

Silicon–carbon compounds (SiC) exhibit good electrical conductivity and chemical stability, making them promising candidates for silicon-based anodes. SiC extends battery lifespan, enhances performance, and addresses volume expansion issues associated with silicon materials. However, its relatively low energy density, moderate lithium-ion embedding, and de-embedding efficiencies result in less efficient charging and discharging processes.

Silicon-based alloys, which form alloy structures with other metal elements like lithium or tin, offer high specific capacities and low charge–discharge potentials. However, they encounter challenges related to volume expansion and cycling stability, requiring further optimization in LIB research.

Currently, silicon is used in a relatively small proportion in anodes for LIBs, typically ranging from 5% to 15%. This is due to silicon's high theoretical specific capacity, along with its tendency to expand in volume during charge and discharge cycles, leading to structural damage to the electrode. To address this challenge, researchers are exploring various methods, such as nanofabrication, surface modification, and the design of porous structures to enhance the stability and lifespan of silicon anodes. With advancements in technology and optimized material design, the incorporation of silicon in anodes may see increased adoption in the future with capacities ≥870 mAh/g [19].

1.5.3 Metal and Alloying Anodes

Metal and alloy anodes are categorized into two main types: pure metal anodes and metal alloy anodes. Pure metal anodes consist of a single metal element, such as lithium (Li) or aluminum (Al), while metal alloy anodes are composed of a combination of two or more metal elements, such as lithium–aluminum alloys and silicon-based alloys. These anode materials offer high-energy densities due to their substantial lithium-ion storage capacity, and they demonstrate efficient performance during battery charging and discharging, thanks to the excellent electrical conductivity of metals.

Table 1.2 Comparison of different performance indices of various anode materials.

Type	Energy density (Wh/kg)	Average voltage (V)	Cycle life	Cost	Safety
Carbon-based materials	200–300	0.1–0.3	Medium	Low	Medium
Silicon-based materials	400–4200	0.4–0.6	High	Medium	High
Metal and alloying anodes	400–900	0.3–0.6	Low	High	Low

Source: Jia et al. [21]/with permission of Elsevier.

For instance, a lithium metal anode shows promise for high-energy-density batteries, boasting a theoretical specific capacity of up to 3860 mAh/g and a very low electrode potential (−3.04 V versus a standard hydrogen electrode). However, metal and alloy anodes undergo volume expansion and contraction during charge–discharge cycles, which can lead to structural damage and reduce the cycling performance of the battery. It is important to note that this volume change phenomenon is not limited to metal and alloy anodes; graphite anodes also undergo similar volume changes during charge–discharge processes.

Metal and alloy anodes typically undergo more severe volume expansion during charge–discharge cycling than graphite anodes. This is because the atomic structure of metals and alloys is more prone to reorganization during these cycles, leading to greater volume changes. Such volume changes can cause structural damage to the battery and reduce its cycling performance.

Therefore, special design considerations are necessary when using metals and alloys as anode materials. For example, the effects of volume change can be minimized by incorporating materials that inhibit expansion or by altering the structural design of the electrode. Furthermore, optimizing the charging and discharging strategies, such as controlling the rates of these processes, can help mitigate the effects of volume change.

Some metal alloy anodes may also encounter the solid solution lithium effect, leading to structural instability and capacity degradation. This phenomenon occurs when lithium ions not only deposit as metallic lithium on the surface of the electrode but also incorporate into the lattice of the alloy, forming a solid solution phase that contains lithium.

Despite the promising electrochemical performance of metal and alloy anodes, a need remains to enhance their performance through structural and material refinements. Addressing issues like volume expansion, structural stability, and the solid solution lithium effect requires optimizing these materials through various methods such as nanotechnology, surface coating, and alloying [20]. Table 1.2 compares different performance indices of various anode materials.

1.6 Electrolytes: Liquid and Solid Electrolytes

Electrolytes, whether liquid or solid, play a pivotal role in the operation of LIBs. They facilitate the movement of ions between the electrodes, allowing for the storage and release of energy. The selection of electrolyte material profoundly influences battery performance, safety, and longevity. This section presents the characteristics and functions of both liquid and solid electrolytes, along with their respective advantages and limitations in LIB technology.

1.6.1 Liquid Electrolytes

The liquid electrolyte is a crucial component in LIBs, typically composed of an organic solvent, lithium salts, and various additives. The organic solvent dissolves the lithium salts and facilitates ion transport throughout the battery. Common organic solvents include dimethyl carbonate and dimethyl acetate, carbon ethers like ethylene glycol dimethyl ether, and nitrile solvents such as propionitrile. These solvents offer good solubility, stability, and high ionic conductivity. Two particularly important solvents in this context are diethyl carbonate (DEC) and ethylene carbonate (EC). While DEC provides advantages such as low viscosity and high ionic mobility, EC is known for its high dielectric constant, aiding in salt dissociation, and contributing to forming a stable solid electrolyte interface (SEI) layer on electrodes. The combination of DEC and EC can enhance the performance and lifespan of LIBs by improving the balance between ionic conductivity and electrochemical stability [22].

Lithium salts function as ion sources, supplying lithium ions for movement during battery operation. Common lithium salts include lithium hexafluorophosphate ($LiPF_6$), lithium tetrafluoroborate ($LiBF_4$), and lithium perchlorate ($LiClO_4$), which dissolve to produce conductive ionic solutions.

Additives are critical in LIB electrolytes by enhancing performance and stability. They serve multiple functions, such as improving electrochemical stability, reducing corrosion and oxidation reactions, inhibiting uncontrolled growth of the SEI, and enhancing the battery's cycle life. While the SEI layer is crucial for the proper functioning of LIBs, excessive growth of the SEI can lead to increased internal resistance and diminished battery performance. Therefore, additives that regulate SEI growth help maintain a thin, uniform SEI layer, essential for stable battery operation over numerous charge and discharge cycles.

Typical additives include film-forming agents, ion-conducting compounds, flame retardants, and overcharge protection agents. Accordingly, these additives work synergistically to ensure the battery operates efficiently and safely within its desired performance range. Researchers like Chen [23] and colleagues from China have highlighted the growing prominence of sodium-ion batteries in large-scale energy storage due to sodium's abundance, cost-effectiveness, and excellent performance. Understanding their chemical properties is crucial for determining their competitive or complementary use alongside lithium-ion technology. Currently, anhydrous organic liquid electrolytes dominate production due to their superior performance, but they require thorough evaluation to ensure optimal electrochemical performance and safety.

1.6.2 Solid Electrolytes

Solid electrolytes are materials used in electrochemical devices, such as LIBs, to replace conventional liquid electrolytes. They offer superior safety, enhanced ionic conductivity, and a broader operating temperature range than their liquid counterparts. Solid electrolytes are typically solid-state materials that enable ionic conduction through defects or channels within their crystal lattice.

These electrolytes exhibit greater stability, addressing safety concerns such as thermal runaway and spontaneous combustion, while providing high ionic conductivity, thermal resistance, and tolerance to elevated temperatures [24].

Solid electrolytes face drawbacks such as low ionic conductivity, challenges with interfacial contact, and susceptibility to oxidation and water vapor. However, the industry's utilization of solid

Table 1.3 Various electrolyte types for comparison.

	Liquid electrolyte	Semisolid electrolyte	Solid-state electrolyte	
Characteristic	**Organic liquid**	**Gel polymer**	**Solid polymer**	**Inorganic solid**
Li+ concentration	Low	Low	High	High
Li+ placement	Unfixed	Fixed	Fixed	Fixed
Conductivity	High	High	Low	Low
Safety	Flammable	Good	Good	Good
Price	High	High	High	Low
Ionic coordination	None	Exist	None	None
Ion exchange number	High	Low	One	One

Source: Wulandari et al. [28]/John Wiley & Sons/CC BY 4.0.

electrolytes has seen progress, with companies like Toyota outlining timelines for the commercialization of solid-state batteries. Scientists are actively researching and refining solid-state electrolyte materials to overcome these technical hurdles [25]. Current research focuses on developing flexible, thin solid electrolyte sheets with a Li+ ion-conducting polymer network to improve interfacial contact [26].

1.6.3 Summary of Electrolyte Comparisons

The electrolyte functions as the pathway for positive and negative ion conduction between the electrodes, while inhibiting electronic conductivity. Therefore, the electrochemical performance of LIBs is closely tied to the characteristics of the electrolyte. Traditionally, LIBs use organic electrolytes because of their excellent stability, wide electrochemical window, and higher operational voltage, which typically exceeds that of aqueous electrolyte batteries by a factor of one [27]. These attributes allow LIBs to achieve high voltage and specific energy properties. However, organic electrolytes exhibit relatively low conductivity and poor thermal stability, which poses safety risks. Therefore, it is crucial to compare the properties of liquid, semisolid, and solid electrolytes. Various types of electrolytes are compared in Table 1.3.

1.7 Separators

The separator in LIBs acts as a nonconductive barrier that allows the passage of lithium ions from the electrolyte. Typically made from polymer materials such as polypropylene (PP), polyethylene (PE), or composite membranes like PP/PE/PP, these separators provide excellent electrical insulation and electrochemical stability. The safety of these separator materials directly impacts overall battery safety. High-performance separators must withstand high temperatures and resist chemical corrosion, remaining stable under harsh operating conditions to address safety concerns related to thermal runaway and explosions.

In LIBs, safety concerns often arise from external impacts that can puncture the battery's internal components, particularly the separator. When a hard object, such as a nail, penetrates the battery, it can damage or collapse the separator, allowing direct contact between the positive and negative electrodes. This contact leads to short-circuiting, which generates significant heat and can ignite the flammable electrolyte, causing fire or even explosion. Traditional separators, when punctured, tend to form large tear holes around the impact site, resulting in extensive areas of short-circuiting and a high risk of thermal runaway.

Advanced separators are now being designed with multilayer structures, consisting of porous substrates and porous layers. The porous layer, applied on the substrate surface, often incorporates inorganic particles that act as binders, enhancing mechanical and thermal stability. Common inorganic particles used in these separators include alumina, silica, magnesia, titanium oxide, and zirconia, among others. These particles provide structural reinforcement, improving resistance to puncture and thermal degradation. The porous substrate itself is typically made from a polymer film, a multilayer polymer film, or a nonwoven fabric. Suitable polymers for the substrate include polyethylene, polypropylene, polyethylene terephthalate, polycarbonate, polyimide, polyethersulfone, and other durable materials. The substrate is engineered to have a high plastic deformation rate, generally around or greater than 40–1800%. This high plasticity ensures that when the separator is punctured, it stretches and conforms around the puncturing object, such as a nail, thereby creating a wrapping effect. This wrapping effect reduces the area of short-circuiting by containing the puncture site, which lowers the risk of internal short circuits and mitigates the likelihood of thermal runaway. The binder polymers in the porous layer can vary widely, including compounds such as vinylidene fluoride-hexafluoropropylene copolymer, polymethyl methacrylate, polyacrylonitrile, polyvinyl acetate, and others. These binders further support the structural integrity of the separator, allowing it to withstand high deformation without tearing. Porosity and thickness are critical parameters of the separator. Typically, the thickness of the porous substrate ranges from 1 to 40 microns, while porosity spans from 10% to 70%, ensuring both electrolyte permeability and mechanical strength. The high plastic deformation rate and structural reinforcement provided by the dual-layer separator design contribute significantly to battery safety. By minimizing the effects of punctures and maintaining separation between the electrodes, this separator design helps prevent large-area short circuits, reduces fire risks, and enhances the overall reliability of LIBs [29].

Another approach in separator technology involves the strategic use of insulating glue layers on electrodes to enhance safety and performance in LIBs. In this LIB design, insulating glue layers are carefully applied to key areas on the electrodes to boost safety and performance, especially under stress conditions like puncturing. The positive and negative leads, which connect the current collectors to external terminals, are each seated in a dedicated recess—a small, carved-out area in the electrode material that allows the leads to be embedded within the electrodes rather than resting on the surface. This recessed placement not only reduces the overall battery thickness but also lowers the risk of misalignment or accidental contact, which can lead to short circuits. A first insulating glue layer is applied to both the upper and lower surfaces of the positive lead, creating a barrier that prevents electrical faults by insulating the lead within the recess. The second insulating glue layer is applied on the positive film, specifically at the section opposite the recess where the negative lead is housed. This additional layer of insulation ensures robust protection at potential points of contact between the electrodes, where the positive and negative leads might otherwise come dangerously close. By designing the glue layers slightly larger than the recesses, the battery achieves complete encapsulation of the lead ends. This encapsulation is especially valuable in mitigating lithium precipitation—a process that can degrade performance and lead to safety issues over time. Furthermore, the glue insulation offers an additional safeguard against nail penetration events, where a sharp object pierces the cell. In such cases, the glue layers help isolate and contain damage, reducing the risk of a thermal runaway reaction that could occur if the electrodes were to make unintended contact. Overall, this design strengthens the internal structure, reduces internal

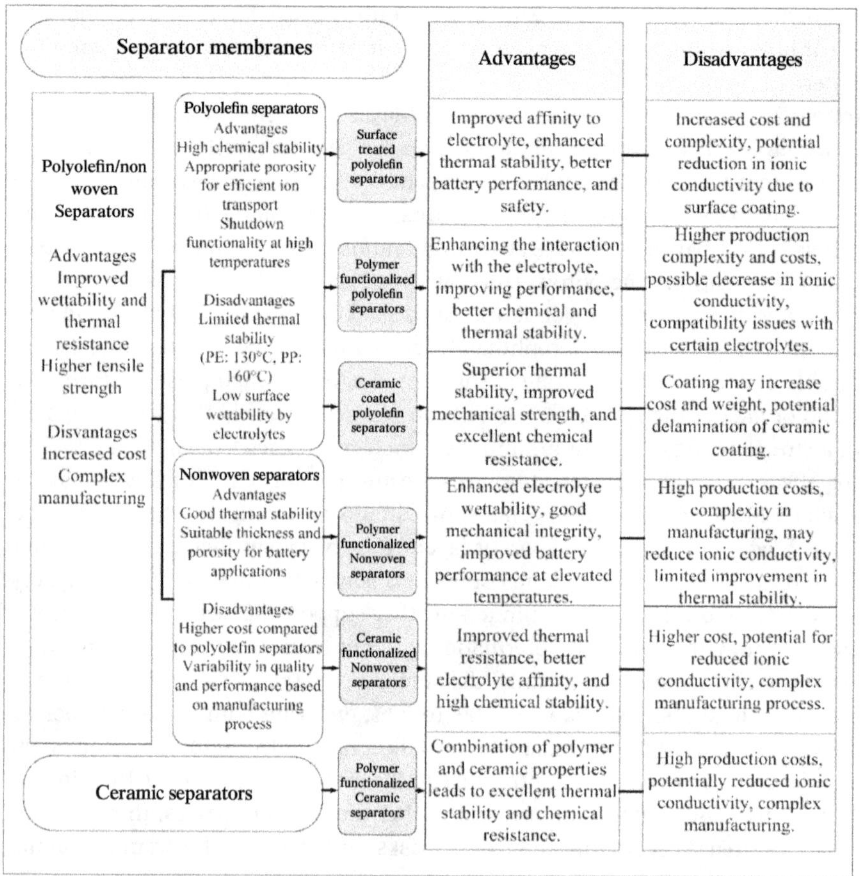

Figure 1.4 The evolutionary classification of separators and the overall advantages and disadvantages of each type of separator. Source: Lingappan et al. [31]/with permission of Elsevier.

resistance, and helps maintain energy density, supporting the battery's longevity and aligning with industry goals for safer, high-performance LIBs [30].

Figure 1.4 summarizes the evolutionary classification of separators and the overall advantages and disadvantages of each type of separator.

To enhance battery safety, researchers are concentrating on developing advanced ceramic separators. These separators, which may be designed as single-layer or multilayer structures, combine the thermal stability and mechanical strength of ceramic materials with the flexibility and low melting temperature of polymer films like polyethylene (PE). This approach aims to combine the high melting temperature and excellent chemical resistance of ceramic layers with the easy processing and cost-effectiveness of polymer films.

Ceramic separators can facilitate the shutdown effect, which enhances the overall safety of batteries through the separators' inherent thermal sensitivity. As the battery heats up due to internal short circuits or external conditions, the ceramic separator material responds by melting or undergoing structural changes upon reaching a critical temperature. This melting or phase transition causes the separator to collapse or close off the ionic transport pathways, effectively stopping ionic transport and halting the battery's electrochemical reactions, preventing further temperature increases and thermal runaway, thereby providing a crucial protection mechanism for battery safety [32].

1.7.1 Polyolefin Separators

Polyolefin separators are essential components in LIBs, preventing physical contact between the anode and the cathode while allowing the transport of ions. There are several types of polyolefin separators, each offering unique features and benefits. Surface-treated polyolefin separators are modified to enhance specific properties such as wettability and adhesion. Accordingly, this treatment improves the separator's compatibility with electrolyte and electrode materials, leading to better battery performance and extended cycle life. Polymer-functionalized polyolefin separators incorporate polymer chains into the polyolefin structure. These polymer chains can provide additional functionalities, such as increased thermal stability and improved mechanical strength. The functional groups on the polymer chains can interact with the electrolyte and electrode materials, potentially enhancing battery performance. Ceramic-coated polyolefin separators feature a thin ceramic layer applied to the substrate's surface. This ceramic coating offers several advantages, including enhanced thermal stability, improved resistance to shrinkage at high temperatures, and reduced penetration of lithium dendrites. The ceramic layer also helps to maintain the separator's structural integrity under extreme conditions, contributing to the overall safety and reliability of the battery [33].

1.7.2 Nonwoven Separators

Nonwoven separators are essential components in modern battery technology, especially in LIBs. They are designed to prevent electrical short circuits between the anode and cathode while facilitating ion transport. Nonwoven separators can be further categorized into polymer-functionalized nonwoven separators and ceramic-functionalized nonwoven separators, each offering unique benefits for enhanced battery performance. Polymer-functionalized nonwoven separators are modified with polymers to improve their properties. These polymers can provide additional functionalities such as increased thermal stability and improved mechanical strength. The functional groups on the polymers can also interact with the electrolyte and electrode materials, potentially enhancing battery performance. Conversely, ceramic-functionalized nonwoven separators feature a ceramic coating that offers several advantages. This ceramic layer can enhance thermal stability, reduce shrinkage at high temperatures, and minimize the penetration of lithium dendrites. The ceramic coating also helps to maintain the separator's structural integrity under extreme conditions, contributing to the overall safety and reliability of the battery [34].

1.7.3 Ceramic Separators

Polymer-functionalized ceramic separators are a type of ceramic separator modified with polymers to enhance their performance and properties. These separators combine the strength and durability of ceramic materials with the flexibility and compatibility of polymers. The polymeric layer is typically applied to the surface of the ceramic separator through methods such as chemical vapor deposition, coating, or dip-coating. Furthermore, this layer can be designed to have specific functional groups that interact with the electrolyte and electrode materials in the battery. One of the key benefits of polymer-functionalized ceramic separators is their improved electrical conductivity. The polymeric layer acts as a conductive pathway between the electrodes, reducing resistance and allowing for faster charge and discharge rates. Additionally, the polymeric layer helps to prevent short circuits and improves safety by providing a barrier between the electrodes and the surrounding environment.

Another advantage of polymer-functionalized ceramic separators is their ability to withstand high temperatures without losing structural integrity. Ceramic materials are known for their excellent thermal stability, which is crucial for batteries in applications such as electric vehicles and power grid storage systems [35].

1.8 Future Trends in Batteries

As the demand for advanced energy storage solutions continues, research and development efforts have expanded beyond traditional lithium-based battery technologies. Over the past decade, significant progress has been made in exploring alternative technologies such as metal-air, sodium-beta, and other metal-ion batteries, paving the way for future trends in battery development [36].

Metal-air batteries, recognized for their high specific energy, use metals like zinc, aluminum, iron, magnesium, or calcium as the anode, with air as the cathode. These batteries can achieve a specific energy storage capacity of up to 600 Wh/kg and an energy density of up to 400 Wh/L [37]. They are available as either electrical or mechanical rechargeable types, with the latter being more prevalent due to easier refueling and recycling. However, they face challenges such as low specific power and electrolyte carbonization. Lithium-air batteries ($Li{-}O_2$), offering extremely high specific energy (~11,700 Wh/kg), are currently being researched for their potential use in electric vehicles [36].

Sodium-, zinc-, and magnesium-ion batteries are emerging as alternatives to LIBs. Sodium-ion batteries are gaining attention due to the abundant availability of sodium, high key performance indicators (KPIs), and energy densities up to 210 Wh/kg [38]. However, Na-ion batteries typically have lower energy density than Li-ion batteries, which means they store less energy for the same weight. Additionally, the development of high-performance sodium-ion (Na-ion) battery materials is ongoing, facing challenges in identifying suitable anodes, cathodes, and electrolytes that can match or exceed the performance of lithium-ion (Li-ion) systems [17]. Zinc-ion batteries, such as the reversible $Zn{-}MnO_2$ battery, show promise for large-scale energy storage in electric vehicles with high voltage and large capacity [39]. In contrast, magnesium-ion batteries offer high specific energy, low cost, eco-friendliness, and safety [40].

Flow batteries convert chemical energy from electroactive materials into electricity through a membrane separating two chemical components. The electrolytes are stored in external tanks and flow through the cell to produce electricity, and there are two main types: redox flow batteries and hybrid flow batteries (e.g., Zn–Cl and Zn–Br systems). Notably, flow batteries offer low energy density, making them unsuitable for portable electronics. However, they have long lifespans, minimal self-discharging, and the ability to discharge fully without affecting the cycle lifespan [41]. Table 1.4 provides a comparison of the different batteries described in this section.

The safety profiles of various battery technologies, including lithium-ion, sodium-ion, metal-air, and flow batteries, differ significantly due to their distinct chemistries and operational conditions. LIBs are known for their high energy density but pose risks such as thermal runaway, which can be triggered by physical damage, overcharging, or exposure to elevated temperatures. This risk is exacerbated by the flammable organic electrolytes present in them. In contrast, sodium-ion batteries offer enhanced safety by leveraging less flammable solvents like propylene carbonate, though they face challenges with high operating temperatures and lower round-trip efficiency [42]. Metal-air batteries are promising but suffer from high charge voltages and the reactivity of oxygen intermediates, which can lead to degradation of the cathode and electrolyte [43]. Flow batteries, conversely, are valued for their safety in stationary applications due to their nonflammable nature, though

Table 1.4 Comparison of the performance of different batteries.

Type	Energy density (Wh/L) [35]	Energy efficiency (%) [35]	Cost	Cycle life	Safety	Environmental effect
Lithium-based batteries	240–300	80–95	High	High	Medium	Sustainable
Sodium-beta batteries	160–250	75–80	High	Medium	Low	Critical
Metal-air batteries	100–300	50–80	Low	Low	Medium	Moderate
Flow batteries	60–90	70–85	Medium	High	High	Sustainable

their electrolytes can be corrosive [44]. As the need for advanced energy storage solutions escalates, efforts are intensifying to overcome the safety challenges inherent in each battery technology.

1.9 Summary

An LIB transfers lithium ions between the anode and cathode through an electrolyte and separator during charge and discharge cycles, while electrons flow through an external circuit. The performance of LIBs is influenced by the choice of cathode materials, which dictate the battery's energy density, cycle life, safety, and cost. As such, several common cathode materials and their properties were discussed. LCO offers high energy density but has high cost and poor safety; LMO is cost-effective and safe but has a shorter cycle life; LFP is comparatively safer with long cycle life but lower energy density; lithium nickel-cobalt-manganese oxide (NMC) provides a balanced mix of high energy density, safety, and cost; and lithium NCA offers high energy density but faces thermal stability challenges. Anode materials, such as carbon-based, silicon-based, and metal/alloying types, each present distinct characteristics and challenges. Carbon-based materials like graphite are cost-effective and stable but have lower energy density. Silicon-based materials offer high capacity but face challenges such as volume expansion. Metal and alloy anodes provide high energy density but suffer from structural instability. Understanding these properties is essential for selecting the most appropriate cathode and anode material for a specific battery application.

Electrolytes in LIBs facilitate ion movement between electrodes, influencing performance, safety, and longevity. Liquid electrolytes, typically composed of organic solvents, lithium salts, and additives, offer high ionic conductivity and stability but face safety risks due to low thermal stability. Solid electrolytes, including oxides, sulfides, and phosphates, provide superior safety and stability, with high ionic conductivity and thermal resistance. However, they encounter challenges such as low ionic conductivity and interfacial contact issues. Ongoing research and development in both liquid and solid electrolytes aim to optimize battery performance and safety, addressing limitations and enhancing the overall efficiency of LIBs.

In LIBs, separators are essential for ion transport and safety, acting as nonconductive barriers while allowing lithium ions to pass between electrodes. Typically composed of polymer materials like polypropylene (PP) and polyethylene (PE) or composite membranes, they ensure electrical insulation and stability. Advanced ceramic separators are being developed to enhance safety by combining the thermal stability and mechanical strength of ceramics with the flexibility of polymers. These separators can incorporate a shutdown effect, where materials within the ceramic layer melt at high temperatures to block ion flow, thus preventing thermal runaway and improving battery safety. Choosing the right combination of materials is essential to enhancing the intrinsic safety of LIBs. The choice of cathode and anode materials not only affects performance but also directly impacts thermal stability, structural integrity, and overall safety.

References

1 Mayemba, Q., Mingant, R., Li, A. et al. (2024). Aging datasets of commercial lithium-ion batteries: a review. *Journal of Energy Storage* 83: 110560.

2 Xia, S., Wu, X., Zhang, Z. et al. (2019). Practical challenges and future perspectives of all-solid-state lithium-metal batteries. *Chem* 5 (4): 753–785.

3 Zhang, Y.C., Briat, O., Boulon, L. et al. (2019). Non-isothermal Ragone plots of Li-ion cells from datasheet and galvanostatic discharge tests. *Applied Energy* 247: 703–715.

4 Zhao, Z., Xu, Z., Chen, J. et al. (2024). A review on functional applications of polyphosphazenes as multipurpose material for lithium-ion batteries. *Journal of Energy Storage* 85: 111049.

5 Sadeghi, A. and Ghaffarinejad, A. (2024). Recent advances of electrode materials based on nickel foam current collector for lithium-based batteries–a review. *Journal of Power Sources* 600: 234275.

6 Guo, R., Wang, F., Rhamdhani, M.A. et al. (2024). Managing the surge: a comprehensive review of the entire disposal framework for retired Lithium-ion batteries from electric vehicles. *Journal of Energy Chemistry* 92: 648–680.

7 Catenaro, E., Rizzo, D.M., and Onori, S. (2021). Experimental analysis and analytical modeling of Enhanced-Ragone plot. *Applied Energy* 291: 116473.

8 Yuxuan, J. (2022). Preparation and electrochemical properties of high-pressure lithium cobaltate cathode materials. Xi'an University of Architecture and Technology. https://doi.org/10.27393/d.cnki.gxazu. 2022.000883.

9 China Science Bulletin (2024, Mar. 12). China's lithium battery 'breakout record' [Online]. https://baijiahao.baidu.com/s?id=1793318849244022241&wfr=spider&for=pc.

10 Energy Scholar (2022, Feb. 15). From LiMn2O4 to partially disordered Li2MnNiO4: progress in lithiated spinel anodes for lithium-ion batteries [Online]. https://zhuanlan.zhihu.com/p/467651714.

11 Li, Z. (2022). Preparation and modification of spinel-type lithium manganate-based cathode materials. China University of Mining and Technology, pp. 001728, 2022.

12 Padhi, A.K., Nanjundaswamy, K.S., and Goodenough, J.B. (1997). Phospho-olivines as positive-electrode materials for rechargeable lithium batteries. *Journal of the Electrochemical Society* 144 (4): 1188. https://doi.org/10.1149/1.1837571.

13 Mados, E., Atar, I., Gratz, Y. et al. (2024). Polymer-based LFP cathode/current collector microfiber-meshes with bi-and interlayered architectures for Li-ion battery. *Journal of Power Sources* 603: 234397.

14 Sina Watch (2024, Feb. 22). Xinda has completed the development of 4C Li-FePO4 battery, and will hold a product launch afterward [Online]. https://k.sina.com.cn/article_1826017320_6cd6d02802001a7ym.html.

15 Qilin, T. (2022). Strukturelles Design und Leistungsstudie von Lithium-Nickel-Kobalt-Mangan-Oxid (NCM811) Kathodenmaterial. Guilin University of Electronic Science and Technology. https://doi.org/10.27049/d.cnki.ggldc.2022.000125.

16 Chen, X., Shen, W., Vo, T.T. et al. (2012). An overview of lithium-ion batteries for electric vehicles//2012 10th International Power & Energy Conference (IPEC). IEEE, pp. 230–235.

17 Lei, F. (2023). Preparation and modification of hollow carbon materials and their performance in lithium-sulfur battery cathode. Jilin University, pp. 007484.

18 Mishra, A.K. and Patial, B.S. (2024). A review on recent advances in anode materials in lithium ion batteries. *Materials Today Electronics* 7: 100089.

19 Kim, D.S., Kim, S.H., and Hong, J.Y. (2023). Influence of oxidation on the electrochemical properties of silicon oxide-based carbon composites for anode materials of lithium-ion batteries. *Carbon Trends* 12: 100279.

20 Hawari, N.H., Huang, X., Butarbutar, L.M. et al. (2024). Unveiling the effect of alloy-type coating for lithium metal plating and stripping on anode-free lithium metal batteries. *Journal of Alloys and Compounds* 975: 172988.

21 Jia, W., Zhang, J., Zheng, L. et al. (2024). Lithium-rich alloy as stable lithium metal composite anode for lithium batteries. *eScience* 4: 100266.

22 Zeng, Q., Lv, Z., Li, S. et al. (2023). Electrolytes for liquid metal batteries. *Materials Research Bulletin* 170: 112586.

23 Lin, Y., Peng, Q., Chen, L. et al. (2024). Organic liquid electrolytes in sodium-based batteries: actualities and perspectives. *Energy Storage Materials* 67: 103211.

24 Sun, J., Yuan, H., Yang, J. et al. (2023). Electrolytes for better and safer batteries: liquid, solid or frameworked, what's next? *Next Materials* 1 (3): 100024.

25 Sina Auto (2023, Jun. 21). Toyota electrification move: all-solid-state batteries on board by 2027 [Online]. https://auto.sina.com.cn/zz/hy/2023-06-21/detail-imyxyyax5432352.shtml.

26 Lee, Y.J., Hong, S.B., Lee, H.J. et al. (2023). Flexible and thin sulfide-based solid electrolyte sheet with Li+-ion conductive polymer network for all-solid-state lithium-ion batteries. *Chemical Engineering Journal* 477: 146983.

27 Lv, J. (2022). Construction of organic solid-state lithium/sodium secondary battery based on butanedinitrile-based electrolyte. Yanshan University. DOI:10.27440/d.cnki.gysdu.2022. 000322.

28 Wulandari, T., Fawcett, D., Majumder, S.B. et al. (2023). Lithium-based batteries, history, current status, challenges, and future perspectives. *Battery Energy* 2 (6): 20230030.

29 Guo, K., Guo, D., Kong, G., and Tao, X. (2021). US Patent 10,964,927 B2, Mar. 30, 2021.

30 Tao, T. and Mo, M.L. (2024). US Patent 11,923,498 B2, Mar. 5, 2024.

31 Lingappan, N., Lee, W., Passerini, S. et al. (2023). A comprehensive review of separator membranes in lithium-ion batteries. *Renewable and Sustainable Energy Reviews* 187: 113726.

32 Whispers (2022, Aug. 09). Lithium-ion Battery Separator Preparation, Testing and Technical Requirements [Online]. https://www.eet-china.com/mp/a152489.html.

33 Yao, S., Zhang, T., Ma, C. et al. (2024). Comparative study of the electrochemical performances of different polyolefin separators in lithium/sulfur batteries. *Materials Research Bulletin* 171: 112604.

34 Kim, Y., Park, B.J., Yu, J.S. et al. Polysiloxane-Coated PI Nonwoven Separators with Higher Thermal and Electrochemical Stability for Lithium Ion Battery Application. Electrochemical Society Meeting Abstracts, The Electrochemical Society, Inc., vol. 2023 (2), pp. 336–336.

35 Jung, Y.C., Kim, S.K., Kim, M.S. et al. (2015). Ceramic separators based on Li+-conducting inorganic electrolyte for high-performance lithium-ion batteries with enhanced safety. *Journal of Power Sources* 293: 675–683.

36 Waseem, M., Ahmad, M., Parveen, A., and Suhaib, M. (2023). Battery technologies and functionality of battery management system for EVs: current status, key challenges, and future prospectives. *Journal of Power Sources* 580: 233349.

37 Yu, T., Fu, J., Cai, R. et al. (2017). Nonprecious electrocatalysts for Li-air and Zn-air batteries: fundamentals and recent advances. *IEEE Nanotechnology Magazine* 11 (3): 29–55.

38 Yabuuchi, N., Kubota, K., Dahbi, M., and Komaba, S. (2014). Research development on sodium-ion batteries. *Chemical Reviews* 114 (23): 11636–11682.

39 Mageto, T., Bhoyate, S.D., Mensah-Darkwa, K. et al. (2023). Development of high-performance zinc-ion batteries: issues, mitigation strategies, and perspectives. *Journal of Energy Storage* 70: 108081.

40 Guo, Q., Zeng, W., Liu, S.L. et al. (2021). Recent developments on anode materials for magnesium-ion batteries: a review. *Rare Metals* 40 (2): 290–308.

41 Doetsch, C. and Pohlig, A. (2020). The use of flow batteries in storing electricity for national grids. In: *Future Energy*, 263–277. Elsevier.

42 Zhao, L., Zhang, T., Li, W. et al. (2023). Engineering of sodium-ion batteries: opportunities and challenges. *Engineering* 24: 172–183.

43 Salado, M. and Lizundia, E. (2022). Advances, challenges, and environmental impacts in metal–air battery electrolytes. *Materials Today Energy* 28: 101064.

44 Sánchez-Díez, E., Ventosa, E., Guarnieri, M. et al. (2021). Redox flow batteries: status and perspective towards sustainable stationary energy storage. *Journal of Power Sources* 481: 228804.

2

Global Suppliers of Battery Raw Materials

Simin Peng, Guanwei Jiang, Yu Zhang, Yulun Zhang, Kianoush Naeli, Virendra Jadhav, Sanjay Tiku, Sahithi Maddipatla, and Lingxi Kong

Lithium-ion batteries (LIBs) have become the cornerstone of modern energy storage, powering everything from smartphones to electric vehicles (EVs). Production of these batteries hinges on the availability of key raw materials, including lithium, cobalt, nickel, manganese, and graphite. Understanding the supply chain dynamics and identifying the primary global suppliers of these materials is crucial for navigating the challenges associated with sourcing and ensuring a stable supply.

2.1 Introduction

The battery manufacturing industry is currently in a phase of rapid growth and transformation. With the increasing worldwide demand for clean energy and electric transportation, there is a growing urgency for high-performance, sustainable, and cost-effective battery technologies. Many countries and companies are ramping up investments to enhance the scale and efficiency of battery production while continuously driving innovation to address the increasingly severe challenges of climate change. However, the industry also faces various challenges, including stability of raw material supply chains, standardization of technologies, and compliance with environmental regulations, amidst advancing technologies and market competition.

Over the years, with the continuous advancement of technology, battery manufacturing has evolved into several highly specialized nodes around the key components, including positive electrode materials, negative electrode materials, electrolytes, separators, binders, conductive additives, and current collectors. Positive electrode materials typically consist of compounds such as lithium cobalt oxide (LCO), lithium manganese oxide (LMO), lithium iron phosphate (LFP), and ternary materials such as lithium nickel manganese cobalt oxides (NMC) and lithium nickel cobalt aluminum oxide (NCA). By comparison, negative electrode materials are commonly composed of carbon, the selection of which significantly impacts the performance and characteristics of the battery.

The electrolyte, whether liquid or solid, facilitates ion transport within the battery, and liquid electrolytes typically comprise solutions or polymers. Next, separators serve as isolating layers between the positive and negative electrodes, preventing direct contact and thus avoiding short circuits and overheating. Importantly, binders and conductive additives are employed to fix electrode materials and enhance the electrode's conductivity, thereby improving stability and performance.

The Safety Challenges and Strategies of Using Lithium-Ion Batteries, First Edition. Edited by Michael G. Pecht.

Current collectors are utilized to collect current from the electrodes and guide it to external circuits, often made of materials such as aluminum foil or copper foil. With all of this in mind, these components collectively form the core structure of the battery, influencing its energy density, cycle life, and safety performance.

2.2 Analysis of Raw Materials

The raw materials for batteries are mainly composed of nickel ore, lithium, cobalt, and graphite, among which nickel ore is mainly divided into sulfide copper nickel ore and oxide nickel ore. These two types undergo distinct ore selection and smelting processes, broadly classified into pyrometallurgical and hydrometallurgical methods. According to a research report released by the China Commercial Industry Research Institute, the global nickel mine production reached 3.3 million tons in 2022, marking a 21% year-on-year increase [1]. Analysts at the China Commercial Industry Research Institute predict that nickel ore production will rise to 3.35 million tons in 2023 and reach 3.5 million tons in 2024. The global nickel production forecast is illustrated in Figure 2.1.

Presently, global manganese ore resources are primarily concentrated in South Africa, China, Australia, and Brazil. According to another report released by the China Commercial Industry Research Institute [2], in 2022, South Africa ranked first with manganese ore reserves reaching 640 million tons, while China ranked second globally with manganese ore reserves amounting to 280 million tons. The estimated global manganese ore reserves are illustrated in Figure 2.2.

Lithium carbonate and lithium hydroxide stand as the primary raw materials utilized in the fabrication of battery cathodes, with major reserves concentrated in South America and Australia. Consequently, they comprise the most prevalent cathode materials in the realm of new energy batteries. Lithium is relatively abundant on Earth, ranking 27th in abundance and comprising approximately 0.0065% of the crust, proving that global lithium resources are ample. According to data from the United States Geological Survey (USGS) [3], sustained exploration endeavors have substantially augmented the known lithium resource base. As of 2022, the global lithium resource total was estimated to be 98 million metric tons, with global reserves reaching 26 million metric tons.

However, the global lithium resources are concentrated. In terms of resource quantity, 53% of global lithium resources are distributed in the “lithium triangle” of South America (Bolivia, Chile, Argentina), followed by Australia, China, and the United States. In terms of reserves, of the

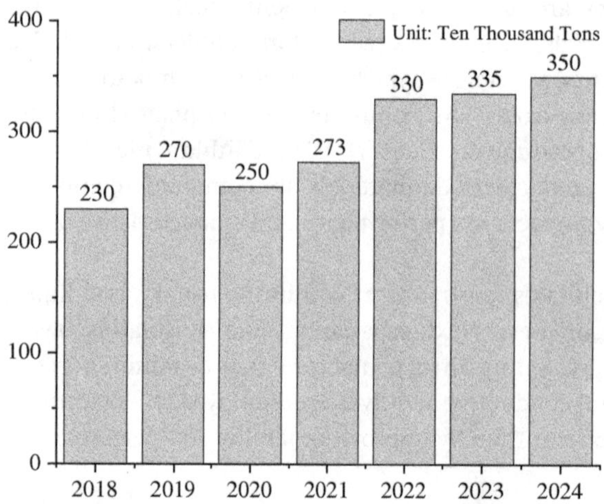

Figure 2.1 Global nickel ore production forecast trend. Source: [1]/NetEase.

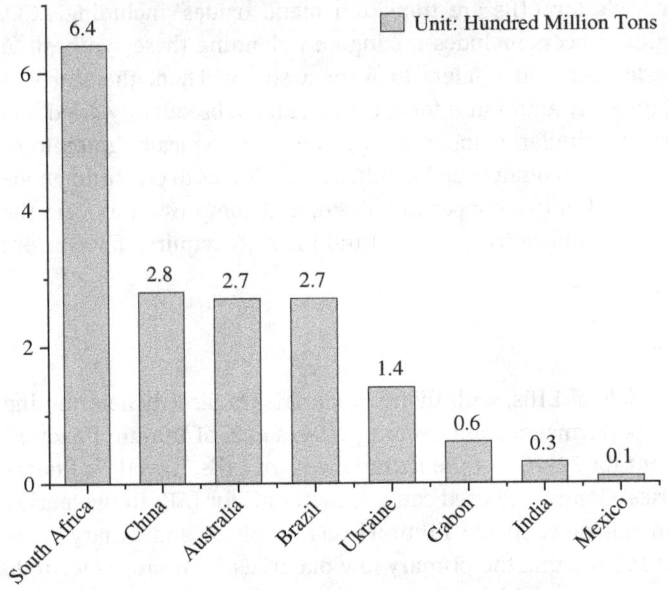

Figure 2.2 Estimates of global manganese ore reserves by country in 2022. Source: [2]/NetEase.

Figure 2.3 Estimated distribution of global lithium reserves in 2022.

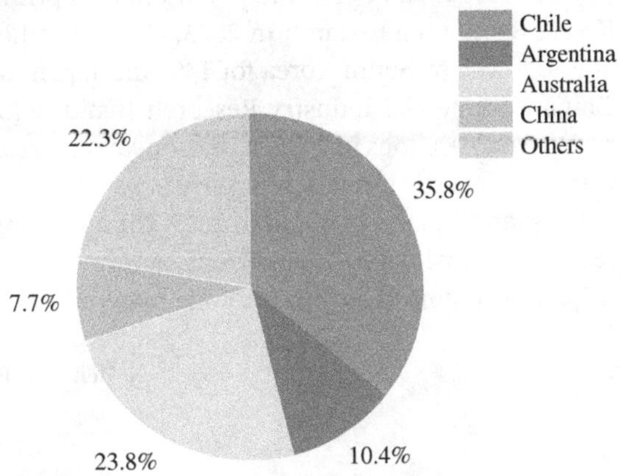

26 million metric tons estimated to exist in 2022, Chile holds the largest lithium reserve at 9.3 million tons, accounting for 35.8% of the total, followed by Argentina with 2.7 million tons (10.4%), Australia with 6.2 million tons (23.8%), and China with 2.0 million tons (7.7%). This distribution is also illustrated in Figure 2.3.

2.3 Battery Cell Component Production

Batteries comprise several highly specialized components, including cathode and anode materials, electrolytes, and separators. Accordingly, the production of these components requires advanced materials chemistry and engineering technologies. The most intricate processing involves forming battery active materials from high-purity chemicals generated during raw material processing, such as lithium hydroxide and nickel sulfate. Further, these materials are processed through specialized synthesis to produce active materials for cathodes and anodes.

The primary cathode active materials for LIBs are transition metal oxides, including LCO, NMC, NCA, and LFP. The production process includes mixing and blending these synthesized active materials with conductive additives and binders to form a slurry. Then, the slurry is uniformly coated onto a metal foil, typically aluminum for cathodes, and subsequently dried and compressed to form the cathode sheet. Similarly, anode active materials, typically graphite or other carbon-based materials, are mixed with binders and conductive additives to create an anode slurry. Following this, the slurry is coated onto a copper foil, dried, and compressed to form the anode sheet. These processes ensure the uniformity and structural integrity required for efficient battery performance.

2.3.1 Positive Electrode Materials

Cathode materials are the core materials of LIBs, with their characteristics directly determining the energy density, cycle life, safety performance, and overall performance of lithium batteries. As a result, cathode materials account for 30–40% of the material cost of LIBs, directly influencing the overall cost of lithium batteries. Commonly used cathode materials for LIBs in the market include lithium cobalt oxide, lithium manganese oxide, lithium iron phosphate, and ternary materials (NMC, NCA), among others. It follows that the primary raw materials for positive electrode materials in batteries include lithium carbonate, nickel sulfate, cobalt sulfate, and lithium ternary materials.

Global production of cathode materials is primarily concentrated in China, Japan, and South Korea. Based on a research in 2023, China's cathode production capacity accounts for 70% globally, followed by South Korea for 14% and Japan for 11% [4]. According to a report released by the China Commercial Industry Research Institute [5], China's ternary materials shipment volume reached 640,000 tons in 2022, a year-on-year increase of 47%. Primarily, market growth is driven by Chinese domestic growth in ternary power batteries and two-way demand from overseas exports. Subsequently, in the first half of 2023, China's shipment volume of ternary cathode sheets for LIBs reached approximately 299,000 tons, accounting for 26% of the global market. To illustrate this, the shipment volume of ternary cathode materials in China is shown in Figure 2.4.

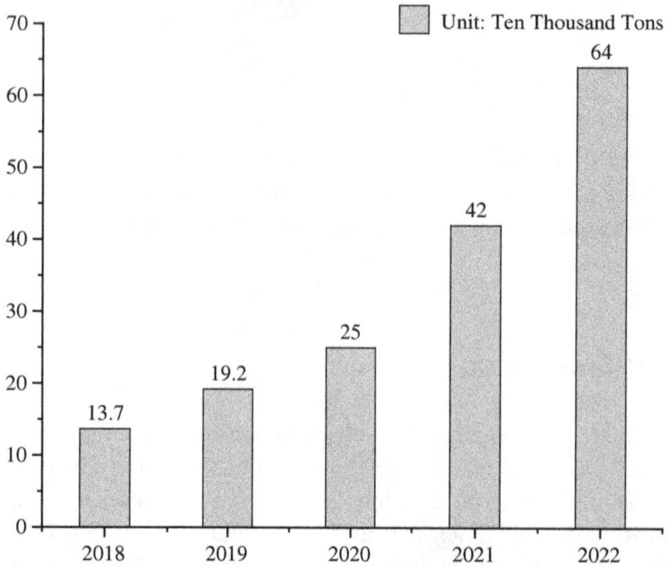

Figure 2.4 Shipment volume of ternary cathode materials in China [5].

The global market for battery-grade lithium carbonate was estimated at 520,000 metric tons in 2023, and it is projected to reach 1.22 million metric tons by 2030, with a compound annual growth rate (CAGR) of 13% from 2023 to 2030. Among these figures, China stands out with a colossal market size for battery-grade lithium carbonate. The total sales volume in the Chinese market was 350,000 metric tons in 2023 and is projected to reach 860,000 metric tons by 2030, with a CAGR of 14% from 2023 to 2030. The lithium carbonate market is compared to China's portion in Figure 2.5.

According to data from the Sinopec Petroleum Industry Research Institute (SPIR), in 2022, among the top 10 global enterprises in terms of shipments of ternary materials, eight were Chinese enterprises; these are illustrated in Figure 2.6.

For lithium cobalt oxide (LCO) cathode materials, the industry relies on a few notable suppliers, including Umicore (headquartered in Brussels, Belgium) [8], Targray (headquartered in Kirkland, Canada) [9], and BASF Shanshan battery materials (headquartered in Changsha, China) [10].

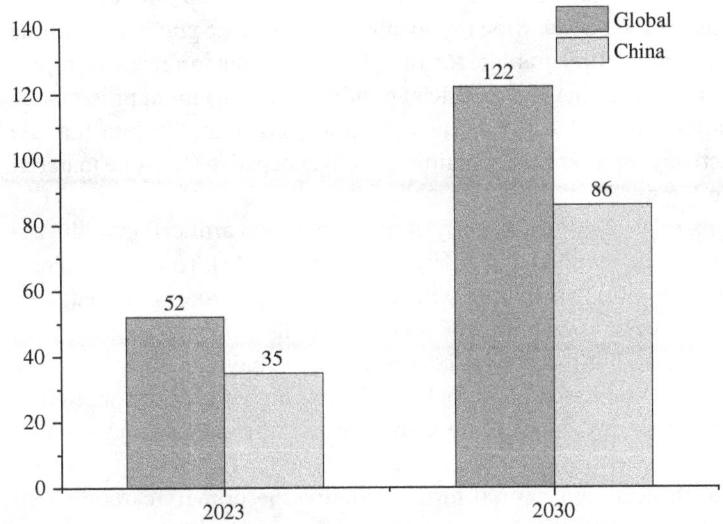

Figure 2.5 Global and China lithium carbonate market sales. Source: Adapted from [6].

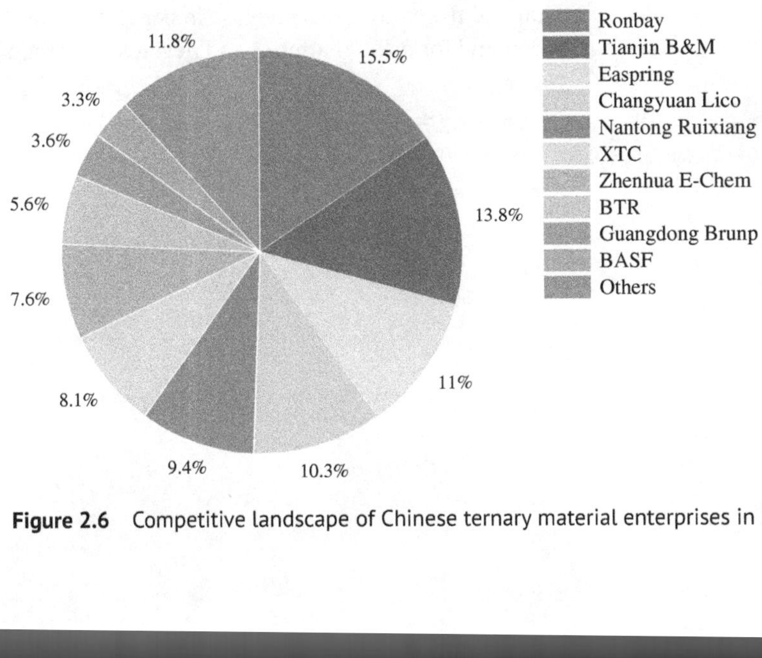

Figure 2.6 Competitive landscape of Chinese ternary material enterprises in 2022 [7].

2.3.2 Negative Electrode Materials

The anode material significantly impacts the initial efficiency and cycling performance of lithium batteries. In essence, it is composed of a mixture of negative electrode active materials (either carbon or noncarbon materials), binders, and additives. Then, this mixture is uniformly coated onto both sides of a copper foil, which is dried and rolled to form a cohesive layer.

Carbon-based negative electrodes include graphite, hard carbon, and soft carbon, further categorized into artificial graphite, natural graphite, and mesophase carbon microspheres. Noncarbon-based negative electrodes comprise lithium titanate, tin-based alloys, and silicon-based alloys. LIB anode materials have evolved alongside technological advancements. Starting from solely artificial graphite, which still dominates, natural graphite and mesophase carbon microspheres are also booming. Additionally, manufacturers utilize soft carbon/hard carbon, amorphous carbon, lithium titanate, silicon-carbon alloys, and various other anode materials.

Considering the cost and performance of the battery, lithium battery manufacturers predominantly choose graphite materials for the anode. Globally, artificial or synthetic graphite accounts for 78% of battery anode materials and natural graphite for 14% [11]. According to a research report on negative electrode materials used in China [12], artificial graphite accounts for approximately 83% of the market, natural graphite about 16%, and other materials, including lithium titanate and silicon-carbon alloys, collectively represent the remaining 1%. The negative electrode material market in China is illustrated in Figure 2.7.

The supply chains for these materials vary significantly. Both natural and artificial graphite are readily available, with substantial production in China, Brazil, and Canada [13]. The supply of natural graphite is largely influenced by mining activities, whereas artificial graphite is derived from petroleum coke and coal tar pitch, necessitating industrial processes for its production. Conversely, lithium titanate is less commonly used but has a more complex supply chain, and it is primarily produced in specialized facilities. Silicon-based alloys, while offering higher capacity, face challenges in large-scale production and supply stability due to their recent introduction and ongoing development.

Graphite, encompassing both artificial and natural forms, remains the primary carbon-based anode material. Artificial graphite is predominantly used in high-capacity vehicle power batteries, high-rate batteries, and medium- to high-end electronic product LIBs. In contrast, natural graphite is mainly utilized in small LIBs and general-purpose electronic product LIBs.

Global trends toward automotive electrification and the continuous increase in penetration rate of new energy vehicles are driving the surge in demand for power batteries and negative electrode

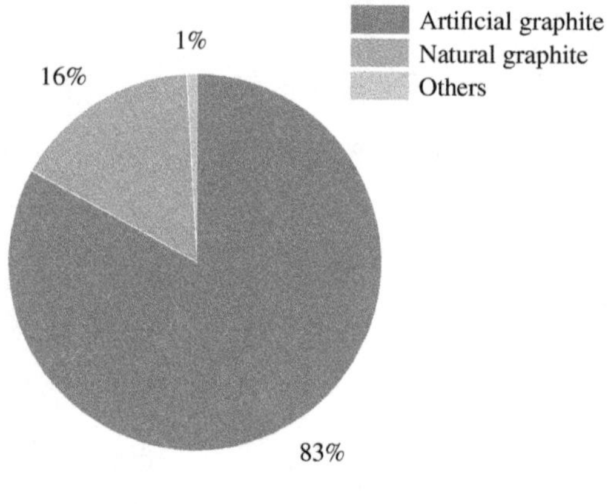

Figure 2.7 Negative electrode materials used in China in 2021.

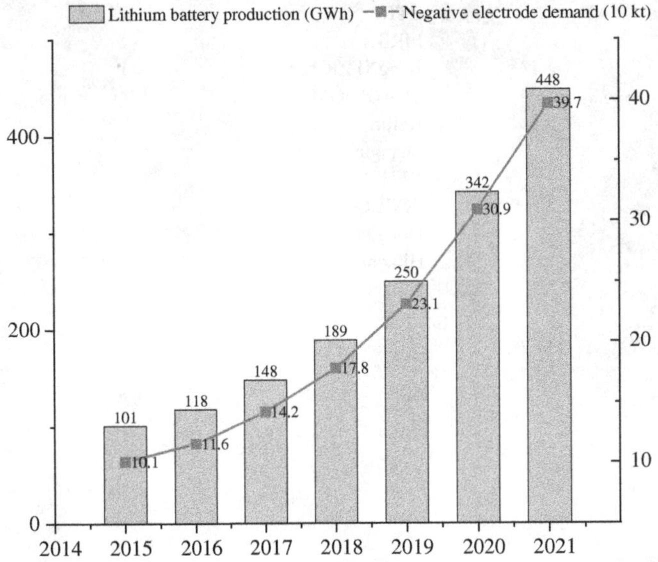

Figure 2.8 Global lithium battery production compared to negative electrode demand. Source: [14]/Zhihu Column.

materials. Globally, lithium battery production reached 342 GWh in 2020, with a demand for negative electrode materials of 309,000 tons. It was estimated that in 2021, the lithium battery production would reach 448 GWh, with a demand for all negative electrodes of 397,000 tons [14]. However, according to the International Energy Agency [15], the global LIB manufacturing capacity was 1.57 TWh in 2022. The estimation for 2025 is 3.97 TWh, which is more than double when compared to the data from 2022. The global production of lithium batteries and the demand for negative electrodes are compared in Figure 2.8.

The negative electrode active materials of batteries mainly include graphite, hard carbon, and silicon-based materials. Graphite is a natural mineral with good conductivity, capable of accommodating the intercalation and deintercalation of lithium ions. According to data released by Xinluo Information in China, the global production of negative electrode materials reached 1.7621 million tons in 2023, with a year-on-year growth rate of 20%. China dominates the production of anodes. Ever back in 2021, the proportion of Chinese negative electrode material production was above 90% [16]. In terms of shipments, the global volume of battery negative electrode materials reached 1.6795 million tons shipped, with China accounting for as much as 95%. Figure 2.9 shows the worldwide companies' share distribution in lithium battery negative electrode materials.

With such a significant market presence, it is crucial to identify the leading suppliers that drive this industry. For graphite anode materials, which continue to be a staple in battery production, key industry suppliers include SGL Carbon (Wiesbaden, Germany) [18], Targray (Kirkland, Canada) [9], and BTR (Shenzhen, China) [19]. Currently, the battery industry is experiencing a surge of innovation, particularly in the development of silicon anode technology. This shift is led by a range of innovative startups, including Sila (Alameda, USA) [20], Enevate (Irvine, USA) [21], Group14 (Seattle, USA) [22], Enovix (Fremont, USA) [23], Amprius (Fremont, USA) [24], among others, who are making significant strides in this field. Most of the top innovative startups are located in the United States.

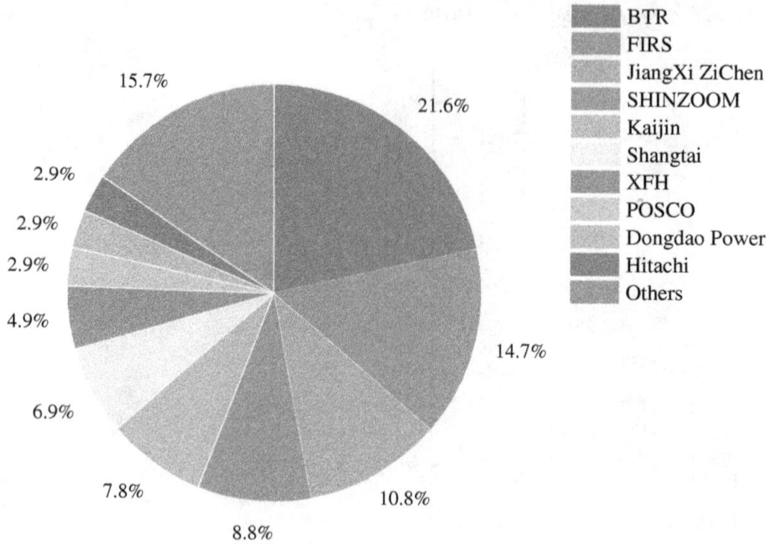

Figure 2.9 Share distribution of global lithium battery negative electrode materials in 2022 [17].

2.3.3 Electrolytes

LIB electrolyte salts are critical components that facilitate ion conduction in LIBs, significantly influencing key performance metrics such as energy density, charge–discharge efficiency, cycle life, and safety. These salts are indispensable to the overall functionality of LIBs.

Electrolyte salts for LIBs are classified based on their chemical properties and application requirements, into inorganic salts, organic salts, and solid electrolytes. Solid electrolytes, in particular, have attracted considerable attention recently due to their high mechanical strength, superior ionic conductivity, and excellent chemical stability. These characteristics make solid electrolytes promising candidates for improving the energy density and safety of LIBs.

In terms of downstream applications, LIB electrolyte salts are utilized in consumer electronics, electric vehicles, and energy storage systems. As of 2023, the consumer electronics sector represents the largest application market, accounting for 57.55% of the total usage [25].

The global electrolyte salt materials market is undergoing rapid development and transformation. In recent years, the demand for high-performance LIBs has been steadily increasing across various sectors, including electric vehicles, wearable devices, and energy storage systems. As a critical component of LIBs, the market demand for electrolyte salt materials has correspondingly expanded. Data indicate that in 2023, the global consumption of LIB electrolyte salt materials reached 118,700 tons, representing a 16.80% increase compared to 2022. Despite the growth in consumption, market revenue in 2023 decreased by 9.98% year on year to $2.125 billion. The market is expected to exhibit a fluctuating growth pattern in the coming years, with revenue initially declining before rising again. Projections suggest that by 2029, the global market revenue will increase to $3.093 billion [26].

Geographically, the Asia-Pacific region dominates the global market for LIB electrolyte salts, with a market share exceeding 80%. Within this region, China stands out as the largest consumer and revenue contributor. According to BECZIS research data, in 2023, China's sales of LIB electrolyte salts amounted to 87,400 tons [27], generating $1.547 billion in revenue, which accounted for 72.80% of the global market share. The top three companies, Guangzhou Tinci Materials Technology,

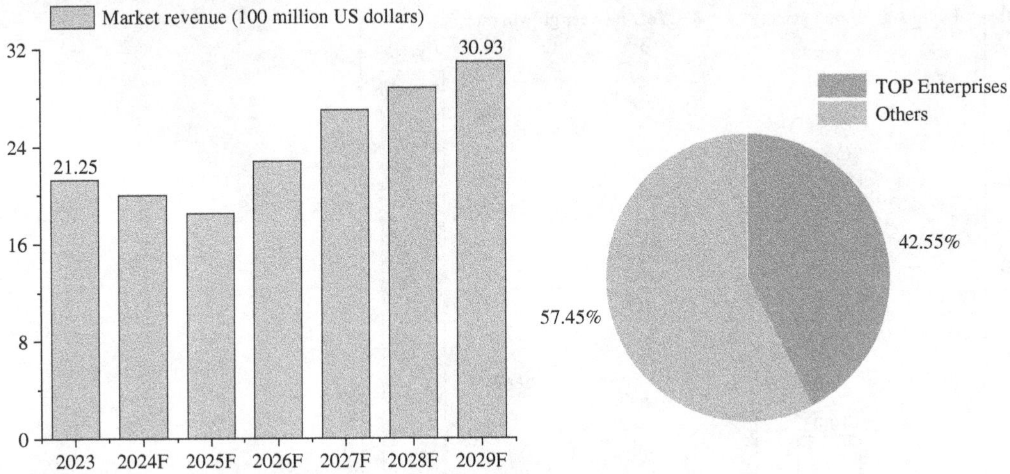

Figure 2.10 Statistical forecast and competitive distribution of the global lithium-ion battery electrolyte salt materials market [25].

Do-Fluoride New Materials, and Jiangsu Xintai Material Technology, dominate the global market with a combined revenue share of 42.55% in 2023. Their respective market shares are 17.22%, 15.87%, and 9.46% (Figure 2.10).

The global electrolyte salt materials market is poised for significant growth, driven by increasing demand for electric vehicles, wearable devices, and energy storage systems. While the total market revenue experienced a decline in 2023 [28], it is expected to recover and grow steadily over the next few years. The competitive landscape remains moderately concentrated, with the key players maintaining substantial market shares.

Electrolyte additives are also essential components in lithium battery electrolytes. They generally do not participate in the electrode reactions of LIBs but can significantly improve various properties of the electrolyte. This enhancement facilitates excellent characteristics such as high energy density, long cycle life, high rate capability, wide temperature range applicability, and high safety.

In 2021, the global market for electrolyte additives for LIBs was approximately $1.5 billion, representing a year-on-year growth of 261.5%. The compound annual growth rate from 2016 to 2021 stood at 42.23%. According to forecasts by the China Industrial Research Institute, the global market for electrolyte additives is expected to reach $1.9 billion by 2022 [29]. The global market for electrolyte additives is illustrated in Figure 2.11.

According to data from EVTank, Chinese enterprises accounted for a remarkable 86% share of the global electrolyte market in the LIB industry in 2021 [30]. In 2022, the global shipment volume of LIB electrolytes reached 998,000 tons, marking a significant year-on-year increase of 67.4%. It is anticipated that by 2025, the shipment volume will increase to 2.037 million tons. The global market distribution of lithium battery electrolytes among various companies in 2021 is shown in Figure 2.12.

Amidst this thriving market, certain companies have emerged as top players in the liquid electrolyte industry. These top players include Mitsubishi Chemical Group (Tokyo, Japan) [31], UBE Corporation (Tokyo, Japan) [32], Shenzhen Capchem Technology (Shenzhen, China) [33], and Guangzhou Tinci Materials Technology (Guangzhou, China) [34], among others.

Solid-state electrolytes are considered the next generation of battery technology. Their non-flammable nature and compatibility with lithium metal anodes could revolutionize battery

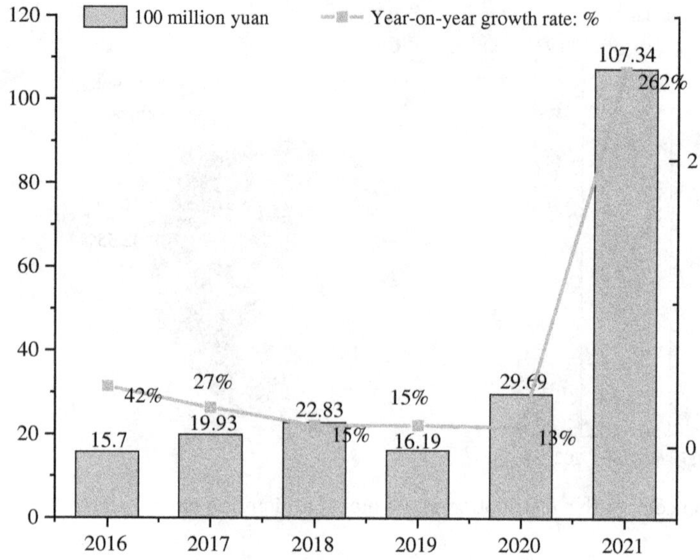

Figure 2.11 Global electrolyte additive market size from 2016 to 2021. Source: [2]/NetEase.

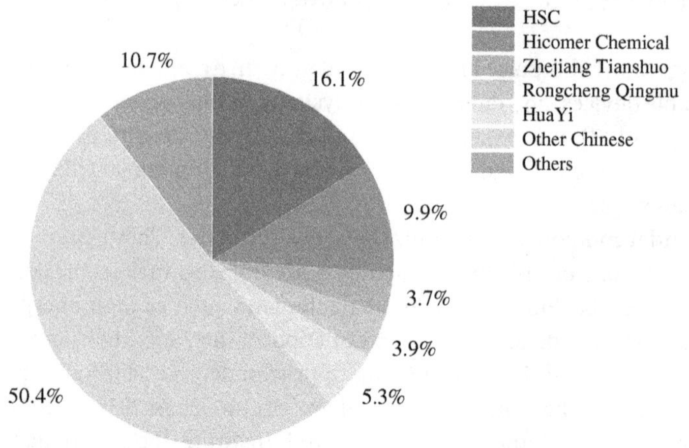

Figure 2.12 Global market share of lithium battery electrolytes in 2021 [30].

capacity and longevity. Yet, these electrolytes face hurdles, such as their lower power output compared to their liquid counterparts, arising from reduced ionic movement, electrode interface issues, and slower electrode reactions. Additionally, production costs for solid-state batteries are currently higher than those for conventional batteries, which may limit their immediate widespread adoption [35]. Several pioneering startups, such as Solid Power (Louisville, USA) [36], SES AI (Woburn, USA) [37], and QuantumScape (San Jose, USA) [38], are advancing the development of solid-state battery electrolytes.

2.3.4 Separators

The battery separator functions as a membrane between the anode and cathode, serving as a crucial component within the LIB cell. In LIBs, the separators establish a barrier to prevent short circuits between the cathode and the anode. Separators are either wound together with the electrodes to form jellyrolls or stacked between the electrodes. For the winding process, the electrodes and separator are under tension pressure, which helps to maintain good contact between different layers. To improve the contact in stacked cells, the cell container is generally used to apply pressure on the different layers.

Lithium-ion cells are fabricated by bonding or laminating layers of electrodes and separators together. To be effective, the properties of the separator need to remain stable during the bonding process. In some instances, separators are coated with artificial layers that help to facilitate the bonding process and improve the contact, thereby reducing interfacial resistance. In the conventional method of producing spirally wound cells, two layers of separators are wound alongside the positive and negative electrodes, resulting in a positive/separator/negative/separator configuration. These components are wound as tightly as possible to ensure optimal interfacial contact, so the separators need to be robust to prevent any contact between the electrodes. Additionally, the separator should not deform or decrease in width, as this could lead to electrode contact.

In the medium term, factors such as declining LIB prices and increasing adoption of electric vehicles are anticipated to drive market growth. On the other hand, the time lag in the supply to the increased demand will create an imbalance that restrains market expansion. Moreover, advancements in battery technology have significantly increased the demand for improvements in separator design. Currently, whether in commercial use or under development, separators have yet to meet the stringent stability and lifespan performance standards required to maintain the efficiency and reliability of battery technologies. It is expected that the gap between increased demand for improved separators and currently less developed separators is likely to create substantial opportunities for the market.

The Asia-Pacific region is expected to be the largest and fastest-growing market, with the majority of demand originating from countries such as China and Japan. To illustrate the growing demand for separators, the global electric vehicle LIB separator market inventory from 2010 to 2023 is shown in Figure 2.13.

With the rapid advancement of the lithium-ion industry, the market size for LIB separators is estimated to be \$5.42 billion in 2024. By 2029, it is anticipated to reach \$12.17 billion, growing at a compound annual growth rate (CAGR) of 17.60% during the forecast period (2024–2029) [39]. The estimated market size of LIB separators in the years 2024 and 2029 is illustrated in Figure 2.14.

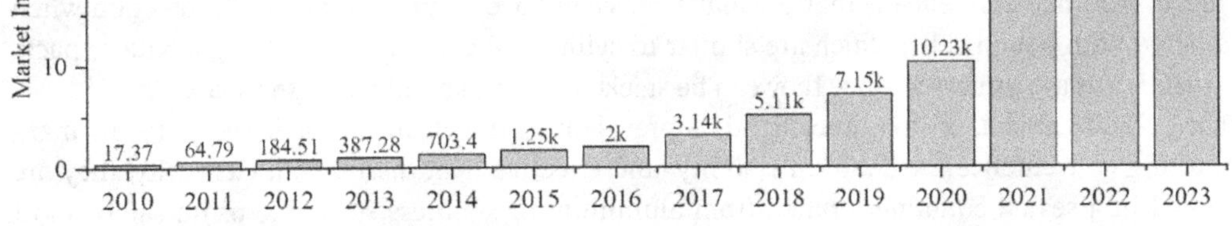

Figure 2.13 Global electric vehicle lithium-ion battery separator market inventory, 2010–2023 (thousands of units). Source: [39]/Mordor Intelligence.

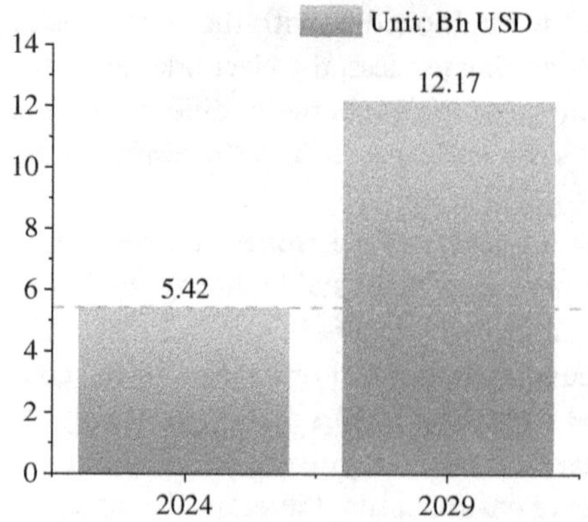

Figure 2.14 Estimated lithium-ion battery separator market size in 2024 and 2029. Source: Adapted from [39].

Manufacturers leading the market in battery separators include Asahi Kasei Corporation (Tokyo, Japan), Celgard (Charlotte, USA), and Cangzhou Mingzhu Plastic Co. Ltd. (Cangzhou, China). These companies are at the forefront of separator technology and production, setting the standard for quality and innovation in the industry.

As the demand for high-performance LIBs continues to rise, the market for battery separators is expected to expand significantly. Consequently, the development of advanced separator materials and technologies will be crucial in meeting the increasing demands of the industry.

2.3.5 Packaging Materials

There are mainly four form factors for LIBs: coin cell, pouch cell, prismatic cell, and cylindrical cell. Among these, the pouch cell is the most widely used packaging enclosure in the consumer electronics industry, due to its low weight (which offers higher cell energy density) and design flexibility for customization.

Coin cells are mostly used in the R&D phase for lab-scale production. Although there exist coin-shaped commercial LIBs, they are usually a short version of cylindrical cells that will be discussed later. Coin cells cannot provide high capacity when compared to other forms due to their limited space for loading active components.

The pouch typically consists of a thin aluminum foil as the middle layer to withstand mechanical stress while acting as a humidity barrier. Additionally, nylon films are used as the outer layer to prevent air penetration, and cast polypropylene (CPP) makes up the inner layer for the melting and sealing of the pouch. These materials must survive not only the mechanical stresses of battery operation but also additional mechanical processes during manufacturing.

Prismatic cells can have either a jellyroll or a stacking electrode-separator assembly, which is similar to pouch cells. However, prismatic cells are usually equipped with a rigid shell that is made of metal. Another difference is that prismatic cells can have additional layers of protection when compared with pouch cells, which are similar to cylindrical cells. When building a battery pack, prismatic cells are preferred since they can be stacked and make full use of the space.

The cylindrical cell, on the other hand, is prevalent in the electric vehicle sector for its manufacturing convenience, mechanical stability, and effective heat dissipation. Generally, they are enclosed in a sealed container crafted from aluminum or stainless steel. To ensure safety, most cylindrical cells incorporate a resealable vent that allows the release of gases when internal pressure surpasses safe levels.

2.4 Battery Management Systems

Batteries comprise assembled cells with varying sizes and shapes. For instance, portable batteries may contain only a few cells, while large industrial batteries consist of hundreds of cells assembled into modules. The proper functioning of these modules and the performance of batteries are managed by sophisticated electronic management systems, commonly known as battery management systems (BMSs). The BMS monitors critical data and processes to prevent batteries from operating outside their safe operating mode.

The relationship between BMSs and LIBs is integral to the efficient and safe operation of these batteries. LIBs, which are widely used in applications ranging from portable electronic devices to large industrial systems, consist of assembled cells that vary in size and shape. Cells are organized into modules, and the proper functioning of these modules, as well as the overall battery performance, is managed by the sophisticated electronic control systems of the BMS.

In 2024, the BMS market size is estimated at $9.30 billion [40], and it is expected to reach $11.79 billion by 2029, growing at a CAGR of 4.85% during the forecast period (2024–2029). Though the market was negatively impacted by COVID-19 pandemic, it has presently returned to pre-pandemic growth levels. The estimated market size for battery management systems in 2024 and 2029 is illustrated in Figure 2.15.

The list of BMS manufacturers includes the integrated circuit (IC) manufacturing powerhouses, such as Texas Instruments (Dallas, USA) [41], STMicroelectronics (Geneva, Switzerland) [42], Analog Devices (Wilmington, USA) [43], and Renesas Electronics (Tokyo, Japan) [44]. In addition to providing the BMS, suppliers must also provide the associated firmware. For redundancy and a faster response time in safety protection of the battery, an independent second-level protection chipset is used along with the BMS. This independent chipset is often supplied by BMS manufacturers, though it is also offered by other IC makers such as Nisshinbo (Tokyo, Japan) [45] and Ricoh (Tokyo, Japan) [46].

Power switches are used to prevent unintended power draws from the battery and to provide isolation from current surges and safety events. Solid-state electronic switches have the best performance (e.g., power delivery, switching time), reliability, and cost impact. Power MOSFET (metal-oxide-semiconductor field-effect transistor) switches are extensively used in battery packs. Because of their efficiency and ability to handle large currents with minimal power loss, power MOSFETs are preferred for high-current applications. Where reliability and longevity are critical, solid-state switches are used, such as in safety circuits and low-current control. In either case, the switch must fully meet the qualification requirements (design, quality, safety) for its usage.

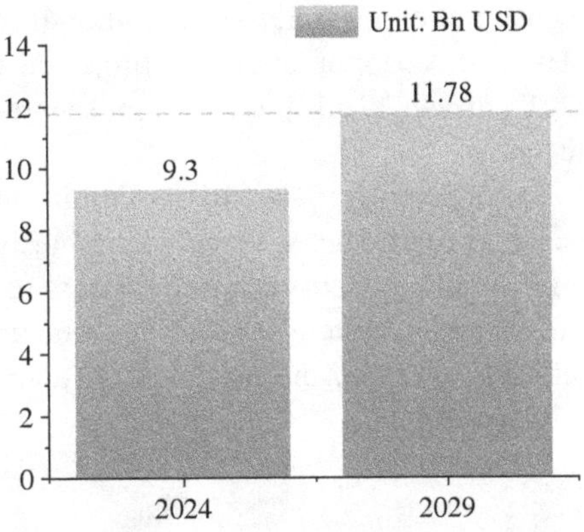

Figure 2.15 Estimated battery management system market size in 2024 and 2029. Source: Adapted from [40].

BMS is connected with cells through connecting elements like cell tabs and bus bars that are used to connect the different cells together in a pack. With this in mind, the pack must have a robust bond between the cell tabs and the printed circuit board (PCB) or between the bus bar and current conduits to function reliably [47]. Consequently, the bonding tools must produce repeatable and reliable bonds. Also, manufacturers must find the best method for the bonding process, considering any possible dissimilarity of metals in the PCB, bus bar, and cell tab. The metal type and its composition must not be altered once the bonding method is established, unless a formal request is made for the battery design or material change and the new arrangement is fully qualified.

2.5 Summary

The battery manufacturing industry is experiencing rapid growth due to rising demand for clean energy and electric transportation. This growth is driven by investments aimed at increasing production efficiency and by innovation to combat climate change. However, the industry faces challenges such as raw material supply stability, technology standardization, and compliance with environmental regulations.

Key components in battery manufacturing include positive and negative electrode materials, electrolytes, separators, binders, conductive additives, and current collectors, each influencing the battery's performance and safety. Raw materials like nickel, manganese, and lithium are critical, with significant global reserves primarily located in South America, Australia, China, and the United States. Positive electrode materials, such as lithium cobalt oxide, lithium manganese oxide, lithium iron phosphate, and ternary materials (NMC, NCA), constitute 30–40% of the battery cost. Following this, primary raw materials include lithium carbonate, nickel sulfate, cobalt sulfate, and lithium ternary materials. Notably, global production is concentrated in China, Japan, and South Korea, with China's ternary materials shipments reaching 640,000 tons in 2022. Furthermore, the battery-grade lithium carbonate market is projected to grow from 520,000 metric tons in 2023 to 1.22 million metric tons by 2030.

The negative electrode in LIBs is primarily composed of graphite, with China dominating production. However, innovations in silicon anode technology are being advanced by companies including Sila and Enevate based in the United States. In 2023, global production of negative electrode materials reached 1.7621 million tons, with China producing 97.3%.

Lithium-ion battery electrolyte salts, crucial for ion conduction and performance, have a global consumption reaching $2.125 billion in 2023, with an expected increase to $3.093 billion by 2029. The LIB separator market is projected to grow from $5.42 billion in 2024 to $12.17 billion by 2029, driven by advancements in separator technology and rising demand for high-performance batteries.

Packaging materials, such as aluminum foil for pouches or aluminum/stainless steel for cylindrical cells, are designed to withstand mechanical stresses and ensure battery safety. The production of these components primarily occurs in regions like China, Japan, and South Korea. The battery management system (BMS) is a sophisticated electronic management tool to ensure the safe and efficient operation of LIBs. The BMS market is projected to grow from $9.30 billion in 2024 to $11.79 billion by 2029.

References

1 Askci·Database (2022). Prediction and Analysis of the Scale and Competitive Pattern of the Global Lithium Battery Electrolyte Additive Market in 2022, 2022 [Online]. https://m.163.com/dy/article/HB71M2JG051481OF.html.

2 A. D. Resource (2023). Research and Analysis on the Industry Chain Graph of China's Lithium Battery Positive Electrode Materials in 2023, 2023 [Online]. https://www.163.com/dy/article/IHQ9SA6505198SOQ.html.

3 I. E. Agency (2022). Global Supply Chains of EV Batteries, IEA, Paris.

4 Indxx, EV Battery Market: Powering the EV Revolution, February 24, 2023, https://www.indxx.com/news-&-research/news-and-research-insider/ev-battery-market-powering-the-ev-revolution (accessed 21 August 2024).

5 L. Beijing Yanjing Bizhi Information Consulting Co. (2023). Analysis Report on Global and Chinese Lithium Battery Positive Electrode Materials Industry, Beijing Yanjing Bizhi Information Consulting Co., Ltd, Beijing, China.

6 D. Research (2023). The global market for battery grade lithium carbonate is expected to continue to grow rapidly, with sales expected to reach 1.22 million tons by 2030 [Online]. https://baijiahao.baidu.com/s?id=1777101360865292129&wfr=spider&for=pc.

7 P. A. Securities (2024). Deep Report on Power & Energy Storage Battery and Electrification Industry, Ping An Securities, Shenzhen, China.

8 Products | Umicore, Umicore [Online]. https://rbm.umicore.com/en/products/ (accessed 6 May 2024).

9 Targray Cathode Materials for Li-ion Battery Manufacturers, Targray [Online]. https://www.targray.com/li-ion-battery/cathode-materials (accessed 6 May 2024).

10 BASF Shanshan Battery Materials Co., Ltd. (BSBM) [Online]. https://catalysts.basf.com/industries/automotive-transportation/battery-materials/global-footprint/basf-shanshan-battery-materials-co-ltd (accessed 6 May 2024).

11 Rystad Energy. Fake it till you make it: Synthetic graphite holds the key to meeting battery demand surge, despite ESG concerns, https://www.rystadenergy.com/news/fake-it-till-you-make-it-synthetic-graphite-holds-the-key-to-meeting-battery-dema (accessed 21 August 2024).

12 P. H. A. Strategy (2024). Analysis of the development speed and prospect prediction of the segmented market for lithium-ion battery negative electrode materials industry [Online]. https://baijiahao.baidu.com/s?id=1792952042728061494&wfr=spider&for=pc.

13 AMC Consultants. Graphite and AMC, https://www.amcconsultants.com/graphite-and-amc (accessed 21 August 2024).

14 H. I. R. I.-H. I. Network (2022). Analysis of the current development status, competitive landscape, and production capacity layout of the global lithium battery negative electrode material industry in 2021 [Online]. https://zhuanlan.zhihu.com/p/497684503?utm_id=0.

15 IEA (2023). World Energy Investment 2023, IEA, Paris. https://www.iea.org/reports/world-energy-investment-2023, Licence: CC BY 4.0.

16 Can the World Make an Electric Car Battery Without China? Agnes Chang and Keith Bradsher, The New York Times, May 16, 2023, https://www.nytimes.com/interactive/2023/05/16/business/china-ev-battery.html (accessed 21 August 2024).

17 S. P. I. O. Research (2023). 2023 Global Lithium Battery Industry Development White Paper, SPIR, Shenzhen, China.

18 Smart Solutions in Graphites & Fiber Composites | SGL Carbon [Online]. https://www.sglcarbon.com/en/ (accessed 3 June 2024).

19 BTR New Material Group Co., Ltd. - btrchina.com [Online]. https://www.btrchina.com/ (accessed 3 June 2024).

20 Sila | Next-Gen Lithium-Ion Battery Materials, Sila [Online]. https://www.silanano.com/ (accessed 6 May 2024).

21 Enevate: Home, Enevate [Online]. https://www.enevate.com/ (accessed 6 May 2024).

22 Group14 | Powering the Silicon Battery Age, Group14 Technologies [Online]. https://www.group14.technology/ (accessed 6 May 2024).

23 Enovix | 100% Active Silicon Lithium-ion Battery - Enovix, Enovix Corporation [Online]. https://www.enovix.com/ (accessed 6 May 2024).

24 Amprius Technologies Silicon Anode Batteries, Amprius Technologies [Online]. https://amprius.com/ (accessed 6 May 2024).

25 M. Monitor (2024). Report on the Growth Trends and Opportunities of the Global and Chinese Lithium ion Battery Electrolyte Salt Materials Market, MARKET MONITOR CO., Changsha, China.

26 Global Lithium-Ion Battery Electrolyte Salt Material Market Revenue May Reach $3093 Million by 2028, Global Market Monitor, May 16, 2024, https://www.linkedin.com/pulse/global-lithium-ion-battery-electrolyte-salt-material-h63ve/ (accessed 21 August 2024).

27 China's Li-Ion Battery Electrolyte Outlook for 2023: Data-Driven Insights, China Chemical Market Insights, September 22, 2023, https://www.linkedin.com/pulse/chinas-li-ion-battery-electrolyte-outlook-2023-market-insights/ (accessed 21 August 2024).

28 Global Lithium Ion Battery Electrolyte Salt Material Market Research Report 2023-Competitive Analysis, Status and Outlook by Type, Downstream Industry, and Geography, Forecast to 2029, Market Research.COM, March, 2023, https://www.marketresearch.com/Maia-Research-v4212/Global-Lithium-Ion-Battery-Electrolyte-33786345/ (accessed 21 August 2024).

29 Zeyuan Yu, Analysis on the development trend of global and Chinese lithium battery electrolyte additives industry in 2023, June 02, 2023, Huaon.com, https://www.huaon.com/channel/trend/900472.html#:~:text=%E6%A0%B9%E6%8D%AE%E6%95%B0%E6%8D%AE%E6%98%BE%E7%A4%BA%EF%BC%8C2021%E5%B9%B4,%E4%B8%8A%E6%B6%A8%E8%87%B3135.56%E4%BA%BF%E5%85%83%E3%80%82 (accessed 21 August 2024).

30 E. C. C. I. R. Institute (2022). Research Report on the Prospects and Investment Opportunities of China's Electrolyte Market, EVTank, China.

31 The Mitsubishi Chemical Group [Online]. https://www.mcgc.com/english/ (accessed 4 June 2024).

32 Electrolyte - UBE Corporation [Online]. https://www.ube.com/contents/en/chemical/battery_materials/electrolytes.html (accessed 4 June 2024).

33 Shenzhen Capchem Technology Co., Ltd. [Online]. https://en.capchem.com/ (accessed 4 June 2024).

34 Tinci Materials Guangzhou [Online]. https://www.tinci.com/ (accessed 4 June 2024).

35 Manthiram, A., Yu, X., and Wang, S. (2017). Lithium battery chemistries enabled by solid-state electrolytes. *Nature reviews Materials* 2 (4): 16103.

36 Solid Power's All-Solid-State Battery Cell Technology, Solid Power Inc. [Online]. https://www.solidpowerbattery.com/all-solid-state-batteries/default.aspx (accessed 6 May 2024).

37 SES AI | The next generation of Li-Metal batteries, SES AI Corporation [Online]. https://ses.ai/ (accessed 6 May 2024).

38 QuantumScape: Building the Best Solid State Battery, QuantumScape [Online]. https://www.quantumscape.com/ (accessed 6 May 2024).

39 Mordor Intelligence, Lithium-Ion Battery Separator Market Size & Share Analysis - Growth Trends & Forecasts (2024–2029) [Online]. https://www.mordorintelligence.com/industry-reports/lithium-ion-battery-separator-market.

40 Mordor Intelligence, Battery Management System Market Size & Share Analysis - Growth Trends & Forecasts (2024-2029) [Online]. https://www.mordorintelligence.com/industry-reports/battery-management-system-market.

41 Battery management ICs, Texas Instruments, 2024 [Online]. https://www.ti.com/power-management/battery-management/overview.html (accessed 14 April 2024).

42 Battery Management ICs, ST Microelectronics, 2024 [Online]. https://www.st.com/en/power-management/battery-management-ics.html (accessed 14 April 2024).

43 Automotive Battery Management Systems, Analog Devices, 2024 [Online]. https://www.analog.com/en/solutions/automotive/electrification-and-powertrain/battery-management-systems-bms.html (accessed 14 April 2024).

44 Battery Fuel Gauge ICs, Renesas, 2024 [Online]. https://www.renesas.com/us/en/products/power-power-management/battery-management/battery-fuel-gauge-ics (accessed 14 April 2024).

45 Nisshinbo Micro Devices Inc. (2018). Li-ion Battery Secondary Protection ICs, Nisshinbo Micro Devices Inc., 1 August 2018 [Online]. https://www.nisshinbo-microdevices.co.jp/en/products/featured-products/r5640_r5641.html (accessed 28 April 2024).

46 Ricoh, Ricoh launches two secondary protection ICs for multi-cell Li-Ion / Li-Polymer battery packs, Ricoh, 30 March 2018 [Online]. https://www.ricoh-europe.com/news-events/news/ricoh-launches-two-secondary-protection-ics-for-multi-cell-li-ion-li-polymer-battery-packs/ (accessed 28 April 2024).

47 Kumar, N., Masters, I., and Das, A. (2021). In-depth evaluation of laser-welded similar and dissimilar material tab-to-busbar electrical interconnects for electric vehicle battery pack. *Journal of Manufacturing Processes* 70: 78–96.

3

Lithium-Ion Cell Manufacturing Process and Form Factors

Simin Peng, Guanwei Jiang, Yuwei Nie, Yu Zhang, Lingxi Kong, and Sahithi Maddipatla

3.1 Lithium-Ion Battery (LIB) Structure Overview

A lithium-ion cell is composed of an anode, a cathode, a porous separator, and an electrolyte that facilitates ion movement between the anode and cathode during charge and discharge cycles. At present, these cells come in various form factors, as shown in Figure 3.1. Cylindrical cells, used in power tools and electric vehicles, feature a jelly roll of anode, cathode, and separator encased in a metal can. Coin cells are small, round, and flat, often found in devices like watches and hearing aids, with a layered anode and cathode inside a small metal casing. Pouch cells are flexible and lightweight, with electrodes and separators stacked or folded within the pouch. They are encased in heat-sealed aluminum plastic film and are mostly used in smartphones and electric vehicles. Prismatic cells, which are rectangular and encased in metal or hard plastic, have electrodes and separators either stacked or wound inside the casing. They offer high capacity in a compact form and are used in laptops, electric vehicles, and energy storage systems.

In a cylindrical battery, the solid components (not including the liquid electrolyte) can be classified into three major parts: the battery cap, jelly roll, and battery can, as shown in Figure 3.2. These three subassemblies contain several other components; they can be manufactured independently and assembled later.

Since the battery cap does not contain active components and is typically premade, the manufacturing process primarily involves the electrode assembly and related steps. The jelly roll structure contains a separator, cathode, and anode. After the jelly roll is made, it is inserted into the battery can, and these two parts are welded together through the anode tab. After adding the electrolyte, the battery cap is placed on top of the battery can to enclose the entire structure. The cap can contain the positive temperature coefficient (PTC) thermistor, the vent disk, and the bottom that makes the battery current interrupt device, providing the venting function. Disassembly of cylindrical batteries from different brands, including MXJO, LG, and Panasonic, shows the myriad design variations (see Figure 3.3).

3.2 Lithium-Ion Battery Manufacturing Process

LIB manufacturing is a complex process that has evolved significantly with the industry's growth. Initially developed for consumer electronics, where hundreds of millions of cells are produced each

The Safety Challenges and Strategies of Using Lithium-Ion Batteries, First Edition. Edited by Michael G. Pecht.

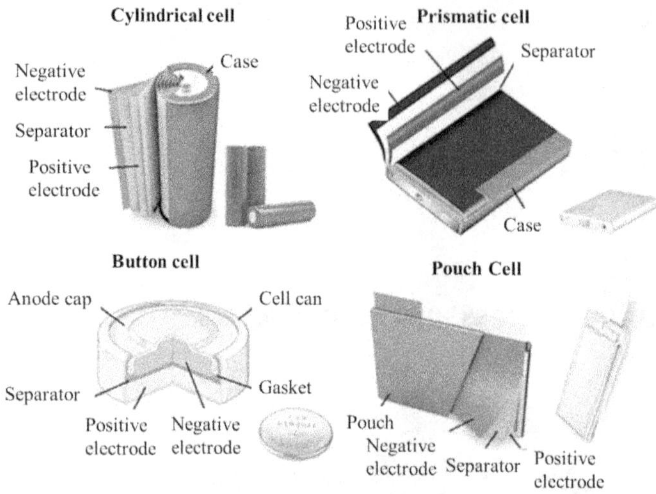

Figure 3.1 Form factors of lithium-ion cells. Source: [1]/Tycorun.

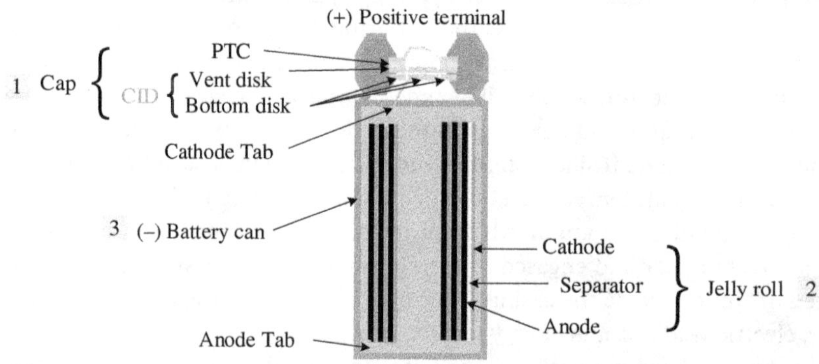

Figure 3.2 Battery subassemblies.

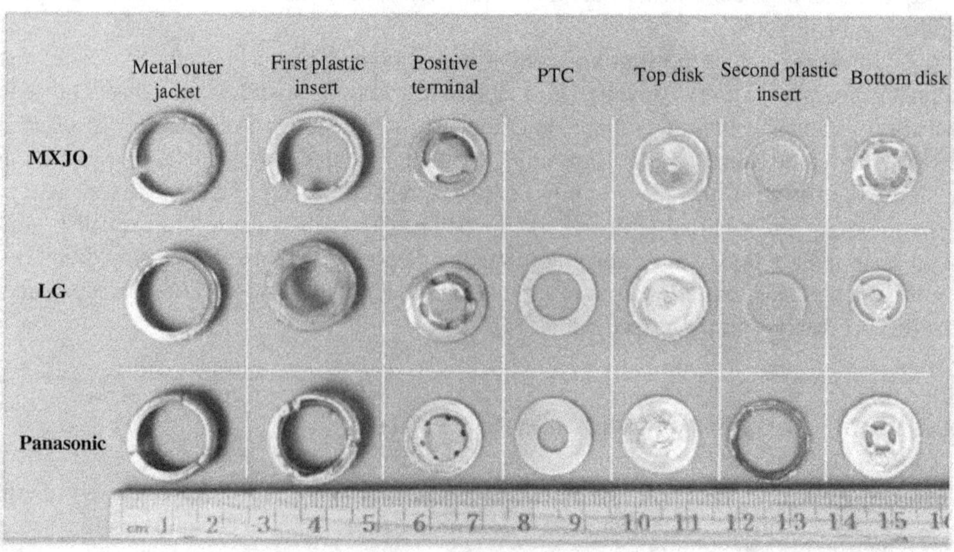

Figure 3.3 Structure of battery caps from various manufacturers. Source: [2]/Elsevier.

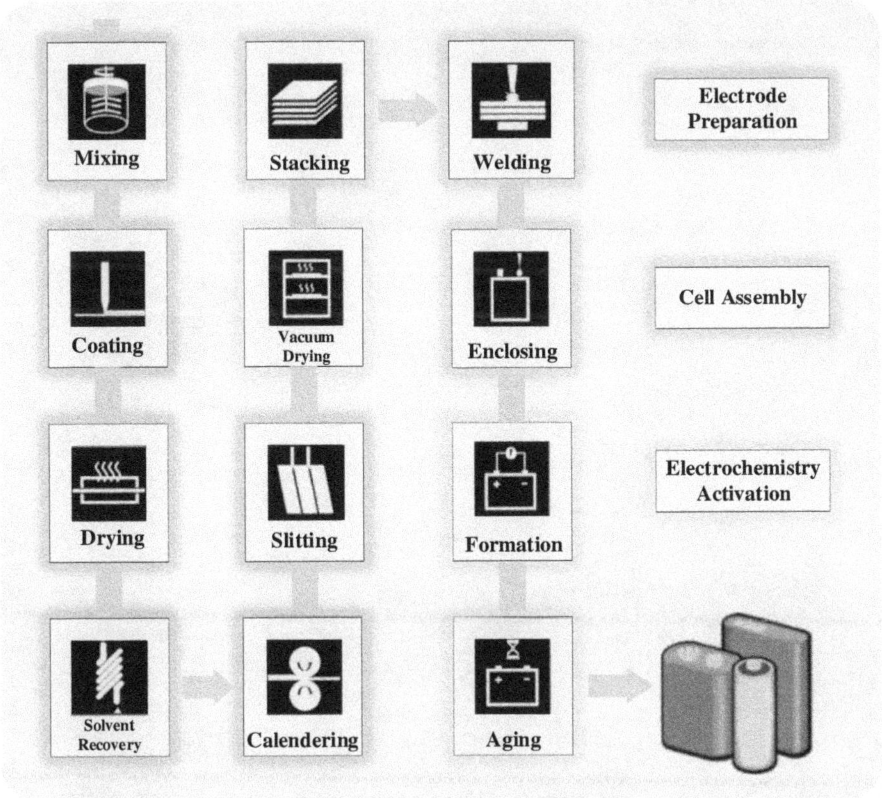

Figure 3.4 Schematic of the LIB manufacturing process. Source: [4]/with permission of Elsevier.

year, the manufacturing techniques have since been refined and adapted for large-scale production. Currently, many billions of cells are produced each year for electric vehicles and energy storage systems.

The production of LIB cylindrical cells involves multiple steps, broadly divided into three main stages: electrode preparation, cell assembly, and battery electrochemistry activation [3, 4]. It begins with electrode preparation, where the active materials are mixed with binders and conductive additives to form a slurry that is coated onto a metal foil. Next, calendaring compresses the coated electrodes to ensure uniform thickness and density, followed by drying to remove residual solvents. Then, during the slitting stage, electrodes are cut into precise widths. Subsequently, the electrodes and separator are layered or rolled into a cylindrical shape in the winding stage.

Assembly comes next, where the electrodes are placed into the casing, followed by electrolyte filling. The cell is then sealed within a casing and fitted with safety features like vents and protective circuits. The final step is formation and testing/aging, where the cell undergoes initial charging and discharging cycles to stabilize the electrochemical properties and ensure it is ready for use. An overview diagram of the manufacturing process is shown in Figure 3.4.

Battery performance is sensitive to the level of residual moisture within the cell. Thus, batteries are typically manufactured in a dry, clean room environment, which is essential for minimizing moisture-related degradation. Dry room manufacturing represents a significant energy demand, consuming 3.9 kWh per cell, or 29% of the total energy usage during manufacturing [4, 5]. The impact of environmental conditions on battery performance underscores the importance of

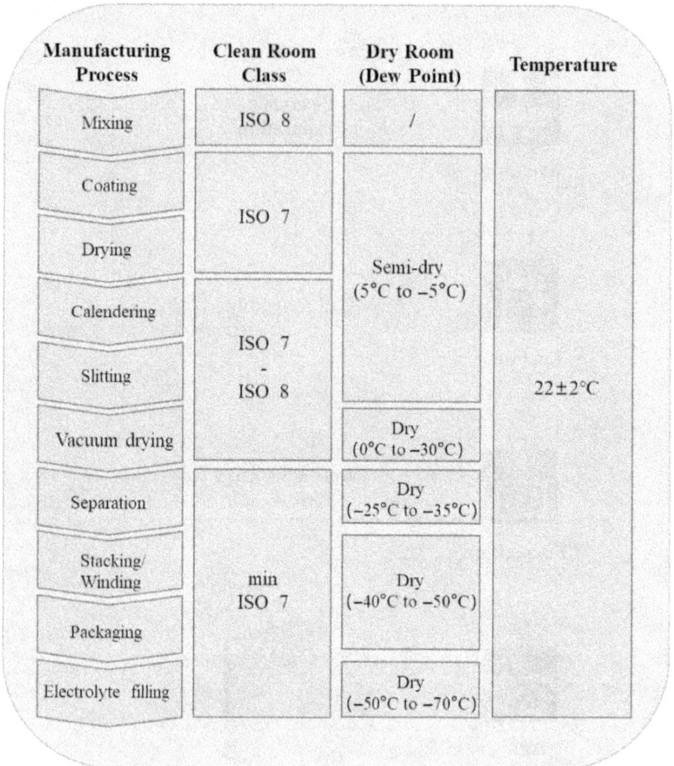

Figure 3.5 Step-by-step dry clean room conditions for LIB manufacturing. Source: Adapted from [5].

controlling them throughout the manufacturing process since moisture could lead to degradation of the electrodes, capacity loss, and gas generation due to side reactions. The rooms for battery manufacturing are typically kept at a dew point of −40 °C although lower temperatures are required for more moisture-sensitive chemicals (such as NMC811 or lithium metal). The clean room designations and environmental conditions for each step, from slurry mixing to electrolyte filling, are depicted in Figure 3.5.

3.2.1 Electrode Sheet Preparation

Regardless of the format (pouch, cylindrical, or prismatic), the first stage in battery manufacturing is electrode preparation. To avoid contamination between materials, production lines are separated: one for the anode and one for the cathode. The anode is generally made of graphite-coated copper foil, while the cathode is composed of aluminum foil coated with materials like LCO, NMC, or NCA. The electrode sheet preparation involves several steps: (a) sintering the raw active materials for the cathode and anode using a furnace; (b) milling the materials into smaller particles; (c) mixing active, conductive, and binder materials into a paste under vacuum; (d) coating the paste onto the current collector to create an electrode; (e) drying the electrodes with a heater; and (f) rolling the electrode to the required thickness using a calendaring machine. Figure 3.6 illustrates the electrode sheet preparation steps.

The electrode preparation process begins with the mixing of the active material, conductive additive, and binder to form a slurry, with a solvent facilitating this blend [7]. For cathodes,

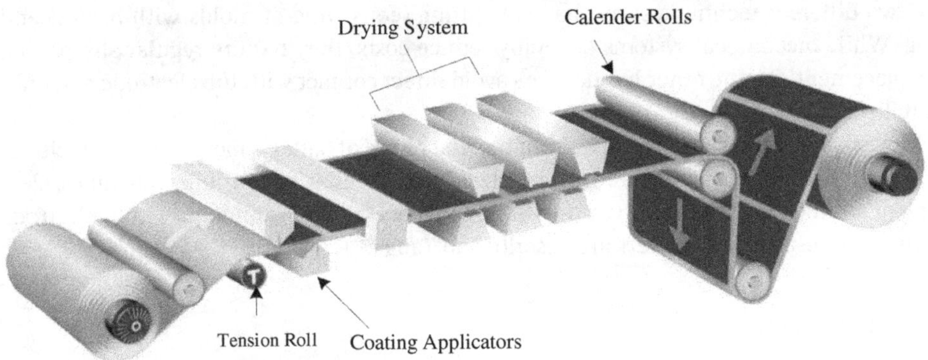

Figure 3.6 Electrode sheet preparation. Source: [6]/Dover Flexo Electronics.

N-methyl pyrrolidone (NMP) dissolves the polyvinylidene fluoride (PVDF) binder, while for anodes, the styrene-butadiene rubber (SBR) binder is dissolved in water with carboxymethyl cellulose (CMC) [8, 9]. This step requires 30 minutes to 5 hours and accounts for approximately 8% of the total manufacturing cost, consuming 0.11 kWh of energy per cell, which is 0.83% of the total energy consumption [4, 5]. Equipment for slurry production comes in two varieties: batch production equipment like planetary mixers and continuous production equipment, which combines basic batching operations and mixing chambers through an automated weighing feeding system.

The slurry is then coated on both sides of the current collector (Al foil for the cathode and Cu foil for the anode) through a pipeline system and dried to evaporate the solvent [10]. This phase can take 35–80 minutes per meter and constitutes 14.96% of the manufacturing cost [4, 5]. The drying process, combined with solvent recovery, allows the active material to adhere to the foil and distribute evenly. It is particularly energy intensive, consuming 6.22 kWh per cell or 46.84% of the total energy [4, 5].

Given the toxicity of common solvents like NMP, a solvent recovery process is essential, especially for the cathode [11], as it represents about 5% of the cost [4, 5]. Markedly, gradual drying is crucial for obtaining good-quality electrodes, requiring ovens with lengths of up to 80 m. The slurry coating, applied on both sides of the foil, can be intermittent or continuous, depending on the size and format of the batteries to be produced. Typically, the width of the strip printed on the roll limits the size of the battery, thus directly affecting the capacity of the production line. Also, the electrode can have uncoated areas required for subsequent assembly, such as for tab welding.

The next step in the electrode preparation process is calendaring, which adjusts the physical properties of the electrodes, such as bonding and density, and it takes 60–100 minutes per meter [4, 5]. It accounts for 5% of the costs and consumes 0.38 kWh of energy per cell (2.86%) [4, 5]. Calendaring encompasses the final treatment of the coated rolls. Similar to the previous step, it is a roll-to-roll process, where the coated rolls pass through two heated rollers to compress the material, ensuring consistent thickness, density, and better adhesion.

Subsequently, the finished electrodes pass through a set of blades and are cut to the required dimensions based on the type of cell (e.g., pouch, cylindrical, prismatic), requiring 80–150 minutes per meter, making up 3% of the costs and consuming 0.71 kWh per cell (5.35%) [4, 5]. Following this, the vacuum drying process removes excess moisture from the electrodes over 12–30 hours, which might otherwise interfere with the electrochemical processes. This step accounts for about 3% of the costs and consumes significant energy, but numbers are not reported [4, 5]. The cutting process can involve two different techniques: mechanical cutting (consisting of molds with blades) and laser cutting. While mechanical systems generally reduce costs, they require regular sharpening and blade replacement. On the other hand, lasers avoid direct contact with the electrode and offer greater flexibility.

The machinery used in the various processes of the first phase of battery manufacturing includes mixers for slurry preparation, coating, and drying machines, calendaring or rolling machines, electrode cutting or slitting machines, and vacuum drying furnaces. The equipment used in electrode manufacturing and respective suppliers are described in Table 3.1.

3.2.2 LIB Cell Assembly

Batteries can be classified as pouch, prismatic, or cylindrical based on their encapsulation method and form factor. From the configuration of electrode sheets, pouch and prismatic batteries can adopt either winding or stacking processes. Conversely, cylindrical batteries, with their curved shape, can only employ winding processes. Although winding processes have had a longer development time, they are mature, cost-effective, and have high yields. However, with the proliferation of electric vehicles, stacking processes have emerged due to their advantages in high volume utilization, structural stability, low internal resistance, and long cycle life. Figure 3.7 shows expanded and fully assembled views of the electrode assembly using the winding (rolling) and stacking processes.

For cylindrical batteries, jelly roll preparation is done via the winding process. Winding involves rolling the positive and negative electrode strips and separator film together at controlled speeds, tensions, and relative positions, allowing for the production of regular-shaped lithium batteries. For the jelly roll winding process, the electrodes and separator form a long ribbon form, with the separator sandwiched between the anode and cathode. Winding the sandwiched layers leads to a jelly roll structure, also known as an LIB cell core. The jelly roll is placed into a battery can with a bottom insulator, and the negative tab is welded to the inner surface of the battery can's bottom.

Stacking processes, on the other hand, layer positive and negative electrode sheets and separator films alternately to form a stack, allowing for the production of regular- or irregular-shaped lithium batteries, providing higher flexibility [14]. Stacking processes achieve the highest structural space

Table 3.1 Equipment used in the manufacture of electrodes and corresponding suppliers.

Equipment	Manufacturer/supplier
Vacuum Mixer	Wuxi Lead Intelligent Equipment, NAURA New Energy, Nagano Automation, TSI
Coating Machine	Shenzhen Yinghe Technology, Wuxi Lead Intelligent Equipment, Katop Automation, NAURA New Energy, Toray Engineering, Hitachi, Nagano Automation, SECI, Jiangmen Kanhoo Industry, Shanghai Putailai
Dryer	Changzhou Ruide Drying Engineering Technology, Noritake, Weiss Technik, Cotes, Shenzhen Dacheng Precision Equipment, Condair
Calendaring or Pressing Machine	Wuxi Lead Intelligent Equipment, NAURA New Energy, Nagano Automation, Dalian Zhiyun Automation
Slitting Machine	Shenzhen Yinghe Technology, Wuxi Lead Intelligent Equipment, NAURA New Energy, Toray Engineering, Nagano Automation
Vacuum Oven	Korea Vacuum, MTI Corporation, Thermo Fisher Scientific, Memmert GmbH

Source: [12]/Customized Energy Solutions India Pvt Ltd.

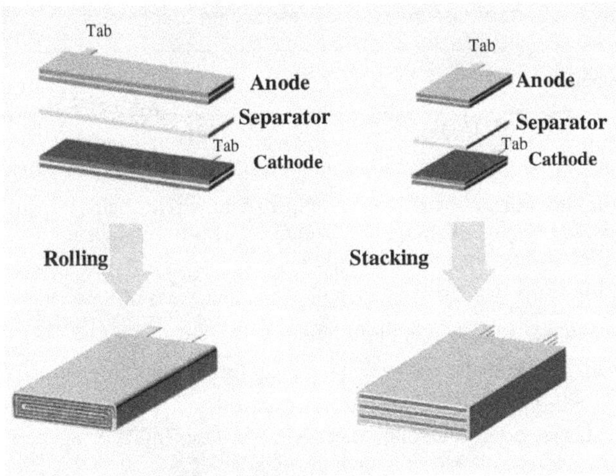

Figure 3.7 Expanded and fully assembled views of lithium-ion cells manufactured using (a) the winding process and (b) the stacking method. Source: [13]/IOP Publishing/CC BY 4.0.

utilization, with the most common method being Z-fold stacking, where the separator is folded onto each electrode layer in a serpentine manner. Alignment is crucial in this process since misalignment may cause electrodes to protrude beyond the separator, leading to short circuits once the battery is completed. Another option is stack lamination, which is similar to single-sheet stacking. However, it combines two components (separator/anode/separator), which are then alternately stacked between cathode layers.

In a pouch cell, stacking represents around 9% of the costs and 5.8% (0.77 kWh per cell) in energy consumption [4, 5]. Regardless of the cell format, tabs are welded to the current collectors, constituting 7% of the costs and consuming 0.25 kWh of energy per cell [4, 5]. The next step is electrolyte filling, wherein the duration depends on the specific cell design and represents around 12% of manufacturing costs [4, 5]. Subsequently, the cell is placed in a vacuum to allow for electrolyte absorption. Then, a top insulator is added to the jelly roll structure, featuring a hole that allows the positive tab to pass through.

The equipment used in the second phase of battery manufacturing includes machines that perform die-cutting, stacking (for pouch cells), winding (for cylindrical and prismatic cells), sealing and tab welding, and electrolyte filling. The machinery used in battery assembly manufacturing and respective suppliers are depicted in Table 3.2.

3.2.3 Sealing of LIBs

After inserting the cell core, a groove is made on the can above the top insulation ring to host the battery cap. The positive tab is welded to the cap, which is then positioned over the groove. Once the cap and cell are aligned, the crimping process is performed [15].

The design of the cylindrical battery incorporates a technique known as "dual sealing." Initially, the cap components are crimped together using the plastic insulation part. This initial crimping ensures a secure attachment of the cap components. Subsequently, during the assembly process, the cap undergoes a second crimping at the groove, which involves bending the edge of the battery can, further enhancing the seal and ensuring the integrity of the battery enclosure.

Table 3.2 Equipment used in the manufacture of battery assembly and corresponding suppliers.

Machine	Manufacturer/supplier
Stacking Machine	Shenzhen Geesun, Wuxi Lead Intelligent Equipment, Toray Engineering, Nagano Automation, Shenzhen Yinghe Technology, Dalian Zhiyun Automation
Winding Machine	Shenzhen Yinghe Technology, Wuxi Lead Intelligent Equipment, Kaido Manufacturing, CKD, Nagano Automation, Shenzhen Geesun Intelligent Technology, KOEM, Dalian Zhiyun Automation
Cell Packaging	Toray Engineering, Manz AG, Shenzhen Yinghe Technology, mPLUS
Tab Welding Machine	mPLUS, Toptec, Wuxi Lead Intelligent Equipment
Electrolyte Filling Machine	Shenzhen Geesun Intelligent Technology, Wuxi Lead Intelligent Equipment, Nagano Automation, Hibar Systems (Tesla)
Cell Assembly Line	Techland, Hitachi, Nagano Automation, Manz AG, Moritani GmbH, Shenzhen Yinghe Technology, Wuxi Autowell Technology, KOEM

Source: [12]/Customized Energy Solutions India Pvt Ltd.

3.2.4 Formation and Testing of LIBs

The formation process entails cells undergoing pre-charging and formation cycles to stabilize the solid electrolyte interphase (SEI) [16, 17]. Furthermore, the cells undergo performance/aging tests, which include high-temperature aging, normal temperature aging, and end-of-life estimation testing, lasting up to 1.5–3 weeks [4, 5]. This crucial phase accounts for the highest cost percentage at 33% and consumes 0.07 kWh per cell [4, 5].

After the formation process and before final sealing, pouch cells typically undergo a degassing process [18]. This step involves removing excess gases that may have accumulated inside the cell during formation. Degassing helps improve the cell's stability and reduces the risk of internal pressure build-up, which can affect performance and safety.

The final sequence, including pre-charging, degassing, forming, and high-temperature aging, among other potential verification steps, may vary depending on the manufacturer and take several weeks. After formation and aging, the battery undergoes end-of-line testing to confirm its performance and capacity. Testing includes visual inspection, capacity measurement, internal resistance measurement, and open circuit voltage testing, ensuring it meets the required performance and safety standards [19]. For example, the newly created battery's performance can be tested using a battery analyzer, such as an impedance tester, to measure the internal resistance. The required equipment necessitates a fully equipped system with channel towers similar to large automated warehouses. As the equipment handles a large number of batteries simultaneously, it has a significant impact on the final size of the production facility. Batteries are ready to use when the final stage is completed.

The formation and aging process constitutes 33% of the total manufacturing process cost. Machines used in this third and final stage of cell manufacturing include battery formation testers, aging cabinets, grading machines, and battery testing machines. Typically, coating, winding, grading, and testing equipment account for 70% of the total cost of lithium-ion cell production equipment although this may vary with the degree of automation. The equipment used during cell finishing and the corresponding suppliers are shown in Table 3.3.

Currently, only a few companies specialize in battery cell manufacturing equipment. The majority of these companies are situated in China, Japan, and South Korea, as depicted in Figure 3.8. However, most of these incumbent suppliers operate at over 95% capacity [20], leaving minimal

Table 3.3 Equipment used in the manufacture of Cell finishing and corresponding suppliers.

Machine	Manufacturer/supplier
Formation Equipment	Sovema Group, Shenzhen Geesun, Nagano Automation, Hanwha Group, HangKe Technology, Shenzhen Yinghe Technology
Aging Cabinet	Arbin Instruments, MACCOR Inc., Chroma ATE Inc., NEWARE Technology Limited
EOL Testing Machine	Sovema Group, HangKe Technology

Source: [12]/Customized Energy Solutions India Pvt Ltd.

room for output expansion. Additionally, these suppliers may prioritize orders from established customers—primarily leading cell manufacturers—over new market entrants from Europe and the United States. Consequently, European battery cell manufacturing companies and electric vehicle (EV) original equipment manufacturers (OEMs) entering the sector will likely encounter a bottleneck in equipment supply, jeopardizing planned production start dates. Thus, securing equipment and supplies emerges as a critical success factor.

European equipment manufacturers have an opportunity to capture a substantial share of the revenue pool by becoming key suppliers to established cell manufacturers expanding into Europe and the United States. Their geographic proximity facilitates the installation, ramp-up, and support of equipment, providing a competitive advantage.

While most battery cell equipment suppliers are located in China, Japan, and South Korea, only a few exist in Europe. The Asia-Pacific region, excluding China, is noted for its coating and cell assembly capabilities. The distribution of key equipment suppliers by region is based on the authors' assessment.

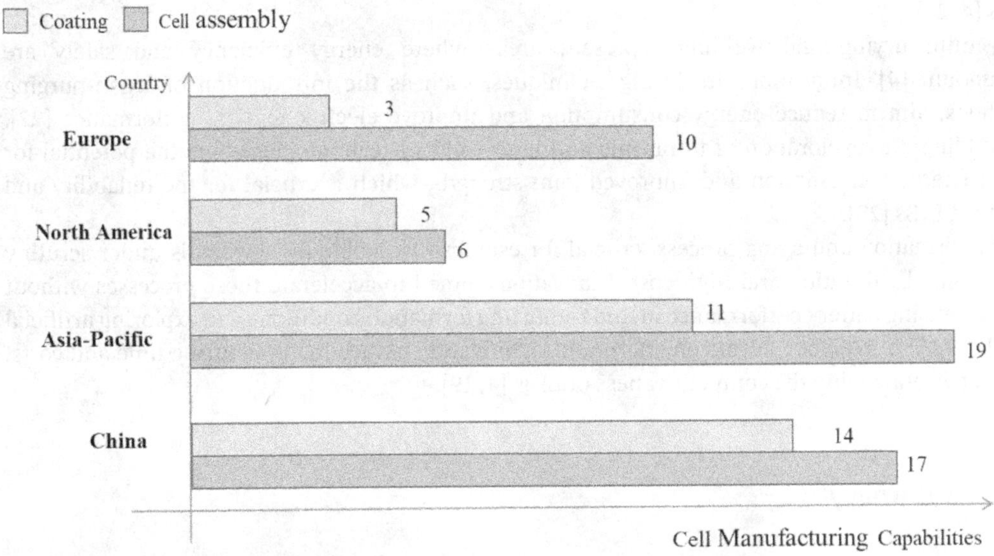

Figure 3.8 Number of key equipment suppliers by region [20].

3.3 Advancements and Refinements in LIB Manufacturing

As the demand for high-performance and cost-effective lithium-ion batteries (LIBs) continues to rise, manufacturers are increasingly focusing on optimizing each stage of the production process to reduce costs and improve efficiency. New methods that can be easily scaled up are emerging, such as high-efficiency mixing, solvent-free coating, and fast formation.

Slurry mixing, accounting for 7.9% of the total manufacturing cost, has been identified as a crucial area for innovation to improve throughput and reduce costs. Hydrodynamic shear mixing (HSM) and advanced ball milling techniques offer promising pathways to enhance mixing efficiency and uniformity [18]. Furthermore, the exploration of ultrasonic mixing and optimized mixing sequences underscores the industry's commitment to improving the electrochemical performance of LIBs [4]. These advancements not only promise higher capacity and energy retention but also aim to reduce solvent usage, thereby lowering both material costs and environmental impact.

Representing approximately 20% of the total cost, the coating and drying stages and solvent recovery are pivotal in the manufacturing process. The transition towards solvent-free and water-based binders represents a significant shift towards reducing the environmental footprint of LIB manufacturing [21]. Innovations like PPG Inc.'s water-based cathode binder [22] and Maxwell Technologies' solvent-free electrode coating technology [23] indicate a move towards more sustainable and cost-effective manufacturing processes. These technologies not only mitigate the use of toxic solvents but also enhance the energy efficiency of the drying process.

Although calendering is a relatively low-cost process, its importance in defining the electrode's physical properties cannot be overstated. Future developments will likely focus on optimizing calendering parameters through advanced simulations and machine learning algorithms [24, 25]. Such innovations will ensure better electrochemical performance and stability while potentially addressing safety concerns related to cell swelling and pressure [24, 25].

As slitting is integral to the roll-to-roll operation, introducing laser cutting technology offers a path to improve edge quality and reduce defects that could lead to short circuits [26]. While traditional methods have been efficient, laser cutting provides unparalleled precision and adaptability to various electrode shapes and sizes, promising to enhance the overall quality and reliability of LIBs [4, 26].

Vacuum drying and welding represent areas where energy efficiency and safety are paramount [4]. Innovations in drying techniques, such as the introduction of argon-purging methods, aim to reduce energy consumption and improve electrochemical performance [27]. In welding, the exploration of ultrasonic and laser welding technologies offers the potential for lower energy consumption and improved joint strength, which is crucial for the reliability and safety of LIBs [28].

The formation and aging process, crucial for establishing stable SEI layers, is under scrutiny for its lengthy duration and high cost. Innovations aimed to accelerate these processes without compromising battery performance include adjusting formation conditions and exploring artificial SEI layers [19, 29]. Such advancements could significantly reduce manufacturing time and costs, ultimately enhancing the competitiveness of LIBs [4, 19].

3.4 Summary

The major steps in LIB manufacturing processes include slurry mixing, coating and drying, solvent recovery, calendering, slitting, winding, vacuum drying, welding, and the formation and aging process. Each of these steps requires a high control of environmental conditions, quality control, and assurance.

References

1 Tycorun. Types and characteristics of lithium ion batteries, Tycorun Batteries [Online]. https://www.tycorun.com/blogs/news/types-and-characteristics-of-lithium-ion-batteries (accessed 9 July 2024).

2 Xu, B., Lee, J., Kwon, D. et al. (2021). Mitigation strategies for Li-ion battery thermal runaway: A review. *Renewable and Sustainable Energy Reviews* 150: 111437.

3 Li, J., Fleetwood, J., Hawley, W.B., and Kays, W. (2021). From materials to cell: state-of-the-art and prospective technologies for lithium-ion battery electrode processing. *Chemical Reviews* 122 (1): 903–956.

4 Liu, Y., Zhang, R., Wang, J., and Wang, Y. (2021). Current and future lithium-ion battery manufacturing. *IScience* 24 (4).

5 Heimes, H.H., Kampker, A., Lienemann, C. et al. (2018). *Lithium-ion battery cell production process*. PEM der RWTH Aachen University.

6 [Online]. https://dfe.com/applications/battery-electrode-tension-control/.

7 Liu, H., Cheng, X., Chong, Y. et al. (2021). Advanced electrode processing of lithium ion batteries: A review of powder technology in battery fabrication. *Particuology* 57: 56–71.

8 Mun, S.C., Jeon, Y.H., and Won, J.H. (2024). Progress and challenges for replacing n-methyl-2-pyrrolidone/polyvinylidene fluoride slurry formulations in lithium-ion battery cathodes. *Progress in Natural Science: Materials International* 34: 194–206.

9 Lee, S., Koo, H., Kang, H.S. et al. (2023). Advances in polymer binder materials for lithium-ion battery electrodes and separators. *Polymers* 15 (23): 4477.

10 Hawley, W.B. and Li, J. (2019). Electrode manufacturing for lithium-ion batteries—Analysis of current and next generation processing. *Journal of Energy Storage* 25: 100862.

11 Bai, Y., Muralidharan, N., Li, J. et al. (2020). Sustainable direct recycling of lithium-ion batteries via solvent recovery of electrode materials. *ChemSusChem* 13 (21): 5664–5670.

12 Gupta, N. Li-ion cell manufacturing: A look at processes and equipment [Online]. https://etn.news/energy-storage/li-ion-cell-manufacturing.

13 Kong, L., Aalund, R., Alipour, M. et al. (2022). Evaluating the manufacturing quality of lithium ion pouch batteries. *Journal of the Electrochemical Society* 169 (4): 040541.

14 RENON (2023). How To Tell If a Lithium Ion Battery Is Bad, 25 Aug 2023 [Online]. https://www.renonpower.com/industry-news/91.html.

15 Maddipatla, S., Kong, L., and Pecht, M. (2024). Safety analysis of lithium-ion cylindrical batteries using design and process failure mode and effect analysis. *Batteries* 10: 76.

16 Schomburg, F., Heidrich, B., Wennemar, S. et al. (2024). Lithium-ion battery cell formation: status and future directions towards a knowledge-based process design. *Energy & Environmental Science* 17 (8): 2686–2733.

17 Kampker, A., Heimes, H., Offermanns, C. et al. (2023). Optimizing the cell finishing process: an overview of steps, technologies, and trends. *World Electric Vehicle Journal* 14 (4): 96.

18 Maiser, E. (2014). Battery packaging - technology review. *AIP Conference Proceedings* 1597 (1): 204–218.

19 Armand, M. et al. (2020). Lithium-ion batteries – current state of the art and anticipated developments. *Journal of Power Sources* 479: 228708.

20 Fleischmann, J., Linder, M., Herring, D. et al. (2022). McKinsey & Company, Unlocking the growth opportunity in battery manufacturing equipment, May 3, 2022.

21 Bryntesen, S.N., Strømman, A.H., Tolstorebrov, I. et al. (2021). Opportunities for the state-of-the-art production of lib electrodes—a review. *Energies* 14 (5): 1406.

22 Wang, H., Zheng, P., Yi, H. et al. (2020). Low-cost and environmentally friendly biopolymer binders for Li–S batteries. *Macromolecules* 53 (19): 8539–8547.

23 Aizat Razali, A., Norazli, S.N., Sum, W.S. et al. (2023). State-of-the-art of solid-state electrolytes on the road map of solid-state lithium metal batteries for E-mobility. *ACS Sustainable Chemistry & Engineering* 11 (21): 7927–7964.

24 Faraji Niri, M., Apachitei, G., Lain, M. et al. (2022). The impact of calendering process variables on the impedance and capacity fade of lithium-ion cells: an explainable machine learning approach. *Energy Technology* 10 (12): 2200893.

25 Tan, C., Ardanese, R., Huemiller, E. et al. (2023). Data-driven battery electrode production process modeling enabled by machine learning. *Journal of Materials Processing Technology* 316: 117967.

26 Dammala, P.K., Dermenci, K.B., Kathribail, A.R. et al. (2023). A critical review of future aspects of digitalization next generation Li-ion batteries manufacturing process. *Journal of Energy Storage* 74: 109209.

27 Orangi, S., Manjong, N., Clos, D.P. et al. (2024). Historical and prospective lithium-ion battery cost trajectories from a bottom-up production modeling perspective. *Journal of Energy Storage* 76: 109800.

28 Grabmann, S., Kriegler, J., Harst, F. et al. (2022). Laser welding of current collector foil stacks in battery production–mechanical properties of joints welded with a green high-power disk laser. *The International Journal of Advanced Manufacturing Technology* 118 (7): 2571–2586.

29 Fichtner, M., Edström, K., Ayerbe, E. et al. (2022). Rechargeable batteries of the future—the state of the art from a BATTERY 2030+ perspective. *Advanced Energy Materials* 12 (17): 2102904.

4

The Lithium-Ion Battery Market and Key Cell Manufacturers

Hayder Ali and Hassan Abbas Khan

In 1800, Alessandro Volta invented the voltaic pile, consisting of alternating zinc and copper discs separated by layers soaked in saltwater [1]. Although it was groundbreaking, this battery was not rechargeable. The first rechargeable battery, introduced in 1859, was the lead-acid battery, which utilizes lead and lead oxide plates submerged in sulfuric acid to facilitate electrochemical conversion and produce electrical energy [1]. This chemistry is low cost, and even though lead is environmentally damaging, lead-acid batteries are still in use today [1].

In 1899, the nickel-cadmium battery was developed, offering a 30–50% higher energy density than its lead-acid counterpart [1, 2]. However, similar to lead-acid batteries, there are environmental concerns, particularly the toxicity of cadmium [2].

By the 1950s, nonrechargeable alkaline batteries emerged, providing enhanced performance and a longer lifespan than previous batteries [3]. Nickel-hydrogen batteries, introduced in the 1970s as a less toxic alternative to nickel-cadmium batteries, are primarily used in spacecraft, satellites, and military applications due to their long cycle life, reliability in extreme conditions, and environmental friendliness [4, 5]. They are also employed in utility systems for load leveling and as backup power, avoiding the environmental and disposal issues associated with lead-acid and nickel-cadmium batteries [5].

The advent of lithium-ion batteries, by researchers including John Goodenough, Stanley Whittingham, and Akira Yoshino, transformed portable electronics due to lithium-ion batteries' superior energy density, absence of memory effects, and minimal self-discharge compared to previous batteries [6]. The 21st century also saw the introduction of numerous other battery types, including lithium-sulfur, sodium-ion, magnesium, and solid-state batteries [7]; however, despite these advancements, lithium-ion batteries have continued to hold significant advantages in terms of energy density, cycle life, operational stability, technology maturity, and cost [8, 9]. The evolution of electric batteries is depicted in Figure 4.1.

Lithium-ion batteries have emerged as a dominant option in portable products [10, 11]. Their applications span from powering the most basic consumer electronics like toothbrushes, smartphones, and laptops to their role in electric vehicles (EVs), energy storage for solar and wind power applications, and even deep space exploration [12, 13]. This versatility has not only marked a milestone in technological advancement, but it also intersects with multiple United Nations sustainable

The Safety Challenges and Strategies of Using Lithium-Ion Batteries, First Edition. Edited by Michael G. Pecht.

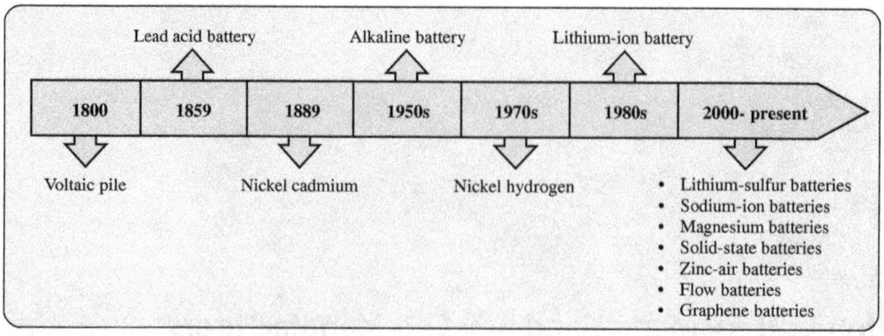

Figure 4.1 Evolution of electric battery technology.

development goals (SDGs), notably SDG 7 (Affordable and Clean Energy), SDG 11 (Sustainable Cities and Communities), and SDG 13 (Climate Action) [14].

4.1 History of Lithium-Ion Battery Commercialization

The historical journey of lithium-ion batteries is fascinating and instructive, providing insights into the evolution of energy storage technology and its societal impact. The conceptual foundation of lithium-ion batteries was laid in the 1970s and 1980s, with the introduction of lithium cobalt oxide (LCO) cathodes in 1980 [15].

Sony's commercialization of the first lithium-ion batteries using LCO in 1991 enabled the rapid development of portable electronic devices and portable power tools [15, 16]. For instance, in 1996, Motorola introduced StarTAC mobile phones with optional lithium-ion batteries, at a time when most phones were limited to lower-capacity nickel-metal hydride (NiMH) batteries [17]. Similarly, in 1997, the Apple Macintosh PowerBook 3400c laptop computer was introduced with 32 Wh lithium-ion batteries, marking a significant advancement in portable electronics power management [18].

The 1990s and early 2000s witnessed advancements in lithium-ion battery technology and its commercial applications. With the commercialization of layered cathode batteries, including nickel cobalt aluminum (NCA), lithium iron phosphate (LFP), nickel manganese cobalt (NMC), lithium manganese oxide (LMO), and lithium titanate (LTO), there was continued diversification and enhancement of battery performance [15]. The Nobel Prize in Chemistry was awarded in 2019 to Goodenough, Whittingham, and Yoshino for contributions to the development of lithium-ion batteries, underscoring the global recognition of lithium-ion batteries' significance [19].

The period from 2010 to 2020 was marked by a shift toward high-energy density chemistries, particularly with NMC and NCA, driven by the burgeoning electric vehicle industry. Notably, Tesla's launch of the Roadster in 2008 and the Model S in 2012, with its range of 426 km, underscored the use of lithium-ion batteries in the EV revolution [20, 21]. Concurrently, Apple's iPhone and MacBook Air, which claimed extended battery life, demonstrated the impact of lithium-ion batteries on consumer electronics [22, 23].

By 2020, the development of high-voltage cathodes and nickel-rich lithium-ion batteries had further advanced the technology [24, 25]. A notable milestone in 2023 was the launch of a condensed battery by CATL, boasting an energy density of up to 500 Wh/kg, which opened up the potential for the electrification of passenger aircraft [26]. A summary of the historical progression of lithium-ion battery development is depicted in Figure 4.2.

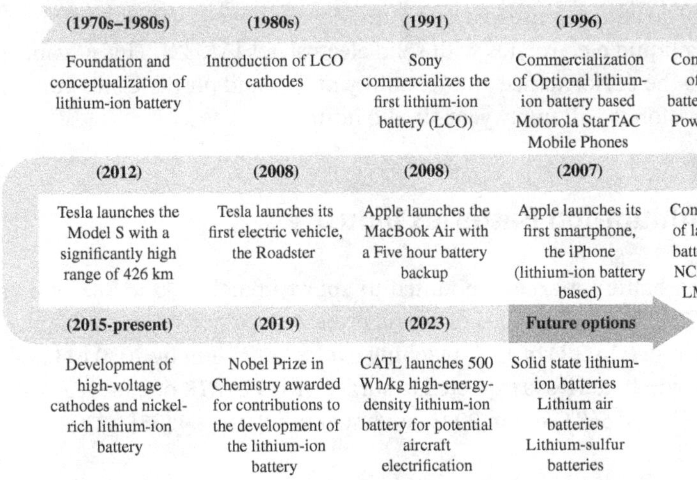

Figure 4.2 Key milestones in the commercial development of lithium-ion batteries.

The evolution of lithium-ion batteries is closely linked to steady improvements in gravimetric and volumetric energy density, which are critical parameters for both consumer electronics and EVs. From an energy density of 80 Wh/kg in 1991, lithium-ion batteries have reached 500 Wh/kg by 2023 [26, 27]. Subsequently, the highest reported energy density of 711 Wh/kg in lab settings is expected to be commercialized by 2025 [28]. Furthermore, the volumetric energy density of 18,650 lithium-ion battery cells increased from 200 Wh/L in 1991 to 888 Wh/L in 2023 [29, 30]. Following this, a volumetric energy density of over 1000 Wh/L has also been reported in lab settings [29]. To emphasize these points, the historical progression of energy density in lithium-ion batteries is illustrated in Figure 4.3.

Before the latest designs can be commercialized, there are challenges to overcome, such as the stability of materials, safety concerns, electrolyte compatibility, and cost [29, 30]. However, advancements in this area have not only enabled the development of more compact and longer-lasting electronic devices, but they have also contributed to the feasibility of EVs with greater range and efficiency.

Looking toward the future, lithium-ion technology is best positioned for further advancements with the development of solid-state, lithium-sulfur, and lithium-air batteries. Solid-state batteries,

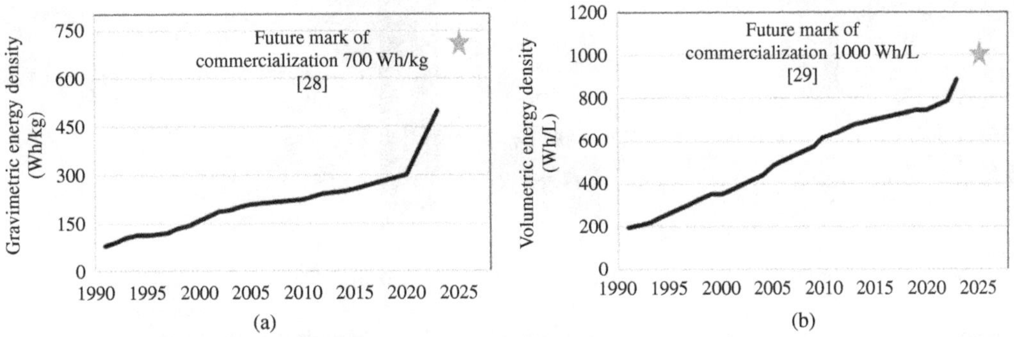

Figure 4.3 Historical progression of lithium-ion battery technology in terms of (a) gravimetric energy density and (b) volumetric energy density.

in particular, promise high energy densities, improved safety, and longer lifespans than traditional lithium-ion batteries by replacing liquid electrolytes with solid electrolytes [31, 32]. This advancement could significantly enhance the performance and reliability of EVs and portable electronics. However, widespread implementation is still many years in the future.

4.2 Expansion of the Lithium-Ion Batteries Industry

As of 2023, the global lithium-ion battery market amounted to approximately $70 to $85 billion [33, 34], with a demand of 558 GWh [35]. This market size is projected to increase significantly, with a compound annual growth rate (CAGR) of 18%, reaching over $250 billion by 2030 [33, 34]. Similarly, the demand for lithium-ion batteries is expected to surge, with a CAGR of 20%, reaching 2000 GWh by 2030, from the baseline of 558 GWh in 2023, as shown in Figure 4.4 [35].

4.3 Geographic Distribution of Battery Manufacturing

According to Bloomberg New Energy Finance's (BloombergNEF's) lithium-ion supply chain rankings [36], China held 77% of the global manufacturing capacity, amounting to 893 GWh in 2022 [36, 37]. This dominance is enabled by China's vertical integration across the EV supply chain, from metal mining to vehicle production, and its status as the largest EV market globally, capturing over half of global sales in 2021 [38]. Major Chinese battery companies include CATL, BYD, and CALB. Additionally, South Korea's LG and Japan's Panasonic have lithium-ion battery production in China [39, 40].

Despite China's dominance, other nations are also advancing in battery manufacturing. Poland contributes 6% to global capacity, with LG's Wrocław gigafactory being the largest in Europe [36, 37]. Another South Korean company, SK Innovation, has a production plant in Poland and plans to establish four more plants in the Silesia region [39].

The United States, though trailing behind China, held 6% of global capacity in 2022, with several major battery manufacturing facilities [37]. Prominent US-based companies include

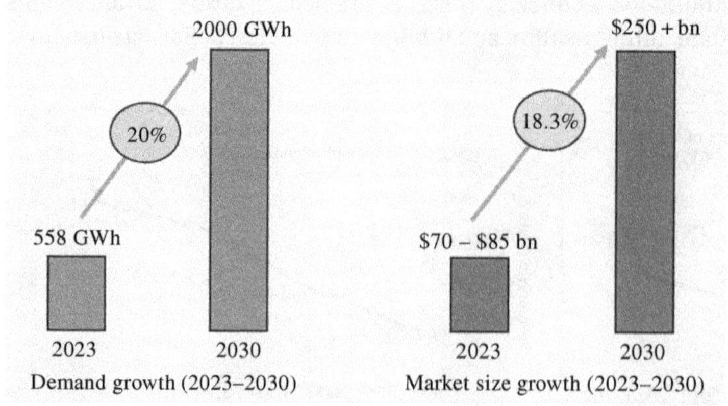

Figure 4.4 Projected demand and market size growth of lithium-ion batteries, 2023–2030. Source: Adapted from [35].

BorgWarner, Microvast, QuantumScape, Sila Nanotechnologies, iM3NY, and Enovix [39, 41]. Tesla and Panasonic jointly operate several battery manufacturing facilities in the United States, including Gigafactory 1 in Nevada, which focuses on producing high-performance lithium-ion batteries for Tesla's EVs [42]. Furthermore, LG operates several production facilities in the United States, notably in Michigan, producing lithium-ion batteries for various applications [43].

Germany, Hungary, Sweden, South Korea, and Japan have also expanded their lithium-ion battery production capacities [39]. With its potential as a large EV market, Germany aims to have 1 million charging stations and 15 million EVs by 2030 [39]. Key manufacturers include VARTA AG (Germany), Samsung (South Korea), and Northvolt (Sweden). Additionally, Hungary is developing its battery manufacturing capacity with plants from Samsung, SK Innovation, and CATL. Sweden's Northvolt produces 14 GWh of battery capacity annually and plans to deliver batteries with an 80% lower carbon footprint, targeting 150 GWh by 2030 [39]. An overview of leading lithium-ion battery-producing countries and their key companies, along with the origin of each company, is presented in Table 4.1.

By 2027, Bloomberg projects that China will hold 69% of global battery manufacturing capacity [36, 37]. The United States is expected to reach 10% [37], driven by incentives such as EV tax credits and investments from both domestic and international companies, such as Ford, General Motors, Toyota, SK Innovation, and LG [44, 45].

Europe is also emerging as a key player in lithium-ion battery manufacturing, projected to reach around 1000 GWh production capacity by 2027 [37]. Investments from domestic and foreign firms, including Volkswagen, CATL, and SK Innovation, underscore the region's commitment to an electric future [46, 47].

The geographic distribution of battery manufacturing reflects efforts across the globe to achieve dominance and self-sufficiency in the transition to electric mobility. China leads the charge, but other nations are gearing up to establish their positions in this evolving landscape. Estimates suggest that the United States will invest $87 billion and Europe $102 billion to establish entirely local supply chains by 2030 [36, 37]. Figure 4.5 illustrates the geographic distribution of battery manufacturing in 2022 and estimated in 2027 [36].

Table 4.1 Leading lithium-ion batteries producing countries and their key companies.

Country	Key lithium-ion battery producers
China	CATL (China), BYD (China), CALB (China), LG (South Korea), Samsung (South Korea), Panasonic (Japan)
United States	BorgWarner (United States), Microvast (United States), QuantumScape (United States), Sila Nanotechnologies (United States), iM3NY (United States), Enovix Corporation (United States), Tesla-Panasonic (United States-Japan), A123 Systems (China), SK Innovation (South Korea), LG (South Korea)
Poland	LG (South Korea), SK Innovation (South Korea)
Germany	VARTA (Germany), Samsung (South Korea), Northvolt (Sweden), CATL (China)
Hungary	CATL (China), Samsung (South Korea), SK Innovation (South Korea)
Sweden	Northvolt (Sweden)
South Korea	LG (South Korea), Samsung (South Korea), SK Innovation (South Korea)
Japan	Panasonic (Japan), Sony Energy Devices (Japan), Toshiba (Japan)

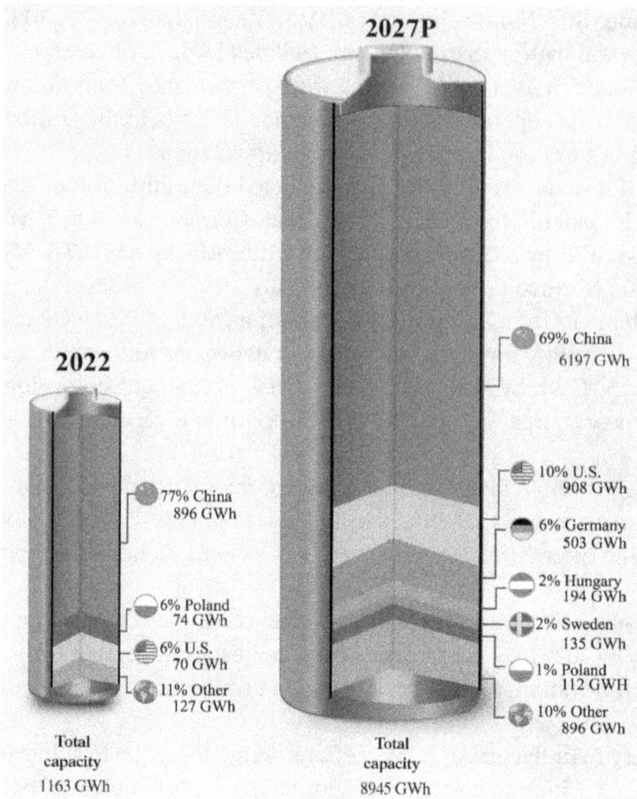

Figure 4.5 Geographic distribution of battery manufacturing in 2022 and estimated in 2027.

4.4 Demand for Batteries

According to Bloomberg [35], EVs—including passenger cars, commercial vehicles, electric two-wheelers, and buses—accounted for 69% of the total lithium-ion battery demand, totaling around 550 GWh during 2022–2023. In contrast, consumer electronics comprised 19%, and stationary storage solutions comprised 12% of the market share, as depicted in Figure 4.6 [35].

According to the International Energy Agency (IEA) Global EV Outlook 2023 [48], lithium-ion battery demand experienced a twelvefold increase from around 43 GWh in 2016 to 550 GWh in 2022, primarily driven by the sale of over 10 million EVs worldwide during that period [48, 49]. This represents a 14% share of the total automobile market [48, 49]. Furthermore, projections for 2030 suggest that EVs will dominate 77% of the lithium-ion battery market, with total demand skyrocketing to approximately 2000 GWh [35].

In 2023, China's demand for lithium-ion batteries reached 312 GWh, representing 57% of the global demand, as shown in Figure 4.7 [48, 49]. This demand spike aligns with an 80% rise in EV sales in 2022 over the previous year, reaching 6 million units [50]. With an installed capacity of 893 GWh, China's production capabilities far exceed its current demand of 312 GWh, providing China with 122% lithium-ion battery self-sufficiency [37, 51]. This capacity is projected to expand to 6197 GWh by 2028 [37, 51].

Europe has witnessed a sharp increase in battery demand, soaring over 26 times from 4.8 GWh in 2016 to 127.7 GWh in 2022, constituting 23% of the global lithium-ion battery demand, as illustrated

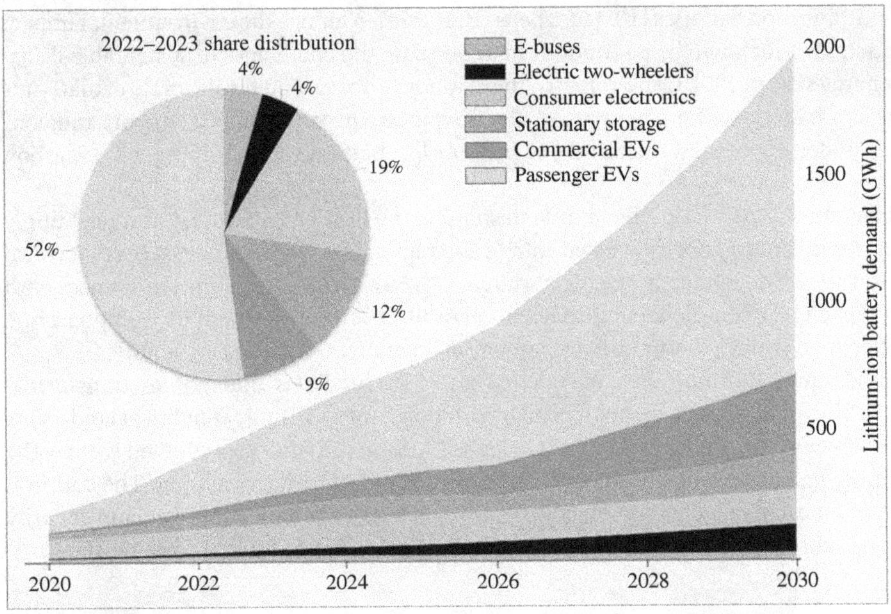

Figure 4.6 Global lithium-ion battery demand based on application (2020–2030).

Figure 4.7 Global lithium-ion battery demand in 2022 based on region.

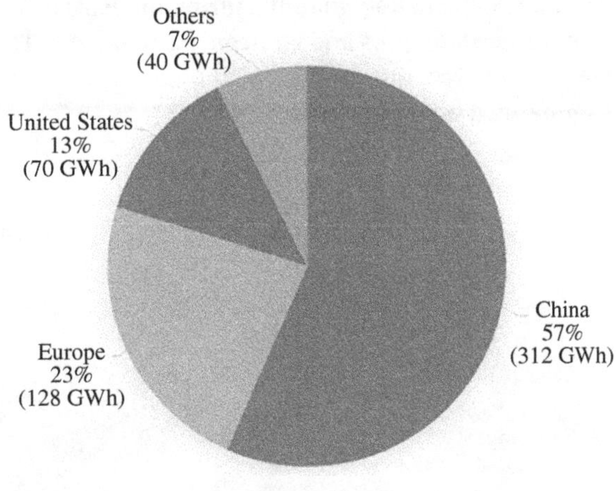

in Figure 4.7 [49]. This surge is attributed to Europe's EV market, the second largest globally after China, with sales of 2.7 million units in 2022 [52]. To put it another way, EVs command approximately 30% of the car market share in China and over 20% in Europe, indicating a strong trend toward sustainable transportation options [53].

CATL leads the battery cell production market, followed by LG, BYD, Panasonic, SK On, and Samsung SDI. Four of these six companies—CATL, LG, SK On, and Samsung SDI—have expanded their manufacturing into Europe, driven by the region's growing demand for energy storage solutions [53]. In 2022, European battery production met 50% of the demand, supported by facilities such as CATL in Germany, LG Energy Solution in Poland, and SK On and Samsung SDI in Hungary [54, 55].

As the lithium-ion battery market continues to evolve, there has been an increased focus on sustainability and the circular economy, leading to innovations in battery recycling and second-life applications for lithium-ion batteries [10, 56]. These efforts aim to reduce the environmental impact of battery production, which will be paramount in addressing the challenges of sustainable transportation and energy storage [57]. Consequently, these innovations extend the life cycle of materials used in lithium-ion batteries. The push toward electrification, driven by policy support and consumer demand, underscores the critical role of lithium-ion batteries in achieving a low-carbon future [52].

Geopolitical factors also play a significant role in shaping the lithium-ion battery market. Supply chain resilience has become a priority, with countries and regions striving to secure access to critical raw materials and reduce dependency on single sources [58, 59]. This strategic shift influences both the production capacity and the global trade patterns of lithium-ion batteries and EVs, emphasizing the need for diversified supply chains and investment in local manufacturing capabilities.

Overall, the landscape of lithium-ion battery demand and production is undergoing a transformative phase, marked by rapid growth, technological innovation, and a shifting geopolitical landscape. As the world moves closer to a sustainable and electrified future, the strategies adopted will lay the groundwork for the resilience and sustainability of the lithium-ion batteries market. The continuous exploration of alternative technologies, coupled with efforts to enhance supply chain security and sustainability, will be crucial in navigating the challenges and opportunities that lie ahead.

4.5 Leading Battery Producers Worldwide

As of 2023, the battery manufacturing industry is dominated by several key players, each with its own unique strengths and strategic partnerships. The top ten battery manufacturers in the world are presented in this section with their current market contributions, with their market shares summarized in Figure 4.8.

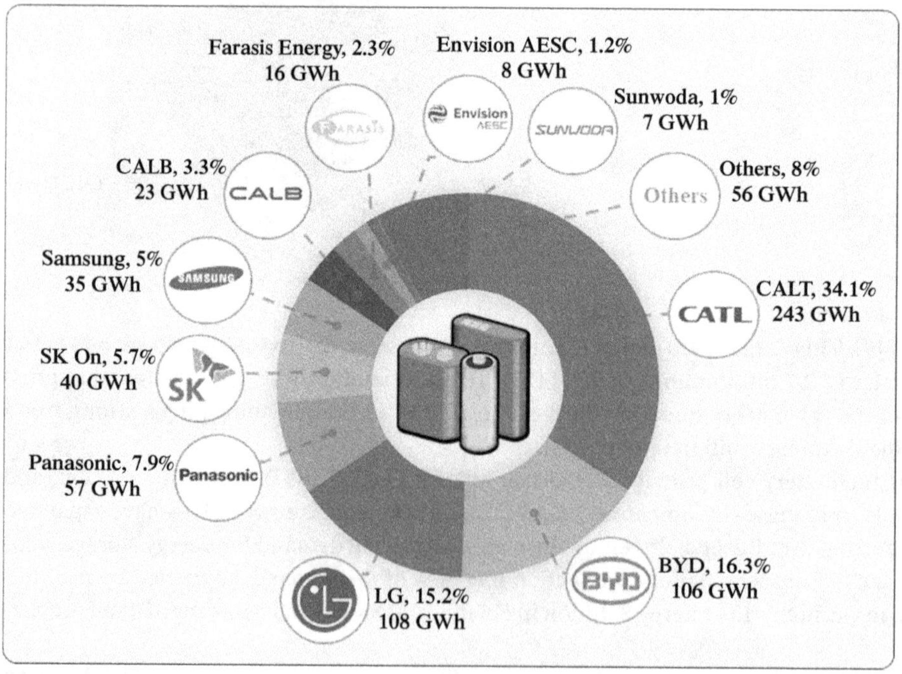

Figure 4.8 Overview of market shares in the battery industry by company in 2023.

4.5.1 Contemporary Amperex Technology Co., Ltd. (CATL)

CATL, headquartered in Ningde, China, is the world's largest battery manufacturer. It holds a 34.1% market share [60, 61] and generated a revenue of $62 billion in 2023 [62]. Specializing in lithium-ion batteries, CATL has established itself as a global primary supplier to numerous EV manufacturers. With a focus on cost competitiveness and technological efficiency, CATL continues to expand its production capacity, with ambitious plans to reach 670 GWh by 2025 [63]. Its ongoing collaborations, such as a $3.5 billion project with Ford Motors, underscore its commitment to driving innovation and meeting the growing demand for sustainable energy solutions [63]. Key products include batteries and their materials and energy storage systems, with significant battery clients including BMW, Ford, Daimler, Hyundai, Honda, Tesla, Toyota, Volkswagen, and Volvo [64]. Although not as prominent as their automotive batteries, CATL also develops batteries for energy storage systems and various portable electronic devices [65]. CATL's lithium-ion battery manufacturing plants are located in China, Germany, and Hungary [66].

Unfortunately, there have been incidents where CATL batteries in EVs have caught fire. CATL batteries in the Aion S model from GAC reportedly caught fire multiple times in 2020 [67]. Furthermore, recalls and safety checks have involved vehicles equipped with CATL batteries [68]. For instance, in 2019, Nio recalled its flagship sport utility vehicles due to a potential issue with the battery packs supplied by CATL [68]. It was discovered that these batteries could short circuit under extreme conditions, attributed to a structural flaw in the battery pack design [68]. Such recall measures are typically precautionary, addressing potential issues that could lead to fires.

In 2023, according to Bloomberg [69], the United States Congress prohibited the Pentagon from procuring batteries produced by six Chinese companies, including CATL, BYD, Envision Energy, EVE Energy, Gotion High-Tech, and Lithium Energy Storage Technology. However, this measure does not extend to commercial purchases by companies such as Ford Motor and Tesla.

4.5.2 BYD Co., Ltd.

From Shenzhen, China, BYD Co., Ltd. has emerged as a major player in the EV industry, overtaking Tesla as the world's largest EV maker [70, 71]. With a revenue of $89 billion and a 16% market share in 2023 [60, 61], BYD's vertically integrated business model sets it apart, allowing for in-house production of batteries and other components [72]. Notably, its battery production in 2023 reached 106.02 GWh [72], and its strategic expansion plans include the construction of multiple battery plants, with a target capacity of 445 GWh by 2024 [63]. Key products include passenger vehicles, batteries, and photovoltaic products, with their primary battery client being their own auto group, BYD Auto [73]. Although EVs are their primary application, BYD also produces batteries for energy storage systems, consumer electronics, and rail transit [74]. BYD has plants located in China and is investing in a new battery assembly plant in Hungary.

Notably, there have been numerous incidents where BYD batteries caught fire, such as in 2024 when a BYD electric car ignited at a Hong Kong charging station [75]. Additionally, there have been reports of fires involving BYD electric buses around the world. For example, in 2020, a BYD electric bus caught fire in Madrid, Spain [76]. These incidents have prompted investigations to identify the causes and mitigate future risks. Although such events are less frequent than with some competitors, BYD has still faced recalls or safety reviews related to battery issues. In 2022, BYD recalled over 60,000 units of its Tang DM-i hybrids equipped with pouch-type batteries, due to a defect in the battery pack that could lead to thermal runaway [77]. According to a 2024 Reuters report, BYD plans to discontinue the use of pouch-type batteries in its hybrid models due to concerns over durability and the risk of fluid leaks [77].

BYD has faced controversies and restrictions globally due to geopolitical tensions and concerns about dependence on Chinese technology. For example, the US Congress has banned the Pentagon from procuring batteries from BYD due to security concerns [69]. India rejected BYD's $1 billion EV joint venture proposal, reflecting caution toward Chinese investments [78]. The European Commission has been investigating Chinese EV manufacturers, including BYD, for unfair subsidies, potentially leading to higher import duties to protect local industries [79].

4.5.3 LG Energy Solution, Ltd.

Based in Seoul, South Korea, LG Energy Solution boasted a 15% market share in 2023 [60, 61], making it the largest non-Chinese battery manufacturer, with a revenue of $26.5 billion [80]. With a rich history of producing lithium-ion batteries since 1999, LG Energy Solution has positioned itself as a leading EV supplier, with battery production in 2023 reaching 111.6 GWh [72]. As it continues to invest in grid-scale energy storage sectors, LG Energy Solution remains a formidable player in the global battery market [63]. Key products include batteries and energy storage systems, with significant clients including Hyundai, Kia, General Motors, Ford, Stellantis, Toyota, Honda, Volkswagen, Renault-Nissan, and Volvo [81]. In addition to producing batteries for EVs, the company is also known for its battery production for consumer electronics and energy storage systems [82]. LG operates its primary lithium-ion battery manufacturing plants in South Korea, China, the United States, Canada, Indonesia, and Poland [83].

Over the years, the company has encountered numerous safety-related incidents. In 2022, the United States National Highway Traffic Safety Administration (NHTSA) started investigating LG's battery cells due to defects that have led to fires in EVs and plug-in hybrids [84]. This review involves over 138,000 vehicles from various manufacturers, including Mercedes-Benz, General Motors, Hyundai, and Stellantis [84]. The incidents led to costly recalls of models like the Chevrolet Bolt EV, which also faced production halts. Relatedly, the NHTSA found that Mercedes-Benz, Chrysler parent Stellantis, General Motors, and Hyundai Motor have issued recalls since 2020 for internal defects in high-voltage batteries that pose a fire risk [84, 85]. Additionally, LG Energy Solution has experienced problems with its energy storage systems. For instance, in 2023, there were reports of fires in LG energy storage units in the United States and Australia [86], leading to the Australian Competition and Consumer Commission (ACCC) proposing a compulsory recall of 5000 LG solar storage batteries in 2024 [87]. Moreover, LG batteries used in consumer electronics have also been recalled, with incidents of fires reported in recent years [88, 89].

4.5.4 Panasonic Holdings Corporation

Panasonic Holdings Corporation, headquartered in Kadoma, Japan, commanded a 7.9% market share in 2023 [60, 61], with a revenue of $60 billion [72, 90], reaching a battery production of 55.8 GWh. Starting in 2011, Panasonic has been producing batteries for Tesla EVs [63]. With initiatives like a $4 billion EV battery plant in Kansas, Panasonic aims to further its position as a primary battery supplier for Subaru Corp., Mazda Motor Corp., Toyota, and Lucid Group [63, 91]. Besides EVs, Panasonic batteries are also used in energy storage systems and consumer electronics [92]. Currently, the company's lithium-ion battery production plants are situated in Japan, the United States, and China [93, 94].

Unfortunately, Panasonic batteries have faced safety concerns in past years. In 2019, an incident was reported where a Tesla vehicle caught fire, involving batteries supplied by Panasonic [95]. Additionally, the company recalled 43,000 battery packs used in notebook computers in 2014 after three incidents of batteries overheating and catching fire [96].

4.5.5 SK Innovation Co., Ltd.

Seoul-based SK Innovation Co., Ltd., with a revenue of $58 billion, held a 5.7% market share in 2023 [60, 97], and their battery production reached 39 GWh [72]. Renowned for its pioneering work in high-energy ternary materials for EV batteries, SK Innovation supplies batteries to major automotive manufacturers. The company continues to experience significant growth, with plans to expand its annual capacity to 150 GWh by 2026 [63]. With a focus on the future, SK is developing new battery technologies, including lithium-iron-phosphate EV batteries [63]. Their key products include batteries, with important battery clients including Ford, Volkswagen, and Hyundai [98]. However, their batteries are also used in various other applications such as energy storage systems, consumer electronics, and industrial tools [99]. Currently, the company has lithium-ion battery production facilities located in South Korea, the United States, Poland, China, and Hungary [100, 101].

In 2021, Hyundai recalled 76,000 Kona EVs, manufactured between 2018 and 2020, following over a dozen reported fires linked to the electric SUV's battery pack, which used SK Innovation batteries [102]. Additionally, in October 2023, the US Labor Department imposed fines totaling $77,200 on SK Battery America for five serious safety violations after workers in a Georgia lithium battery plant suffered potentially permanent respiratory damage during a fire [103].

SK Innovation faced a serious legal issue in the United States when it was initially banned from importing, producing, and selling electric vehicle batteries for ten years [104, 105]. This was due to claims that it had stolen trade secrets from its competitor, LG Energy Solution. However, SK Innovation settled with LG to avoid the ban by agreeing to pay $1.8 billion [104, 105].

4.5.6 Samsung SDI Co., Ltd.

A division of Samsung Electronics Co., Ltd. and based in South Korea, Samsung SDI Co., Ltd. held a 5% market share in 2023, a revenue of $17 billion [60, 106], and battery production reaching 27.9 GWh [72]. Leveraging its experience in energy storage systems, Samsung SDI supplies batteries to Ford, BMW, Audi, Volkswagen, Volvo, and Hyundai [107]. With ambitious expansion plans, including a $2.5 billion battery plant in the United States, Samsung SDI is poised to capitalize on the growing demand for EV batteries and renewable energy solutions [63]. Additionally, they are also a renowned manufacturer of batteries for consumer electronics, as well as for energy storage systems and industrial tools [108]. Current production facilities are situated in South Korea, China, the United States, Japan, Germany, Vietnam, India, Hungary, and Malaysia [109].

Samsung has faced several high-profile battery incidents, particularly with its consumer electronics. The most prominent was the Samsung Galaxy Note 7 smartphone in 2016 [110, 111], which became infamous as reports of it catching fire or exploding due to battery defects emerged [110, 111]. This resulted in a global recall and the eventual discontinuation of the model, significantly affecting Samsung's reputation and financial standing [110, 111].

4.5.7 CALB Group Co., Ltd.

Operating out of Jiangsu, China, CALB Group Co., Ltd., with a revenue of $3 billion, has emerged as a prominent producer of lithium-ion power batteries. The company boasted a 3.3% market share in 2023 [60, 112], with battery production reaching 16.74 GWh [72]. Their strategic partnerships with automotive manufacturers like Xpeng, Leapmotor, Geely Automobile, and GAC Aion highlight its growing influence in the EV ecosystem. With plans to expand its total capacity to 500 GWh by

2025, CALB Group is well positioned to meet the escalating demand for EV batteries and energy storage solutions [63]. Key products include batteries and their materials, with large battery clients including Xpeng, Leapmotor, Geely Automobile, and GAC Aion [113]. In addition to EVs, CALB's batteries are used in energy storage systems, consumer electronics, industrial settings, and marine applications [114]. The company has manufacturing facilities primarily in China, with significant expansions planned in Europe [115].

Specific incidents involving CALB batteries have not been widely publicized or reported in major news sources or industry reports at the same scale as other major battery manufacturers. However, CALB, like all battery manufacturers, is subject to the inherent risks associated with lithium-ion batteries, such as thermal runaway and potential fire hazards.

In 2022, CALB faced a legal setback related to battery patent infringement. Notably, the company lost a lawsuit brought by its competitor CATL, resulting in a court order for CALB to stop selling certain products and pay damages [116].

4.5.8 Farasis Energy (Gan Zhou) Co., Ltd.

Based in China, Farasis Energy (Gan Zhou) Co., Ltd. focuses on the research, development, and commercialization of lithium-ion power batteries for EVs. With a revenue of $8.8 billion in 2023 [117], Farasis had a 2.3% market share [60, 61], and their battery production reached 5.58 GWh [72]. The company has partnered with automakers like Jiangling Motors Electric Vehicle, Daimler, GAC Aion, FAW, Volkswagen, Dongfeng Motor, JMC, and Mitsubishi [118, 119]. In addition to manufacturing batteries for EVs, Farasis also produces batteries for energy storage systems [120]. The company's lithium-ion battery manufacturing plant is situated in China, with plans to construct a 20 GWh facility in Turkey [121].

4.5.9 Envision AESC

Envision Automotive Energy Supply Corporation (Envision AESC), based in Zama, Japan, is a joint venture of Nissan Motor Company, Ltd., NEC Corporation, and NEC Tokin Corporation. They commanded a 1.2% market share in 2023 [60, 61], a revenue of less than $1 billion [72, 122], and battery production reached 11.16 GWh [72]. The company's strategic partnerships with Renault, BMW, and Nissan highlight its expertise in providing power batteries for EVs [63, 123]. Besides EVs, Envision Automotive Energy Supply Corporation (Envision AESC) batteries are also used in energy storage systems [124]. With an eye toward the future, Envision plans to expand its annual capacity to 25 GWh by 2030 [63]. In addition to China, the company operates battery manufacturing facilities in the United Kingdom and Japan, and it is planning to construct a new plant in the United States [125, 126].

In 2023, the United States imposed bans on six Chinese lithium-ion battery manufacturers over security concerns [69, 127]. According to Bloomberg, the list includes Envision Energy [69, 127], but the restriction only applies to the US government and not commercial transactions.

4.5.10 Sunwoda Electric Battery Co., Ltd.

A subsidiary of Sunwoda Electronic Co., Ltd., based in Shanghai, China, Sunwoda Electric Battery held a 1% market share in 2023 [60, 61], revenue of $8.8 billion [72, 128], and battery production reaching 5.58 GWh [72]. With partnerships including automakers like Daimler, BAIC Group, Guangzhou Automobile, Volkswagen, Dongfeng Motor, and Great Wall Motor [129], Sunwoda is poised for growth, as evidenced by its plans to build multiple battery factories with significant production capacities [63]. Outside their EV batteries, Sunwoda also makes batteries for consumer electronics and energy storage systems [130]. In addition to China, the company has production facilities in India, Vietnam, Hungary, and Morocco [131]. Notably, there have been no widely reported safety concerns involving Sunwoda batteries.

4.6 Battery Suppliers and Their Market Clients

The relationship between battery manufacturers and their clients is pivotal in several key industries, especially as the demand for EVs grows. Lithium-ion battery manufacturers supply batteries to a diverse range of clients, including EV manufacturers, consumer electronics companies, and firms involved in energy storage solutions. Significantly, 69% of lithium-ion battery demand in 2023 consisted of batteries used in EVs.

As examples of the importance of relationships with EV manufacturers, the top Chinese company CATL and Japanese company Panasonic manufacture batteries for Tesla vehicles. In addition to Tesla, CATL also supplies batteries to BMW, Daimler, and Ford, among others. LG Chem supplies batteries to many EV makers, such as Ford, General Motors, Honda, and others, yet BYD manufactures batteries for their own EVs. Figure 4.9 shows the relationship between EV battery suppliers and EV makers, where dark blue denotes a relationship and light gray indicates the absence of one.

Batteries are utilized in various applications beyond EVs, including mobile phones and laptops. Many lithium-ion battery manufacturers supply batteries to these sectors although detailed disclosures are often limited. Notably, Amperex Technology Limited (ATL) secured about one-third of the mobile phone battery market share in 2023 [132]. Established in 1999, ATL is recognized for producing lithium polymer batteries primarily for consumer electronics [133]. In 2011, its parent company at the time, TDK Corporation, divested its automotive battery division, resulting in the formation of CATL. Consequently, CATL became an independent entity focused on automotive and energy storage batteries [133]. Table 4.2 displays the relationships between battery manufacturers and their clients in the mobile phone and laptop sectors.

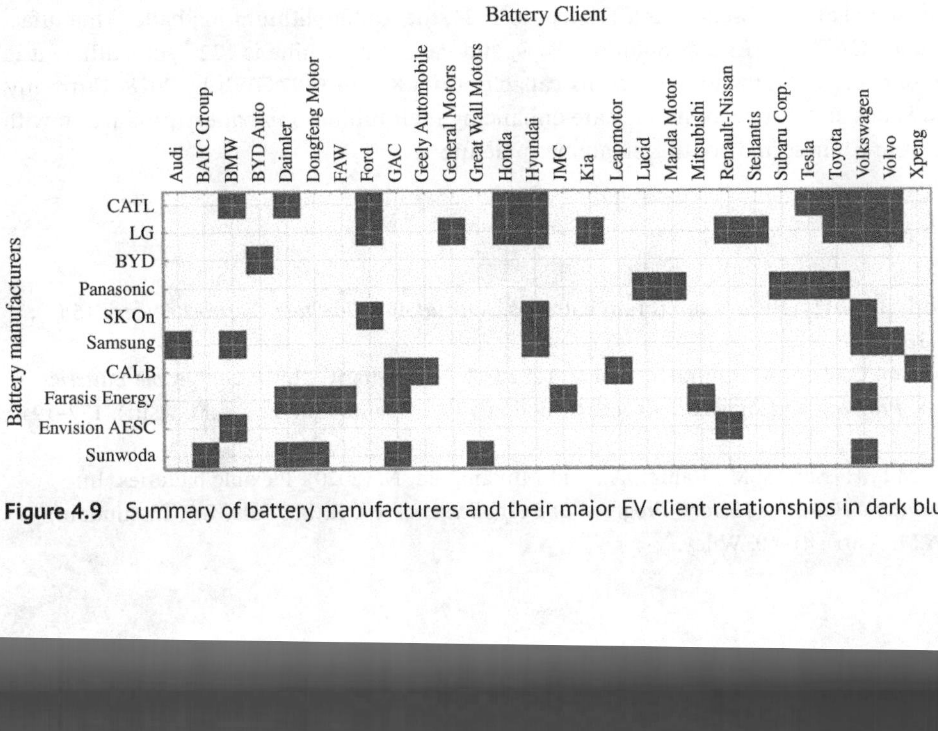

Figure 4.9 Summary of battery manufacturers and their major EV client relationships in dark blue.

Table 4.2 Overview of battery manufacturers and their key clients in mobile phones and laptops/tablets.

Lithium-ion batteries manufacturer	Mobile phone companies	Laptop/tablets companies
LG [134, 135]	Apple, LG	HP, LG
BYD [136–138]	Apple, Huawei	Apple, Dell
Panasonic [139]	–	Panasonic, Lenovo
Samsung [140]	Samsung	Apple, Samsung
Sony [141]	Sony	Dell, Sony
Sunwoda [139, 142, 143]	Apple, Huawei	Lenovo
ATL (Amperex Technology Limited) [139, 140, 144–147]	Apple, Xiaomi, Oppo	Apple, HP, Dell, Lenovo

4.7 Summary

The lithium-ion battery market size is around $80 billion and has evolved significantly since the first commercial release of lithium-ion batteries in 1991. In particular, the batteries play a crucial role in powering contemporary consumer electronics, EVs, and renewable energy storage systems. Overall, the market size is expected to grow to more than $230 billion by 2030.

While lithium-ion batteries are used in smartphones, laptop computers, and many other portable consumer products, the increased demand for lithium-ion batteries is predominantly driven by the EV market. In 2023, the EV market accounted for 69% of global lithium-ion battery consumption and is expected to grow further due to the automotive sector's shift toward electrification and the implementation of zero-carbon policies. Consumer electronics account for 19% of the demand, while stationary energy storage systems comprise 12%.

There are now many thousands of battery manufacturers in the world. Notably, over 60% of global lithium-ion battery production capacity is owned by Chinese companies (some big and many small ones), followed by South Korea at 28%, and Japan at 8%. The remaining capacity is distributed among the United States, Sweden, and France. In 2024, the leading lithium-ion battery manufacturers included CATL, LG Energy Solution, BYD, and Panasonic. China is 122% self-sufficient in battery production and aims to increase its capacity from 893 to 6197 GWh by 2028. Currently, the United States and European nations are enhancing their lithium-ion battery production with assistance from Chinese and South Korean manufacturers.

References

1 Scrosati, B. (2011). History of lithium batteries. *Journal of Solid State Electrochemistry* 15 (7): 1623–1630.

2 Jeyaseelan, C., Jain, A., Khurana, P. et al. (2020). Ni-Cd Batteries. In: *Rechargeable Batteries: History, Progress, and Applications* (ed. R. Boddula, R.P. Inamuddin, and A.M. Asiri), 177–194. Wiley.

3 Khan, M.I., Hassan, M.M., Rahim, A., and Muhammad, N. (2020). Flexible batteries. In: *Rechargeable Batteries: History, Progress, and Applications* (ed. R. Boddula, R.P. Inamuddin, and A.M. Asiri), 41–60. Wiley.

4 Pathak, A.D., Saha, S., Bharti, V.K. et al. (2023). A review on battery technology for space application. *Journal of Energy Storage* 61: 106792.

5 Bernard, P. and Lippert, M. (2015). Nickel-cadmium and nickel–metal hydride battery energy storage. In: *Electrochemical Energy Storage for Renewable Sources and Grid Balancing* (ed. P.T. Moseley and J. Garche), 223–251. Elsevier https://www.sciencedirect.com/book/9780444626165/electrochemical-energy-storage-for-renewable-sources-and-grid-balancing.

6 Tyutyunnik, V.M. (2021). Another breakthrough in power supply technology–lithium-ion batteries: 2019 Nobel Prize winners in Chemistry John Goodenough, Stanley Whittingham, and Akira Yodhino. *Journal of Advanced Materials and Technologies* 6 (3): 163–166.

7 Lou, S., Zhang, F., Fu, C. et al. (2021). Interface issues and challenges in all-solid-state batteries: lithium, sodium, and beyond. *Advanced Materials* 33 (6): 2000721.

8 Ohno, S. and Zeier, W.G. (2021). Toward practical solid-state lithium-sulfur batteries: challenges and perspectives. *Accounts of Materials Research* 2 (10): 869–880.

9 Biemolt, J., Jungbacker, P., van Teijlingen, T. et al. (2020). Beyond lithium-based batteries. *Materials* 13 (2): 425.

10 Ali, H., Khan, H.A., and Pecht, M.G. (2021). Circular economy of Li Batteries: technologies and trends. *Journal of Energy Storage* 40: 102690.

11 Park, S., Jeong, S.Y., Lee, T.K. et al. (2021). Replacing conventional battery electrolyte additives with dioxolone derivatives for high-energy-density lithium-ion batteries. *Nature Communications* 12 (1): 838.

12 Ali, H., Beltran, H., Lindsey, N.J., and Pecht, M. (2023). Assessment of the calendar aging of lithium-ion batteries for a long-term—Space missions. *Frontiers in Energy Research* 11: 1108269.

13 Peng, J., Meng, J., Wu, J. et al. (2023). A comprehensive overview and comparison of parameter benchmark methods for lithium-ion battery application. *Journal of Energy Storage* 71: 108197.

14 United Nations (UN) (2015). Sustainable Development Goals (SDGs). United Nations. https://sdgs.un.org/goals (accessed 16 March 2024).

15 Liu, W., Placke, T., and Chau, K. (2022). Overview of batteries and battery management for electric vehicles. *Energy Reports* 8: 4058–4084.

16 Wang, J., Ma, J., Zhuang, Z. et al. (2024). Toward direct regeneration of spent lithium-ion batteries: a next-generation recycling method. *Chemical Reviews* 124 (5): 2839–2887.

17 Motorola (1999). StarTAC User Guide. Motorola. https://ringtones.specialtyansweringservice.net/wp-content/uploads/2014/08/manuals/motorola-startac-userguide.pdf (accessed 18 March 2024).

18 Apple (2017). Macintosh PowerBook 3400c/180: technical Specifications. Apple. https://support.apple.com/en-gb/112136 (accessed 17 March 2024).

19 The Nobel Prize (2019). The Nobel Prize in Chemistry 2019. The Nobel Prize. https://www.nobelprize.org/prizes/chemistry/2019/summary/ (accessed 18 March 2024).

20 Perkins, G. and Murmann, J.P. (2018). What does the success of Tesla mean for the future dynamics in the global automobile sector? *Management and Organization Review* 14 (3): 471–480.

21 Ulrich, L. (2013). Top 10 tech cars: slenderized. *IEEE Spectrum* 50 (4): 34–41.

22 Apple (2008). Apple Introduces MacBook Air—The World's Thinnest Notebook. Apple. https://www.apple.com/newsroom/2008/01/15Apple-Introduces-MacBook-Air-The-Worlds-Thinnest-Notebook/ (accessed 18 March 2024).

23 Apple (2007). Apple Reinvents the Phone with iPhone. https://www.apple.com/uk/newsroom/2007/01/09Apple-Reinvents-the-Phone-with-iPhone/ (accessed 18 March 2024).

24 Xiang, J., Wei, Y., Zhong, Y. et al. (2022). Building practical high-voltage cathode materials for lithium-ion batteries. *Advanced Materials* 34 (52): 2200912.

25 Xu, J., Cai, X., Cai, S. et al. (2023). High-energy lithium-ion batteries: recent progress and a promising future in applications. *Energy & Environmental Materials* 6 (5): e12450.

26 CATL (2023). CATL launches condensed battery with an energy density of up to 500 Wh/kg, enables electrification of passenger aircrafts. CATL. https://www.catl.com/en/news/6015.html (accessed 18 March 2024).

27 Placke, T., Kloepsch, R., Dühnen, S., and Winter, M. (2017). Lithium ion, lithium metal, and alternative rechargeable battery technologies: the odyssey for high energy density. *Journal of Solid State Electrochemistry* 21: 1939–1964.

28 Li, Q., Yang, Y., Yu, X., and Li, H. (2023). A $700\,W{\cdot}h{\cdot}kg^{-1}$ rechargeable pouch type lithium battery. *Chinese Physics Letters* 40 (4): 048201.

29 Chen, F., Han, J., Kong, D. et al. (2021). $1000\,Wh\,L^{-1}$ lithium-ion batteries enabled by crosslink-shrunk tough carbon encapsulated silicon microparticle anodes. *National Science Review* 8 (9): nwab012.

30 PR Newswire (2022). Enpower Greentech Achieved Breakthrough in Cylindrical Batteries. PR Newswire. https://www.prnewswire.com/news-releases/enpower-greentech-achieved-breakthrough-in-cylindrical-batteries-301667423.html (accessed 14 May 2024).

31 Zheng, Y., Shen, Y., Guo, J. et al. (2024). Recent advances in solid-state lithium batteries based on anode engineering. *Nano Research Energy* 3 (3): e9120118.

32 Wang, L., Mukherjee, A., Kuo, C.Y. et al. (2024). High-energy all-solid-state lithium batteries enabled by Co-free $LiNiO_2$ cathodes with robust outside-in structures. *Nature Nanotechnology* 19 (2): 208–218.

33 Market.us (2023). Global Lithium Ion Battery Market. Market.us. https://market.us/report/lithium-ion-battery-market/ (accessed 19 March 2024).

34 Precedence Research (2023). Lithium-ion Battery Market. Precedence Research. https://www.precedenceresearch.com/lithium-ion-battery-market (accessed 30 April 2024).

35 Bloomberg (2021). How Far Can Lithium-ion Batteries Go. Bloomberg. https://www.bloomberg.com/news/storythreads/2021-09-16/how-far-can-lithium-ion-batteries-go (accessed 1 March 2024).

36 BloombergNEF (2022). China's Battery Supply Chain Tops BNEF Ranking for Third Consecutive Time, with Canada a Close Second. BloombergNEF. https://about.bnef.com/blog/chinas-battery-supply-chain-tops-bnef-ranking-for-third-consecutive-time-with-canada-a-close-second/ (accessed 14 March 2024).

37 Elements (2023). Battery Manufacturing Capacity by Country. Elements. https://elements.visualcapitalist.com/chinas-dominance-in-battery-manufacturing/ (accessed 20 March 2024).

38 Wang, H., Feng, K., Wang, P. et al. (2023). China's electric vehicle and climate ambitions jeopardized by surging critical material prices. *Nature Communications* 14 (1): 1246.

39 Finance Yahoo (2023). Lithium Battery Production By Country: Top 12 Countries. Finance Yahoo. https://finance.yahoo.com/news/lithium-battery-production-country-top-183050554.html (accessed 15 May 2024).

40 Reuters (2023). LG Chem with China's Huayou to make battery materials in Indonesia, Morocco. Reuters. https://www.reuters.com/technology/lg-chem-partners-with-huayou-group-build-joint-lfp-cathode-plant-morocco-2023-09-24/ (accessed 21 May 2024).

41 Yahoo Finance (2023). Top 12 Battery Manufacturers in USA. Yahoo Finance. https://finance.yahoo.com/news/top-12-battery-manufacturers-usa-173335709.html (accessed 21 May 2024).

42 Reuters (2023). Panasonic to boost battery output at Tesla's Nevada Gigafactory by 10%. Reuters. https://www.reuters.com/business/autos-transportation/panasonic-boost-battery-output-teslas-nevada-gigafactory-nikkei-2023-06-05/ (accessed 15 May 2024).

43 American Progress (2023). LG Energy Battery Plant Expansion Project. American Progress. https://www.americanprogress.org/article/lg-energy-battery-plant-expansion-project/ (accessed 15 May 2024).

44 CNN (2022). Honda, LG announce $4.4 billion US battery factory for EVs. CNN. https://edition.cnn.com/2022/08/29/business/honda-lg-electric-vehicles/index.html (accessed 15 March 2024).

45 Reuters (2023). Ford-SK venture to get $9.2 billion US loan for battery plants. Reuters. https://www.reuters.com/markets/asia/ford-sk-joint-venture-set-receive-92-billion-us-government-loan-battery-plants-2023-06-22/ (accessed 15 March 2024).

46 FDI Intelligence (2022). CATL announces Hungary's largest ever investment. FDI Intelligence. https://www.fdiintelligence.com/content/news/catl-announces-hungarys-largest-ever-investment-81374 (accessed 15 March 2024).

47 Merics (2023). Watts the plan, Europe – Chinese battery investments on or off? Merics. https://merics.org/en/comment/watts-plan-europe-chinese-battery-investments-or (accessed 20 March 2024).

48 International Energy Agency (IEA) (2023). Global EV Outlook. International Energy Agency. https://iea.blob.core.windows.net/assets/dacf14d2-eabc-498a-8263-9f97fd5dc327/GEVO2023.pdf (accessed 21 March 2024).

49 International Energy Agency (IEA) (2023). Trends in batteries. International Energy Agency. https://www.iea.org/reports/global-ev-outlook-2023/trends-in-batteries (accessed 1 March 2024).

50 Bloomberg (2022). China Electric Car Sales Forecast to Hit Record 6 Million. Bloomberg. https://www.bloomberg.com/news/articles/2022-08-09/china-s-july-car-sales-rise-20-on-demand-for-electric-vehicles (accessed 20 March 2024).

51 Goldman Sachs (2022). The U.S. and Europe Are Moving Toward Battery Self-Sufficiency. Goldman Sachs. https://www.goldmansachs.com/intelligence/pages/the-us-and-europe-are-moving-toward-battery-self-sufficiency.html (accessed 20 March 2024).

52 International Energy Agency (IEA) (2023). Electric Vehicles. International Energy Agency. https://www.iea.org/energy-system/transport/electric-vehicles (accessed 15 March 2024).

53 EV Markets Reports (2023). Europe's EV Boom: Powering the Battery Revolution. EV Markets Reports. https://evmarketsreports.com/europes-ev-boom-powering-the-battery-revolution/ (accessed 20 March 2024).

54 EV Markets Reports (2023). Europe's Battery Supply to Ramp Up by 2030. EV Markets Reports. https://evmarketsreports.com/europes-battery-supply-to-ramp-up-by-2030/ (accessed 20 March 2024).

55 SK On Hungary (2023). About us. SK On Hungary. https://skonhungary.com/about-us/ (accessed 21 May 2024).

56 Ali, H., Khan, H.A., and Pecht, M. (2022). Preprocessing of spent lithium-ion batteries for recycling: need, methods, and trends. *Renewable and Sustainable Energy Reviews* 168: 112809.

57 Bridge, G. and Faigen, E. (2022). Towards the lithium-ion battery production network: thinking beyond mineral supply chains. *Energy Research & Social Science* 89: 102659.

58 Hu, X., Wang, C., Lim, M.K. et al. (2023). Critical systemic risk sources in global lithium-ion battery supply networks: static and dynamic network perspectives. *Renewable and Sustainable Energy Reviews* 173: 113083.

59 Sanchez-Lopez, M.D. (2023). Geopolitics of the Li-ion battery value chain and the Lithium Triangle in South America. *Latin American Policy* 14 (1): 22–45.

60 Clean Technica (2024). Top 10 Battery Producers In The World — 2023. Clean Technica. https://cleantechnica.com/2024/01/19/top-10-battery-producers-in-the-world-2023-provisional-data/ (accessed 30 April 2024).

61 Statista (2023). Global market distribution of lithium-ion battery makers in 2023. Statista. https://www.statista.com/statistics/235323/lithium-batteries-top-manufacturers/ (accessed 30 April 2024).

62 Companies Market Cap (2023). CATL. Companies Market Cap. https://companiesmarketcap.com/catl/revenue/ (accessed 20 March 2024).

63 Yahoo Finance (2023). 12 Biggest Battery Manufacturers in the World. Yahoo Finance. https://finance.yahoo.com/news/12-biggest-battery-manufacturers-world-153020817.html (accessed 1 March 2024).

64 BBC (2020). Tesla battery supplier CATL says new design has one million-mile lifespan. BBC. https://www.bbc.com/news/technology-52966178.amp (accessed 19 March 2023).

65 CATL (2023). Energy Storage Solutions. CATL. https://www.catl.com/en/ess/ (accessed 30 April 2024).

66 CATL (2023). Company Profile. CATL. https://www.catl.com/en/about/profile/ (accessed 15 May 2024).

67 Business Korea (2020). CATL Batteries Catch Fires. Business Korea. https://www.businesskorea.co.kr/news/articleView.html?idxno=50937 (accessed 30 April 2024).

68 Yicai Global (2019). CATL Supplied Batteries That Prompted Nio's Mass Car Recall. Yicai Global. https://www.yicaiglobal.com/news/catl-supplied-batteries-that-prompted-nio-mass-car-recall (accessed 30 April 2024).

69 Bloomberg (2024). US to Ban Pentagon From Buying Batteries From China's CATL, BYD. Bloomberg. https://www.bnnbloomberg.ca/us-to-ban-pentagon-from-buying-batteries-from-china-s-catl-byd-1.2024262 (accessed 15 May 2024).

70 The Guardian (2024). China's BYD overtakes Tesla as top-selling electric car seller. The Guardian. https://www.theguardian.com/environment/2024/jan/02/chinas-byd-overtakes-tesla-as-top-selling-electric-car-seller (accessed 14 March 2024).

71 Companies Market Cap (2023). BYD. Companies Market Cap. https://companiesmarketcap.com/byd/revenue/ (accessed 18 March 2024).

72 Clean Technica (2023). Top 10 Battery Producers In The World. Clean Technica. https://cleantechnica.com/2023/09/18/top-battery-producers-in-the-world/ (accessed 1 March 2024).

73 BYD (2020). BYD's New Blade Battery Set to Redefine EV Safety Standards. BYD. https://en.byd.com/news/byds-new-blade-battery-set-to-redefine-ev-safety-standards/ (accessed 12 March 2024).

74 BYD (2023). Energy Storage System. BYD. http://www.bydenergy.com/solution-cn?locale=EN (accessed 30 April 2024).

75 SCMP (2024). BYD electric car catches fire at Hong Kong charging station, expert says short circuit could be the cause. SCMP. https://www.scmp.com/news/hong-kong/society/article/3249963/byd-electric-car-catches-fire-hong-kong-charging-station-expert-says-short-circuit-could-be-cause (accessed 30 April 2024).

76 Sustainable Bus (2021). EMT Madrid goes with BYD, again. New order for 20 e-buses. Sustainable Bus. https://www.sustainable-bus.com/electric-bus/emt-madrid-byd-electric-buses/ (accessed 30 April 2024).

77 Reuters (2024). Exclusive: BYD to stop making pouch batteries for its hybrid EVs on leakage concerns. Reuters. https://www.reuters.com/business/autos-transportation/byd-stop-making-pouch-batteries-its-hybrid-evs-leakage-concerns-sources-2024-01-17/ (accessed 30 April 2024).

78 Times of India (2023). Modi govt rejects Chinese EV maker BYD's \$1b mfg plant. Times of India. https://timesofindia.indiatimes.com/auto/news/modi-govt-rejects-chinese-ev-maker-byds-1b-mfg-plant-proposal-in-india/articleshow/102036056.cms (accessed 15 May 2024).

79 MIT Technology Review (2023). Europe is about to crack down on Chinese electric cars. MIT Technology Review. https://www.technologyreview.com/2023/09/26/1080293/europe-chinese-ev-investigation-subsidy/ (accessed 15 May 2024).

80 Companies Market Cap (2023). LG Energy Solution. Companies Market Cap. https://companiesmarketcap.com/lg-energy-solution/revenue/ (accessed 20 March 2024).

81 Business Korea (2023). LG Energy Solution to Supply Batteries to Top 5 Automakers, including Toyota. Business Korea. https://www.businesskorea.co.kr/news/articleView.html?idxno=202934 (accessed 18 March 2024).

82 LG (2023). LG Energy Storage System. LG. https://www.lg.com/us/ess/home (accessed 30 April 2024).

83 LG (2024). Global Network. LG. https://news.lgensol.com/company-overview/fast-facts/ (accessed 16 May 2024).

84 Electrive (2022). NHTSA investigates LG ES battery recalls. Electrive. https://www.electrive.com/2022/04/06/nhtsa-investigates-lg-es-battery-recalls/ (accessed 21 May 2024).

85 Chevrolet (2020). Bolt EV and Bolt EUV Recall Information. Chevrolet. https://www.chevrolet.ca/en/bolt-recall (accessed 30 April 2024).

86 9news (2023). 'Can catch fire at any time': energy storage batteries recalled after home burns down. 9news. https://9news.com.au/national/lg-battery-product-recall-home-energy-storage-systems-catch-fire-without-warning/0dc878eb-45b3-4474-9982-a3e895d67944 (accessed 30 April 2024).

87 Australian Competition and Consumer Commission (ACCC) (2024). Proposed compulsory recall of dangerous LG solar storage batteries. ACCC. https://www.accc.gov.au/media-release/proposed-compulsory-recall-of-dangerous-lg-solar-storage-batteries (accessed 30 April 2024).

88 Fortune (2016). Samsung Reportedly Looks to a New Battery Supplier After the Note 7 Disaster. Fortune. https://fortune.com/2016/12/19/samsung-galaxy-note-7-batteries/ (accessed 25 August 2024).

89 Manufacturing (2017). New LG phone influenced by Samsung's Note 7 troubles. Manufacturing. https://www.manufacturing.net/home/news/13115301/new-lg-phone-influenced-by-samsungs-note-7-troubles (accessed 25 August 2024).

90 Companies Market Cap (2023). Panasonic. Companies Market Cap. https://companiesmarketcap.com/panasonic/revenue/ (accessed 14 March 2024).

91 Bloomberg (2040). Panasonic to Make Upgraded EV Battery as Soon as This Year. Bloomberg. https://www.bloomberg.com/news/articles/2024-01-14/panasonic-to-make-upgraded-electric-car-battery-as-early-as-2024 (accessed 19 March 2024).

92 Panasonic (2024). Battery Storage. Panasonic. https://na.panasonic.com/us/energy-solutions/battery-storage/ (accessed 30 April 2030).

93 Panasonic (2018). Production and Shipments of Automotive Lithium-ion Batteries. Panasonic. https://news.panasonic.com/global/press/en180313-3 (accessed 16 May 2024).

94 Reuters (2023). Panasonic plans multiple new battery factories in North America by 2030. Reuters. https://www.reuters.com/business/panasonic-plans-new-battery-production-capacity-north-america-by-2030-2023-05-18/ (accessed 16 May 2024).

95 Business Insider (2019). Tesla's investigation confirms a Model S that caught fire on video in Shanghai had a faulty battery module. Business Insider. https://www.businessinsider.com/tesla-battery-fire-shanghai-update-investigation-findings-2019-7 (accessed 30 April 2024).

96 Reuters (2014). Panasonic to recall 43,000 PC batteries after three catch fire. Reuters. https://www.reuters.com/article/us-panasonic-batteries-recall/panasonic-to-recall-43000-pc-batteries-after-three-catch-fire-idUKKBN0E80NV20140528/ (accessed 30 April 2024).

97 Companies Market Cap (2023). SK Innovation. Companies Market Cap. https://companiesmarketcap.com/sk-innovation/revenue/ (accessed 20 March 2024).

98 Bloomberg (2024). Ford Battery Partner SK On Sees Strong EV Growth Beyond 2024. Bloomberg. https://www.bloomberg.com/news/articles/2024-01-10/ford-battery-partner-sk-on-sees-strong-ev-growth-beyond-2024 (accessed 19 March 2024).

99 SK (2024). Our Companies. SK. https://eng.sk.com/companies (accessed 30 April 2024).

100 SK Innovation Hungary (2024). About Us. SK Innovation Hungary. https://skonhungary.com/about-us/ (accessed 16 May 2024).

101 The Korea Economic Daily (2023). SK IE Technology mulls adding Lithium-ion Batteries factory in US, says SKIET president. The Korea Economic Daily. https://www.kedglobal.com/batteries/newsView/ked202302280019 (accessed 16 May 2024).

102 The Verge (2021). Hyundai recalls Kona EVs over battery fire risk. The Verge. https://www.theverge.com/2021/2/24/22299121/hyundai-kona-electric-recall-battery-fire-lg (accessed 30 April 2024).

103 Reuters (2024). US says SK Battery faces $77,200 in fines over employee health hazards. Reuters. https://www.reuters.com/world/us/us-says-sk-battery-faces-77200-fines-over-employee-health-hazards-2024-04-08/ (accessed 30 April 2024).

104 Reuters (2021). U.S. ITC says SK Innovation didn't infringe LG Energy Solution patents. Reuters. https://www.reuters.com/article/idUSKBN2BO428/ (accessed 30 April 2024).

105 Bloomberg (2021). Ford EV F-150 Battery Supplier Loses Fight Over U.S. Imports. Bloomberg. https://www.bloomberg.com/news/articles/2021-02-10/lg-chem-wins-u-s-import-ban-on-sk-innovation-ev-batteries (accessed 16 May 2024).

106 Samsung (2024). Samsung Electronics Announces Fourth Quarter and FY 2023 Results. Samsung. https://news.samsung.com/global/samsung-electronics-announces-fourth-quarter-and-fy-2023-results (accessed 18 March 2024).

107 The Korea Economic Daily (2022). Samsung SDI in talks with global automaker for battery joint venture. The Korea Economic Daily. https://www.kedglobal.com/ev-batteries/newsView/ked202201280005 (accessed 20 March 2024).

108 Samsung (2024). Energy Storage System. Samsung. https://www.samsungsdi.com/ess/index.html (accessed 30 April 2024).

109 Samsung (2024). Global Network. Samsung. https://www.samsungsdi.com/about-sdi/global-network.html (accessed 16 May 2024).

110 CNET (2017). Samsung Galaxy Note 7 recall: here's what happens now. CNET. https://www.cnet.com/tech/mobile/samsung-galaxy-note-7-return-exchange-faq/ (accessed 30 April 2024).

111 Forbes (2017). Samsung Explains Note 7 Battery Explosions, And Turns Crisis Into Opportunity. Forbes. https://www.forbes.com/sites/maribellopez/2017/01/22/samsung-reveals-cause-of-note-7-issue-turns-crisis-into-opportunity/?sh=5d8881924f12 (accessed 30 April 2024).

112 WSJ (2023). CALB Group Co. Ltd. WSJ. https://www.wsj.com/market-data/quotes/HK/XHKG/3931/financials (accessed 14 March 2024).

113 Battery Tech Online (2024). CALB Powers Ahead with Revolutionary 'U' Structure Battery. Battery Tech Online. https://www.batterytechonline.com/ev-batteries/calb-powers-ahead-with-revolutionary-u-structure-battery (accessed 15 March 2024).

114 CALB (2023). Energy Storage Systems. CALB. https://www.calbusainc.com/energy-storage-systems (accessed 30 April 2024).

115 Cnevpost (2022). Chinese battery maker CALB to build production base in Portugal. Cnevpost. https://cnevpost.com/2022/11/03/calb-to-build-production-base-in-portugal/ (accessed 16 May 2024).

116 Bestmag (2022). CALB loses battery patent infringement lawsuit brought by CATL. Bestmag. https://www.bestmag.co.uk/calb-loses-battery-patent-infringement-lawsuit-brought-by-catl/ (accessed 15 May 2024).

117 Simply Wall St (2024). Farasis Energy (Gan Zhou) Full Year 2023 Earnings: Revenues Beat Expectations, EPS Lags. Simply Wall St. https://simplywall.st/stocks/cn/capital-goods/shse-688567/farasis-energy-gan-zhou-shares/news/farasis-energy-gan-zhou-full-year-2023-earnings-revenues-bea (accessed 20 March 2024).

118 Farasis Energy (2023). Battery Goes Orange. Farasis Energy. https://www.farasis-energy.com/app/uploads/2023/08/Farasis_Brochure_Battery_Goes_Orange_2023_US.pdf (accessed 24 March 2024).

119 Energy Trend (2022). Farasis Energy Plans to Build a 24 GWh Production Site for Manufacturing LFP Energy Storage Batteries," Energy Trend. https://www.energytrend.com/news/20221011-30029.html (accessed 15 March 2024).

120 Farasis Energy (2024). Battery Goes Orange. Farasis Energy. https://www.farasis-energy.com/en/ (accessed 30 April 2024).

121 Farasis Energy (2024). Focus on Sustainability and Performance. Farasis Energy. https://www.farasis-energy.com/en/products/ (accessed 16 May 2024).

122 Growjo (2022). Envision USA Revenue and Competitors. Growjo. https://growjo.com/company/Envision_USA (accessed 15 March 2024).

123 Just Auto (2022). Envision AESC to Supply EV Cells to BMW Spartanburg. Just Auto. https://www.just-auto.com/deals-dashboards/envision-aesc-to-supply-ev-cells-to-bmw-spartanburg/ (accessed 14 March 2024).

124 Envision Group (2024). Envision Smart Storage. Envision Group. https://www.envision-group.com/en/smartstorage.html (accessed 1 May 2024).

125 Auto Economic Times (2020). China's Envision Eyes France for New EV Battery Factory. Auto Economic Times. https://auto.economictimes.indiatimes.com/news/auto-components/chinas-envision-eyes-france-for-new-ev-battery-factory/78092262 (accessed 16 May 2024).

126 Asia Nikkei (2023). EV Battery Maker Envision AESC to Build 2nd Plant in U.S.Asia Nikkei. https://asia.nikkei.com/Business/Automobiles/EV-battery-maker-Envision-AESC-to-build-2nd-plant-in-U.S (accessed 16 May 2024).

127 Trend Force (2024). Six Companies, Including BYD and CATL, Are Included in the U.S. Procurement Ban List. Trend Force. https://www.trendforce.com/news/2024/01/22/news-six-companies-including-byd-and-catl-are-included-in-the-u-s-procurement-ban-list/ (accessed 16 May 2024).

128 MarketScreener (2023). Sunwoda Electronic Co., Ltd. Reports Earnings Results for the Half Year Ended June 30, 2023. MarketScreener. https://www.marketscreener.com/quote/stock/SUNWODA-ELECTRONIC-CO-LTD-11318066/news/Sunwoda-Electronic-Co-Ltd-Reports-Earnings-Results-for-the-Half-Year-Ended-June-30-2023-44714336/ (accessed 15 March 2024).

129 Battery Tech Online (2024). Sunwoda EVB Pursues IPO, Green Initiatives. Battery Tech Online. https://www.batterytechonline.com/ev-batteries/sunwoda-evb-pursues-ipo-green-initiatives (accessed 14 March 2024).

130 Sunwoda Energy (2024). Sunwoda Energy. Sunwoda Energy. https://www.sunwodaenergy.com/ (accessed 1 May 2024).

131 Sunwoda (2024). About Us. Sunwoda. https://en.sunwoda.com/about.html#industrial (accessed 16 May 2024).

132 Tech Insights (2023). Smartphone Battery Market Share Q1 2023: ATL Captures 45 Percent Share Despite Revenue Decline. Tech Insights. https://www.techinsights.com/blog/smartphone-battery-market-share-q1-2023-atl-captures-45-percent-share-despite-revenue-decline (accessed 1 May 2024).

133 Aceon Group (2022). The History of CATL: An Asian Giant. Aceon Group. https://www.aceongroup.com/news/the-history-of-catl-an-asian-giant/ (accessed 21 May 2024).

134 KOHA (2017). LG Will Supply Apple with Batteries for the iPhone 9. KOHA. https://www.koha.net/en/tech/33380/LG-will-supply-Apple-with-batteries-for-the-iPhone-9 (accessed 1 May 2024).

135 EEPower (2005). LG Chem Signs Battery Supply Contract with Hewlett-Packard. EEPower. https://eepower.com/news/lg-chem-signs-battery-supply-contract-with-hewlett-packard/# (accessed 1 May 2024).

136 Pandaily (2022). BYD Upgrades Shanghai Battery Factory to Supply Apple. Pandaily. https://pandaily.com/byd-upgrades-shanghai-battery-factory-to-supply-apple/ (accessed 1 May 2024).

137 Huawei (2023). BYD and Huawei Successfully Deploy Smart Factory Solution. Huawei. https://e.huawei.com/en/case-studies/solutions/enterprise-network/byd (accessed 1 May 2024).

138 EEPower (2004). Simplo Partners with BYD to Capture Dell Battery Orders. EEPower. https://eepower.com/news/simplo-partners-with-byd-to-capture-dell-battery-orders/ (accessed 1 May 2024).

139 Lenovo (2023). Lenovo Manufacturing Sites and Suppliers. Lenovo. https://static.lenovo.com/ww/docs/sustainability/list-of-lenovo-mfg-sites-and-suppliers-27oct.pdf (accessed 5 May 2024).

140 The Lec (2022). Samsung SDI to Apply EV Battery Production Technique to Smartphone Battery. The Lec. https://www.thelec.net/news/articleView.html?idxno=3995 (accessed 25 April 2024).

141 The New York Times (2006). Dell to Recall 4.1 Million Batteries Made by Sony. The New York Times. https://www.nytimes.com/2006/08/15/technology/15iht-dell.2487518.html (accessed 25 April 2024).

142 Apple (2017). Supplier Clean Energy. Apple. https://images.apple.com/environment/pdf/Apple_Supplier_Clean_Energy_Program_Update_October_2017.pdf (accessed 30 April 2024).

143 Nikkei Asia (2023). Huawei Supplier Shares Jump Amid China Chip Self-Sufficiency Push. Nikkei Asia. https://asia.nikkei.com/Spotlight/Supply-Chain/Huawei-supplier-shares-jump-amid-China-chip-self-sufficiency-push (accessed 1 May 2024).

144 Tech Insights (2023). Amperex Technology Limited (ATL) BLP973 Lithium-Ion Battery with Battery Health Engine (Oppo Reno 9 Pro) Battery Characterization. Tech Insights. https://www.techinsights.com/blog/amperex-technology-limited-atl-blp973-lithium-ion-battery-battery-health-engine-oppo-reno-9 (accessed 1 May 2024).

145 Xiaomi (2024). Redmi Note 13 FAQ. Xiaomi. https://www.mi.com/global/support/faq/details/KA-91513/ (accessed 1 May 2024).

146 Dell (2022). MSDS for Dell Batteries. Dell. https://i.dell.com/sites/csdocuments/Legal_Docs/en/amperex_sds_2022.pdf (accessed 1 May 2024).

147 HP (2009). Material Safety Datasheet. HP. https://h20195.www2.hp.com/v2/GetDocument.aspx?docname=c07078536 (accessed 1 May 2024).

5

Lithium-Ion Battery Cell and Pack Design Considerations

Yulun Zhang, Kianoush Naeli, Virendra Jadhav, and Sanjay Tiku

The design of lithium-ion cells encompasses mechanical, chemical, and safety considerations. Battery pack design involves configuring cells to meet the voltage, capacity, and power requirements of specific applications. Series and parallel configurations are tailored to balance performance and reliability, supported by a battery management system (BMS) that monitors and regulates cell conditions and the charging and discharging of the pack. With these ideas in mind, electrical and mechanical assembly should ensure seamless integration of cells within the pack, incorporating thermal management solutions to maintain safe operating temperatures.

5.1 Cell Design Considerations

Battery cell qualification is a critical process that involves several stages to ensure the cells' safety, reliability, and performance before they are integrated into products. The extensive timeline for qualification is due to the depth of testing and analysis each cell undergoes at every stage. Particularly, cycle life tests, which have repeated sequences of battery charge and discharge cycles, are among the most time-consuming tests for battery cell qualification. Despite employing accelerated life testing methods, such as using elevated temperatures and increased charge-discharge rates, the process still demands a significant duration, often lasting several months, to complete a substantial number of full charge-discharge cycles.

This evaluation is essential to maintain the high standards of quality and safety associated with battery cells. Additionally, the testing period for this qualification is extended by the need for thorough audits, detailed action reviews, rigorous on-site factory and production line audits, and the meticulous development of product deliverables. This ensures that each battery cell reaching the market has undergone exhaustive scrutiny and meets established standards for performance and safety.

5.1.1 Mechanical Structure

In a battery cell, the cathode and anode are assembled using either a rolling or stacking method, with a separator in between to prevent short circuits while enabling ion flow. The active materials are applied onto current collectors and then wound into a spiral structure known as a jelly roll, or multiple electrode and separator layers are piled up in what is called the stacking process. The assembly of the electrodes and separators is essential for the cell's ion flow and overall performance.

The Safety Challenges and Strategies of Using Lithium-Ion Batteries, First Edition. Edited by Michael G. Pecht.

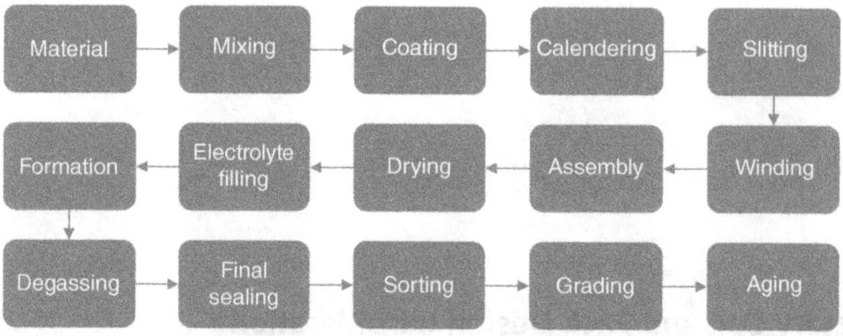

Figure 5.1 Simplified overview of the Li-ion battery cell manufacturing process chain. Source: Fichtner et al. [1]/John Wiley & Sons/CC BY 4.0.

Starting with the raw materials, the manufacturing process for a battery cell spans approximately four weeks and includes several stages, as shown in Figure 5.1 [2]. It starts with material inspection to ensure the quality of incoming materials, followed by mixing active materials and combining them with solvents and binders. Then, this mixed slurry is coated onto current collectors, which are basically thin sheets of aluminum or copper. Subsequently, the coated collectors are compressed into rolls of desired thickness after drying. At this stage, with active battery ingredients coated on them, the anode and cathode sheets are ready for slitting into strips to fit the size of a specific battery design. Next, through either winding or stacking, the anode, cathode, and separator are assembled along with the outer pouch packaging (e.g., with Mylar film). Before electrolyte infusion into the pouch holding the layered structure for the cell, it is vacuum dried to remove any moisture from the assembly. After electrolyte infusion and primary sealing, the cell is ready for activation.

Cell activation is initiated by charging and discharging the cell to form the solid electrolyte interphase (SEI) layer, followed by degassing to remove any unwanted gases created during SEI formation. Following this, the cell is finally sealed to make it airtight, and then it is folded and cut to the required size. Once the final product is ready, binning (essentially sorting and grading) is used to classify cells based on performance criteria, followed by outgoing quality control (OQC) for final checks before shipping. During the entire process, the integrity of the mechanical structure and manufacturing considerations are paramount to ensure the cell's safety and functionality. The mechanical design must also account for the battery's placement and clearance within the device, ensuring that the dimensions are maintained during alignment and assembly.

5.1.2 Chemical Architecture

The chemical design of battery cells is a sophisticated process that involves the careful selection and combination of materials to create a cell that is both efficient and safe. The cathode and anode materials must be chosen for their ability to store and release energy effectively, while the electrolyte needs to facilitate the flow of ions between the two electrodes during charging and discharging cycles. The separator plays a crucial role in preventing short circuits while allowing ion flow. Together, these components work in harmony to produce a battery cell that powers our devices and vehicles.

Electrode loading is a critical aspect of battery cell design that significantly impacts the cell's capacity and energy density. As such, it involves the precise application of active material onto the cathode and anode current collectors. The process requires accurate control over the material's thickness and uniformity to ensure the cell's optimal performance and durability. Higher electrode loadings can enhance the cell's energy density, but they can also present challenges such as maintaining the structural integrity of the electrodes and ensuring efficient ion transport. Achieving a proper balance between cathode and anode loading is essential to prevent issues like lithium plating, which can occur if the anode loading is too high compared to the cathode.

The electrolyte dosage in a battery cell is another key factor that influences the cell's operational efficiency and lifespan. Electrolyte volume within the cell is typically measured in grams per ampere-hour (g/Ah), which helps to determine the optimal amount of electrolyte needed based on the cell's capacity. It is a pivotal cell design parameter, as insufficient electrolyte can accelerate capacity degradation, while too much electrolyte might result in leakage or cell protrusion.

5.1.3 Safety Architecture: TCO

The battery thermal cutoff (TCO) is a critical safety component integrated into battery cells to mitigate the risk of overheating, excessive discharge current, and potential safety hazards. It works by increasing resistance when the temperature goes above a set limit, which cuts off the current to prevent damage or fire.

The TCO, a bimetallic plate, is typically placed directly on the battery tab, sitting between the battery cell electrode and its connection to the pack. In the event of excessive current or heat exposure, the TCO's bimetallic plate responds by snapping upwards, breaking the main contact arm's connection, and halting the current flow. This interruption persists until the temperature normalizes, and the bimetallic plate, upon cooling, snaps back into place, reestablishing the connection, as shown in Figure 5.2 [3]. Regarded as a secondary line of defense, TCOs are typically employed alongside other protective devices within the pack-level protection circuit module, such as fuses and gas gauges.

When selecting a TCO, it is important to consider several critical parameters: the current limit, temperature range, form factor, resistance, and the robust connection to the electrode. The choice of TCO must align with the battery's specific characteristics, encompassing its chemistry, capacity, and design. To ensure consistent performance throughout the battery's operational life, the TCO should be subjected to thorough testing.

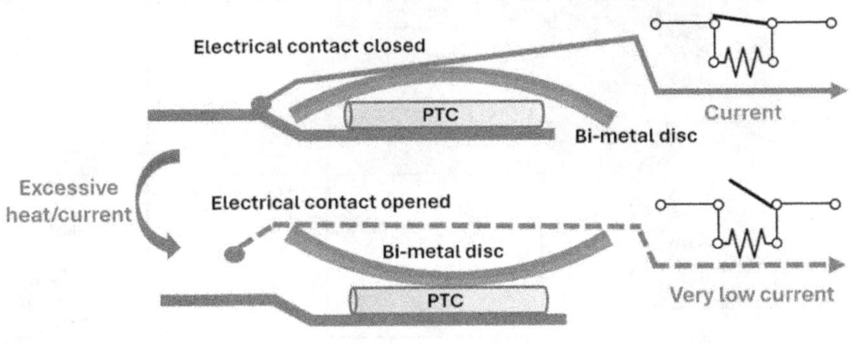

Figure 5.2 How battery cell thermal cutoff provides protection to the circuit. Source: Adapted from [3]/BOURNS.

5.2 Pack Design Considerations

A battery pack is a set of cells assembled as a single unit to store and deliver energy reliably and safely. The relationship between cell, pack, and the host device in a handheld product is shown in Figure 5.3. A battery pack includes the battery cell (i.e., the charge tank), the monitoring and control system, safety protection mechanisms, the enclosure, and connections to the outside world. Importantly, a pack is expected to have more outputs than just the positive and negative power lines; it must also send the battery operation status data (e.g., temperature). Figure 5.4 shows a simplified block diagram for the different components of a battery pack.

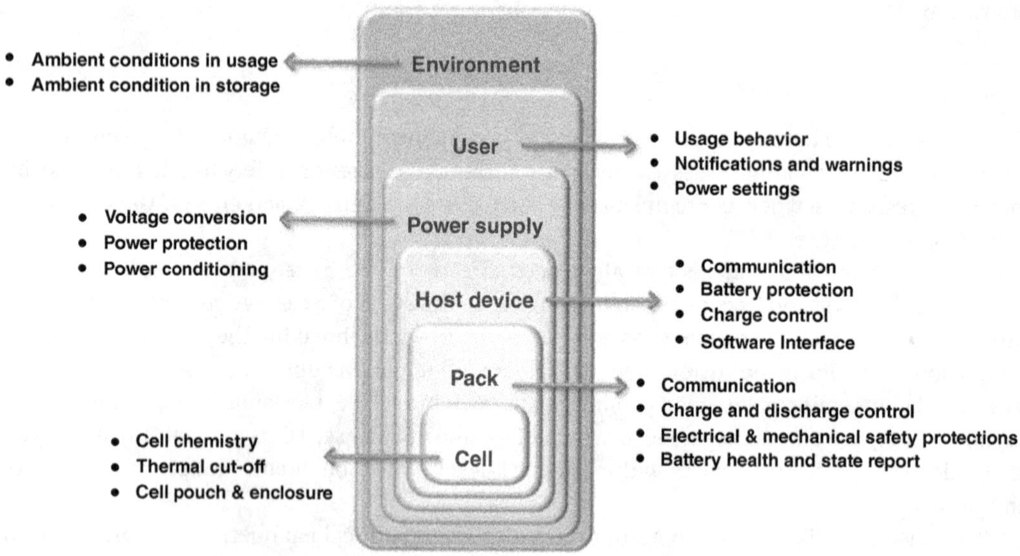

Figure 5.3 Conceptual diagram of a host device and its user. Source: Adapted from [4]/IEEE.

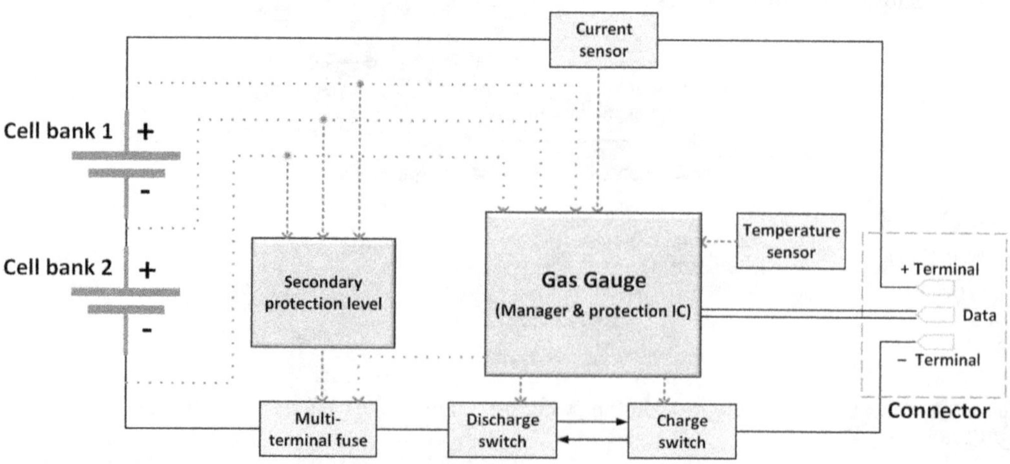

Figure 5.4 A simplified battery pack block diagram.

5.2.1 Cell Configurations in a Pack

In making a pack, the cells can be connected in parallel, in series, or in combinations of both. The cells in a parallel connection will form a bank, also known as a module (Figure 5.5). The cells in a bank can have different capacities. This aspect must be considered during pack design, particularly if their capacity difference causes a significant current imbalance between them. The capacity of the bank is approximately the sum of individual cell capacities. It is not exactly equal to the sum because the input resistance in each cell can create a voltage drop on the cell during charge and discharge events, preventing it from utilizing its full charge capacity. For the same reason, there may need to be a limit in the capacity difference between the cells in a bank since the current flow to each cell is proportional to its capacity.

The cells or the banks in series form a configuration called a string (Figure 5.6). The charge capacity of the string is determined by the single cell or bank that has the smallest charge capacity among the elements in the string. This is because all cells and banks in a series configuration are bound to hold the same amount of charge to meet Kirchhoff's current law (KCL) if there is no other intervention.

Figure 5.7 illustrates the two approaches used in making cell configurations of a pack. In a parallel-first approach, the banks are made first, and then they are put in a single string. In a series-first approach, the cells are first connected in strings, and then the strings are put in parallel together. Importantly, the notation for the cell configuration in a pack is "xPyS," where x refers to the number of parallel cells and y is the number of series elements. Under those circumstances, the notation for a parallel-first pack starts with "xP," but the nomenclature for pack definition

Figure 5.5 Parallel configuration of cells (bank or module).

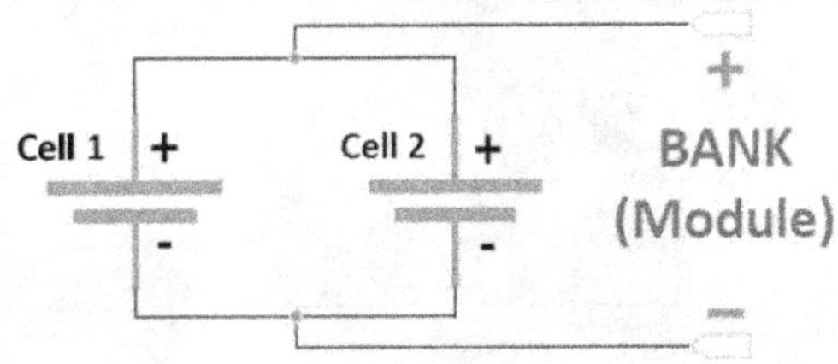

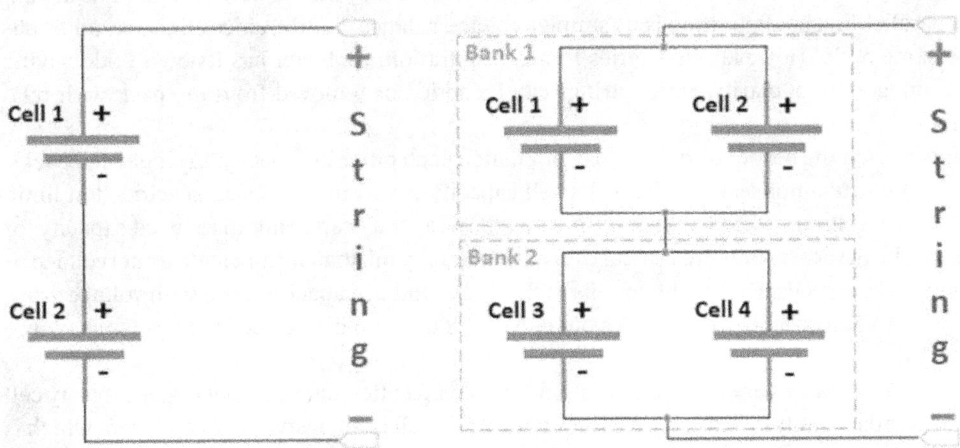

Figure 5.6 Cells (left) or banks (right) in series form a string.

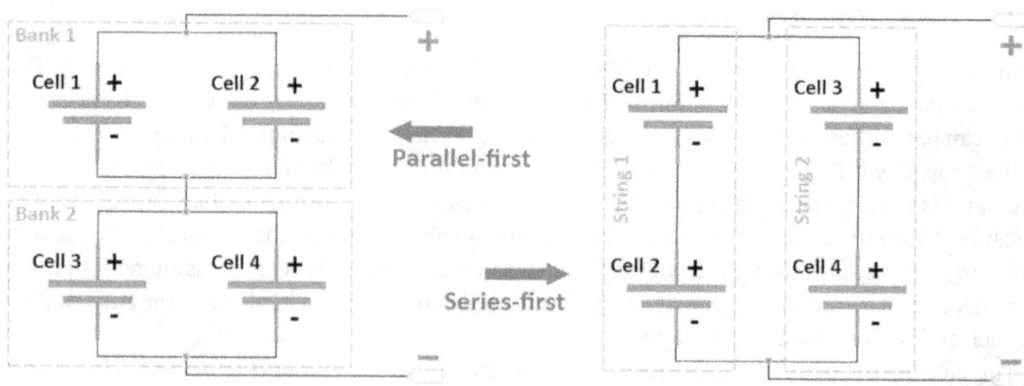

Figure 5.7 Two approaches in making a battery pack: parallel-first (left) and series-first (right).

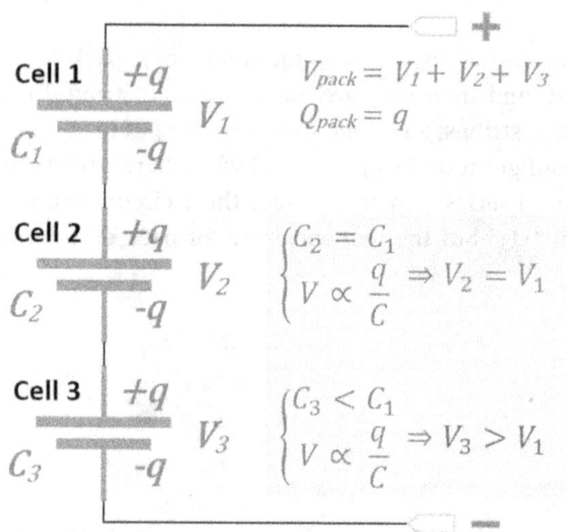

Figure 5.8 Relation between capacity and voltage of a pack and its cells.

must be clearly identified as sometimes the flipped notation of "ySxP" is used instead. The advantage of a parallel-first configuration is its simpler voltage balance control, since all the parallel cells have the same node. However, in a series-first configuration, each cell has its own node, giving it the advantage of modularity. Series strings can be added or removed from the pack with relative ease.

In practice, even the best-quality cells may not match each other in capacity. In single-cell packs, this does not create a noticeable issue if the cell capacity is within the design specification limit. However, when cells are placed in series, a mismatch in capacity can result in reduced capacity for the pack, in the best case. In more critical cases, the capacity mismatch can create undervoltage or overvoltage risks for cells. For a series of cells with C1, C2, and C3 capacities, and with voltage drops of V1, V2, and V3, respectively, the pack voltage V_{pack} is the sum of each cell voltage, as shown in Figure 5.8 [5].

If a string has at least two cells in series with different capacities, charging the larger capacity cell to its full potential results in charging the smaller capacity cell to an overvoltage condition, which is a safety risk. Similarly, discharging the pack can drive the smallest capacity cell to an undervoltage condition. Therefore, the purpose of making a string is to ensure that the cells or banks in series have minimum variation in relative capacity. Otherwise, either the extra capacity of larger cells is wasted, or the pack will be at risk of overvoltage or undervoltage situations.

Variations in cells' capacities can have multiple root causes. Of course, manufacturing and processing variations in cell production can affect the final capacity. Also, differences in the cell inventory regarding storage time or conditions can alter the cell capacity since the amount of cell capacity loss is proportional to the duration of storage and corresponding temperature history. In a similar fashion, the temperature difference that is created over the battery pack by an uneven thermal profile of the host device (e.g., a laptop) also affects capacity degradation. Since cells with different temperatures age differently, and over time, the pack will lose its cell-capacity balance. Unfortunately, this issue cannot be controlled by pack design or through manufacturing controls. The only potential way to mitigate this is through thermal design at the device level that ensures similar temperature profiles for different cells.

5.2.2 Battery Management System (BMS)

The cell is the source of energy in the battery. The amount of energy that the cell can store, its energy capacity, is proportional to the amount of charge that it can hold. The variation of charge over time is the current, which can be sourced out (discharging) or sent into the battery (charging). While simple batteries (e.g., coin cells) completely rely on the host device for regulating charge and discharge operations, such a task in more complex applications (e.g., phones, laptops, and electric vehicles) is largely assigned to the battery itself. Such battery performance monitoring is conducted by the BMS in the battery pack. In other words, the pack is formed by a combination of the BMS and a cell or multiple cells. For simplicity, this definition of a pack will be used in this chapter, but in electric vehicles, the pack can be a superset of an intermediate stage, which is called a battery module [6].

The brain of a battery pack is a microcontroller unit (MCU), often referred to as a gas gauge (GG) or fuel gauge (FG). The gas gauge has various functions, including communication, running pack operations, health reporting, and safety monitoring [4, 5, 7, 8]. The pack communicates with the host device (e.g., laptop) through the gas gauge, and it provides auxiliary information like the battery identification descriptors and maximum charge voltage. Additionally, the gauge may also transmit other unique parameters, such as battery type, serial number, and production information.

Pack operation duties of the GG include commanding the units in the BMS (e.g., power switches, fuses, etc.), controlling the charge and discharge current flow, and balancing the charge in cells. Furthermore, GG monitors and reports various battery health metrics. This responsibility includes reporting the remaining state of charge (RSOC) and conducting battery life prognosis through the battery state of health (SOH). GG also monitors and predicts potential battery failure by monitoring the voltage, capacity, and impedance of the cells, prevents the impact of incoming noise on the battery operation, and records the battery status and lifetime events (e.g., runtime, maximum discharge current, minimum operating temperature, etc.). GG functions related to safety include recording and regulating the battery temperature and maintaining the battery limits of voltage and current versus temperature for both charge and discharge tasks [4, 9]. GG also executes safety procedures, prevents over-temperature, overvoltage, and overcurrent incidents, and enables the reactivation of an overdischarged battery in a safe manner.

A reliable GG and a well-designed BMS can be tailored to salvage a cell with poor or degraded quality. However, the benefit of the GG is not limited to this case. The GG function is essential for the safety and reliability of multicell packs too, particularly when there is a variation in the cells' capacities. To some extent, the gas gauge can mitigate the capacity difference between the elements of a string through voltage or charge balancing [5]. In this operation, the GG transfers charge in a string from cells that are at a higher voltage to cells that are at a lower voltage. This way, all the cells in series will be approximately at the target voltage at the end of the charging step. In doing so, the gas gauge must monitor the voltage of each of the series elements. It must be noted that the typical GG can handle balancing of one string only. This means, if the pack design utilizes multiple strings, it must have as many GGs as the number of strings.

For cost considerations in consumer electronics, each pack most commonly uses a single gas gauge. That is why, in the common pack configuration, there is most often a series of banks (parallel-first) instead of a bank of strings (series-first), as illustrated in Figure 5.7. The other benefit of a parallel-first approach is that the standard deviation of the sum of capacities distribution, as in the banks' capacity distribution, is smaller than the standard deviation in individual cells' distribution. The summation acts like a low-pass filter, removing the effects of extreme highs and lows. In other words, the highs and lows will cancel each other when summed together. That said, when a pack is made as a series of banks (parallel-first), it will lose the modularity benefit. On the other hand, in a pack configuration with strings in parallel (series-first), the pack capacity can be easily increased by simply adding more strings, with the least amount of modification to the BMS. Also, in this configuration, the battery function will not come to a halt if one string fails, and such a damaged string can be easily replaced. In large-scale installations of batteries (e.g., battery energy storage systems), this type of modularity is more important than in consumer electronics. Overall, as a best practice in pack manufacturing, even with a highly capable gas gauge, the series elements should be matched in such a way that they have the smallest difference in capacity. Moreover, for the same reason, the cells are recommended to all be taken from the same manufacturing lot, to minimize the impact of aging from inventory storage.

The gas gauge controls the flow of charge and discharge current by utilizing power switches. Importantly, power switches are selected based on the application and size of the pack. The most common type utilized in portable devices is the solid-state switch [5], but for large packs electromechanical relays can be an option too. Examples of solid-state switches are power metal-oxide-semiconductor field-effect transistors (MOSFETs), insulated-gate bipolar transistors (IGBTs), and switchable diodes. In consumer electronics, given the relatively low power operation and size limitations, power MOSFETs are the best choice. Conversely, electromechanical relays may become useful for high-current applications, in which the size and switching time are not critical. Nevertheless, the cycle life and reliability of electromechanical relays typically are smaller than those of solid-state switches, mainly because of failures in moving parts and the contacts in relays [10].

The other power components that are used in a pack for BMS protection itself are the transient voltage suppressors (TVSs), e.g., Zener diodes. They are used for protection against electrostatic discharge (ESD) and electrical overstress (EOS) events.

The gas gauge triggers the charge and discharge power switches (Figure 5.4) when the current or voltage limits are not met. In doing so, it directly monitors the voltage of each bank and reads the pack current on its power line. However, additional protection components besides a GG are needed in a pack mainly for two reasons: redundancy and response time. The reliability of the system increases by adding more independent components that can deliver the same function, especially since the time it takes for the GG to react to a safety event may not be short enough. Specifically, thermal runaway events in the lithium cobalt oxide (LCO) chemistry can happen in a fraction of a second [8]; therefore, a safety component with a faster response time than the GG is needed. With this in mind, a secondary-level protection unit, independent of the gas gauge, addresses the issues. It must permanently disable the pack, such as by triggering a fuse, if the safety limits are exceeded [5, 7].

The gas gauge needs to monitor the battery temperature and current in evaluating the SOH and guarding the safety limits [11]. For this purpose, temperature sensors are used as part of the BMS. The battery temperature is measured by thermistors, which are resistors with a distinct negative (NTC) or positive (PTC) temperature coefficient. Importantly, a pack may utilize multiple thermistors. As best practice, the thermistors must be placed as close as possible to the cells with the highest and lowest temperatures in the pack, considering the thermal profile of the host device when operating.

The most common method for the GG to monitor the battery current is by reading the voltage drop over a small resistor that is placed in series in the main power line. An alternative method is to use the Hall effect [11], which is more suitable if the voltage drop on the series resistor affects the battery output (e.g., high load applications). However, for a GG to reliably execute its function, robust firmware is needed, and, notably, the firmware is usually coded by the gas gauge manufacturer. Additionally, the functionality of the BMS can be augmented in coordination with the host device software, which controls the battery charger unit and the system power load [12].

5.2.3 Electrical Assembly

The different components of the BMS were described in the previous section, from the perspective of functionality. But in the physical world, they must be placed and routed together, for example, on a printed circuit board, a flex, and through connectors. Different considerations for the design of the many elements of the electrical assembly are discussed in this section.

A printed circuit board assembly (PCBA) holds most of the components and creates connections between them through solder joints and traces. The most important aspects to be considered while designing a battery pack PCBA are its power handling capability, thermal dissipation properties, and ability to prevent leakage currents. High resistance of the power lines in the PCBA can result in a high voltage drop and loss of available capacity in the presence of high current loads; thus, the trace design for power lines is important. Localized or overall heat generated in the PCBA in the presence of load can push the cells' temperatures beyond the safe limits, which requires heat dissipation strategies and design elements to prevent hot spots and enable rapid spreading of heat generated during heavy loads.

Leakage currents result in battery discharge, even when it is not in use. They create a risk to the pack's shelf life and reduce the battery's runtime. Notably, leakage currents can be a result of poor circuit or layout design, inadequate isolation in the printed circuit board, or high ambient humidity. To prevent the impact of humidity, the correct materials need to be chosen for the printed circuit board layers. Also, covering sensitive regions of the board with a humidity protection layer can help.

Other relevant considerations for PCBA design include ESD protection, moisture protection, and reliability of the board-level and secondary interconnects, including anisotropic conductive filament (ACF) connections for flexible PCBs and solder joints. ESD events can happen through the handling of the pack, for example, in the factory line or at the repair site, or even by interaction with the host device. Since a battery is an active component, moisture condensation or water spills can result in corrosion or dendritic growth, potentially causing shorts. Effective practices in blocking moisture condensation and damage include using a protective conformal layer to cover the critical components and prevent void creation in the water-permeable fillings on the board. The conformal layer must have insignificant water permeability, preventing the water molecules from reaching the metal layers in the PCB or the components assembled on it. Otherwise, if water can diffuse into the layer, the voids in the layer may trap the water vapor. The water vapor molecules can change their phase to liquid as temperature swings occur [13]. Pads for the cell or flex connection must result in a reliable bond, and all solder connections must be designed to withstand the fatigue caused by temperature swings during use.

The electrical tracing to the cell tabs (i.e., the cell electrodes' terminals) can be through the PCBA, or through more flexible routings, known as flexes and bus bars. The decision to pick the optimal tracing approach depends on the shape of the pack and the cost. In either case, the resistance of the path is the main concern. To handle large currents with a flexible routing approach, bus bars may be a better option than flex, because bus bars have thicker conductive layers. Regardless of the chosen approach, the output of the pack can be delivered through pads or a connector that is attached to the PCBA or flexible circuit.

Robust connections are required between the pack and the host device and between the cell and the rest of the pack. Given that, a failure in connections can cause a disruption to the battery output. Also, the mechanical strength of the connection is important, as a weak connection can result in electrical discontinuity, which leads to voltage loss or imbalance and, consequently, the failure of the pack. Furthermore, a low electrical resistance of the connection is critical, which will minimize the voltage drop on the connection under high power loads. With all of this in mind, common methods for connecting the cell tabs with the PCBA pads, or with the flex, include spot welding, laser welding, and ultrasound bonding [14]. Welding of the entire tab surface area or soldering may not be preferred because of the associated high thermal loading during assembly, which can damage the cell. There are both destructive and nondestructive methods for monitoring the weld strength [15], such as a peel test (destructive) [16] and visual inspection (nondestructive) [17].

The pack connection to the system can be either removable or permanent; however, removable connections are mostly used. For example, when repairability is a concern, a permanent connection is not desirable. In particular, a removable connection can be accomplished through custom-made designs or commercial off-the-shelf (COTS) connectors, e.g., latch-lock connectors or zero insertion force connectors (ZIF). Importantly, the electrical resistance and mechanical retention of connectors must not deteriorate over the life of the device, especially if the user needs to remove and reinsert the battery pack frequently (e.g., in the case of charging the pack in an external unit). Other concerns for the connectors include corrosion at the contact region (e.g., galvanic corrosion between dissimilar metals) or fretting corrosion [18].

5.2.4 Mechanical Assembly

All the different components of the pack must be put together not only in a complete electrical circuit, but also in a tightly controlled mechanical design. This is to facilitate the assembling of packs as independent modules in the packing factory, also known as the pack house, before integrating them into the host device like a laptop. If the pack is expected to be handled outside of the host device manufacturing line, then the mechanical assembly is usually formed around a frame skeleton or inside a hard enclosure, which can be made of plastic or metals. In all cases, the internal components of the pack are placed together using a mechanical retention mechanism or adhesive, and they are often wrapped by insulating tapes. To that end, the mechanical assembly design for packs needs to meet specific requirements related to structural integrity and safety.

From a structural standpoint, the assembly should be able to support the pack weight and hold all components of the pack within the expected clearances. Mechanical assembly design needs to minimize component movements during the handling of the pack or during usage in the host device. Also, the design should accommodate the cell thickness change within acceptable margins through the life of the pack. Additionally, the mechanical structure should be electrically insulated from the battery and have thermal stability over the pack and system operation temperature range. Safety considerations require the mechanical structure to be nonflammable to impede or prevent the spread of fire. It is also required to be robust enough to not break into shards in abuse scenarios—the shards can penetrate cell pouches and create safety concerns like a short circuit between the electrodes or electrolyte leakage.

5.3 OEM Device Design Considerations

Typically, mobile devices use Li-ion batteries for portability and high-energy-density considerations. As such, mechanical considerations such as compression (e.g., putting your laptop in a backpack between books or stacking a bunch of laptops for storage) or mechanical shock or drop events (e.g., dropping your phone while getting out of the car) are important in product design. With this in mind, the device enclosure plays a critical part in protecting the battery during normal usage. Proper care must be taken to ensure that any deformations caused by mechanical events do not adversely affect the battery. Similar considerations are necessary for the inclusion of battery packs or modules in an electric vehicle (EV) design to isolate them from heat, vibration, or shock stresses during vehicle operation, charging and discharging, or during outdoor parking or storage.

One common method for the integration of batteries into device enclosures in consumer electronics is to use pressure-sensitive adhesive (PSA) tape to attach the battery to the enclosure. This method is targeted for achieving the thinnest possible device while maximizing the battery size, since the adhesive takes little space in the thickness direction and can provide excellent bonding to the enclosure. Consequently, this method relies on automated assembly to integrate the battery into the device. While the battery is well protected during normal use, this method can be challenging to remove the battery pack from the device during rework, repair, or replacement. For regular high-strength PSA tapes, the temperature must be raised to make the adhesive joint weak, or lowered to make the adhesive joint brittle before the battery can be safely removed from the enclosure. This can pose challenges outside of the factory, as the means to do it the right way may not be easily accessible or cost-effective.

An improvement to standard PSA is stretch release PSA, which allows removal of the tape by pulling on a tab. If the mechanical architecture of the device allows easy access to the pull tabs, this can be used effectively. In fact, a trained person can have a very high rate of success without damaging the battery. However, a layperson may still break the tabs and essentially default to a situation of standard PSA.

Another method is to use a frame in which the battery is installed, and the frame is then integrated into the chassis, usually by screws. Thus, this is a common method for the integration of battery banks in electric vehicle designs, since this method enables manual assembly of battery packs into the device. It also provides ease of replaceability in the field. The frame method, however, has an impact on the overall capacity of the battery, as some of the in-plane space is taken up by the frame. Care needs to be taken to ensure that mechanical integration of pouch cells into the frame accounts for potential new failure modes involving the cells running into the frame during shipping, assembly, or usage of the device.

5.3.1 Device Functional and Performance Requirements

There are several device-level functional design requirements for batteries, including a capacity determined by the desired runtime claim and energy consumption rate, power delivery determined by the system performance requirements, and quick charging time determined by battery chemistry and cell design. The physical size of the battery is influenced by the capacity requirement and the technology used. In a typical battery pack involving multiple cells, some part of the volume is dedicated to, and thereby lost to, battery electronics and mechanical structure, which are non-energy-storing parts of the battery. Design teams try hard to reduce the volume and weight of these components, while ensuring that the safety is not compromised and the design meets the overall cost budget.

5.3.2 Enclosure Design for Battery Protection

Safety is an overarching requirement for all batteries. For a Li-ion pouch cell-based battery pack, the mechanical enclosure provides critical protection from mechanical, handling, and environmental damage. As such, strict design guidelines must be followed for enclosure design. Mobile products, such as laptops, phones, and electric vehicles, experience vibration, shock, and drop events as part of regular use or abuse. Product designers need to evaluate this and determine what would be considered regular use and ensure that during these events, the battery is well protected. This involves understanding the relative motion of the battery, enclosure, and any other structural elements in the device. As such, finite element analysis (FEA) can be an effective tool to simulate different scenarios, and experimental data can be collected on critical cases. While a typical FEA considers the cells as solid elastic bodies, it must be recognized that they are jelly rolls or stacked layers suspended in a polymeric gel, requiring adjustments to be made to account for the solid-fluid interactions.

Lithium-ion battery cells tend to expand and contract during normal charging and discharging operations. Further, as the batteries age, there can be cumulative gas generation that gets trapped in the cells, leading to an expanded battery. If sufficient allowance is not provided for this expansion, the cells may push against other parts of the device. For devices with displays near the battery packs, like phones and laptops, artifacts may appear on the display resulting from deformation of display internal structures caused by battery swelling. Additionally, if a swollen battery pushes against a sharp feature inside the device, it may leak. Consequently, care must be taken to ensure that no sharp features can contact the battery even after it swells.

Special care should be taken to ensure that any electrical component next to the cells is not able to contact the cells in scenarios like compression or drop events. The terminals of individual cells need to be protected with insulation to prevent shorting with external sources. The same precautions should be taken on PCBs used for connecting the cells to make a pack, and good practice involves not having any exposed traces on the PCB and placing all of the routings in internal layers only.

5.3.3 Replacement and Reworkability

The ability of end users to replace an old battery within their device is an important consideration. In truth, prior to the introduction of Li-ion cells, this was the norm. Sustainability and cost-conscious people regard the lack of battery replaceability in a device as an obvious downside, and several government regulations attempt to address this [19, 20]. To this end, the industry is working on battery replaceability while also facing the challenges of handling pouch cells safely and the associated risk of counterfeits. Some of the original equipment manufacturers (OEMs) like Apple and HP have already made battery replacement feasible at authorized service providers to mitigate the safety aspects [21, 22]. Also, service provider companies such as iFixit and Batteries Plus provide full-service repair including battery replacement for several PCs as well as phones and tablets.

The need to make replaceable battery packs drives some design choices as well. While attaching the battery pack to the enclosure by using PSAs provides the most space for battery capacity, it is the least desirable option from a removability perspective. It is necessary to either raise the temperature to make the adhesive more ductile or lower the temperature to make the adhesive very brittle, before the pack can be removed from the enclosure. On the flip side, using a frame to support the cells makes it easy to replace the battery pack with commonly available tools, at the expense of precious real estate inside the device.

5.3.4 BMS and Smart Charging

The BMS is a critical part of lithium-ion batteries. At the device level, at the very least, the lowest and highest voltages, the charging profile, and the maximum discharge current must be managed to ensure safe operations of the battery. These parameters are also influenced by ambient temperature. For example, it may be okay to discharge a battery below freezing temperature, but there may not be a safe way to charge it due to the possibility of lithium plating.

Smart charging refers to any of a number of ways to ensure a longer lifespan of the battery. In its simplest form, a user is given a choice to charge the battery only up to 80–90% of the maximum charge capacity. Importantly, this keeps the battery materials in the Goldilocks region by avoiding more stressful regions above 80%. It is known that above a certain voltage, particularly for LCO chemistry (around 4.2V), the cell materials can be highly reactive, and more time spent in this region can be detrimental to the life of the battery, reducing the battery capacity faster as compared to when the battery is used at lower charge levels [23]. A more advanced scheme of smart charging limits the max charging voltage based on usage patterns. For example, in mobile phones, battery charging above 80% can be delayed until early morning even if the phone is plugged in for charging the previous night. As the battery chemistry evolves and device makers get additional information on the usage patterns of their devices, smart charging will continue to evolve [24].

5.3.5 Usage Patterns and Telemetry

Usage patterns of a device can have a significant impact on the design and lifespan of the battery. For example, in the case of mobile applications like a phone or an electric vehicle, it is typical to charge the battery at night and drain it during the day [25]. In the case of a device like a laptop or tablet, the usage profiles can have quite different scenarios. It has been observed that a laptop may be plugged into an AC wall outlet for most of the day while connected to an external keyboard, mouse, and display [26]. In this case, the battery is not cycled but kept at a high charge state for a long time. It is also possible that the same device is sometimes used in a typical mobile-like manner and sometimes in a desktop manner. Consequently, this can influence the battery design for a laptop. While cycle life (i.e., charge-discharge cycles) is the main criterion for mobile applications, the behavior of the battery for a long duration at high voltage is an important additional criterion for laptops and tablets.

Another factor that plays into battery design for integration into a device is the thermal profile of the device. Phones mostly consist of a single cell, but laptops, tablets, and EV batteries have multiple cells—which may or may not be identical. In addition, laptops have high-power CPUs and GPUs, which act as sources of heat inside the device. If the heat-spreading within the device is not well designed, this can result in one cell being hotter than others, leading to a reduced lifespan of the overall pack. While lowering hot spots within the system by spreading heat quickly across the device is a good practice in any case, it becomes an important need for battery reliability. Heat-spreading materials, such as graphene sheets, are often necessary in addition to the heat sinks and fans typically used [27, 28]. Graphene sheets are used due to their high in-plane thermal conductivity and heat-spreading ability at a very low weight. While metal-based heat spreaders are used to move heat away from heat-generating components (like CPU) quickly, they pose a risk of mechanical damage to the battery if they come in direct contact with a pouch cell battery. In a battery pack consisting of multiple cells, it is important to minimize the temperature difference between different cells, and graphene provides a great option to achieve heat-spreading with minimal risk of mechanical damage. Also, metals are much heavier and are not preferred in mobile devices. In the case of EVs, during extreme use scenarios like a hot day in Arizona or a freezing night in Alaska, additional thermal management systems are used to maintain batteries in the most optimal operational temperature environments. This has a penalty, though, in terms of diverting some stored energy and lowering the range of the EVs.

5.4 Summary

A cell is the basic unit of a battery that stores and delivers electrical energy. Battery cells consist of five main components: cathode, anode, electrolyte, separator, and packaging materials. The choice of the right combination of materials is determined by the requirements for the cell and corresponding mechanical structure, chemical architecture, and safety architecture for long and safe operation.

A battery pack is a set of components assembled as a single unit to store and deliver energy reliably and safely. More specifically, a battery pack includes the battery cell, monitoring and control system, safety protection mechanisms, enclosure, and connections to the outside world. Cells can be connected in parallel and series combinations to make a pack. Connecting cells in series increases the voltage of the pack. Connecting cells in parallel increases the capacity of the pack. The advantage of a parallel-first configuration is its simpler voltage balance control, whereas a series-first configuration has the advantage of modularity. Also, series strings can be added or removed from the pack with less complexity.

The battery performance monitoring is conducted by the BMS in the battery pack. The brain of the battery pack is a MCU often referred to as a gas gauge or fuel gauge. As such, the gas gauge has various functions, including communication, running pack operations, health reporting, and safety monitoring. Different components of the BMS are placed and routed together on a printed circuit board, a flex, and through connectors. A mechanical structure for the pack provides structural integrity during external mechanical stresses and enables it to meet regulatory and other safety requirements.

At the OEM device level, battery design is mainly focused on overall integration schemes from electrical and mechanical perspectives. Device-level functional design requirements for batteries include the capacity determined by the battery runtime claim, power delivery determined by the system performance requirements, and charging time determined by battery chemistry and cell design. Usage patterns of a device can also have an impact on the design and lifespan of the battery (how many charge-discharge cycles are required, how long might the battery be subjected to storage or calendar aging, what is the anticipated C-rate of the battery). Additionally, proper installation of batteries into devices, protection of the cells during device assembly, and customer use have become critical functions of the device enclosure. For example, automated handling, installation, and inspection processes are now necessary for manufacturers to avoid inadvertent damage to pouch cells. For sustainability and legislated circular economy requirements, reuse and recycling of materials used for lithium-ion batteries must be considered in terms of safety, as these are becoming more common practices.

References

1 Fichtner, M., Edström, K., Ayerbe, E. et al. (2022). Rechargeable batteries of the future—the state of the art from a BATTERY 2030+ perspective. *Advanced Energy Materials* 12 (17): 2102904.

2 Kwade, A., Haselrieder, W., Leithoff, R. et al. (2018). Current status and challenges for automotive battery production technologies. *Nature Energy* 3 (4): 290–300.

3 Bourns (2017). Overcoming circuit protection challenges in lithium-ion battery packs. https://www.bourns.com/docs/technical-documents/technical-library/circuit-protection/application-notes/mini-breakers---overcoming-circuit-protection-challenges-in-lithium-ion-based-portable-electronics-application-note.pdf. (accessed 22 April 2024).

4 IEEE Standard 1725-2021 (2021). *IEEE standard for rechargeable batteries for mobile phones.* IEEE Power and Energy Society.

5 Barsukov, Y. and Qian, J. (2013). *Battery Power Management for Portable Devices.* Norwood, MA: Artech.

6 Thaler, A. and Watzenig, D. (2014). *Automotive Battery Technology.* New York: Springer Cham.

7 Arora, A., Lele, S., Medora, N., and Souri, S. (2019). *Lithium-Ion Battery Failures in Consumer Electronics.* Artech.

8 Andrea, D. (2020). *Lithium-Ion Batteries and Applications: A Practical and Comprehensive Guide to Lithium-Ion Batteries and Arrays, from Toys to Towns, Volume 1, Batteries.* Artech.

9 Japan Electronics and Information Technology Industries Association; Battery Association of Japan (2007). *A Guide to the Safe Use of Secondary Lithium Ion Batteries in Notebook-Type Personal Computers.* Japan Electronics and Information Technology Industries Association; Battery Association of Japan.

10 Saleem, M.M. and Nawaz, H. (2019). A systematic review of reliability issues in RF-MEMS switches. *Micro and Nanosystems* 11 (1): 11–33.

11 Tan, X., Vezzini, A., Fan, Y. et al. (2023). *Battery Management System and its Applications.* Singapore: Wiley.

12 Volta Foundation, "2023 Battery Report," [Online], Available https://volta.foundation/battery-report-2023 (accessed 28 August 2024).

13 Nikolayev, V.S., Beysens, D., Gioda, A. et al. (1996). Water recovery from dew. *Journal of Hydrology* 182: 19–35.

14 Brand, M.J., Schmidt, P.A., Zaeh, M.F., and Jossen, A. (2015). Welding techniques for battery cells and resulting electrical contact resistances. *Journal of Energy Storage* 1: 7–14.

15 Kumar, N., Masters, I., and Das, A. (2021). In-depth evaluation of laser-welded similar and dissimilar material tab-to-busbar electrical interconnects for electric vehicle battery pack. *Journal of Manufacturing Processes* 70: 78–96.

16 Shin, S., Nam, S., Yu, J. et al. (2021). Ultrasonic metal welding of multilayered copper foils to nickel-plated copper sheet in lithium-ion battery cell. *Metals* 11 (1195): 8.

17 McGovern, M.E., Bruder, D.D., Huemiller, E.D. et al. (2023). A review of research needs in nondestructive evaluation for quality verification in electric vehicle lithium-ion battery cell manufacturing. *Journal of Power Sources* 561: 1–15.

18 Zwicker, M., Moghadam, M., Zhang, W., and Nielsen, C. (2020). Automotive battery pack manufacturing – a review of battery to tab joining. *Journal of Advanced Joining Processes* 1: 100017.

19 NCSL (2023). National conference of state legislatures. https://www.ncsl.org/technology-and-communication/right-to-repair-2023-legislation (accessed 21 May 2025).

20 The European Parliament and The Council of the European Union (2023). Regulation (EU) 2023/1542 of the European Parliament and of the Council. https://eur-lex.europa.eu/legal-content/EN/TXT/PDF/?uri=CELEX:32023R1542 (accessed 21 May 2025).

21 Apple. iPhone Battery Replacement. https://support.apple.com/iphone/repair/battery-replacement (accessed 28 August 2024).

22 HP Notebook PCs – Purchasing a replacement battery. https://www.hp.com/us-en/shop/cv/battery-replacement?cjdata=MXxOfDB8WXww&jumpID=af_gen_nc_ns&utm_medium=af&utm_source=cj&utm_campaign=Microsoft+Shopping+%28Bing+Rebates%2C+Coupons%2C+etc.%29&utm_content=5250933_Microsoft+Shopping+%28Bing+Rebates%2C+Coupon (accessed 28 August 2024).

23 Qin, B., Liu, Z., Zheng, J. et al. (2015). Single-ion dominantly conducting polyborates towards high performance electrolytes in lithium batteries. *Journal of Materials Chemistry A* 7773–7779.

24 Microsoft (2024). Smart charging on surface. https://support.microsoft.com/en-us/surface/smart-charging-on-surface-dda48e48-950b-421c-a436-c48c55e158c4 (accessed 21 May 2025).

25 Banerjee, N., Rahmati, A., Corner, M.D. et al. (2007). *Users and Batteries: Interactions and Adaptive Energy Management in Mobile Systems, UbiComp 2007: Ubiquitous Computing. UbiComp 2007, Lecture Notes in Computer Science*. Berlin: Springer-Verlag.

26 Jones, M.E., Wei, B., and Hung, D. (2013). Laptop energy-saving opportunities based on user behaviors. *Energy Efficiency* 6: 425–431.

27 Zuev, S., Prokhorov, D., Maleev, R. et al. (2021). Application of graphene in the cooling system of a personal electronic computer. *Russian Microelectronics* 50: 404–411.

28 Nkansah, A., Wang, N., Ye, L. et al. (2018). Reliability study on high thermally conductive graphene film as heat spreader in electronics cooling applications. IMAPS Nordic Conference on Microelectronics Packaging (NordPac), Oulu, Finland.

6

Design and Process Failure Modes and Mechanisms

Sahithi Maddipatla, Saurabh Saxena, and Michael G. Pecht

Degradation is unavoidable during the life of lithium-ion batteries, including storage. The performance of batteries is typically measured in terms of capacity (mAh) and internal resistance (Ω). Both of these metrics degrade over time (capacity decreases and internal resistance increases) due to electrochemical, thermal, and mechanical processes within the battery [1]. Additionally, batteries can suffer from catastrophic and abrupt failures due to internal defects and accidental abuse during manufacturing, assembly, or usage.

Lithium-ion batteries are intrinsically unsafe due to the materials required to make them function as intended—the flammable electrolyte, oxidizing cathode material, and polymeric separator material with low thermal stability. These are complex systems produced through multistep manufacturing processes. Manufacturing processes affect the trajectory of degradation and the risk of catastrophic events.

Battery safety evaluation requires assessing the battery's design and manufacturing process along with experimental testing. With that in mind, performance tests are usually required by various standards organizations, including Underwriters Laboratories (UL) and the International Electrotechnical Commission (IEC). However, these tests are conducted only on a small portion of the manufactured batteries and may not be sufficient to reveal all manufacturing defects. Furthermore, when used in products, a monitoring and management system, known as a battery management system (BMS), is required to prevent any abusive operation, such as overcharging or overheating. The BMS also plays a critical role in managing the degradation of batteries.

Despite performance and safety tests from standards and the monitoring and control systems, there are still numerous incidents of battery performance issues and exothermic thermal events within products (e.g., in phones, computers, and electric vehicles) made by well-known companies. These issues often originate from defects that arise in the manufacturing process, which further evolve during storage and cycling operations, eventually becoming sources of failures later. Understanding the role of design choices and the manufacturing processes in battery failures is crucial for ensuring battery safety.

6.1 Introduction

The performance of Li-ion batteries is typically characterized by capacity and internal resistance parameters. Capacity refers to the amount of charge (in ampere-hours) a battery can deliver during

The Safety Challenges and Strategies of Using Lithium-Ion Batteries, First Edition. Edited by Michael G. Pecht.

a complete discharge event. The battery's internal resistance is the sum of the ohmic resistance due to battery materials, the charge-transfer resistance, and the resistance associated with the slow diffusion of lithium in the solid electrode particles. Capacity is related to the battery's energy storage capability, while resistance is associated with the battery's power capability. Both parameters change as the battery ages, leading to a gradual capacity reduction and increased resistance. In many applications, end of life is the time or number of cycles when the battery capacity is reduced to 70% or 80% of the initial value.

The gradual degradation of the battery while reaching its end of life is common in all applications deploying Li-ion batteries. This degradation occurs due to various failure mechanisms operating at the material level. As a battery is charged-discharged, certain undesirable side reactions occur inside the battery, consuming the active lithium ions and reducing the battery's capacity. The electrode active materials also undergo repeated volume change due to the insertion and removal of lithium ions, leading to deterioration and, in extreme cases, disintegration of the electrode structure. Additionally, temperature cycling caused by charge-discharge exacerbates the failure mechanisms further. Manufacturing quality issues, such as uneven electrode coatings, the presence of foreign particles, and an undeveloped solid electrolyte interphase (SEI) layer, can also accelerate the rate of degradation [2, 3].

Lithium-ion batteries can fail abruptly, leading to overheating, fires, or explosions. When the battery temperature rises beyond a certain point due to a short circuit or overheating, a process known as thermal runaway starts. In this process, self-sustaining exothermic reactions like SEI layer decomposition, anode-electrolyte decomposition, electrolyte decomposition, and cathode decomposition coupled with internal short circuit continue to increase the battery temperature and internal pressure beyond a point of no recovery, leading to the violent release of gases and battery fires [4]. Catastrophic failures typically occur under abusive or overstress conditions, such as overcharging, overheating, and short circuits, among others. However, they can also occur due to the accumulation of damage over time and manufacturing defects. In the case of battery packs, thermal runaway can propagate from one cell to another, leading to a violent chain of fires and explosions.

The number of applications of Li-ion batteries is continuously rising, with a corresponding increase in demand for them. These batteries are used in smartphones, smartwatches, tablets, laptops, e-cigarettes, hoverboards, e-bikes, electric cars, and grid storage. Unfortunately, the rising demand has resulted in many poor-quality batteries entering the market, thus increasing the potential for battery safety issues. The number of safety events has been steadily growing, as evidenced by news reports from various countries [5–8]. Hence, battery designers and manufacturers must address the safety issues of lithium-ion batteries to reduce the frequency of battery failures and their consequences.

Failure mode and effect analysis (FMEA) is an engineering-structured analysis of the risks of potential failures [9]. It systematically investigates each design element or manufacturing process to determine how it can cause system the failures (battery). In the following sections, the design and process FMEA will be discussed, elaborating on the underlying mechanisms that are responsible for battery safety issues[1].

---

1 This paragraph and subsequent sections consist of excerpts, figures, and tables which have been reprinted from Maddipatla, S.; Kong, L.; Pecht, M. Safety Analysis of Lithium-Ion Cylindrical Batteries Using Design and Process Failure Mode and Effect Analysis. *Batteries* 2024, *10*, 76, Copyright 2024 by the authors. Licensee MDPI, Basel, Switzerland. This article is an open access article distributed under the terms and conditions of the Creative Commons Attribution (CC BY) license (https://creativecommons.org/licenses/by/4.0/).

6.2 Failure Mechanisms in Li-Ion Batteries

Failure mechanisms are central to the physics-of-failure (PoF) approach to reliability assessment of systems, subsystems, and components. Failure mechanisms are the "processes by which physical, electrical, chemical, and mechanical stresses induce failures" [10]. These mechanisms describe the fundamental way in which a device or component can fail. On the other hand, failure modes are defined as how a failure is physically observed.

Li-ion batteries contain five major components: negative electrode (anode), positive electrode (cathode), separator, electrolyte, and current collectors. The electrodes contain active material mixed with conductive additives and binders and coated on a current collector. The binder acts as a glue to hold this structure together. Ideally, only active material should participate in the battery redox reactions, which allow the gain and loss of electrons to generate electricity. The purpose of the electrolyte is to serve as a medium for Li^+ ion transfer between the two electrodes and as an electronic insulator to prevent self-discharge. The separator serves as an electrically insulating layer between the two electrodes while allowing the flow of Li^+ ions between them. The designs and materials of these components play a critical role in battery safety.

Commercial lithium-ion batteries are available in four main form factors: cylindrical, pouch, prismatic, and coin. Li-ion batteries' design, assembly, and manufacturing process can vary from manufacturer to manufacturer and is proprietary. However, the basic principles of cell design, materials, and manufacturing are usually the same and are well-known. A cylindrical battery, a popular form factor, is used here to discuss the design and process impact on battery safety. Cylindrical batteries are composed of a rolled-up assembly called a jelly roll, which includes the anode, cathode and separator sheets tightly wound together and connected with electrical tabs. A schematic of a cylindrical lithium-ion battery is shown in Figure 6.1.

6.2.1 Negative Electrode (Anode)

The anode active materials used, including graphite, silicon, germanium, and lithium titanate, show good thermal stability compared to the cathode active material [11]. Graphite in the form

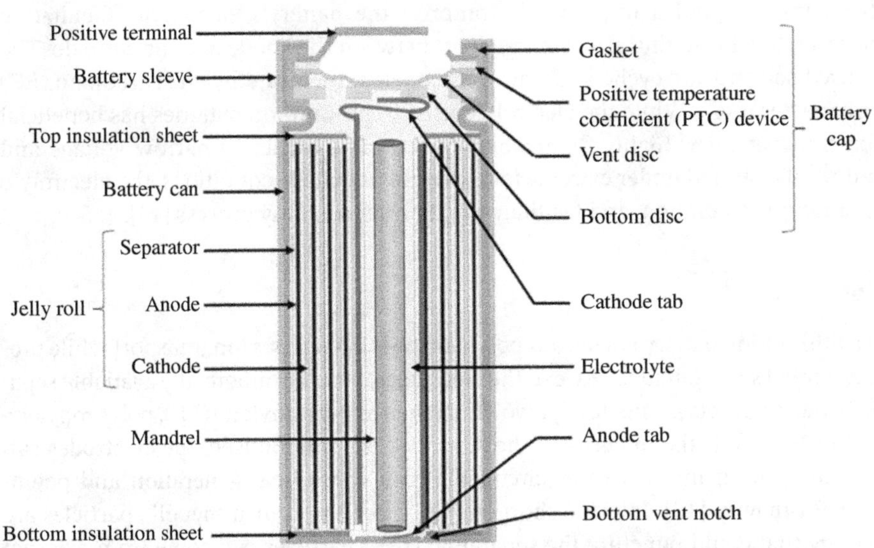

Figure 6.1 Cylindrical battery structure.

of powder, combined with binder material, is widely used [12]. The cell's performance is affected by parameters like purity, particle size, particle size distribution, particle shapes, particle porosity, the crystalline phase of carbon, and degree of compaction [13, 14].

Lithium plating and dendrite formation are potentially dangerous mechanisms that occur on the negative electrode. If the battery is charged rapidly or at low temperatures, lithium metal can deposit on the surface of the graphite anode rather than intercalating into it. Over time, these deposits can develop into dendritic structures that may puncture the separator and initiate internal short circuits between the anode and cathode. This situation can lead to catastrophic failure and is often difficult to detect until a short circuit occurs. Overcharging of Li-ion batteries beyond their manufacturer-specified end-of-charge voltage or 100% state of charge (SoC) and design defects, such as misbalancing of capacity or the mass ratio of the anode and cathode, can also lead to lithium plating and dendrite formation mechanisms [15, 16].

6.2.2 Positive Electrode (Cathode)

The popular cathode active materials operate on an intercalation principle; such materials include lithium cobalt oxide (LCO), lithium iron phosphate (LFP), spinels like lithium manganese oxide (LMO), mixed metal oxides like lithium nickel cobalt aluminum oxide (NCA), and lithium nickel manganese cobalt oxide (NMC). Notably, the thermal stability of the battery is affected by the composition of the cathode active material, with the nickel content playing an important role [17]. A high nickel content contributes to a higher capacity at the expense of safety characteristics. The cathode materials are susceptible to thermally driven decomposition, particularly if the cell is overcharged. They undergo a phase transition and release oxygen in an overcharge state, along with heat generation and electrolyte decomposition, which can lead to a thermal runaway [18]. The amount and type of gases generated are directly proportional to the cathode materials' oxidation (oxygen-releasing) capability, where LCO shows the highest oxidation capability [19].

6.2.3 Electrolyte

An organic solvent mixture and a lithium salt comprise the battery's electrolyte, facilitating lithium-ion transport while inhibiting electron transport between the anode and the cathode. The most commonly used solvents are cyclic and linear carbonates, such as ethylene carbonate (EC) and dimethyl carbonate (DMC). While the electrolyte used in lithium-ion batteries has beneficial properties for ion transport, it is highly flammable and unstable outside a narrow voltage and temperature window. If operated under extreme temperature or voltage conditions, the electrolyte can decompose, cause gas generation, and facilitate the thermal runaway process [15].

6.2.4 Separator

The separator in a lithium-ion battery is a porous polymer sheet that allows ion transport while preventing internal electrical short circuits between the electrodes. Most commercially available separator materials in battery markets are made of polyolefins, such as polyethylene (PE), polypropylene (PP), and hybrid PP/PE [20]. If the integrity of the separator is compromised, the electrodes can make contact, resulting in an internal short circuit that can cause heat generation and potentially lead to thermal runaway [15]. Internal short circuits can also occur if metallic particles are present in the battery that could puncture the separator. These particles can come from external contamination during assembly, lithium dendrite growth, and copper dissolution (from a negative current collector). Unfortunately, separator pore closure can also aid localized lithium plating [21]. The separator is usually wider than the electrodes to prevent short circuits at the edges; however, the separator will melt, shrink, and collapse under high temperatures, leading to internal short circuits.

6.2.5 Current Collectors

Thin foils of copper and aluminum are used as current collectors for the anode and cathode materials. A tab is welded to the current collector, which acts as a bridge that connects the electrode to the external circuit. The material, location, and number of tabs affect the cell's performance in terms of uniform current distribution, heat generation, and ohmic resistance [22]. During overdischarge, an abusive operating condition, the copper used for the anode current collector can oxidize, leading to copper dissolution. Subsequently, copper ions can get deposited on graphite during the next charge operation, leading to the potentially dangerous formation of copper dendrites [15].

6.2.6 Battery Cap Structure

A cylindrical battery is sealed with a cap located on the top of the cathode tab; a typical cylindrical battery cap structure is illustrated in Figure 6.2. The cap provides an electrical connection between the cathode tab and the positive terminal, enabling the transfer of electrical current between the battery's electrodes and the external circuit. Also, the cap consists of a nonconductive plastic insert or gasket to insulate the positive terminal from the battery can. The battery cap is vital in a cylindrical battery because it often consists of safety devices to protect the cell from thermal runaway and explosion [23]. For example, the cap comprises conductive parts that include the positive temperature coefficient (PTC) thermistor and the bottom and top disk that act as a current interrupt device (CID) [24]. Another function of the top disk is to act as the safety vent, which releases the internal pressure of the battery by allowing gases generated inside to be released in a controlled manner.

The CID is designed to disconnect the battery's internal current flow in case of excessive internal pressure, preventing thermal runaway. When the internal pressure of a cylindrical lithium-ion battery reaches a predetermined level, typically between 1.0 and 1.2 MPa, the top disk will be pushed upward to break the weak point, which is the welded connection between the central point of the top disk and the bottom disk. As a result, the electrical pathway between the current collector and the external load will be disconnected [25]. The notch in the vent disc opens when the internal cell pressure increases beyond a limit, allowing the gases to escape using the through holes in the bottom disc and the exhaust holes in the top terminal plate. For additional venting, a bottom vent is added to the battery can to prevent sidewall rupture [25].

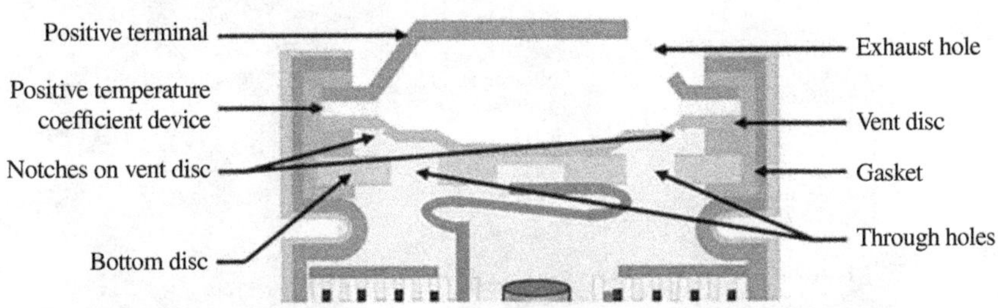

Figure 6.2 Typical cylindrical battery cap structure.

6.3 Lithium-Ion Cell Manufacturing Process

The manufacturing and assembly of a lithium-ion battery involve the precise fabrication of battery cans/casings and caps, the preparation of the electrode stack, its assembly into a jelly roll structure or layered stack, tab welding and assembly into battery can/casing, and the sealing of the battery to ensure no leakage [26, 27]. These steps, along with thorough quality checks, contribute to the production of reliable batteries. There are three major steps in battery manufacturing: electrode preparation, cell assembly, and formation (Figure 6.3).

The electrode preparation process begins with drying and mixing active materials (graphite, lithium cobalt oxide, LFP), conductive additives (e.g., carbon black), and a binder (e.g., polyvinylidene fluoride [PVDF]) in a solvent to form a slurry. Key quality parameters include slurry homogeneity, particle size, purity, and viscosity, all influenced by mixing sequence, filtering, shear forces, and blending time [26]. The slot-die coating process is the most common method for coating the slurry mixture onto the current collectors [28]. Other techniques, such as spray coating or doctor-blade coating, may also be used. The coating thickness ranges from 70 to 350 μm [22]. The slurry pump's coating speed, width, and precision define the coating's thickness accuracy, homogeneity, and surface quality (blowholes, particles) [26].

Post-coating, the electrode is dried to remove solvents, ensuring control over residual humidity, surface finish, and drying speed [26]. Calendering follows to compress the electrode, improving density, porosity, and adhesion; parameters like line speed (30–100 m/min), roller diameter (600–1000 mm), and line load (500–1000 N/m) are monitored to prevent fractures [29, 30]. The electrode sheet is then slit and vacuum-dried (12–30 hours) to eliminate moisture, as any remaining humidity could lead to hydrogen fluoride (HF) generation [26, 31]. Tabs are welded onto the current collector ends, usually by ultrasonic or laser welding, minimizing contact resistance and thermal stress to avoid heat buildup [29, 32]. Insulation tape covers the tab to prevent electrical shorts [22].

The anode electrode, separator, and cathode electrode are stacked together and rolled up to form a jelly roll. During the winding process, a center pin can be used to prevent deformation of the electrode assembly. This center pin, also called a deformation prevention core, is removed

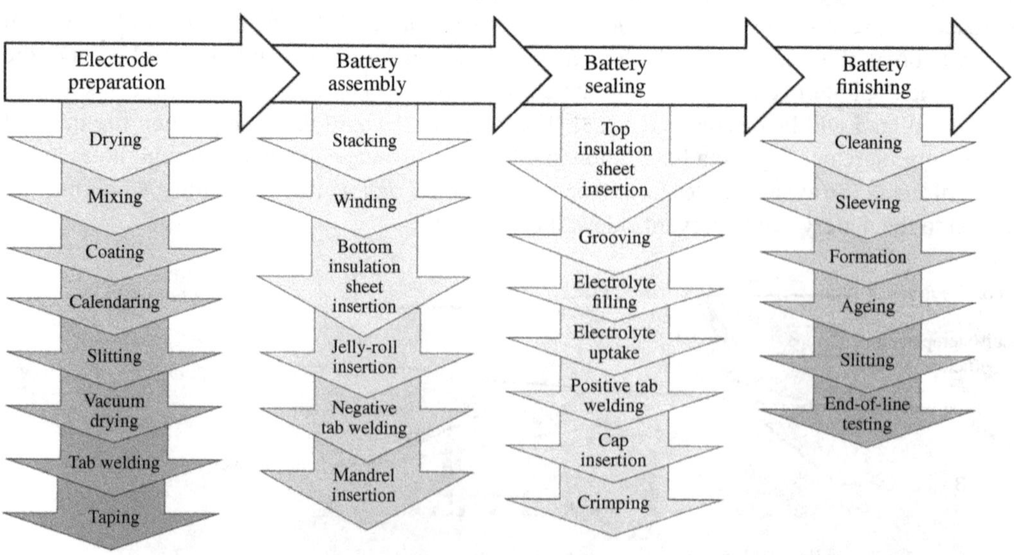

Figure 6.3 Cylindrical cell manufacturing process.

once the winding process is complete [33]. It should be noted that winding speed, web tension, and web edge control influence the quality of the jelly roll. An adhesive tape is used to secure the jelly roll in its tightly wound configuration, ensuring that the layers of electrodes and the separator remain firmly in place during subsequent assembly processes. The jelly roll and a bottom insulator are inserted into the battery can, and the negative tab is welded to the internal surface at the bottom of the can. This is a challenging step because the weld should not penetrate the battery can [34].

For cell assembly, anode, separator, and cathode layers are wound into a jelly roll, stabilized with adhesive tape, and inserted into a battery can with a negative tab weld. Process control over winding speed, web tension, and edge alignment is critical for a secure structure [33, 34]. A mandrel aids stability, and a groove is added to secure the cap without deforming the electrodes [35]. The electrolyte is filled post-drying, followed by crimping to secure the cap, with precise pressure control to avoid can deformation. Finally, the battery is insulated and sleeved [36].

A mandrel is inserted to provide mechanical stability to the cell and facilitate pressure relief by offering an unobstructed route for the fluidized material and gases to migrate from the base to the cell's positive terminal [35]. Then, a top insulator is applied to the jelly roll structure; the insulator contains a hole that allows the positive tab to go through it. A groove is made on the can above the top insulation ring to host the battery cap, but the grooving speed and depth need to be precisely controlled to avoid electrode deformation. After another drying process to remove the remaining moisture, the electrolyte is filled into the battery before the final sealing. Although the separator of a lithium-ion battery has a porous structure, an electrolyte uptake/wetting step is performed to facilitate the infusion of the electrolyte and wet the entire jelly roll structure to ensure a homogenous distribution of the electrolyte [36]. Once the cap and cell are aligned, the crimping process can begin. This involves the crimping machine applying pressure to the edges of the cap, causing it to deform and grip the top edge of the battery. The pressure is carefully controlled to ensure that the cap is securely attached to the battery but not so much that it causes damage or deformation of the battery. Subsequently, the battery is cleaned and then wrapped with a sleeve made of insulating material and then undergoes the formation process and aging.

In the formation process, an SEI layer is developed over several low-current charge-discharge cycles, critical for preventing lithium plating and electrolyte decomposition. Typically, the formation process involves three to five charge-discharge cycles at low currents (usually C/20), followed by rest, which can take several days or weeks [37]. High-current and high-temperature formation cycles produce a porous SEI layer that cannot prevent electrolyte decomposition due to contact of the electrolyte with the anode surface [38]. High currents can also lead to lithium plating, reducing cell safety [3]. Aging tests assess self-discharge, while end-of-line testing confirms visual integrity, capacity, internal resistance, and open-circuit voltage to ensure compliance with performance and safety criteria [26].

6.4 Role of the Design and Manufacturing Process in Battery Safety

Cylindrical lithium-ion batteries are complex systems with multistep manufacturing processes. This introduces the possibility of diverse failure modes that can harm the reliability and safety of the battery. Tables 6.1 and 6.2 list all the design and process parameters influencing cell safety. During cell manufacturing, control of these parameters is essential for building reliable and safe lithium-ion cells.

An integrated safety analysis consolidates design failure modes and effect analysis (DFMEA) and process failure modes and effects analysis (PFMEA). DFMEA provides a structured methodology to evaluate and address potential failure modes in various components and aspects of cylindrical

Table 6.1 Design parameters influencing battery safety.

Element	Influencing parameters
Anode material	Purity, particle size, particle size distribution, particle porosity, crystalline phase of carbon, degree of compaction, surface area, coating thickness, surface roughness, conductivity of coating material, binder type, binder-to-active-material ratio
Cathode material	Purity, particle size, particle size distribution, particle porosity, thermal stability, degree of compaction, surface area, coating thickness, surface roughness, binder type, binder-to-active-material ratio, conductivity of the binder
Electrolyte	Salt-to-solvent ratio, thermal stability of solvent, amount and composition of additives, moisture content, viscosity, conductivity, corrosive properties, melting point
Current collectors	Mechanical strength, corrosion resistivity, conductivity, thickness, contact resistance, adhesion strength
Separator	Mechanical strength, thermal and dimensional strength, permeability, porosity, chemical structure, surface energy with electrolyte and electrode materials, thickness, uniformity of pore size, tensile strength
Electrode tabs	Material, location, number of tabs, thickness, corrosion resistance
Current interrupt device	Activation pressure or temperature, response time, corrosion resistivity
Safety vents	Activation pressure, geometry
Positive temperature coefficient device	Thermal mass, heat dissipation coefficient, trip time, material composition, resistance stability, corrosion resistance

Table 6.2 Process parameters influencing battery safety.

Process step	Influencing parameters
Drying	Temperature profile, humidity, drying speed, foil pretension, drying time
Mixing	Mixing and dispersing sequence, shear forces of the equipment, blending time, mixing temperature, mixing speed, humidity
Coating	Coating speed, coating width, precision of the slurry pump, humidity
Calendering	Line speed, roller diameters, line load, calendering temperature, roller alignment
Laser slitting	Laser power, scanning speed, laser beam quality, focal spot size
Tab welding	Current profile, weld force, weld duration, electrode alignment
Winding	Winding speed, web tension
Grooving	Grooving speed, grooving depth, tool, and groove alignment
Electrolyte wetting	Wetting time, ambient humidity
Crimping	Crimping force, crimping speed, crimping electrode alignment
Formation	Charging current, voltage range, temperature, formation duration

lithium-ion batteries, including materials selection and design. PFMEA focuses on potential failure modes caused by manufacturing and assembly process deficiencies. Tables A1 and A2 summarize DFMEA and PFMEA for cylindrical cells.

The integrated DFMEA–PFMEA shows that localized heating and a short circuit increase the risk of thermal runaway increased gas generation due to moisture or electrolyte leakage increases the risk of explosion. Notably, manufacturing and assembly defects in the safety devices also reduce the safety of the battery. The following section describes each of the mechanisms that can affect the safety of the battery.

6.4.1 Internal Short Circuit

Internal short circuits in lithium-ion batteries can result in thermal runaway, as they generate heat that initiates a sequence of exothermic reactions [35]. Internal short circuits can occur due to lithium dendrites, separator holes, burrs on the electrodes/tabs, and metallic contaminants. Additionally, poor puncture or tensile strength or insufficient separator thickness can increase the risk of tear [15]. Proper insulation by the separator is required during the alignment of electrode sheets and assembly of the battery. The design of wider separators with high thermal resistance is suggested to avoid a short circuit due to separator shrinkage when exposed to high temperatures [41].

Ensuring proper capacity alignment between the negative and positive electrodes is critical to designing lithium-ion batteries that can operate safely. This is achieved by maintaining a proper negative/positive (N/P) ratio. The N/P ratio is defined as the ratio of the reversible capacity of the negative electrode and positive electrode, and it is controlled between 1.03 and 1.2 [11]. A low N/P ratio can cause the anode potential to drop to less than 0 V versus Li/Li^+ during charging, which could lead to lithium plating on the surface of the anode. Conversely, a high N/P ratio reduces the battery's energy density because it requires more anode material, which adds weight and volume without proportionally increasing the overall energy storage capacity.

Researchers have employed numerous methods to evaluate, detect, and study lithium plating, including analyzing voltage plateau signals [42], measuring cell thickness [43], and creating simulation models [44–46]. However, despite these efforts, accurately predicting the likelihood of internal short circuits resulting from lithium plating remains difficult. This further increases the importance of mitigating the failure modes during design and manufacturing that potentially cause lithium plating.

6.4.2 Localized Heating

Heat is generated in the battery due to entropy change and Joule heat [47]. When the heat generation rate exceeds that of dissipation, it results in a temperature rise. Poor-quality materials and manufacturing processes, or an inhomogeneous distribution of materials, can cause uneven temperature distribution due to different heat generation and heat dissipation conditions in the electrode [11]. Subsequently, localized heating increases the risk of thermal runaway due to the initiation of exothermic side reactions.

Reducing the porosity or increasing the electrode's thickness can lead to an increase in ion concentration and potential gradient, which can influence the generation of Joule heat [48]. Also, electrode particle fracture due to calendering and the vacuum drying process can cause local hotspots. Therefore, it becomes essential to ensure that design factors like porosity and electrode thickness are appropriately selected and tested. Thermal simulations are generally conducted using lumped model [49], but including local hotspots increases the accuracy of model prediction and aids in a better design of protection limits [50].

6.4.3 Increased Gas Generation

Gases generated in the battery increase the internal pressure and, in some situations, can cause the battery to vent or rupture, eventually leading to thermal runaway due to the reaction of hot, flammable gases from the battery with oxygen [51]. Increased side reactions involving electrolyte decomposition elevate gas generation in lithium-ion batteries. The poor thermal stability of cathodes and an improper formation process for SEI layer growth on anodes can lead to electrolyte decomposition that produces gases [17, 52]. Cathode material composition can impact the amount of gas generation during electrical abuse conditions like overcharge due to electrolyte decomposition at the cathode surface [53]. Coatings like transition metal oxide nanoparticles on the cathode have shown the potential to enhance stability and prevent side reactions during overcharge [54]. Electrode porosity should also be optimized to facilitate Li^+ ion diffusion and minimize overpotential [55].

The presence of moisture in a lithium-ion battery can generate gas through a series of chemical reactions with the electrolyte and lithium salts. When moisture enters the cell, it reacts with lithium salt to produce HF and other byproducts. However, the generated HF is highly reactive and can react with the organic solvents in the electrolyte or the electrode materials, leading to more gas generation [56]. This moisture can result from manufacturing defects and poor sealing due to improper design or lack of quality control during manufacturing. Notably, the water content present in the anode during the manufacturing process should drop from about 1000 ppm to about 200 ppm after vacuum drying, emphasizing the importance of the drying steps in the process [31]. Poor thermal stability of the gasket in the battery cap can lead to sagging during temperature cycling, leading to moisture ingress and electrolyte leakage [57].

Electrode crosstalk refers to a phenomenon wherein the byproducts generated at one electrode initiate adverse side reactions on the opposing electrode, leading to exothermic reactions and gas release [58]. This occurrence often stems from the dissolution of transition metals, the magnitude of which is influenced by factors such as cathode composition, electrolyte formulation, and the formation of a stable SEI layer [59]. Introducing aluminum doping into the transitional metal cathode material can mitigate the dissolution process [60] while incorporating appropriate electrolyte additives can suppress active material corrosion and oxygen evolution [58, 61].

6.4.4 Malfunctioning of Safety Devices

When a lithium-ion battery goes into thermal runaway, the energy stored within the battery is often released in a matter of milliseconds [24]. Improper design and manufacturing processes that can compromise the functionality of the safety devices increase the risk of cell failures or hazardous incidents.

The PTC device is a temperature-sensitive resistor that limits current flow when the battery's temperature exceeds a specified threshold, protecting the battery from overheating. However, improper design of the PTC can increase the battery's internal resistance, thereby increasing thermal loss [62]. The PTC's thermal mass and the heat transfer coefficient affect its response time, and choosing an incorrect PTC material can cause the device to fail, leading to uncontrolled current flow and a higher risk of thermal runaway [63].

If the safety vent, another safety device in a cylindrical battery, becomes obstructed and the internal pressure is not released in a timely manner, the pressure may continue to build up and cause the battery case to rupture or even lead to an explosion [64, 65]. To mitigate this risk, a mandrel and bottom vent are considered in certain models of cylindrical cells. As a result, this design enhancement increases the venting efficiency and reduces the thermal impact of a single battery rupture in a battery pack [25]. Without an internal mandrel, the electrode assembly can collapse, blocking the gas flow and increasing the risk of uncontrolled cell venting.

It is essential to choose the proper activation pressure to ensure the CID is activated only when there is a risk of thermal runaway or explosion and not during normal operation. Manufacturing issues, such as improper assembly, misalignment, or defects in the CID components, can cause it to fail or activate prematurely, compromising the battery's safety and performance. Also, the activation of the CID can be affected by the welding connection. If the welding connection is too strong, the top disk may not break the welded connection, preventing the CID from activating when needed. The trigger pressure of the safety vent in cylindrical batteries is typically higher than that of the pressure-responsive CID [66], and the vent's opening area can determine the flow rate and burst pressure during venting [67]. Therefore, considering all potential failure modes due to improper venting, and simulating them to verify the design, can help improve the battery's safety.

6.5 Summary

Improving lithium-ion battery safety involves optimizing design, ensuring rigorous manufacturing quality control, and integrating protective features to manage potential safety issues. Given the complex factors contributing to failures—such as battery chemistry, design specifications, operational conditions, and intended applications—it is challenging to develop a universal ranking of causes. However, key areas of concern can be identified, corresponding to critical design and manufacturing processes that can exacerbate short circuits, localized heating, and abnormal gas generation within cells.

The integrated FMEA-based safety analysis helps identify design parameters that need to be considered while modeling a particular failure mechanism to predict an onset of failure. Identifying the process parameters and failure causes associated with these failures can guide designing quality checks in the manufacturing process. These quality checks, implemented as part of in-process quality control, can improve safety. This analysis can also be utilized for a post-facto root cause analysis to determine the most probable explanation for a failure.

Appendix

Table A1 DFMEA for cylindrical lithium-ion battery.

Component	Potential failure mode	Potential failure cause	Effect on battery safety
Cathode	Poor thermal stability Poor overcharge tolerance High reactivity to electrolyte	Incorrect material selection; incorrect design parameters; poor-quality material from the supplier	Increased gas generation and risk of explosion and thermal runaway
	Metal contamination		The presence of metal particles can cause nucleation sites for the formation of lithium dendrites, increasing the risk of short circuit
	Improper N/P ratio		The N/P ratio is not maintained, which can lead to lithium plating, increasing the risk of an internal short circuit
	Large particle size		This can result in a tear of the separator, causing an internal short circuit
	Small specific area		This can result in severe electrochemical polarization, increasing the temperature at the cathode and the risk of thermal runaway

(Continued)

Table A1 (Continued)

Component	Potential failure mode	Potential failure cause	Effect on battery safety
	Delamination of cathode sheet		Nonuniform current distribution resulting in localized heating, increasing risk of thermal runaway
	High compaction density	Wrong settings for calendering during design	This can result in fractures, causing nonuniform current distribution, resulting in localized heating, which increases the risk of thermal runaway
Anode	Following similar failure modes (with similar failure causes and effects on safety) as cathode are observed: poor thermal stability; metal contamination; large particle size; small specific area; improper N/P ratio; delamination of anode sheet		
Electrolyte	Poor overcharge tolerance Poor thermal stability Highly corrosive	Incorrect material selection; incorrect design parameters; poor-quality material from the supplier	Increased gas generation and risk of explosion
	Lack of electrolyte		Increased electrochemical polarization due to improper soaking and nonuniform current distribution, increasing risk of lithium plating and internal short circuit
	Excess electrolyte		Improper seal, causing electrolyte leakage, leading to the formation of a flammable gas mixture, and increasing the risk of explosion
	Micro short circuit		Increasing risk of an internal short circuit
	Low ionic conductivity High freezing point		This can result in severe electrochemical polarization, increasing the temperature and the risk of thermal runaway
Current collector	Low tensile strength	Incorrect material selection; incorrect design parameters; poor-quality material from the supplier	Metal foil fracture, causing nonuniform current distribution, resulting in localized heating, increasing risk of thermal runaway
Separator	High shutdown temperature	Incorrect material selection; incorrect design parameters; poor-quality material from the supplier	Delay in separator shutdown at elevated temperatures, increasing risk of thermal runaway
	High shrinkage Low puncture strength Low thickness Low tensile strength		Increased risk of an internal short circuit, leading to thermal runaway
	High porosity		Reduction in mechanical strength, increasing risk of tear and internal short circuit
	Nonuniform pore size distribution		Nonuniform current distribution resulting in localized heating, lithium plating, and increasing risk of thermal runaway

Table A1 (Continued)

Component	Potential failure mode	Potential failure cause	Effect on battery safety
Cap	High PTC base resistance	Incorrect material selection; incorrect design parameters; poor-quality material from the supplier	Increased internal resistance, causing a reduction in capacity
	High PTC temperature inflection point		Delay in functioning of PTC, increasing risk of thermal runaway
	High CID and vent activation pressure		Delay in activation of CID, increasing risk of internal pressure buildup, causing thermal runaway or explosion
	Insufficient air flow rate		Increased risk of internal pressure buildup, causing thermal runaway or explosion
	Low gasket thermal stability		Improper seal, causing electrolyte leakage, leading to the formation of a flammable gas mixture, increasing the risk of explosion
	High gasket diffusion coefficient		
	Low insulation ring thermal stability		
	Mismatched dimensions of PTC, insulation ring, vent disk, bottom disk, and gasket		
Tab	Improper hardness	Incorrect material selection; incorrect design parameters; poor-quality material from the supplier	The tab could cut through the separator, causing an internal short circuit and increasing the risk of thermal runaway
	Improper location and number of tabs		Nonuniform current distribution results in localized heating and lithium plating, increasing the risk of thermal runaway

Source: Maddipatla et al. [39]/MDPI/CC BY 4.0.

Table A2 PFMEA for manufacturing a cylindrical lithium-ion battery.

Process	Substep	Potential failure mode	Potential failure cause	Effect on battery safety
Electrode preparation	Mixing	Wrong material chosen for mixing	Error during procurement; wrong labeling	N/P ratio[a)] is not maintained, which can lead to lithium plating, increasing the risk of an internal short circuit [26]
		Wrong composition of materials for the mixture	Error during measurement before mixing	
		Presence of moisture	Improper storage of raw material; improper warehouse humidity conditions	Increased gas generation causes poor safety due to the risk of explosion [40]

(Continued)

Table A2 (Continued)

Process	Substep	Potential failure mode	Potential failure cause	Effect on battery safety
		Presence of metal contaminants	Procurement of poor-quality material; improper storage of raw material	The presence of metal particles can cause nucleation sites for the formation of lithium dendrites, increasing the risk of short circuit [11]
		Presence of dust contaminants		The creation of discontinuities in electrode structure due to dust can result in local hotspots, increasing the risk of thermal runaway [11]
		The presence of solid content in the mixture	Short blend time setting	This can result in a tear of the separator, causing an internal short circuit
		Insufficient viscosity of the mixture	Improper blend time setting; improper temperature of mixture	Improper coating of the mixture causes nonuniform current distribution and peeling of the coated film, which results in localized heating, increasing the risk of thermal runaway [16]
	Coating	Nonuniform coating Improper surface finish—presence of holes or voids	Improper viscosity of the mixture; improper alignment of slot die; uneven flow rate of slot die; foil surface is uneven	Nonuniform current distribution results in localized heating and lithium plating, increasing the risk of thermal runaway [17]
		Improper surface finish—nonuniform dispersion or presence of agglomerates	Improper blend time during mixing	Nonuniform current distribution, resulting in localized heating, increasing risk of thermal runaway Increase in electrical polarization
		Improper dimensions of coat—cathode width out of lower limit	Improper setting of equipment	The N/P ratio increases, reducing the energy density of the battery
		Improper dimensions of coat—anode width out of lower limit		The N/P ratio is lowered, which can lead to lithium plating, increasing the risk of an internal short circuit
	Calendering	Thickness below lower limit; reduced electrode porosity	Improper setting of gap and force between the rollers	The low porosity of electrodes increases their diffusion resistance due to slow kinematics, which can result in lithium plating, increasing the risk of an internal short circuit
		Thickness above the upper limit; increased electrode porosity		Cycle life performance degradation; jelly roll diameter above upper limit, making assembly difficult

Table A2 (Continued)

Process	Substep	Potential failure mode	Potential failure cause	Effect on battery safety
		Warping of electrode sheets		Nonuniform current distribution resulting in localized heating, increasing risk of thermal runaway
		Increased surface roughness	The uneven surface of the rollers	
		Occurrence of fractures in the material	Improper setting of gap and force between the rollers	Cracks in the cathode result in a low N/P ratio, which can lead to lithium plating, increasing the risk of an internal short circuit. Fractures lead to increased moisture sorption, leading to gas generation, increasing the risk of explosion
	Slitting	Improper width—cathode too narrow or anode too wide	Improper setting of equipment	The N/P ratio is lowered, which can lead to lithium plating, increasing the risk of an internal short circuit
		Improper width—cathode too wide or anode too narrow		
		Improper width—separator too wide		Increases the risk of an internal short circuit
		Improper width—separator too narrow		
		Improper height of electrodes and separator		
		Presence of burrs	Wear of slitting knife; unclean edge of slitting knife	
		Improper geometry of the cutting edges	Improper setting of equipment	
		Presence of metallic foreign particles	Unclean equipment and workshop conditions	The presence of metal particles can cause nucleation sites for the formation of lithium dendrites, increasing the risk of short circuit
	Vacuum drying	Presence of moisture	Improper setting of room humidity level; insufficient setting of drying time	Increased gas generation, causing poor safety due to the risk of explosion
		Occurrence of fractures in the material	Improper setting of room temperature and drying time	Cracks in the cathode result in a low N/P ratio, which can lead to lithium plating, increasing the risk of an internal short circuit. Fractures lead to increased moisture sorption, leading to gas generation, increasing the risk of explosion

(Continued)

Table A2 (Continued)

Process	Substep	Potential failure mode	Potential failure cause	Effect on battery safety
	Tab welding	Insufficient weld strength and improper weld tension	Improper setting of equipment; improper weld position	Increased contact resistance causes increased Joule heating, creating thermal hotspots and increasing the risk of thermal runaway
		Presence of burrs or protrusions		This can result in a tear of the separator, causing an internal short circuit
		Overwelding of tabs		It can damage the electrode sheet, causing an internal short circuit
		Presence of dust contaminants	Unclean equipment and workshop conditions	The creation of discontinuities in electrode structure due to dust can result in local hotspots, increasing the risk of thermal runaway
	Taping	Poor coverage of tab	Improper position settings in equipment	This can result in a tear of the separator, causing an internal short circuit
		Poor adhesion of the tape	Poor-quality procurement; presence of contaminants in the workshop	This can result in the tear of the separator and improper current distribution, resulting in localized hotspots, causing an internal short circuit
		Presence of dust contaminants	Unclean equipment and workshop conditions	The creation of discontinuities in electrode structure due to dust can result in local hotspots, increasing the risk of thermal runaway
Cell assembly	Stacking and winding	The presence of holes in separators	Poor quality of separator	Internal short circuit
		Improper positioning of electrodes and separator	Operator fault	
		Improper rolling of jelly roll—tight winding	Improper tension settings in equipment; improper removal of winding rod	
		Improper rolling of jelly roll—loose winding		Nonuniform current distribution resulting in localized heating, increasing risk of thermal runaway
		Improper rolling of jelly roll—winding spiral		
		Improper rolling of jelly roll—center collapse		Can scrape the electrode layers while welding the bottom tab, increasing the risk of an internal short circuit

Table A2 (Continued)

Process	Substep	Potential failure mode	Potential failure cause	Effect on battery safety
	Bottom insulation sheet and jelly roll insertion	The diameter of the bottom insulation ring is small	Operator fault; manufacturing defects	Short circuit between negative terminal and positive electrode
		Improper alignment of jelly roll—inclined	Improper alignment of equipment	Weld rod can scrape the electrode layers while welding the bottom tab, increasing the risk of an internal short circuit
	Bottom tab welding	Insufficient weld strength and improper weld tension	Improper setting of equipment; improper weld position	Increased contact resistance causes increased Joule heating, creating thermal hotspots and increasing the risk of thermal runaway
		Presence of burrs or protrusions		This can result in a tear of the separator, causing an internal short circuit
		Overwelding of tabs to can		Can damage the can, resulting in an improper seal, causing electrolyte leakage, leading to the formation of a flammable gas mixture, increasing the risk of explosion
		Electrode damage due to weld rod	Improper alignment of equipment	Weld rod can scrape the electrode layers while welding the bottom tab, increasing the risk of an internal short circuit
Cell sealing	Top insulation; sheet insertion	The diameter of the top insulation ring is small	Operator fault; manufacturing defects	Short circuit between negative terminal and positive terminal
		Improper alignment of insulation sheet	Improper alignment of equipment	
	Grooving	Improper groove height—low	Improper settings in equipment	Electrode deformation can cause an internal short circuit
		Improper crimping diameter—low		
		Improper groove height—high		Movement of jelly roll, increasing risk of an internal short circuit
		Improper crimping diameter—high		Improper seal, causing electrolyte leakage leading to the formation of a flammable gas mixture, increasing the risk of explosion

(Continued)

Table A2 (Continued)

Process	Substep	Potential failure mode	Potential failure cause	Effect on battery safety
		Presence of dust contaminants	Unclean equipment and workshop conditions	The creation of discontinuities in electrode structure due to dust can result in local hotspots, increasing the risk of thermal runaway
	Electrolyte filling	The presence of contaminants in the electrolyte	Procurement of poor-quality material; improper storage of raw material	Increased gas generation, causing poor safety due to the risk of explosion
		Presence of moisture	Improper setting of room humidity level	
		Increased electrolyte quantity	Improper settings in equipment	Electrolyte leakage leads to the formation of a flammable gas mixture, increasing the risk of explosion
		Reduced electrolyte quantity		Inhomogeneous distribution of electrolyte, causing nonuniform current distribution, increasing risk of lithium plating and internal short circuit
	Positive tab welding	Insufficient weld strength and improper weld tension	Improper setting of equipment; improper weld position	Increased contact resistance causes increased Joule heating, creating thermal hotspots and increasing the risk of thermal runaway
		Presence of burrs or protrusions		This can result in a tear of the separator, causing an internal short circuit
	Cap insertion	Improper cap diameter—low	Manufacturing defects	Improper seal, causing electrolyte leakage leading to the formation of a flammable gas mixture, increasing the risk of explosion
		Improper cap diameter—high		
		Misalignment of cap	Improper alignment of equipment	
	Crimping and sealing	Improper crimping force—low	Improper equipment settings and alignment	Improper seal increasing the risk of explosion and leakage
		Misaligned crimp		
		Improper crimping force—high		Electrode deformation, causing an internal short circuit

Table A2 (Continued)

Process	Substep	Potential failure mode	Potential failure cause	Effect on battery safety
Cell finishing	Sleeving	Improper sleeve thickness—low	Manufacturing defects	Insufficient wear resistance, causing an external short circuit
		Improper sleeve thickness—high		Poor heat dissipation, increasing risk of thermal runaway
		Improper sleeve damaged or punctured	Improper loading into equipment	External short circuit
	Formation	Unstable SEI layer formation	High currents during the formation cycle; short formation cycles	Increased gas generation, reduced capacity, and risk of thermal runaway due to continuous SEI growth

a) N/P ratio is the ratio of negative electrode capacity to positive electrode capacity.
Source: Maddipatla et al. [39]/MDPI/CC BY 4.0.

References

1 Broussely, M., Biensan, Ph., Bonhomme, F. et al. (2005). Main aging mechanisms in Li ion batteries. *Journal of Power Sources* 146 (1–2): 90–96. https://doi.org/10.1016/j.jpowsour.2005.03.172.

2 Mohanty, D., Hockaday, E., Li, J. et al. (2016). Effect of electrode manufacturing defects on electrochemical performance of lithium-ion batteries: cognizance of the battery failure sources. *Journal of Power Sources* 312: 70–79. https://doi.org/10.1016/j.jpowsour.2016.02.007.

3 Mao, C., An, S.J., Meyer, H.M. III, et al. (2018). Balancing formation time and electrochemical performance of high energy lithium-ion batteries. *Journal of Power Sources* 402: 107–115. https://doi.org/10.1016/j.jpowsour.2018.09.019.

4 Maleki, H., Deng, G., Anani, A., and Howard, J. (1999). Thermal stability studies of Li-ion cells and components. *Journal of the Electrochemical Society* 146 (9): 3224–3229. https://doi.org/10.1149/1.1392458.

5 CNN Business (2024). Lithium-ion battery fires are happening more often. Here's how to prevent them. https://edition.cnn.com/2023/03/09/tech/lithium-ion-battery-fires/index.html (accessed 28 May 2024).

6 BBC (2024). Norfolk Fire and Rescue Service sees increase in battery fires. https://www.bbc.com/news/articles/cl4p6mn0md6o (accessed 28 May 2024).

7 CTV News (2024). E-bike battery fires in Canada becoming more common. https://www.ctvnews.ca/sci-tech/lithium-ion-battery-fires-on-the-rise-across-canada-fire-chiefs-warn-1.6842177 (accessed 28 May 2024).

8 Down To Earth (2024). Trial by 'fire': A wake-up call for India's electric vehicle strategy. https://www.downtoearth.org.in/news/renewable-energy/trial-by-fire-a-wake-up-call-for-india-s-electric-vehicle-strategy-83333 (accessed 28 May 2024).

9 Bartolomé, E. and Benítez, P. (2022). Failure mode and effect analysis (FMEA) to improve collaborative project-based learning: case study of a Study and Research Path in mechanical engineering. *International Journal of Mechanical Engineering Education* 50 (2): 291–325. https://doi.org/10.1177/0306419021999046.

10 Hu, J.M., Barker, D.B., Dasgupta, A., and Arora, A.K. (1992). Role of failure-mechanism identification in accelerated testing. *Annual Reliability and Maintainability Symposium 1992 Proceedings*, 181–188. https://doi.org/10.1109/ARMS.1992.187820.

11 Wu, X., Song, K., Zhang, X. et al. (2019). Safety issues in lithium ion batteries: materials and cell design. *Frontiers in Energy Research* 7: 65. https://doi.org/10.3389/fenrg.2019.00065.

12 Mikolajczak, C., Kahn, M., White, K., and Long, R.T. (2011). *Lithium-Ion Batteries Hazard and Use Assessment*, Springer Briefs in Fire. Boston, MA: Springer. ISBN: 978-1-4614-3485-6.

13 Ogumi, Z. and Inaba, M. (2002). Carbon anodes. In: *Advances in Lithium-Ion Batteries* (ed. W.A. Van Schalkwijk and B. Scrosati), 79–101. Boston, MA: Springer. ISBN: 978-0-306-47356-2.

14 Nazri, G.A. and Yebka, B. (2000). Reactivity and safety aspects of carbonaceous anodes used in lithium-ion batteries—correlation of structural parameters and reactivity. In: *Materials for Lithium-Ion Batteries* (ed. C. Julien and Z. Stoynov), 161–178. Dordrecht, The Netherlands: Springer. ISBN: 978-0-7923-6651-5.

15 Hendricks, C., Williard, N., Mathew, S., and Pecht, M. (2015). A failure modes, mechanisms, and effects analysis (FMMEA) of lithium-ion batteries. *Journal of Power Sources* 297: 113–120. https://doi.org/10.1016/j.jpowsour.2015.07.100.

16 Kasnatscheew, J., Placke, T., Streipert, B. et al. (2017). A tutorial into practical capacity and mass balancing of lithium ion batteries. *Journal of the Electrochemical Society* 164 (12): A2479–A2486. https://doi.org/10.1149/2.0961712jes.

17 Bak, S.-M., Hu, E., Zhou, Y. et al. (2014). Structural changes and thermal stability of charged $LiNi_xMn_yCo_zO_2$ cathode materials studied by combined *in situ* time-resolved XRD and mass spectroscopy. *ACS Applied Materials & Interfaces* 6 (24): 22594–22601. https://doi.org/10.1021/am505526s.

18 Ma, L., Nie, M., Xia, J., and Dahn, J.R. (2016). A systematic study on the reactivity of different grades of charged $Li[Ni_xMn_yCoz]O_2$ with electrolyte at elevated temperatures using accelerating rate calorimetry. *Journal of Power Sources* 327: 145–150. https://doi.org/10.1016/j.jpowsour.2016.07.071.

19 Kong, W., Li, H., Huang, X., and Chen, L. (2005). Gas evolution behaviors for several cathode materials in lithium-ion batteries. *Journal of Power Sources* 142 (1–2): 285–291. https://doi.org/10.1016/j.jpowsour.2004.11.018.

20 Zhou, H., Fear, C., Parekh, M. et al. (2022). The role of separator thermal stability in safety characteristics of lithium-ion batteries. *Journal of the Electrochemical Society* 169 (9): 090521. https://doi.org/10.1149/1945-7111/ac8edf.

21 Cannarella, J. and Arnold, C.B. (2015). The effects of defects on localized plating in lithium-ion batteries. *Journal of the Electrochemical Society* 162 (7): A1365–A1373. https://doi.org/10.1149/2.1051507jes.

22 Yao, X.-Y. and Pecht, M.G. (2019). Tab design and failures in cylindrical li-ion batteries. *IEEE Access* 7: 24082–24095. https://doi.org/10.1109/ACCESS.2019.2903581.

23 Balakrishnan, P.G., Ramesh, R., and Prem Kumar, T. (2006). Safety mechanisms in lithium-ion batteries. *Journal of Power Sources* 155 (2): 401–414. https://doi.org/10.1016/j.jpowsour.2005.06.019.

24 Kong, L., Li, C., Jiang, J., and Pecht, M.G. (2018). Li-ion battery fire hazards and safety strategies. *Energies* 11 (9): 2191. https://doi.org/10.3390/en11092191.

25 Xu, B., Kong, L., Wen, G., and Pecht, M.G. (2021). Protection devices in commercial 18650 lithium-ion batteries. *IEEE Access* 9: 66687–66695. https://doi.org/10.1109/ACCESS.2021.3071271.

26 Heimes, H., Kampker, A., Lienemann, C. et al. (2018). *Lithium-Ion Battery Cell Production Process*. PEM der RWTH Aachen University.

27 Liu, Y., Zhang, R., Wang, J., and Wang, Y. (2021). Current and future lithium-ion battery manufacturing. *iScience* 24 (4): https://doi.org/10.1016/j.isci.2021.102417.

28 Schmitt, M., Baunach, M., Wengeler, L. et al. (2013). Slot-die processing of lithium-ion battery electrodes—Coating window characterization. *Chemical Engineering and Processing* 68: 32–37. https://doi.org/10.1016/j.cep.2012.12.007.

29 Kwade, A., Haselrieder, W., Leithoff, R. et al. (2018). Current status and challenges for automotive battery production technologies. *Nature Energy* 3 (4): 290–300. https://doi.org/10.1038/s41560-018-0108-3.

30 Huttner, F., Diener, A., Heckmann, T. et al. (2021). Increased moisture uptake of NCM622 cathodes after calendering due to particle breakage. *Journal of the Electrochemical Society* 168 (9): 090539. https://doi.org/10.1149/1945-7111/ac1c53.

31 Kosfeld, M., Westphal, B., and Kwade, A. (2023). Moisture behavior of lithium-ion battery components along the production process. *Journal of Energy Storage* 57: 106174.

32 Yang, N., Zhang, X., Shang, B.B., and Li, G. (2016). Unbalanced discharging and aging due to temperature differences among the cells in a lithium-ion battery pack with parallel combination. *Journal of Power Sources* 306: 733–741.

33 Kim, J.K., Saito, A., and Kim, Y.K. (2008). Jelly-roll type electrode assembly, lithium secondary battery having the same, and method for manufacturing the same. US Patent 7,442,465, issued October 28, 2008.

34 Lee, J., Kim, T., Jung, H., and Shin, B. (2018). Method for welding metal tab of electrode layer for cable battery and electrode manufactured thereby. US Patent 9,947,913, issued April 17, 2018.

35 Finegan, D.P., Darcy, E., Keyser, M. et al. (2017). Characterising thermal runaway within lithium-ion cells by inducing and monitoring internal short circuits. *Energy & Environmental Science* 10 (6): 1377–1388.

36 Davoodabadi, A., Jin, C., Wood, D.L. III, et al. (2020). On electrolyte wetting through lithium-ion battery separators. *Extreme Mechanics Letters* 40: 100960.

37 An, S.J., Li, J., Du, Z. et al. (2017). Fast formation cycling for lithium ion batteries. *Journal of Power Sources* 342: 846–852. https://doi.org/10.1016/j.jpowsour.2017.01.011.

38 Bhattacharya, S. and Alpas, A.T. (2012). Micromechanisms of solid electrolyte interphase formation on electrochemically cycled graphite electrodes in lithium-ion cells. *Carbon* 50 (15): 5359–5371.

39 Maddipatla, S., Kong, L., and Pecht, M. (2024). Safety analysis of lithium-ion cylindrical batteries using design and process failure mode and effect analysis. *Batteries* 10: 76.

40 Wood, D.L., Li, J., and An, S.J. (2019). Formation challenges of lithium-ion battery manufacturing. *Joule* 3 (12): 2884–2888.

41 Huang, X. (2011). Separator technologies for lithium-ion batteries. *Journal of Solid State Electrochemistry* 15 (4): 649–662.

42 Gao, S., Feng, X., Lu, L. et al. (2019). An experimental and analytical study of thermal runaway propagation in a large format lithium ion battery module with NCM pouch-cells in parallel. *International Journal of Heat and Mass Transfer* 135: 93–103.

43 Feng, X., Fang, M., He, X. et al. (2014). Thermal runaway features of large format prismatic lithium ion battery using extended volume accelerating rate calorimetry. *Journal of Power Sources* 255: 294–301.

44 Parhizi, M., Ahmed, M.B., and Jain, A. (2017). Determination of the core temperature of a Li-ion cell during thermal runaway. *Journal of Power Sources* 370: 27–35.

45 Akolkar, R. (2014). Modeling dendrite growth during lithium electrodeposition at sub-ambient temperature. *Journal of Power Sources* 246: 84–89.

46 Arora, P., Doyle, M., and White, R.E. (1999). Mathematical modeling of the lithium deposition overcharge reaction in lithium-ion batteries using carbon-based negative electrodes. *Journal of the Electrochemical Society* 146 (10): 3543.

47 Oh, K.-Y. and Epureanu, B.I. (2016). A novel thermal swelling model for a rechargeable lithium-ion battery cell. *Journal of Power Sources* 303: 86–96.

48 Zhao, H., Yang, Q., Yuca, N. et al. (2016). A convenient and versatile method to control the electrode microstructure toward high-energy lithium-ion batteries. *Nano Letters* 16 (7): 4686–4690.

49 Feng, X., He, X., Ouyang, M. et al. (2015). Thermal runaway propagation model for designing a safer battery pack with 25 Ah $LiNi_xCo_yMn_zO_2$ large format lithium ion battery. *Applied Energy* 154: 74–91.

50 Kim, G.-H., Pesaran, A., and Spotnitz, R. (2007). A three-dimensional thermal abuse model for lithium-ion cells. *Journal of Power Sources* 170 (2): 476–489.

51 Mishra, D., Zhao, P., and Jain, A. (2022). Thermal runaway propagation in Li-ion battery packs due to combustion of vent gases. *Journal of the Electrochemical Society* 169 (10): 100520.

52 Bryngelsson, H., Stjerndahl, M., Gustafsson, T., and Edström, K. (2007). How dynamic is the SEI? *Journal of Power Sources* 174 (2): 970–975.

53 Arai, J., Okada, Y., Sugiyama, T. et al. (2015). *In situ* solid state ^7Li NMR observations of lithium metal deposition during overcharge in lithium ion batteries. *Journal of the Electrochemical Society* 162 (6): A952.

54 Lee, H., Ren, X., Niu, C. et al. (2017). Suppressing lithium dendrite growth by metallic coating on a separator. *Advanced Functional Materials* 27 (45): 1704391.

55 Beuse, T., Fingerle, M., Wagner, C. et al. (2021). Comprehensive insights into the porosity of lithium-ion battery electrodes: a comparative study on positive electrodes based on $LiNi_{0.6}Mn_{0.2}Co_{0.2}O_2$ (NMC622). *Batteries* 7 (4): 70.

56 Langklotz, U., Schneider, M., and Michaelis, A. (2012). Water uptake of tape-cast cathodes for lithium ion batteries. *Journal of Ceramic Science and Technology* 4 (2): 69–76.

57 Kim, S.J., Lee, J.J., and Ku, C.H. (2009). Cap assembly for preventing gasket sag, and cylindrical secondary battery having the same. EU Patent 2475026B1, issued December 31, 2009. Available online: Google Patents (accessed February 17, 2024).

58 Song, Y., Wang, L., Sheng, L. et al. (2023). The significance of mitigating crosstalk in lithium-ion batteries: a review. *Energy & Environmental Science* 16 (5): 1943–1963.

59 Zhan, C., Wu, T., Lu, J., and Amine, K. (2018). Dissolution, migration, and deposition of transition metal ions in Li-ion batteries exemplified by Mn-based cathodes–a critical review. *Energy & Environmental Science* 11 (2): 243–257.

60 Li, W., Kim, U.-H., Dolocan, A. et al. (2017). Formation and inhibition of metallic lithium microstructures in lithium batteries driven by chemical crossover. *ACS Nano* 11 (6): 5853–5863.

61 Yim, T., Woo, S.-G., Lim, S.H. et al. (2015). 5V-class high-voltage batteries with over-lithiated oxide and a multi-functional additive. *Journal of Materials Chemistry A* 3 (11): 6157–6167.

62 Asare, E., Evans, J., Newton, M. et al. (2016). Effect of particle size and shape on positive temperature coefficient (PTC) of conductive polymer composites (CPC)—a model study. *Materials & Design* 97: 459–463.

63 Pesaran, A.A., Kim, G.-H., Smith, K., and Darcy, E. (2008). Designing safe lithium-ion battery packs using thermal abuse models. National Renewable Energy Laboratory.

64 Kong, L., Hu, X., Gui, G. et al. (2020). Computed tomography analysis of Li-ion battery case ruptures. *Fire Technology* 56: 2565–2578.

65 Yao, X.-Y., Kong, L., and Pecht, M.G. (2020). Reliability of cylindrical li-ion battery safety vents. *IEEE Access* 8: 101859–101866.

66 Xu, B., Lee, J., Kwon, D. et al. (2021). Mitigation strategies for Li-ion battery thermal runaway: a review. *Renewable and Sustainable Energy Reviews* 150: 111437.

67 Ferreira, S.R., Hargather, M., and Mier, F.A. (2017). Determining the internal pressure in 18650 format lithium ion batteries under thermal abuse. No. SAND2017-10799C. Albuquerque, NM (United States): Sandia National Lab. (SNL-NM).

7

Thermal Runaway and Mitigation Strategies

Simin Peng, Yue Shen, Genkai Xia, Sahithi Maddipatla, Lingxi Kong, Weiping Diao, and Michael G. Pecht

Lithium-ion batteries, while popular for their high energy density and long cycle life, are not without safety issues. One significant concern is thermal runaway, a process where the battery overheats uncontrollably, potentially leading to fires or explosions. As a result, lithium-ion batteries are designed with multiple safety features to prevent and mitigate potential hazards caused by overheating, overcharging, short-circuiting, etc. These safety mechanisms can be inherent to the battery's materials and design or based on protections added to the battery system.

7.1 Thermal Runaway in Lithium-Ion Batteries

Thermal runaway in lithium-ion batteries is a series of chain reactions characterized by an uncontrollable rise in temperature [1]. Heat is inevitably generated due to the electrochemical reactions within a battery [2]. Under normal conditions, this heat can be dissipated, keeping the battery temperature under control. However, heat generation increases significantly during abnormal operations, such as overcharging, external heating, or physical damage like nail penetration and crush. The natural heat dissipation mechanisms become insufficient, leading to a rapid temperature rise.[1]

Components within lithium-ion batteries are susceptible to decomposition when temperatures approach 130 °C. For instance, the solid-electrolyte interphase (SEI) on the anode begins to decompose at around 80 °C [3], the electrolyte starts breaking down at 100–120 °C [4], and polyethylene (PE) and polypropylene (PP) separators collapse at about 135 °C [5]. Electrolyte breakdown releases flammable gases like H_2, CH_4, C_2H_6, and C_2H_4 [6]. If the separator collapses, internal short circuits may occur, releasing stored electrical energy, heat, and gas, thus exacerbating the temperature increase and promoting further chain reactions until thermal runaway ensues.

Lithium-ion batteries are prone to thermal runaway under conditions of rising temperatures, overcurrent loads, excessive internal pressures, and overvoltage. The impacts of thermal runaway vary depending on the cell format; cylindrical and prismatic cells might rupture or explode, while pouch cells are more likely to swell or catch fire. When cells are configured in series or

1 This paragraph and subsequent sections consist of excerpts and figures that have been reprinted from Xu et al. [1], with permission from Elsevier.

parallel within battery packs, thermal runaway in a single cell can propagate to adjacent cells, potentially causing catastrophic failures. Various safety strategies are employed at both the cell and package levels to address this challenge. At the cell level, specialized safety features, like positive temperature coefficient (PTC) elements, current interrupt devices (CIDs), safety vents, and protection circuits, mitigate hazards. Supplementary safety features include shutdown separators, metal-polymer current collectors, electrolyte additives, and safer electrolytes. At the package level, thermal runaway safety measures are categorized into passive and preventive protections, such as ventilation structures, fire-retardant materials, and battery management systems (BMSs) [1].

7.2 Safety Mechanisms and Mitigation Strategies in Lithium-Ion Batteries

There are various mechanisms to ensure the safety of lithium-ion batteries. Among these are CIDs, PTCs, venting mechanisms, flame retardants, and protection circuits. However, these safety mechanisms are not present in all lithium-ion batteries. Their inclusion often comes with trade-offs, such as increased manufacturing costs and potential reductions in battery performance due to added weight or reduced energy density. Furthermore, implementing these safety features can vary significantly depending on the cell format. For instance, cylindrical cells might incorporate different safety mechanisms than pouch or prismatic cells. Several mitigation strategies have been developed to enhance the safety and reliability of lithium-ion battery packs. The primary approaches include BMS optimization and battery thermal management/cooling systems.

7.2.1 Current Interrupt Devices (CID)

A current interrupt device (CID) functions as a safety mechanism designed to mitigate risks associated with excessive internal pressure or temperature. It breaks the electrical connection within a battery when pressure or temperature surpasses a specific threshold. CIDs are generally categorized into two types based on their response mechanism: pressure-responsive CIDs and temperature-responsive CIDs [1].

The pressure-responsive CID structure for cylindrical cells is depicted in Figure 7.1. According to Figure 7.1a, the central point forms the only electrical link between the metallic foil and the top disk, while the second plastic insert insulates other areas. When the internal pressure in the battery escalates to approximately 145 psi [7], the top disk deforms outward, forcing the central point upward and breaking the connection between the metallic foil and the top disk. This disconnect halts the rise in internal pressure by stopping the current flow. The CID may not operate correctly if the weld strength between the metallic foil and the top disk is too strong or too weak. In cylindrical cells, the CID is a one-time protective device that permanently disables the battery once activated. With a single cell's voltage less than 5 V, a CID can sever the electrical current to prevent thermal runaway under overcharge or external short-circuit conditions.

Exposure to elevated temperatures can potentially lead to internal short circuits in batteries containing components with lower decomposition thresholds. Thermal fuses and shape memory alloys are designed to disrupt current flow under high temperatures, categorizing them as temperature-responsive CIDs. A thermal fuse, typically constructed from materials such as Sn, Bi, In, or Pb, operates within a temperature range of 85–120 °C, ideally around 95 °C ± 5 °C. Integrated

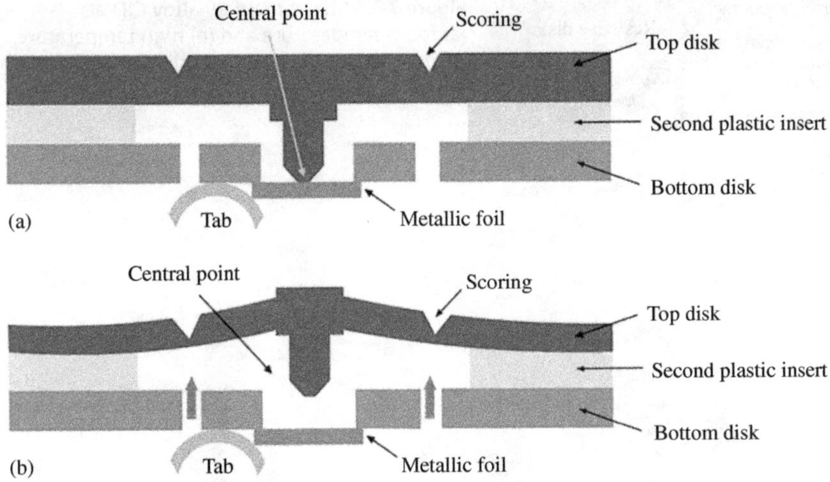

Figure 7.1 A pressure-responsive current interrupt device (CID) structure in a cylindrical cell. Source: Xu et al. [1]/with permission of Elsevier.

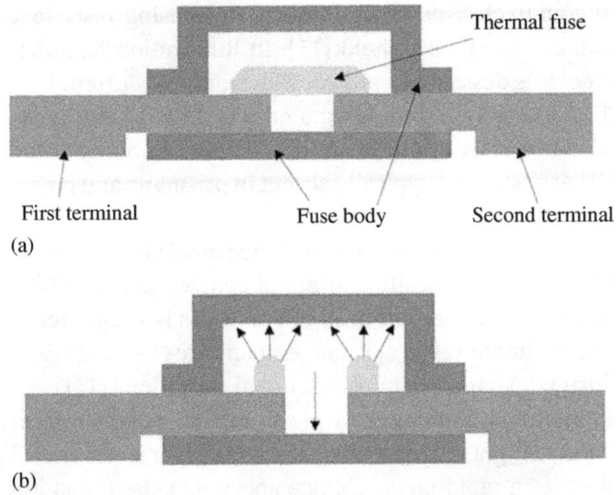

Figure 7.2 A thermal fuse design in a prismatic battery under (a) normal conditions and (b) high temperatures. Source: Xu et al. [1]/with permission of Elsevier.

into batteries, this fuse melts at a predetermined temperature, preemptively breaking the electrical circuit to halt thermal runaway, as shown in Figure 7.2 [8].

Shape memory alloys provide a reversible solution as a temperature- or magnetic-field-driven CID. In the event of temperature spikes from overcharging or short circuits, a disk spring within the shape memory alloy bends backward, effectively cutting off the electrical circuit to prevent further temperature escalation. As the temperature normalizes, the disk spring returns to its original configuration, restoring the electrical connection [9]. This temperature-responsive design modifies the shape of a plate inside the battery as the temperature rises, disrupting the electrical circuit to prevent undesirable internal pressure levels [10]. The plate reverts to its original shape at normal or predetermined temperatures, as illustrated in Figure 7.3.

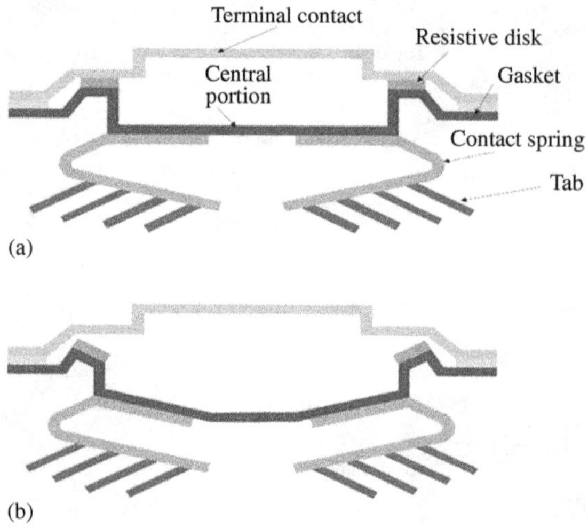

Figure 7.3 Shape memory alloy CID at (a) room temperature and (b) high temperature. Source: Xu et al. [1]/with permission of Elsevier.

7.2.2 Positive Temperature Coefficient (PTC)

A PTC component is defined as having the characteristic of significantly increasing resistance when it reaches a specific temperature and, or current threshold [11]. In lithium-ion batteries, PTC thermistors serve as resettable passive devices designed to protect cells from overcurrent conditions, such as external short circuits. These thermistors are integrated into Li-ion batteries to safeguard individual cells by increasing resistance to rising temperatures. While commonly found in commercial cylindrical cell designs, PTC thermistors are typically absent in prismatic and pouch batteries [12, 13].

During normal operation, current flows from the cathode's current collector through the positive tab, metallic foil, bottom disk, top disk, PTC thermistor, positive terminal contact, and finally to the external load. The PTC material used for circuit protection in Li-ion batteries is a conductive polymer, which exhibits low resistance in the milliohm range at room temperature (1–10 mΩ) and undergoes significant resistance changes during operation to nonoperational conditions [14].

The PTC thermistor is typically a thin annulus composed of a conductive polymer layer sandwiched between two metal annular disks, often incorporating a PE polymer layer mixed with carbon black [15]. This polymer undergoes a rapid phase change above its glass-transition temperature, sharply expanding in volume and increasing its resistance due to the greater distance and reduced conduction pathway between embedded carbon black particles. Upon cooling below its glass-transition temperature, the polymer reverts to a crystalline state, restoring conductivity. When exposed to temperatures above 100 °C from external short circuits, the conductive polymer in the PTC thermistor transitions to a high-resistance state, reducing the battery's current load [16]. Upon returning to operational temperatures, the polymer resumes conductivity, restoring the battery's function. The typical temperature-resistance characteristic of PTC thermistors is illustrated in Figure 7.4.

PTC thermistors are crucial components embedded in the caps of most commercial cylindrical batteries, where they effectively mitigate hazards like external short circuits or overcurrent situations at the single-cell level. However, delays in the PTC's response to internal hot spots within the battery, such as at the SEI layer and electrodes, may occur due to their placement distance [17]. To address this, proposals have been made to integrate PTC components directly into electrodes, significantly increasing electrode resistance when batteries overheat. Feng et al. demonstrated

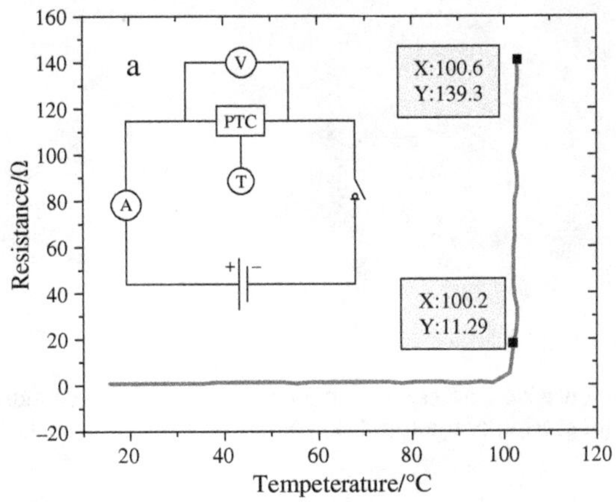

Figure 7.4 The electric circuit and the resistance-temperature curve of a PTC from a commercial cylindrical lithium-ion cell. Source: Xu et al. [1]/with permission of Elsevier.

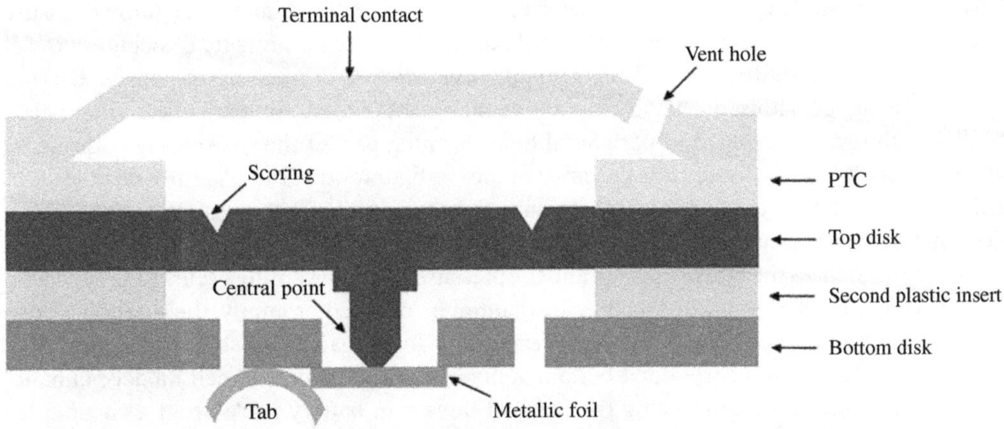

Figure 7.5 Example of the positive temperature coefficient thermistors in a cylindrical cell. Source: Xu et al. [1]/with permission of Elsevier.

this through their novel epoxy-carbon PTC electrode design [18]. Naturally, aside from lithium batteries, PTC technology finds applications in other battery types that need protection against overheating, such as nickel-cadmium batteries and nickel-metal hydride batteries. Figure 7.5 shows the structure of PTC thermistors in a cylindrical cell [1].

7.2.3 Venting Mechanisms

Gas buildup during thermal runaway can elevate a battery's internal pressure, potentially causing the battery case to rupture if the pressure is not adequately released. Safety vents are integral to alleviating internal pressure by venting out generated gases. These vents are strategically designed weak points on the battery casing to expel gases safely. According to the International Electrotechnical Commission (IEC) 62133 standard, "venting" is defined as the controlled release of excessive internal pressure from a cell or battery to prevent rupture or explosion [19]. The Institute of Electrical and Electronics Engineers (IEEE) 1725 standard specifies that cells must incorporate a reliable vent design or mechanism, such as foil, edge, seam, or score, to minimize projectile risks and retain cell contents to ensure safety [11].

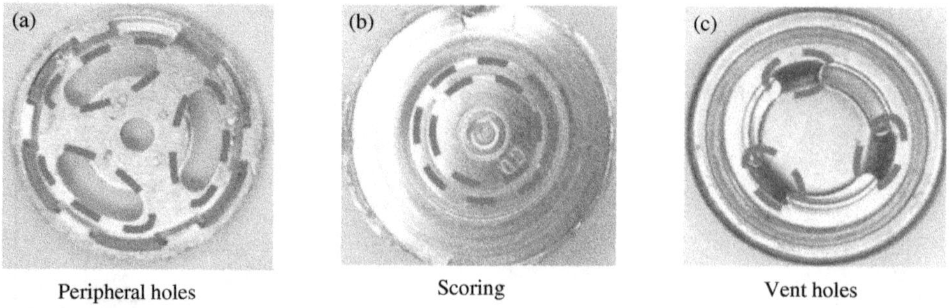

Figure 7.6 Safety vent in a commercial 18650 cylindrical battery: (a) peripheral holes of the bottom disk, (b) scoring of the top disk, and (c) vent holes. Source: Xu et al. [1]/with permission of Elsevier.

Safety vents are crucial for preventing complete battery rupture when internal pressures reach critical levels, such as 113–128 psi (0.85 MPa) for prismatic cells [20] and 406 psi (2.8 MPa) for 18650 cylindrical cells [21]. Various preset scorings or notches in the different cell formats ensure that safety vents open at specified pressure thresholds [1]. The trigger pressure of safety vents, set at 406 psi for 18650 cylindrical cells [21], generally exceeds that of pressure-responsive CIDs at 145 psi [7]. Positioned within the battery cap, the vent consists of three components: a bottom disk, a top disk with preset scoring, and peripheral holes forming part of the gas venting pathway, as shown in Figure 7.6. Each battery design varies in the configuration of these peripheral holes.

In the event of sudden excessive internal pressure, the prescored top disk ruptures, allowing gases or electrolytes to vent through these holes, thereby preventing a fire or explosion [22]. The size of the vent opening influences the rate of gas flow and the pressure at which venting occurs [23]. However, if a vent becomes blocked and internal pressure cannot be released promptly, the escalating pressure may rupture the battery case or lead to an explosion. To mitigate this risk, manufacturers like LG Chem and Sony have incorporated bottom vents in certain cylindrical cell models, enhancing venting efficiency and minimizing the thermal impact in battery packs from a single cell's rupture [1].

7.2.4 Flame Retardants

Flame retardants are additives used to mitigate or halt the combustion of materials, often incorporated into components like electrolytes and diaphragms to enhance battery safety. They impede the spread of fire during abnormal conditions, such as overcharging, overdischarging, or external short circuits. Flame retardants act through mechanisms like inhibiting chemical reactions, thermal decomposition isolation, heat absorption, and inerting.

The electrolyte in Li-ion batteries consists of a nonaqueous organic solvent solution containing inorganic lithium salt and additives [4]. This solvent, typically a blend of cyclic and acyclic carbonate solvents, is chosen for its stability at the battery's operating voltage, generally above 3 V. However, acyclic solvents like diethyl carbonate (DEC), dimethyl carbonate (DMC), and ethyl methyl carbonate (EMC) are flammable due to their low flash points—15–33 °C [22]. During thermal runaway, the electrolyte can emit flammable gases, exacerbating the risk of battery fires as the temperature rises.

Efforts to enhance the thermal stability of electrolytes include adding flame retardants (FRs) that surpass the flash points of these acyclic solvents. FRs like triethyl phosphate (TEP), trimethyl phosphate (TMP), and hexamethylcyclophosphate (HMPN) have been studied. However, their introduction can impact battery performance due to high viscosity or electrode instability [24, 25]. Researchers have synthesized alternatives like tris(2,2,2-trifluoromethyl) phosphate (TFP), bis(2,2,2-trifluoromethyl) methyl phosphate (BMP), and (2,2,2-trifluoromethyl) diethyl phosphate (TDP), which maintain electrolyte conductivity and improve thermal stability without significantly affecting battery cycling performance [25]. Effective FR concentrations typically remain below 20%, with optimal values reported for specific FRs such as 10 wt% BMP-PF6, cresyl diphenyl phosphate (CDP), and 4-isopropyl phenyl diphenyl phosphate (IPPP) [26, 27].

An alternative approach involves embedding FRs within the battery separator, as demonstrated by Liu et al., who developed a thermal-triggered, flame-retardant, nonwoven separator. This separator features a core-shell structure with a triphenyl phosphate (TPP) core and a poly(vinylidene fluoride-hexafluoropropylene) (PVDF-HFP) shell. PVDF-HFP ensures FR containment and compatibility with electrolytes, releasing TPP at around 160 °C during early combustion stages without compromising battery electrochemical reactions [28]. This design effectively enhances flame-retardant capabilities while preserving battery performance integrity.

Although flame retardants can help reduce the risk of fires, the efficacy of flame retardants hinges significantly on factors such as type, dosage, and compatibility with other battery components, underscoring the importance of careful selection and formulation in battery design and production. Additionally, for many of the common flame retardants, companies have generally been concerned with how the flame retardants poorly affect capacity and performance.

7.2.5 Shutdown Separators

The separator typically comprises polymers, although paper and gels have also been used. Positioned between the positive and negative electrodes, the separator prevents direct electrode contact and facilitates the movement of lithium ions essential for battery operation. The separator is designed from insulating materials and has a porous structure that absorbs liquid electrolyte [4].

Single-sheet polymer separators contribute to battery safety by undergoing a process known as separator shutdown when exposed to temperatures near their melting points. The melting point is typically around 160 °C for polypropylene (PP) and 140 °C for polyethylene (PE) separators [29]. During the shutdown, the separator's pores close, temporarily halting ion flow and electrochemical reactions, thereby increasing battery impedance. However, maintaining separator integrity is crucial; if the separator shrinks or fails, direct electrode contact can occur, leading to internal short circuits.

Triple-layered polymer separators are designed with PE sandwiched between two PP layers, offering a wider temperature margin for shutdown functionality. The PE middle layer functions as the primary shutdown mechanism by melting at around 130 °C, effectively stopping ion flow to prevent overheating. With a higher melting point of about 165 °C, the PP outer layers provide mechanical strength and structural integrity, resisting shrinkage and tearing under mechanical stress or thermal fluctuations. This design ensures the separator can operate safely over a wider temperature range than single-layer separators, offering controlled shutdown and reducing the risk of thermal runaway, fires, or explosions. Alternatively, ceramic materials can be added to polymer separators to create ceramic/polymer composites that withstand higher temperatures and exhibit greater thermal stability than pure polymer separators [30].

However, the shutdown function can provide little protection under rapid localized overheating, overcharging, or internal short circuits. Although the shutdown separator has been implemented in cylindrical cells, the batteries went into thermal runaway under overcharge conditions [31]. The possible reason was the inhomogeneity of the temperature distribution, causing an incomplete separator shutdown [31]. Other researchers pointed out that the shutdown feature would fail if rapid localized overheating caused separator shrinkage or even melting [32, 33]. Thus, the separator shutdown function can provide a limited degree of protection for a single battery or battery package consisting of a small number of batteries [1].

7.2.6 Metal-Polymer Current Collectors

Current collectors serve as substrates for the active material coating and as conductive paths for electrons out of a cell. Conventionally, the anode uses copper (Cu), and the cathode uses aluminum (Al) as material for the current collector. The design of the current collector's thickness involves a trade-off between increasing the energy density and maintaining structural integrity during manufacturing and charge–discharge cycling [34, 35]. Under mechanical impact (e.g., automobile accidents), internal short circuits that connect the anode and the cathode can happen, and short-circuiting the two current collectors leads to the most severe consequences [34].

To reduce the risk of thermal runaway, the metallic current collectors can be replaced with thermoplastic polymer foils that are metalized on both sides. The metal layers are typically 0.5–1.0 μm [36]. When an internal short-circuit event like nail penetration occurs, the thermoplastic polymer substrate will melt and shrink away from the nail due to the heat generated. The melting causes the active material to lose support and split apart, disrupting the electric conduction path created by the short circuit and reducing the chances of thermal runaway. Using metal-polymer foils also saves the metal material, decreases the weight, and increases the specific energy [36].

7.2.7 Protection Circuitry and Battery Management System

Protection circuitry is designed to safeguard against various operational risks, including overcharge, overdischarge, overcurrent, and overtemperature conditions [37]. IEEE 1725–2011 (IEEE standard for rechargeable batteries for cellular telephones) identifies protection circuitry as essential for mitigating risks such as overvoltage, undervoltage, electrostatic discharge, and excessive current for individual cells and battery packs [11]. Protection circuitry operates based on preset voltage, current, and temperature thresholds, where sensors compare these parameters to predetermined limits. Upon exceeding these thresholds, metal-oxide-semiconductor field-effect transistors (MOSFETs) are activated to disable the connection between the battery and system load or charging circuitry, preventing further risks.

Different types of protection circuits are tailored to specific needs, such as Takahashi et al.'s design for prismatic cells. This design includes a voltage monitor and a fuse-switch unit to detect and disconnect circuits when voltage exceeds safe levels, as shown in Figure 7.7 [38].

Overdischarge poses risks like reduced capacity and internal short circuits. To counter this, protection circuits often employ devices like Zener diodes that act as switches to disconnect the battery load when voltage drops to critical levels, as depicted in Figure 7.8. When the battery voltage rises above the discharge cut-off level, the Zener diode reconnects the circuit, ensuring battery protection against overdischarge incidents.

Protection circuitry provides hardware-based safety mechanisms where the voltage/current/temperature thresholds are not programable and serve as the last layer of protection at the battery level. On top of the protection circuitry, battery management algorithms are designed to optimize battery performance and safety. BMS optimization is essential for ensuring the safe and efficient operation of lithium-ion battery packs. An optimized BMS employs various strategies, primarily focusing on temperature monitoring and alert systems, energy management and balancing, and

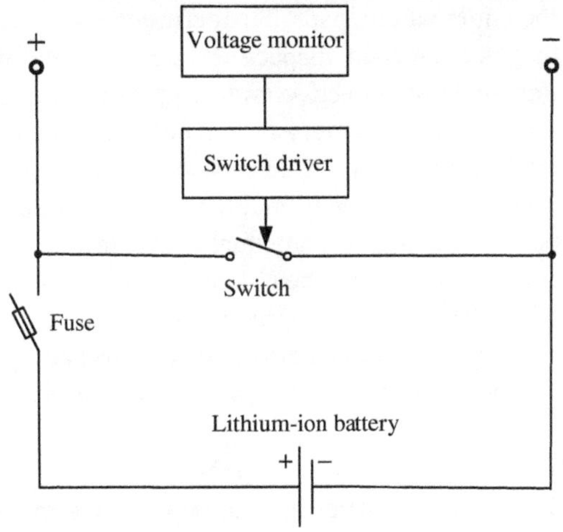

Figure 7.7 Overcharge protection circuit in Li-ion batteries. Source: Xu et al. [1]/with permission of Elsevier.

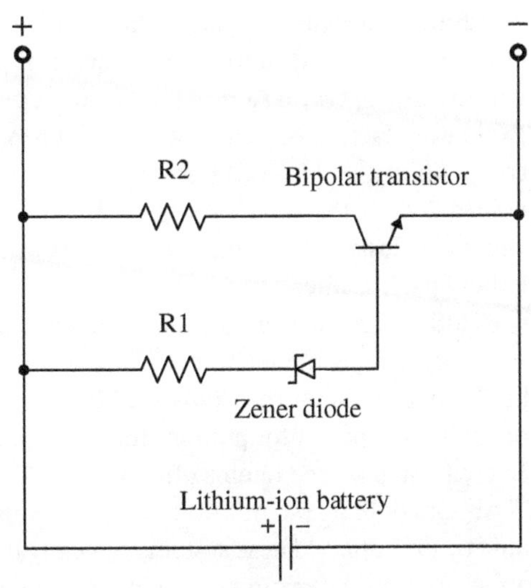

Figure 7.8 Overdischarge protection circuit in Li-ion batteries. Source: Xu et al. [1]/with permission of Elsevier.

fault diagnosis and maintenance. By continuously monitoring key parameters and applying proactive management techniques, an optimized BMS enhances battery performance and significantly reduces the risks associated with overheating and cell failure.

Temperature monitoring and warning systems are essential for BMS optimization, particularly in preventing thermal runaway in lithium-ion battery packs. However, many BMS configurations do not measure the temperature of each cell, relying instead on sensors positioned across the battery pack. When temperatures exceed predetermined safety thresholds, the BMS promptly issues alerts and takes appropriate actions, such as reducing the charging rate, halting discharge, or activating cooling systems. This early warning mechanism enables real-time detection of potential thermal runaway risks, allowing for proactive measures that significantly reduce the likelihood of fires or explosions. Additionally, the BMS can log temperature changes for later analysis, aiding in fault diagnosis [39].

The BMS is designed to detect abnormal conditions in real time, such as short circuits, overvoltage, overheating, and internal faults. When an issue is detected, the BMS automatically isolates the faulty cell to prevent the problem from escalating and compromising the safety of the entire battery pack. Furthermore, the BMS provides periodic maintenance recommendations to guide users in inspecting and maintaining the battery, ensuring it operates optimally. By implementing effective fault diagnosis and maintenance strategies, the BMS not only enhances the safety of the battery system but also prolongs its overall lifespan, thereby significantly reducing the likelihood of thermal runaway [39].

Due to variations in manufacturing and operating conditions, individual cells within a battery pack may experience unequal charge levels, leading to some cells being overcharged or deeply discharged, increasing the risk of thermal runaway. Energy balancing techniques dynamically adjust cells' charge and discharge rates to achieve uniform energy distribution. This approach not only enhances the overall performance of the battery pack but also extends its lifespan, effectively reducing the risk of overheating and failure due to cell imbalances.

7.2.8 Battery Thermal Management Systems

Battery thermal management systems (BTMS) are crucial for ensuring the safety and performance of lithium-ion batteries, particularly in high-power or high-energy-density applications such as electric vehicles and energy storage systems. An effective cooling system helps maintain the battery within a safe operating temperature range, preventing overheating and the risk of thermal runaway. Evaluating factors beyond thermal performance, including scalability, practicality, energy usage, compatibility, and overall system cost, is crucial. It is generally advised that the BTMS should not exceed 20% of the total battery pack weight [40]. Additionally, thermal management structures should be designed to avoid interference with other components, such as wiring, bus bars, and the battery pack casing.

BTMS designs can be classified in several ways, such as passive versus active, parallel versus series, or by specific techniques like heating or cooling, using liquid, air, thermoelectric coolers (TEC), phase change materials (PCMs), or heat pipes (HP) [40]. Active systems, which typically rely on additional power for pumps and fans, are commonly applied in liquid- and air-cooling methods. In contrast, passive systems often utilize PCMs or HPs to dissipate heat from the battery surface.

An air-cooling system lowers battery temperature by circulating air to dissipate the heat generated by the battery. These systems typically include fans, heat sinks, and ducting. The fan circulates air around the battery, removing hot air and introducing cooler air for effective cooling. Heat sinks enhance the contact area between the air and the battery, improving cooling efficiency. The advantages of an air-cooled system include simplicity, low cost, and ease of maintenance. However, its limitations include relatively weak heat dissipation capability and significant dependence on ambient temperature [41].

A liquid-cooling system effectively lowers battery temperature by circulating a coolant that carries away heat generated by the battery. These systems typically consist of a coolant, a pump, a radiator, and piping. The pump drives the coolant to flow, allowing it to absorb heat from the battery and dissipate it to the external environment through the radiator. Liquid-cooling systems offer high cooling efficiency and precise temperature control, effectively preventing battery overheating and enhancing safety and service life [42].

A PCM cooling system manages battery temperatures by absorbing and storing heat through a material that undergoes a phase transition, usually from solid to liquid. As the battery generates heat, the PCM absorbs it, melting in the process and thereby regulating temperature without requiring additional energy input. Key components include the PCM and housing structures designed to contain the material and maximize surface contact with the battery. PCM systems offer passive, energy-efficient temperature control with enhanced uniformity. Still, they can be limited by their heat storage capacity and the need for cooling cycles to reset the PCM to its solid state [40].

On the other hand, a heat pipe (HP) cooling system uses a closed-loop pipe containing a working fluid to transfer heat away from the battery rapidly. As the battery heats up, the working fluid in the HP vaporizes at the heat source, moves along the pipe, and then condenses at a cooler section where heat is released. This process recycles as the condensed fluid returns to the heat source through capillary action or gravity, creating an efficient, continuous heat transfer cycle. While heat pipes have excellent thermal conductivity, they are usually combined with liquid cooling or forced air to help release heat more effectively at the condensation end [40].

7.3 Safety Mechanisms Used in Cells with Different Form Factors

Lithium-ion batteries are available in various form factors, each designed to meet specific needs in size, shape, capacity, and performance. The main form factors include cylindrical, prismatic, pouch, and coin cells, each incorporating different safety features to suit their applications.

Coin cells, or button cells, are small, round batteries typically used in low-power devices such as watches, hearing aids, and small medical devices. Their compact size and shape make them ideal for limited-space applications. Despite their small size, some commercial coin cells incorporate several safety features, such as pressure relief valves and PTC devices, to manage overcurrent conditions and prevent short-circuiting.

Cylindrical cells are the most common and widely recognized form factor, often seen in devices like power tools, electric vehicles, and consumer electronics. They are typically denoted by numbers indicating their dimensions, such as 18650 or 21700. Cylindrical cells are favored for their high mechanical stability and standardized sizes compared to other form factors, facilitating easy integration into various applications. Their design often includes PTCs, CIDs, and venting mechanisms to safely release built-up gases and prevent pressure buildup that could lead to explosions.

Prismatic cells are rectangular and often encased in hard or semirigid packaging, making them suitable for automotive applications and large-scale energy storage systems. These cells offer higher energy density per unit volume than cylindrical cells, allowing for more efficient use of space. Their rectangular shape also provides flexibility in design integration. They often include CIDs and safety vents.

Pouch cells use a flexible, laminated foil casing, allowing for various shapes and sizes. This form factor is widely used in smartphones, tablets, portable devices, and electric vehicles. Pouch cells' lightweight and flexible nature maximizes energy density and offers high design flexibility for custom shapes and sizes. However, due to their less robust casing, pouch cells require additional protective measures, such as safety coatings and encapsulation. They often employ nonflammable electrolytes and protective layers to enhance safety and prevent combustion. Table 7.1 comprehensively compares the various lithium-ion battery form factors—cylindrical, prismatic, pouch, and coin cells.

Table 7.1 Comparison of safety mechanisms in different Li-ion battery form factors.

Safety mechanism	Coin	Cylindrical	Prismatic	Pouch
PTC	Present in some commercial cells	Commonly present	Not present	Not present
CID	Present in some commercial cells	Commonly present	Commonly present	Not present
Safety vents	Not present	Commonly present	Commonly present	Not present

7.4 Summary

The number of battery safety incidents is expected to rise due to the ongoing expansion of the global Li-ion battery market. To address this challenge, safety strategies increasingly focus on preventing hazards before thermal runaway occurs. Protection mechanisms inherent in Li-ion batteries include PTC thermistors, CIDs, safety vents, protection circuitry, shutdown separators, metal-polymer current collectors, electrolyte flame-retardant additives, and safe electrolyte chemistries.

Most major battery manufacturers, such as Panasonic, Ningde Times, and Samsung, prioritize safety by incorporating a full array of safety mechanisms like CIDs, PTCs, exhaust ports, and flame retardants into their products. However, some companies do not. For example, LG Chem does not use a PTC in most of its cylindrical cells. Several factors may be at play for companies that do not integrate safety features into their cells, including cost, weight, and capacity loss (e.g., a PTC will slightly reduce the cell capacity). A company that uses such batteries must consider the trade-offs before making a decision, and cost should not be the only criterion if there is potential for safety risks.

In addition to cell-level safety features, battery pack-level strategies for mitigating thermal runaway include BMS optimization and battery thermal management. Effective active cooling methods, such as gas and liquid, and passive cooling methods using PCMs and HPs help dissipate heat to maintain safe operating temperatures and prevent the cells from reaching critical thresholds. Meanwhile, BMS optimization enables precise monitoring and control of cell voltages, currents, and temperatures across the pack, ensuring balanced operation and early fault detection. Together, these strategies enhance battery safety by actively managing thermal risks and improving the overall reliability of Li-ion battery systems.

References

1 Xu, B., Lee, J., Kwon, D. et al. (2021). Mitigation strategies for Li-ion battery thermal runaway: a review. *Renewable and Sustainable Energy Reviews* 150: 111437.

2 Choudhari, V.G., Dhoble, A.S., and Sathe, T.M. (2020). A review on effect of heat generation and various thermal management systems for lithium ion battery used for electric vehicle. *Journal of Energy Storage* 32: 101729.

3 Spotnitz, R. and Franklin, J. (2003). Abuse behavior of high-power, lithium-ion cells. *Journal of Power Sources* 113 (1): 81–100.

4 Kong, L., Li, C., Jiang, J. et al. (2018). Li-ion battery fire hazards and safety strategies. *Energies* 11 (9): 2191.

5 Feng, X., Ouyang, M., Liu, X. et al. (2018). Thermal runaway mechanism of lithium ion battery for electric vehicles: a review. *Energy Storage Materials* 10: 246–267.

6 Spinner, N.S., Field, C.R., Hammond, M.H. et al. (2015). Physical and chemical analysis of lithium-ion battery cell-to-cell failure events inside custom fire chamber. *Journal of Power Sources* 279: 713–721.

7 Ould Ely, T., Kamzabek, D., and Chakraborty, D. (2019). Batteries safety: recent progress and current challenges. *Frontiers in Energy Research* 7: 71.

8 Pechini, M.P. (1967). Washington, DC: US patent and trademark office. US Patent, vol. 3, pp. 697, 1967.

9 Jani, J.M., Leary, M., Subic, A. et al. (2014). A review of shape memory alloy research, applications and opportunities. *Materials & Design (1980–2015)* 56: 1078–1113.

10 Toshizou, K. and Toshiharu, H. (1995). Safety device for use in secondary battery. U.S. Pat No. US5766793A, 1995.

11 Standard for rechargeable batteries for cellular telephones - redline. pp. 1725–2011, 2011.

12 Stephens, D., Shawcross, P., and Stout, G. et al. Lithium-ion battery safety issues for electric and plug-in hybrid vehicles. Washington, DC, USA: National Highway Traffic Safety Administration, vol. 2017, pp. 261.

13 Jeevarajan, J. (2014). Safety of commercial lithium-ion cells and batteries. In: *Lithium-Ion Batteries*, 387–407. Elsevier, vol. 2014.

14 Kim, J.I., Kang, P.H., and Nho, Y.C. (2004). Positive temperature coefficient behavior of polymer composites having a high melting temperature. *Journal of Applied Polymer Science* 92 (1): 394–401.

15 Asare, E., Evans, J., Newton, M. et al. (2016). Effect of particle size and shape on positive temperature coefficient (PTC) of conductive polymer composites (CPC)—a model study. *Materials & Design* 97: 459–463.

16 Manzo, M.A., Brewer, J.C., and Bugga, R.V. et al. (2010). NASA Aerospace Flight Battery Program: Generic Safety, Handling and Qualification Guidelines for Lithium-Ion (Li-Ion) Batteries; Availability of Source Materials for Lithium-Ion (Li-Ion) Batteries; Maintaining Technical Communications Related to Aerospace Batteries (NASA Aerospace Battery Workshop).

17 Zhong, H., Kong, C., Zhan, H. et al. (2012). Safe positive temperature coefficient composite cathode for lithium ion battery. *Journal of Power Sources* 216: 273–280.

18 Feng, X.M., Ai, X.P., and Yang, H.X. (2004). A positive-temperature-coefficient electrode with thermal cut-off mechanism for use in rechargeable lithium batteries. *Electrochemistry Communications* 6 (10): 1021–1024.

19 Arterburn, D., Ewing, M., and Prabhu, R. et al. (2017). FAA UAS Center of Excellence task A4: UAS ground collision severity evaluation.

20 Kim, J.H., Lee, K.H., Ko, D.C. et al. (2017). Design of integrated safety vent in prismatic lithium-ion battery. *Journal of Mechanical Science and Technology* 31: 2505–2511.

21 Young, K.H., Koch, J.M., Wan, C. et al. (2017). Cell performance comparison between C14-and C15-predominated AB2 metal hydride alloys. *Batteries* 3 (4): 29.

22 Kong, L., Hu, X., Gui, G. et al. (2020). Computed tomography analysis of li-ion battery case ruptures. *Fire Technology* 56: 2565–2578.

23 Ferreira, S.R., Hargather, M., and Mier, F.A. (2017). Determining the internal pressure in 18650 format lithium ion batteries under thermal abuse. Albuquerque, NM, United States: Sandia National Lab. (SNL-NM).

24 Xu, K., Zhang, S., Allen, J.L. et al. (2002). Nonflammable electrolytes for Li-ion batteries based on a fluorinated phosphate. *Journal of the Electrochemical Society* 149 (8): A1079.

25 Xu, K., Ding, M.S., Zhang, S. et al. (2003). Evaluation of fluorinated alkyl phosphates as flame retardants in electrolytes for Li-ion batteries: I. Physical and electrochemical properties. *Journal of the Electrochemical Society* 150 (2): A161.

26 Choi, J.A., Sun, Y.K., Shim, E.G. et al. (2011). Effect of 1-butyl-1-methylpyrrolidinium hexafluorophosphate as a flame-retarding additive on the cycling performance and thermal properties of lithium-ion batteries. *Electrochimica Acta* 56 (27): 10179–10184.

27 Wang, Q., Ping, P., Sun, J. et al. (2010). Improved thermal stability of lithium ion battery by using cresyl diphenyl phosphate as an electrolyte additive. *Journal of Power Sources* 195 (21): 7457–7461.

28 Liu, K., Liu, W., Qiu, Y. et al. (2017). Electrospun core-shell microfiber separator with thermal-triggered flame-retardant properties for lithium-ion batteries. *Science Advances* 3 (1): e1601978.

29 Venugopal, G., Moore, J., Howard, J. et al. (1999). Characterization of microporous separators for lithium-ion batteries. *Journal of Power Sources* 77 (1): 34–41.

30 Orendorff, C.J. (2012). The role of separators in lithium-ion cell safety. *The Electrochemical Society Interface* 21 (2): 61.

31 Roth, E.P., Doughty, D.H., and Pile, D.L. (2007). Effects of separator breakdown on abuse response of 18650 Li-ion cells. *Journal of Power Sources* 174 (2): 579–583.

32 Arora, P. and Zhang, Z. (2004). Battery separators. *Chemical Reviews* 104 (10): 4419–4462.

33 (2012). *Batteries for Sustainability: Selected Entries from the Encyclopedia of Sustainability Science and Technology*. Springer Science & Business Media.

34 Pham, M.T., Darst, J.J., Walker, W.Q. et al. (2021). Prevention of lithium-ion battery thermal runaway using polymer-substrate current collectors. *Cell Reports Physical Science* 2 (3).

35 Diao, W., Xu, B., and Pecht, M. (2021). Charging induced electrode layer fracturing of 18650 lithium-ion batteries. *Journal of Power Sources* 484: 229260.

36 Gruhn, H., Krüger, T., Mund, M. et al. (2023). Challenges in contacting metal–polymer current collectors in pouch cells. *Journal of Manufacturing and Materials Processing* 7 (6): 219.

37 Wang, Y., Han, Z., and Shi, J. (2014). One cell Lithium-ion battery protection IC. *2014 9th IEEE Conference on Industrial Electronics and Applications*, IEEE, vol. 2014, pp. 1427–1432.

38 Takahashi, K., Fujimura, H., Takahashi, K., and Yanai, Y. (2011). U.S. Patent Application. vol. 12/857, pp. 951, 2011.

39 Ou, Y., Zhou, P., Hou, W. et al. (2024). Smart materials for safe lithium-ion batteries against thermal runaway. *Journal of Energy Chemistry* 94: 360–392.

40 Olabi, A.G., Maghrabie, H.M., Adhari, O.H.K. et al. (2022). Battery thermal management systems: recent progress and challenges. *International Journal of Thermofluids* 15: 100171.

41 Dhuchakallaya, I. and Saechan, P. (2024). Numerical study of a pressurized gas cooling system to suppress thermal runaway propagation in high-energy-density lithium-ion battery packs. *Journal of Energy Storage* 101: 113916.

42 Wu, C., Sun, Y., Tang, H. et al. (2024). A review on the liquid cooling thermal management system of lithium-ion batteries. *Applied Energy* 375: 124173.

8

Battery Qualification

Rashed A. Islam

Battery qualification is a process that involves tests and assessments conducted on battery cells, packs, and entire products to verify the batteries' performance, reliability, and safety adherence to predefined standards and specifications. Passing battery qualification standards is crucial for ensuring safety, yet it does not guarantee complete safety. While qualification tests assess whether the battery design meets specific criteria and standards, they are conducted on a limited number of samples and may not represent all manufacturing variability or defects. Qualification, therefore, evaluates the design but does not ensure safety in all scenarios. Additionally, qualification tests often overlook specific application conditions, which may lead to issues or premature failures in real-world use. Developing more comprehensive, application-specific qualification criteria is essential to ensure batteries meet performance and safety requirements.

Battery qualification is vital in addressing known failure modes and mechanisms and identifying additional failure modes and mechanisms. By subjecting batteries to electrical, thermal, mechanical, and abuse testing, manufacturers can assess how batteries perform under different operating conditions and identify failure modes and mechanisms that may occur during real-world use [1]. Such a proactive approach allows manufacturers to implement design improvements and safety features to mitigate or prevent these failure modes, thus enhancing the overall reliability and safety of batteries. Additionally, battery qualification provides valuable data and insights that enable continuous improvement in battery technology, leading to advancements in performance, durability, and safety.

8.1 Key Performance Metrics

Battery failure modes can arise due to various factors, and understanding these failure modes and the failure mechanisms behind them is crucial for improving battery design, safety, and reliability. Early identification of critical failure modes during the development process allows for proactive measures to address them effectively.

The predominant failure modes identified in the literature include thermal runaway, internal short circuits, overcharging and overdischarging, capacity fade, dendritic formation, electrolytic breakdown, voltage instability, and the presence of foreign materials such as impurities and contaminants [2–5]. These modes of failure can be caused by numerous failure mechanisms.

The Safety Challenges and Strategies of Using Lithium-Ion Batteries, First Edition. Edited by Michael G. Pecht.

Capacity, efficiency, cycle life, and voltage stability are some of the key battery metrics that collectively provide a valuable assessment of battery performance and are crucial for selecting, designing, and optimizing battery systems for a wide range of applications. Each metric is described in more detail throughout this section.

8.1.1 Capacity

Battery capacity is typically measured in terms of ampere-hours (Ah) or milliampere-hours (mAh) and is represented by this formula:

$$Q = xnF$$

where Q relates to the quantity of electric charge, x represents the number of moles of reaction, n signifies the number of electrons transferred per mole of reaction, and F denotes the Faraday constant. However, battery capacity is commonly expressed in terms of mass rather than the number of moles. Therefore, a modified formula can be used as follows:

$$Q = \frac{nF}{M_r}$$

where, M_r denotes the molecular mass. This yields the capacity in units of ampere-hours per gram (Ah/g).

Several methodologies exist to determine the rated capacity, which denotes the maximum charge storage and delivery capability under specific conditions, and manufacturers typically derive this rating through standardized testing procedures. One prevalent approach, coulomb counting, involves tracking the total charge entering or leaving the battery over time, obtained by integrating the current flow. Discharge testing offers another approach, wherein the battery undergoes full charging before discharge at a constant current, until reaching a predetermined voltage cutoff, with the total delivered charge indicating the capacity [6]. Alternatively, voltage versus state of charge (SoC) curves present graphical depictions of the battery voltage across various charge states, facilitating capacity estimation through curve analysis. Additionally, impedance spectroscopy has emerged as a method to assess battery health by measuring the impedance changes at different frequencies, reflecting capacity variations.

Peukert's law describes the interrelation among discharge current, capacity, and discharge time. By adhering to specified discharge currents and employing Peukert's law, analysts can calculate the battery capacity, assess adherence to performance specifications, and optimize battery utilization in various applications. Peukert's law is as follows:

$$t = H\left(\frac{C^K}{(IH)^K}\right)$$

where t is the actual discharge time in hours, H is the rated discharge time in hours, I is the actual discharge current in amperes, C is the rated capacity in ampere-hours, and K is the dimensionless Peukert constant (usually a value from 0.95 to 1.1 [7]).

8.1.2 Efficiency

Battery efficiency is typically measured by comparing the amount of energy stored during charging to the amount of energy released during discharging. The efficiency of a battery system can be calculated using the following formula:

$$\text{Efficiency} = \frac{\text{Energy Discharged}}{\text{Energy Charged}} \times 100\%$$

One factor that contributes to losses during charging and discharging operations is the coulombic efficiency, which quantifies the ratio of charge removed during discharge to charge added during charging. This also considers losses stemming from internal resistance, side reactions, and other inefficiencies inherent to the battery.

Voltage efficiency is equally critical, accounting for losses attributed to voltage drops and internal circuitry losses experienced during both charging and discharging cycles. Additionally, energy losses manifesting as heat, or other forms of wasted energy, result from resistance within the battery, self-discharge phenomena, and inefficiencies inherent in the charging process. Also, heat dissipation is a notable consideration, as the heat generated during charging and discharging activities signifies energy losses within the system, thereby warranting inclusion in efficiency calculations. By comprehensively assessing these efficiency factors, battery analysts can gain insights into the overall performance and effectiveness of battery systems, facilitating informed decision-making and optimization strategies for diverse applications.

Efficiency measurements may also consider factors such as the temperature of the battery, the rate of charging and discharging, and the depth of discharge. Notably, the efficiency of a battery system may also vary depending on its state of charge and operating conditions.

8.1.3 Battery Cycle Life

Battery cycle life assessment involves subjecting the battery to repeated charge and discharge cycles under controlled conditions until it reaches the end of its operational life. These cycling tests are crucial during both manufacturing and reliability assessments to comprehensively evaluate performance degradation.

The process includes several key steps. First, the initial capacity of the battery is measured to establish a baseline for comparison. Next, a specific cycling protocol is defined, including parameters like depth of discharge (DoD), charging rate, discharging rate, and temperature, tailored to the battery chemistry and application. The battery is then subjected to repeated charge and discharge cycles according to the defined protocol. Periodic capacity measurements are taken during cycling to monitor degradation, often involving discharging the battery to a specified cutoff voltage and then measuring the delivered charge.

Finally, end-of-life criteria are established, based on factors like percentage decrease in capacity or number of cycles. These criteria vary across applications, with consumer electronics typically accepting capacity degradation to around 80% and reduced cycle life, while medical devices may require more stringent criteria within the range of 10–20%. Similarly, electric vehicles and grid energy storage systems have their own specific criteria for capacity degradation, influencing their ability to store and discharge energy effectively. Such meticulous testing ensures batteries meet performance requirements across diverse applications.

Cycle life estimation is a crucial aspect of battery evaluation, providing insights into how long a battery can sustain its performance under repeated charge-discharge cycles before reaching the end of its useful life. This estimation is based on the observed degradation rate and defined end-of-life criteria. For consumer electronics like smartphones and laptops, lithium-ion batteries typically endure approximately 300–500 charge–discharge cycles before their performance begins to degrade significantly. In contrast, electric vehicles (EVs) and hybrid electric vehicles (HEVs) require batteries engineered for higher cycle life, ranging from 1000 to 2000 cycles, depending on battery chemistry and vehicle usage patterns. Grid energy storage systems demand even greater durability due to frequent charge–discharge cycles, with batteries designed to withstand anywhere from 2000 to 8000 cycles or more, depending on the application and chemistry. With renewable energy systems such as solar or wind energy storage, batteries must support intermittent energy generation and are designed for extensive cycling, ranging from 3000 to 10,000 cycles or more. Medical device batteries, catering to critical healthcare applications, vary widely in cycle life, with designs ranging from 500 to 2000 cycles or more, tailored to the device's specific requirements and intended use in healthcare settings.

8.1.4 Voltage Stability

Voltage stability in batteries is evaluated through various methods aimed at ensuring consistent and reliable voltage output under diverse conditions. Open circuit voltage (OCV) measurement, conducted when the battery is not under load, provides insights into its stability during idle periods. Conversely, voltage under load assessments involves applying a known load to monitor stability during operation. Transient voltage response tests evaluate how quickly the battery stabilizes after abrupt load changes, while voltage regulation assessments gauge its ability to maintain stable voltage output despite varying loads. Temperature dependence evaluations involve thermal cycling tests to observe voltage stability across temperature ranges. Long-term stability testing involves continuous operation monitoring over extended periods to assess real-world performance. Dynamic load testing simulates usage scenarios to evaluate voltage stability under varying load conditions. Collectively, these methods provide a comprehensive understanding of a battery's voltage stability and help identify potential performance issues.

Safe limits for parameters such as capacity, voltage, internal resistance, cycle life, temperature, and safety features ensure optimal battery performance and reliability, with typical safe limits varying based on battery type (lithium ion) and design (pouch or cylindrical).

- Capacity: maintaining at least 80% of the initial rated capacity after a certain number of cycles
- Voltage: operating within a specified range, such as 3.0–4.2 V for lithium-ion batteries
- Internal resistance: remaining within a certain range to ensure efficient energy transfer.
- Cycle life: achieving a predetermined number of cycles without significant capacity loss.
- Temperature: operating within a specified temperature range, often between −20 °C and 60 °C for consumer batteries
- Safety: meeting industry standards for safety features and passing relevant safety tests to prevent hazards like thermal runaway or explosion

8.2 Battery Qualification Process

The battery qualification process is a comprehensive effort that navigates through several critical stages to ensure the reliability, safety, and performance of batteries across various applications. It begins with design validation, where rigorous attention is paid to ensuring that the battery design aligns with specified requirements [8]. This involves considerations such as design capacity, form factor, and the integration of safety features to meet industry standards. Performance testing follows, where the battery's capabilities are rigorously evaluated against predefined metrics. Parameters such as capacity, efficiency, cycle life, and voltage stability are scrutinized to assess the battery's overall performance and suitability for its intended application. Safety assessments form another crucial aspect of the qualification process. Through a series of tests, potential safety risks such as thermal runaway, short circuits, or overcharging are identified and addressed to ensure the battery's safe operation [9].

The battery validation test proceeds through three distinct phases. First, at the cell level, individual cell testing is conducted to assess electrical, thermal, and mechanical characteristics. Safety assessments, including abuse testing like overcharge and thermal stress tests, are performed to gauge a cell's resilience under adverse conditions. When qualifying cell-level batteries, a thorough assessment of battery performance parameters is vital. Capacity evaluation involves measuring energy storage accuracy, consistency across cycles, and adherence to manufacturer specifications. Efficiency assessment covers energy conversion during charging and discharging, considering losses and operating conditions. Voltage stability monitoring ensures consistent output, even under varying loads or rapid charging. Evaluating cycle life involves determining degradation over charge-discharge cycles and simulating real-world usage conditions. Safety considerations like thermal stability and overcharge protection are also crucial. With all of that said, comprehensive testing ensures batteries meet performance, reliability, and safety standards for their intended applications.

Moving to the pack level, cells are assembled into a cohesive battery pack. Here, comprehensive performance testing evaluates an entire pack's charge-discharge cycles, voltage stability, output voltage range, charge–discharge current, operation temperature, and thermal management capabilities. Safety assessments at this level include measures for overcharge and overdischarge protection to ensure a pack's safe operation [9]. When qualifying pack-level batteries, thorough evaluation is essential to ensure performance, reliability, and safety. Key aspects include verifying the battery capacity and uniformity across cells, measuring energy efficiency during charging and discharging, monitoring voltage stability under various conditions, assessing battery cycle life through accelerated aging tests, and inspecting mechanical integrity for protection against external factors. Additionally, considering safety features like thermal management and overcurrent protection is crucial. In sum, comprehensive testing and analysis of these factors ensure that battery packs meet performance standards, reliability requirements, and safety regulations for their intended applications.

At the product level, a battery pack is integrated into the final product. End-to-end testing is conducted to simulate real-world application scenarios, ensuring that a product operates seamlessly in practical environments. System-level performance testing and safety assessments provide valuable insights into a product's overall functionality and safety. At the product level, integrating a battery pack marks a crucial step, demanding thorough end-to-end testing to mimic real-world scenarios and guarantee seamless operation. Capacity is scrutinized to ensure it meets or exceeds energy demands, optimizing the battery pack's ability to store and deliver electrical energy efficiently. Efficiency evaluations focus on energy conversion during charging and discharging, striving for minimal losses to extend operational times and enhance overall efficiency. Monitoring voltage stability across different loads and discharge cycles is essential to maintain consistent performance and safeguard the product's electronics from voltage fluctuations. Assessing cycle life involves determining the number of charge-discharge cycles before significant capacity degradation occurs, typically through accelerated aging tests, providing insights into the battery pack's durability and lifespan. Moreover, safety assessments encompass system-level testing, evaluating thermal management, overcurrent protection, and measures against thermal runaway. These evaluations ensure the product meets regulatory standards and user expectations, delivering on functionality, reliability, and safety.

Throughout each stage, meticulous analysis and documentation of results are conducted to provide a comprehensive understanding of a battery's performance, safety, and suitability for its intended application. This multifaceted approach to battery qualification ensures that batteries meet stringent quality standards and perform reliably in diverse real-world scenarios (Figure 8.1).

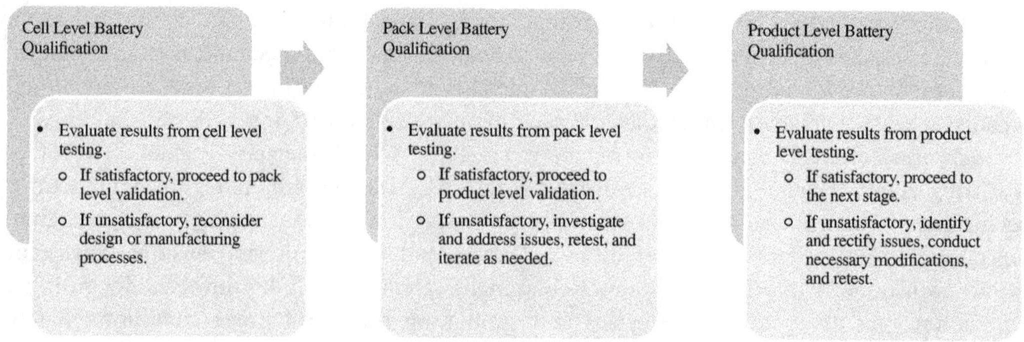

Figure 8.1 End-to-end battery qualification process.

8.3 Battery Qualification Testing Protocols

Testing protocols play a pivotal role in the battery qualification process, serving to validate the reliability and performance of batteries under various conditions. These protocols encompass a range of tests across different domains [10].

Electrical tests involve assessing parameters like voltage, current, and resistance to ensure the electrical integrity of the battery. This includes evaluating a battery's ability to deliver power consistently and efficiently throughout its life. Electrical tests assess a battery's ability to store and deliver electrical energy efficiently. These tests help to identify any irregularities or inefficiencies in energy storage and conversion processes, directly impacting capacity and efficiency. Electrical tests also evaluate voltage output under various load conditions, ensuring stability. Further, identifying voltage fluctuations or inconsistencies helps to maintain stable voltage levels, contributing to voltage stability. Battery protection tests evaluate the effectiveness of built-in protection mechanisms such as overcharge protection, overdischarge protection, and short-circuit protection [11, 12].

Safety tests encompass various aspects, including electrical, thermal, and mechanical safety. Certifying compliance with safety standards and regulations mitigates risks of performance degradation, ensuring consistent capacity, efficiency, voltage stability, and battery cycle life.

Battery performance tests focus on assessing a battery's performance under different operating conditions, including charge-discharge cycles, energy efficiency, and capacity retention. Performance tests evaluate a battery's ability to meet specified performance criteria under different operating conditions. Assessing performance parameters such as charge–discharge rates and energy delivery helps to validate capacity and efficiency ratings. Performance tests provide insights into a battery's voltage response under dynamic load conditions, contributing to voltage stability assessments.

Thermal tests are conducted to evaluate a battery's thermal management capabilities, heat dissipation efficiency, and resistance to thermal runaway. These tests simulate extreme temperature conditions to assess a battery's stability and safety under adverse thermal scenarios. High temperatures can degrade battery efficiency and shorten cycle life due to increased internal resistance and accelerated chemical reactions. Thermal tests also evaluate the battery's response to overheating scenarios, helping to ensure safety by identifying any potential risks of thermal runaway or fire hazards.

Mechanical tests verify the structural integrity of a battery and its components, ensuring resistance to vibrations, shocks, and other mechanical stresses. They assess the physical integrity of a battery, including the casing, interconnections, and terminals. Importantly, ensuring mechanical robustness prevents damage that could compromise safety or lead to performance degradation. This includes tests for vibration resistance, shock resistance, and impact resistance, among others. Battery abuse testing, a subset of mechanical tests, specifically focuses on subjecting batteries to extreme conditions, such as impact, crushing, and puncture, to assess their resistance to physical damage.

Battery environmental testing evaluates a battery's response to environmental factors like humidity, salt exposure, liquid ingress, dust ingress, altitude vibration, and shock, which can also impact mechanical integrity. To ensure clarity and distinction between test categories, it is essential to make them mutually exclusive, emphasizing the unique aspects of each test type while maintaining comprehensive coverage of mechanical evaluations [9]. These tests simulate the range of environments and potential abusive conditions that batteries may encounter during their lifecycle. Charge-discharge cycling tests assess a battery's performance under normal operating conditions, while storage tests evaluate its stability and integrity during storage. Together, these tests help to identify any potential issues related to performance degradation, self-discharge, or storage-induced damage, assessing a battery's reliability and safety over time. Environmental tests expose a battery to various conditions that simulate real-world usage environments and help assess its durability, directly impacting cycle life. It also helps to evaluate a battery's response to environmental stressors, ensuring it can withstand adverse conditions without posing safety risks.

8.3.1 Cell-Level Qualification

In the meticulous design process of lithium-ion battery cells, several critical factors must be carefully considered and addressed to ensure optimal performance, safety, and efficiency. Beginning with chemistry selection, the choice of battery chemistry must align precisely with an application's requirements, considering energy density, power density, cycle life, and safety aspects (Figure 8.2).

Form factor optimization follows, where the physical dimensions of a cell are tailored to fit seamlessly into the intended application, considering size, shape, and weight. Determining the voltage and capacity requirements of a cell is crucial, ensuring it meets the power demands of the application while aligning with overall system specifications.

Thermal management mechanisms, such as internal cooling structures and conductive materials, are then integrated to regulate temperature effectively. Safety features are meticulously incorporated at the cell level, including thermal cutoffs and pressure relief vents, safeguarding against potential hazards like overcharging and short circuits. Additionally, achieving the desired cycle life entails careful electrode material selection and separator design to enhance long-term durability. Energy efficiency is optimized by minimizing internal resistance, thus reducing energy losses during charge and discharge cycles.

Materials selection prioritizes conductivity, stability, and compatibility to enhance performance and safety. Reliable manufacturing processes are implemented to ensure consistency and quality in cell production, controlling variables such as the electrode coating thickness and electrolyte filling. Electrode and separator designs are optimized for efficient charge and discharge processes, considering factors like porosity and material composition. Effective electrolyte management strategies prevent dendrite formation and improve ion conductivity [13].

Failure mode and effect analysis (FMEA) is conducted to identify potential failure modes and implement preventive measures [14]. Environmental impact is carefully considered to optimize recyclability and minimize a battery's environmental footprint. Finally, robust testing

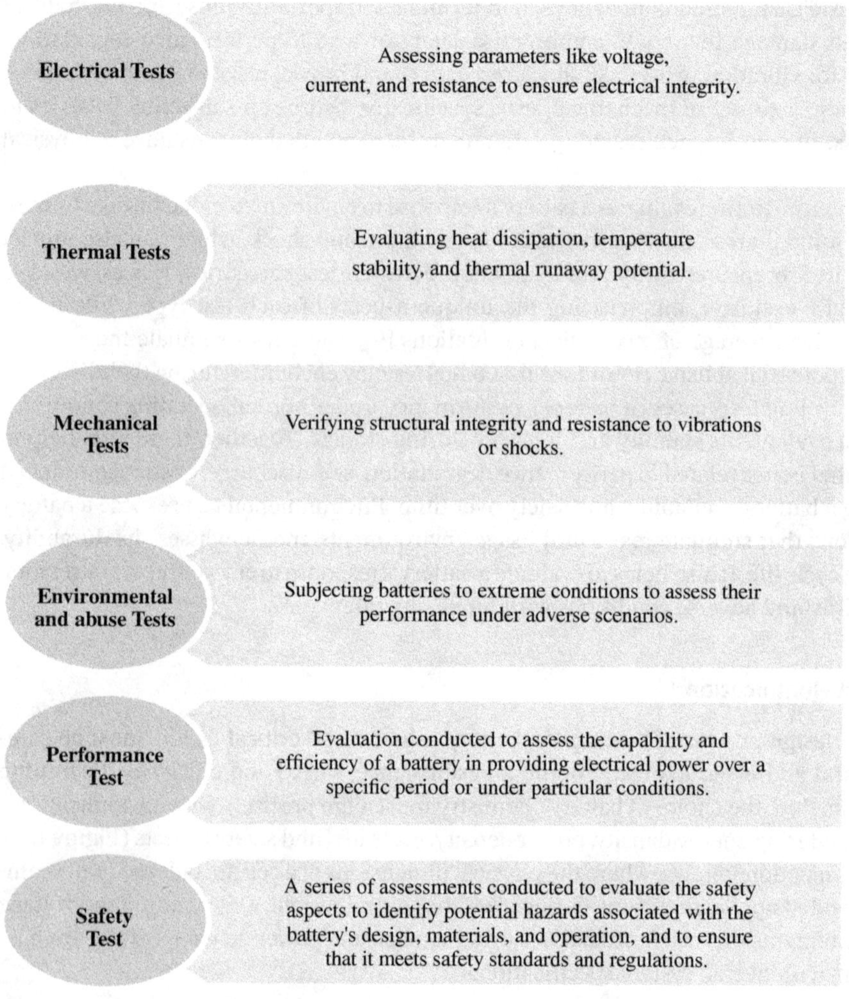

Figure 8.2 Battery qualification test protocol.

and validation protocols are implemented to ensure cells meet specified standards through comprehensive electrical characterization, safety testing, and environmental assessments. Through attention to these key design guidelines, lithium-ion batteries can be designed to meet the stringent demands of various applications while prioritizing safety, reliability, and sustainability [15].

By adhering to cell-level battery design guidance points, industries can design or choose cells that are optimized for specific applications, ensuring a balance between performance, safety, and longevity. When conducting cell-level testing, the general test specifications in Table 8.1 should be adhered to; however, specifications may also vary depending on the specific applications and requirements [16–20].

Table 8.1 Cell-level battery test guidelines.

Test name	Test condition	Sample size
	Electrical test	
Battery voltage and internal resistance test	Measure voltage and impedance between positive and negative terminals in an open circuit while the FET is in the "on" state. Measure ACIR (Alternating Current Internal Resistance) and DCIR (Direct Current Internal Resistance). The test results are considered valid regardless of the specific section in which they are obtained. Passing criteria: ACIR and DCIR may involve comparing the measured resistance values against predetermined thresholds or limits. These thresholds can be based on factors such as battery chemistry, design specifications, intended application, or performance standards.	5–10
External short-circuit test	Conducting an external short-circuit test involves subjecting a rated charged cell to ambient temperatures of 25 °C and 55 °C in separate tests. In each scenario, a test sample is short-circuited by connecting positive and negative terminals using a wire with resistance between 60 and 100 mΩ for a duration of 1 hour. Passing criteria: No smoke, no fire, and no explosion	5–10
	Mechanical test	
Safety-of-free-fall test	Charge the battery to full capacity at an ambient temperature of 20 °C ± 5 °C. Following this, subject the battery to six random drops from a height of 1.0–1.5 m onto a hard (concrete/granite/quartz) floor at the same temperature. Passing criteria: No smoke, no fire, and no explosion	5–10
Crush test	The battery crush test involves applying a force of 10–15 kN to a rated charged cell, focusing on crushing the cell longitudinally. Passing criteria: No smoke, no fire, and no explosion	5–10
Impact test	This test requires placing a 13–15 mm diameter bar across the center of the sample and dropping an 8–10 kg weight from a height of 750–1000 mm onto the sample, with the impact side being longitudinal. Passing criteria: No smoke, no fire, and no explosion	5–10
Vibration test	Charge the battery pack to full capacity. Subject the battery pack to vibrations ranging from 7 to 18 Hz with a constant peak acceleration of 1 g. Apply vibrations with a constant amplitude of 0.8 mm and a peak acceleration increasing from 1 g to 8 g. Further, subject the battery pack to vibrations up to 500 Hz with a constant peak acceleration of 8 g. Repeat the entire vibration sequence for 15 minutes per set, completing a total of 12 sets over 3 hours. Vibrate the battery pack in three directions: X, Y, and Z. After completion of the test, allow the sample to rest for 2 hours before inspection. Passing criteria: No temperature rise, smoke, fire, or explosion observed during the test. The battery's appearance exhibits no abnormalities or signs of damage. Mass loss does not exceed 0.1% of the original mass. The open circuit voltage of the battery remains above 90% of its initial value throughout the test.	5–10

(Continued)

Table 8.1 (Continued)

Test name	Test condition	Sample size
	Environmental test	
High-temperature storage test	The fully charged battery undergoes a 168-hour storage period at 75 °C in a high-temperature chamber, followed by a 2-hour storage period at room temperature. Passing criteria: The battery's appearance should show no abnormalities, such as peeling labels, uneven clearances between the label and the case, waving or warping of the case, or any shaking noise. These indicators serve as visual cues for structural integrity and overall quality. Second, the battery must meet specified electrical characteristics, ensuring that it operates within acceptable parameters. By fulfilling both appearance and electrical criteria, the battery demonstrates its reliability and suitability for use.	5–10
Low-temperature storage test	The fully charged battery undergoes a 168-hour storage period at −40 °C in a low-temperature chamber, followed by a 2-hour storage period at room temperature. Passing criteria: The battery's appearance should show no abnormalities, such as peeling labels, uneven clearances between the label and the case, waving or warping of the case, or any shaking noise. These indicators serve as visual cues for structural integrity and overall quality. Second, the battery must meet specified electrical characteristics, ensuring that it operates within acceptable parameters. By fulfilling both appearance and electrical criteria, the battery demonstrates its reliability and suitability for use.	5–10
High-temperature and high-humidity storage test	After a full charge, the battery is stored in a thermostatic chamber at 60 °C and 90% humidity, or 85 °C and 85%, for 168 hours, followed by a 2-hour storage period at 25 °C and 50% humidity. Passing criteria: The battery's appearance should show no abnormalities, such as peeling labels, uneven clearances between the label and the case, waving or warping of the case, or any shaking noise. These indicators serve as visual cues for structural integrity and overall quality. Second, the battery must meet specified electrical characteristics, ensuring that it operates within acceptable parameters. By fulfilling both appearance and electrical criteria, the battery demonstrates its reliability and suitability for use.	5–10
Thermal shock test	Following a full charge, the battery undergoes 100–500 cycles (depending on the application) of alternating exposure: −40 °C for 30 minutes, followed by 75 °C for 30 minutes. The cell is then removed during the final −40 °C conditions to confirm dewfall. Subsequently, the battery is stored at room temperature for 2 hours. Passing criteria: The battery's appearance should show no abnormalities, such as peeling labels, uneven clearances between the label and the case, waving or warping of the case, or any shaking noise. These indicators serve as visual cues for structural integrity and overall quality. Second, the battery must meet specified electrical characteristics, ensuring that it operates within acceptable parameters. By fulfilling both appearance and electrical criteria, the battery demonstrates its reliability and suitability for use.	5–10

Table 8.1 (Continued)

Test name	Test condition	Sample size
Temperature humidity cycling test	The test involves 14 cycles of high temperature/high humidity (HT/HH) at 60 °C and 90% ± 2% relative humidity, or 85 °C and 85% ± 2% relative humidity, coupled with a low-temperature (LT) phase at −20 °C or −40 °C (uncertain relative humidity. Each cycle comprises specific temperature and humidity transitions over a 12-hour period, including ramps, dwells, and temperature variations. HT/HH and LT specifications will depend on the application. Passing criteria: The battery's appearance should show no abnormalities, such as peeling labels, uneven clearances between the label and the case, waving or warping of the case, or any shaking noise. These indicators serve as visual cues for structural integrity and overall quality. Second, the battery must meet specified electrical characteristics, ensuring that it operates within acceptable parameters. By fulfilling both appearance and electrical criteria, the battery demonstrates its reliability and suitability for use.	5–10
Altitude test	The test involves subjecting the system to altitude conditions of 20,000 ft., with a −20 °C temperature for a 16-hour dwell period and a ramp rate of 1000 ft./min. Subsequently, the system experiences another cycle at 20,000 ft., with a 35 °C temperature for a 16-hour dwell period, again with a 1000 ft./min ramp rate. Repeat these two steps for a total of 10 cycles. Passing criteria: The battery's appearance should show no abnormalities, such as peeling labels, uneven clearances between the label and the case, waving or warping of the case, or any shaking noise. These indicators serve as visual cues for structural integrity and overall quality. Second, the battery must meet specified electrical characteristics, ensuring that it operates within acceptable parameters. By fulfilling both appearance and electrical criteria, the battery demonstrates its reliability and suitability for use.	5–10
	Battery cell performance test	
Battery cycling test	Conduct 300–5000 (depending on the application) charge–discharge cycles on a battery, tracking cycle count versus battery performance. Maintain temperatures at both the highest and lowest operational levels of the product. If the discharge rate is excessively low, conduct the test at both pack and cell levels as needed. Passing criteria: The battery shows compliance with battery specifications postassessment. Additionally, the battery's appearance and structure must exhibit no anomalies, including peeled labels, irregular clearances between the label and the case, waving or warping of the case, or any audible shaking noise. Furthermore, the battery must demonstrate satisfactory electrical characteristics. Fulfilling these requirements ensures both the physical integrity and electrical performance of the battery after testing.	5–10
Battery deep discharge test	Discharge the battery entirely, then subject it to 30–50 cycles of charge–discharge after high temperature and high humidity (60 °C and 90% RH for 168 hours) and 30–50 cycles after low-temperature storage (−40 °C and uncontrolled humidity for 168 hours) to assess compliance. Passing criteria: The battery shows compliance with battery specifications postassessment. Additionally, the battery's appearance and structure must exhibit no anomalies, including peeled labels, irregular clearances between the label and the case, waving or warping of the case, or any audible shaking noise. Furthermore, the battery must demonstrate satisfactory electrical characteristics. Fulfilling these requirements ensures both the physical integrity and electrical performance of the battery after testing.	5–10

(Continued)

Table 8.1 (Continued)

Test name	Test condition	Sample size
	Safety test	
Overcurrent protection test	This test focuses on overcurrent protection with or without the thermal cutoff (TCO). Connect the battery cell to the current source and configure it to deliver a controlled current exceeding the cell's rated maximum continuous discharge current. Ensure that the current source can deliver the desired current level accurately and safely. Gradually increase the current flowing through the battery cell until it reaches or exceeds the maximum continuous discharge current specified by the manufacturer. Monitor the cell's voltage, temperature, and any other relevant parameters using the data logging device. Passing criteria: The battery must adhere to the specified current value and delay time for overcurrent protection operation as outlined in the requirement specification. The protective mechanisms should function as intended, providing effective safeguards. Additionally, there should be no signs of overheating, voltage drop, smoke emission, ignition, or damage to any components during the operation of these protections.	5–10
Overcharge test	After battery charging, conduct a 24-hour test under specified conditions. The test is conducted at room temperature, with the current set at twice the recommended maximum consecutive charge current (mA). For voltage, it depends on the recommended charge voltage: • If the recommended charge voltage is 18 V or less, the minimum voltage is lower than twice the maximum charge voltage of the battery pack or 22 V. • If the recommended charge voltage exceeds 18 V, the minimum voltage is 1.2 times the maximum charge voltage. Passing criteria: The requirement stipulates that batteries must not exhibit any instances of fire, explosion, or smoke. This criterion is met if the batteries remain intact and do not catch fire for a period of seven days following the test.	5–10
Overcharge protection test	The test aims to validate overcharge protection for cells. Cells are set at 4.0 V, and the voltage is slowly increased to confirm the charge field effect transistor (FET) turns off. Additionally, the reset overcharge protection is tested by decreasing cell voltage to verify the charge FET turns on. The entire process is repeated 100 times for thorough assessment. Passing criteria: The overvoltage protection functions and resets according to the specified parameters for both the set voltage and delay time, ensuring normal operation as per the specifications.	5–10
Overdischarge protection test	The test aims to verify the overdischarge protection for cells. All cell voltages are set to 3 V, and the total voltage is slowly decreased to confirm the discharging FET turns off. To test the reset condition, cell voltages are slowly increased until the discharging FET turns on. The entire process is repeated 100 times, with the flexibility to change reset conditions if necessary, ensuring comprehensive evaluation. Passing criteria: The overdischarge protection reliably detects and resets according to the specified conditions outlined in the specifications.	5–10

8.3.2 Pack-Level Qualification

Designing battery packs involves integrating individual cells into a cohesive and efficient energy storage system. In the intricate process of designing a battery pack, several crucial elements must be meticulously considered and integrated to ensure optimal performance, safety, and reliability. Beginning with cell selection, individual cells are chosen based on the application's specific requirements, prioritizing factors such as energy density and cycle life. Thermal management systems are then implemented to regulate temperature effectively within the pack, utilizing methods like cooling plates or liquid cooling to enhance safety.

Mechanical integrity is upheld through robust enclosure designs that withstand mechanical stresses and impacts. Voltage and capacity matching ensures uniformity among cells, optimized through balancing systems. Safety features are integrated, including protection circuits and thermal sensors, to prevent hazardous events. Pack configuration and battery management systems are optimized for efficiency and health monitoring, respectively. Electrical integration minimizes resistances, while serviceability considerations facilitate easy maintenance. Encapsulation and sealing safeguard against environmental factors, while form factor optimization maximizes integration possibilities. Electromagnetic compatibility and reliability testing ensure compliance with standards and durability under various conditions, while regulatory compliance guarantees adherence to safety regulations and industry standards. Attention to these key aspects will help to ensure that the battery packs are engineered to meet the reliability and safety demands of diverse applications. The pack-level battery tests are outlined in Table 8.2 [21, 22].

Table 8.2 Pack-level battery test guideline.

Test item	Test condition	Sample size
	Electrical test	
ESD test	The electrostatic discharge (ESD) test involves subjecting the device to both contact and air discharges at specified voltages. For contact discharge, the voltage is set at ±10 kV, while for air discharge, it is ±15 kV. Each discharge method is applied 20 times, with 10 discharges in each polarity. The test is conducted on every defined location of the battery pack to assess its susceptibility to electrostatic discharge events and ensure adequate protection against ESD-induced damage. Passing Criteria: No smoke, no fire, and no explosion. The pack must meet specified electrical characteristics to ensure proper operation within acceptable parameters.	3–5
Battery voltage and internal resistance test	Measure voltage and impedance between positive and negative terminals in an open circuit while the FET is in the "on" state. Measure the alternating current internal resistance (ACIR) and direct current internal resistance (DCIR). The test results are considered valid regardless of the specific section in which they are obtained. Passing criteria: ACIR and DCIR may involve comparing the measured resistance values against predetermined thresholds or limits. These thresholds can be based on factors such as battery chemistry, design specifications, intended application, or performance standards.	5–10

(Continued)

Table 8.2 (Continued)

Test item	Test condition	Sample size
External short-circuit test	Conducting an external short-circuit test involves subjecting a rated charged cell to ambient temperatures of 25 °C and 55 °C in separate tests. In each scenario, a test sample is short-circuited by connecting positive and negative terminals using a wire with resistance between 60 and 100 mΩ for a duration of 1 hour. Passing criteria: No smoke, no fire, and no explosion. The pack must meet specified electrical characteristics to ensure proper operation within acceptable parameters.	5–10
	Thermal test	
Temperature and humidity (3 corner) test	The testing procedure comprises different levels tailored to specific applications, each characterized by varying temperature conditions. In level 1 testing, temperatures range from 0 °C to 45 °C, with high-temperature (HT) and low-temperature (LT) limits defined accordingly. Level 2 testing elevates the temperature parameters slightly, with HT reaching 50 °C and LT set at −5 °C. For more rigorous assessments, level 3 testing involves temperatures of 60 °C for HT and −10 °C for LT. Each level is conducted over a 24-hour period per corner, with the number of cycles ranging from 5 to 30, depending on the product's anticipated lifespan and application requirements. These tests serve to gauge the performance and resilience of the product under diverse temperature conditions, aiding in ensuring its reliability and suitability for its intended use. Passing criteria: Must meet the capacity requirement to deliver expected energy levels, ensure voltage stability for consistent power output, operate within specified temperature ranges to prevent overheating, demonstrate durability through charge–discharge cycles, integrate safety features like overcharge protection, meet environmental regulations, and pass functional tests for compatibility with intended applications.	3–5
Altitude test	In operational conditions, the battery is subjected to a 17,000 ft. altitude, with temperatures ranging from −0 °C to 45 °C for a dwell time of 16 hours, using a ramp rate of 2000 ft./minute. Passing criteria: Must meet capacity, efficiency, and voltage stability requirements.	3–5
	Mechanical test	
Safety-of-free-fall test	Charge the battery pack to full capacity at an ambient temperature of 20 °C ± 5 °C. Following this, subject the battery to six drops (two drops per orientation) from a height of 1.0–1.5 m onto a hard (concrete/granite/quartz) floor at the same temperature. Passing criteria: No physical damage to the battery cells, casing, or internal components. No deformation, punctures, or ruptures in the battery cells or casing. No thermal runaway, no fire, and no explosion.	3–5
Crush test	The battery pack is subjected to a crushing force to simulate impact or compression. The battery crush test involves applying a force of 50–200 kN to a rated charged cell, focusing on crushing the cell longitudinally. Passing criteria: No electrolyte leakage, no internal short circuits, or no thermal runaway. No damage to internal connections, circuitry, or safety features within the battery pack. No fire or explosion.	3–5

Table 8.2 (Continued)

Test item	Test condition	Sample size
Vibration test	Charge the battery pack to full capacity. Subject the battery pack to vibrations ranging from 7 to 18 Hz with a constant peak acceleration of 1 g. Apply vibrations with a constant amplitude of 0.8 mm and a peak acceleration increasing from 1 g to 8 g. Next, subject the battery pack to vibrations up to 500 Hz with a constant peak acceleration of 8 g. Repeat the entire vibration sequence for 15 minutes per set, completing a total of 12 sets over 3 hours. Vibrate the battery pack in three directions: X, Y, and Z. After completion of the test, allow the sample to rest for 2 hours before inspection. Passing criteria: No temperature rise, smoke, fire, or explosion observed during the test. No fatigue, no cracking, or no loosening of connections. No heat generation or thermal runaway if not adequately dissipated.	3–5
	Environmental test	
Temperature and humidity cycling test	The testing process involves various levels tailored to specific applications, each characterized by distinct temperature and humidity conditions. In level 1 testing, conditions entail a high-temperature/high-humidity (HT/HH) environment at 45 °C and 90% relative humidity (RH), while the low-temperature (LT) environment is set at −10 °C with uncontrolled relative humidity (UNC RH). Level 2 testing escalates the temperature and humidity parameters, with HT/HH reaching 60 °C and LT set at −20 °C, both with UNC RH. Level 3 testing further intensifies conditions, with HT/HH at 75 °C and LT at −40 °C, both at UNC RH. Each testing level involves a 2-hour dwell period followed by a 4-hour transfer time, totaling 12 hours per cycle. The number of cycles varies across levels, with 8, 16, or 42 cycles conducted depending on the product's anticipated lifespan and application requirements. These tests aim to assess the product's performance and durability under diverse temperature and humidity conditions, ensuring its reliability and suitability for its intended use. Passing criteria: No corrosion of battery terminals, connectors, or internal components, which could compromise electrical connections and conductivity. No leakage of electrolyte. Degradation of capacity and performance must be within acceptable limits.	3–5
Thermal cycling test	The testing process entails various levels tailored to specific applications, each characterized by distinct temperature ranges. In level 1 testing, temperatures span from −20 °C (LT) to 45 °C (HT), providing a moderate environment for assessment. Level 2 testing intensifies conditions, with temperatures ranging from −40 °C (LT) to 60 °C (HT), while level 3 further escalates to a wider range of −40 °C (LT) to 85 °C (HT). For the battery management system (BMS) board, testing mirrors level 3 conditions. Each test cycle involves a 30-minute dwell period followed by a 7.5-minute transfer time, totaling 75 minutes per cycle. The number of test cycles varies from 300 to 1000, depending on the product's anticipated lifespan and application requirements. These tests aim to evaluate the product's performance and resilience under different temperature extremes, ensuring its reliability and suitability for various operational environments. Passing criteria: No deformation in the casing and leakage of electrolyte. Degradation of capacity and performance must be within acceptable limits. Causing internal short circuits and potential thermal runaway.	3–5

(Continued)

Table 8.2 (Continued)

Test item	Test condition	Sample size
Low-temperature storage test	The test specification level varies based on the intended application, with each level offering a distinct temperature range and duration. Level 1 testing, suited for indoor devices, subjects the product to temperatures as low as −20 °C for 168 hours. Level 2 testing extends to both indoor and outdoor applications, exposing the device to a more rigorous environment with temperatures dipping to −40 °C for 500 hours. For products intended for outdoor use, level 3 testing further intensifies the conditions, extending the duration to 1000 hours while maintaining the same temperature range of −40 °C. These tests serve to assess the product's durability and performance under varying temperature extremes, ensuring its reliability in diverse operational settings. Passing criteria: No diminished performance, capacity loss, and potential damage due to increased internal resistance, material contraction, and reduced efficiency in cold conditions.	3–5
High-temperature and high-humidity storage test	The test specifications vary according to the application requirements, with different levels offering distinct temperature and humidity conditions. Level 1 testing subjects the product to 45 °C and 90% relative humidity (RH) for a duration of 168 hours. For more stringent evaluations, level 2 testing increases the temperature to 60 °C and maintains the same humidity level for an extended duration of 500 hours. Level 3 testing further intensifies the conditions by raising both temperature and humidity to 85 °C and 85% RH, respectively, also for 500 hours. It is imperative that the testing chamber does not condense during these rigorous evaluations to ensure an accurate and reliable assessment of the product's performance under such environmental stressors. Passing criteria: No corrosion of battery terminals, connectors, or internal components, compromising electrical connections and conductivity. No leakage of electrolyte. Degradation of capacity and performance must be within acceptable limits.	3–5
Salt mist	Per IEC standard 60068-2-52, severity (3): The test procedure involves subjecting the units under test (UUTs) to a series of salt spray periods and subsequent humidity storage cycles. Each salt spray period lasts 2 hours at a temperature of 35 °C, followed by a 22-hour humidity storage period at 40 °C and 90% relative humidity. After four salt spray periods, there is a 72-hour storage period at a temperature of 23 °C with relative humidity maintained between 45% and 55%. The salt mist solution consists of 5% NaCl by weight, with a pH range of 6.5–7.2. Passing criteria: No corrosion of battery terminals, connectors, or internal components, compromising electrical connections and conductivity. No leakage of electrolyte. Degradation of capacity and performance must be within acceptable limits.	3–5
Altitude test	In a nonoperational condition, the battery undergoes testing at 20,000 ft. altitude, with temperatures varying from −20 °C to 60 °C for a dwell time of 16 hours, also employing a ramp rate of 2000 ft./minute. Passing criteria: No deformation in the casing and leakage of electrolyte. Degradation of capacity and performance must be within acceptable limits. Causing internal short circuits and potential thermal runaway.	3–5

Table 8.2 (Continued)

Test item	Test condition	Sample size
Dust test	Per IEC standard 60,529-IP6X: The dust test involves an 8-hour duration, with dust continuously blown into the chamber, followed by a minimum settlement period of 2 hours to ensure complete settling before opening the chamber. During the test, the units undergo a 2-hour orientation period in the four worst-case orientations, which are more prone to dust leakage. The dust used for the test should either be Arizona dust or talcum powder with particles smaller than 75 μm. Approximately 2 kg of dust per cubic meter of the chamber volume is recommended, with the relative humidity maintained below 30% if the chamber has humidity control capabilities. Passing criteria: Must maintain electrical performance, mechanical integrity, safety features, reliability, sealing effectiveness, and functional operation in dusty environments.	3–5
Liquid ingress test	The liquid ingress test, as per IEC standard 60529, categorizes levels of protection against liquid ingress. IPX2 denotes suitability for indoor applications, IPX4 for both indoor and outdoor use, and IPX6 for outdoor applications. IPX7 and above signify complete protection against liquid ingress. The test involves checking the unit's functionality immediately after clearing water and again after a 48-hour period to ensure sustained performance under varying conditions of liquid exposure. Passing criteria: No corrosion of battery terminals, connectors, or internal components, compromising electrical connections and conductivity. No leakage of electrolyte. Degradation of capacity and performance must be within acceptable limits.	3–5
Highly accelerated life test	The test aims to identify vulnerabilities in the design by subjecting the product to various stress conditions. These conditions include assessing the high, low, and nominal voltage margins of the product under maximum load. Additionally, the product undergoes cold step stress, ranging from 20 °C to the lower design limit or −90 °C, and hot step stress, from 30 °C to the upper design limit or 100 °C. Rapid thermal transition testing involves moving the product between the lower and upper test limits. Vibration step stress testing ranges from 5 g to the design limit of 80 g. Finally, a combined test involves a temperature range from the lower to the upper test limit and vibration steps from the design limit (or 80 g). These conditions collectively assess the product's performance under various stressors to uncover potential weaknesses in the design.	3–5
	Battery pack performance test	
Battery cycling test	300–5000 charge–discharge cycles depending on product life and application. Temperature should be held at the highest and lowest operating temperature of the product. If the discharge rate is too low, then perform at the cell level. Passing criteria: Must maintain capacity, voltage stability, internal resistance, cycle life, safety performance, and functional operation.	3–5
Battery deep discharge test	The test process involves several steps to assess battery performance. Initially, the battery is drained completely. Following this, the battery undergoes either heat soak or cold storage for one week at level 2 intensity. Subsequently, the battery is subjected to 30 cycles of charge and discharge after the heat soak or cold storage to evaluate its performance under these conditions. This sequence of steps aims to simulate real-world scenarios and stressors to determine the battery's reliability and durability. Passing criteria: Must maintain capacity, voltage recovery, internal resistance, safety performance, cycle life, and functional operation.	3–5

(Continued)

Table 8.2 (Continued)

Test item	Test condition	Sample size
	Safety test	
Short-circuit test	Connect the digital multimeter and high-current shunt in series with the battery pack to monitor the current flow during the test. Prepare heavy-duty cables with alligator clips to create a short circuit across the terminals of the battery pack. Ensure that the short-circuiting device, if used, can deliver a controlled short circuit and can be activated remotely or safely from a distance. Carefully connect the heavy-duty cables with alligator clips across the positive and negative terminals of the battery pack to create a short circuit. Monitor the current flow through the digital multimeter and high-current shunt to ensure that the short circuit is established. Passing criteria: No abnormal audible noise, visible sparks, smoke emission, or temperature rise. Monitor the current flow and voltage drop across the battery pack to assess its ability to limit the short-circuit current and prevent thermal runaway or other safety hazards.	3–5
Overcurrent protection test	This test focuses on overcurrent protection, differentiated by software and hardware methods. For software-based protection, the current value and delay time are scrutinized using an oscilloscope to ensure they meet specifications. The same procedure applies to hardware-based protection, with the examination repeated 100 times. Notably, due to limitations in checking hardware overcurrent protection in the charge direction for battery packs, execution by the PCB is permitted. Passing criteria: The battery pack must adhere to the specified current value and delay time for overcurrent protection operation as outlined in the requirement specification. The protective mechanisms should function as intended, providing effective safeguards. Additionally, there should be no signs of smoke emission, ignition, or damage to any components during the operation of these protections.	3–5
Overcharge test	After battery charging, conduct a 24-hour test under specified conditions. The test is conducted at room temperature, with the current set at twice the recommended maximum consecutive charge current (mA). For voltage, it depends on the recommended charge voltage: • If the recommended charge voltage is 18 V or less, the minimum voltage is lower than twice the maximum charge voltage of the battery pack or 22 V. • If the recommended charge voltage exceeds 18 V, the minimum voltage is 1.2 times the maximum charge voltage. Passing criteria: The requirement stipulates that batteries must not exhibit any instances of fire, explosion, or smoke. This criterion is met if the batteries remain intact and do not catch fire for a period of seven days following the test.	3–5
Overcharge protection test	The test aims to validate overcharge protection for cells. Cells are set at 4.0 V, and the voltage is slowly increased to confirm the charge field effect transistor (FET) turns to the "off" state. Additionally, the reset overcharge protection is tested by decreasing cell voltage to verify the charge FET turns to the "on" state. The entire process is repeated 100 times for thorough assessment. Passing criteria: The overvoltage protection functions and resets according to the specified parameters for both the set voltage and delay time, ensuring normal operation as per the specifications.	3–5
Over-discharge protection test	The test aims to verify the overdischarge protection for cells. All cell voltages are set to 3 V, and the total voltage is slowly decreased to confirm the discharging FET turns to the "off" state. To test the reset condition, cell voltages are slowly increased until the discharging FET turns to the "on" state. The entire process is repeated 100 times, with the flexibility to change reset conditions if necessary, ensuring comprehensive evaluation. Passing criteria: The overdischarge protection reliably detects and resets according to the specified conditions outlined in the specifications.	3–5

8.3.3 Product-Level Qualification

Designing batteries at the product level requires a comprehensive approach to address various considerations for optimal performance, safety, and reliability. Application-specific design entails tailoring the battery to meet the unique requirements and constraints of the intended application, encompassing factors like power demand and environmental conditions. System integration ensures the smooth incorporation of the battery into the overall product design, accounting for spatial requirements and thermal management.

Safety considerations are paramount, with robust features such as protection circuits and fail-safe mechanisms to mitigate potential hazards. Thermal management systems are designed to regulate battery temperature effectively during operation, enhancing safety and longevity. Mechanical compatibility ensures the battery can withstand environmental stresses, while form factor and packaging optimization ensure seamless integration and aesthetics. Voltage and power requirements are aligned with the product's needs, emphasizing energy efficiency and usability. User interface and experience are prioritized, providing clear feedback and intuitive interaction elements. Serviceability is considered for ease of battery replacement, and regulatory compliance is ensured through adherence to safety standards. A sophisticated battery management system monitors performance, and life cycle analysis assesses sustainability factors.

Product-level battery validation varies significantly depending on the application, due to differences in requirements, operating conditions, safety standards, and performance expectations. Validation for consumer, electric vehicle, medical, aerospace and defense, energy storage, and industrial applications will be discussed in this section.

For consumer electronics, validation of battery performance encompasses critical factors such as energy density, cycle life, charging time, and safety under real-world usage conditions. This process involves a range of tests aimed at ensuring reliability across various consumer electronic devices like smartphones, laptops, and wearables. Operational tests include evaluations under extreme conditions such as high-/low-temperature storage, humidity, and altitude variations, simulating everyday usage scenarios. Storage tests focus on assessing battery performance under varying temperature and humidity conditions, along with thermal cycling and altitude simulations. Battery performance tests involve system-level cycling, deep discharge assessments, and ingress tests including salt spray and IPxx (dust and liquid) tests. Mechanical tests evaluate the battery's ability to withstand impacts, vibrations, pressure, and shocks, ensuring durability and safety in diverse environments. Through these rigorous validation processes, battery reliability and performance are confirmed, instilling confidence in their use across a wide range of consumer electronics [23].

For electric vehicles (EVs), aspects such as energy efficiency, power output, temperature regulation, and fast-charging capabilities are prioritized. Testing procedures encompass evaluations under high-stress conditions, including rapid acceleration and sustained high-power output, along with assessments of vehicle-level durability over thousands of charging cycles. Operational tests simulate extreme environmental conditions such as high-/low-temperature storage, humidity variations, thermal cycling, and altitude changes. Storage tests focus on assessing battery performance under different temperature and humidity levels, as well as thermal cycling and altitude variations. Battery performance tests include cycling, deep discharge, and rapid charge-discharge cycling assessments. Additionally, ingress tests evaluate the vehicle's resistance to external elements through cyclic salt spray and IPxx tests, as well as shock, impact, and mechanical vibration assessments under both operational and nonoperational conditions. Through these rigorous validation processes, EV manufacturers ensure the reliability, durability, and safety of their vehicles across diverse operating conditions [24].

For medical devices, validation is of paramount importance, placing a strong emphasis on safety, reliability, and long-term performance, given the critical implications of battery failure for patients' well-being. Rigorous testing protocols are implemented to uphold stringent quality control standards. This includes battery cycle life testing, which assesses the battery's durability under various conditions such as high-/low-temperature storage, humidity variations, thermal cycling, and altitude changes. Safety testing procedures involve evaluations for overcharge and overdischarge scenarios, as well as vibration and mechanical shock testing to ensure the device's resilience. Additionally, transportation testing evaluates the device's performance over extended periods of storage, ensuring its reliability during transit and storage. By subjecting medical devices to thorough validation processes encompassing these key aspects, manufacturers prioritize patient safety and confidence in the device's performance across diverse operating environments [25].

For aerospace and defense, validation focuses on extreme environmental conditions, including temperature variations, high vibration, and radiation exposure. Testing includes qualification for aerospace standards such as DO-160 and MIL-STD-810, which involve rigorous testing to ensure reliability and safety in aerospace and defense applications.

For energy storage systems, validation focuses on critical factors including energy density, efficiency, cycle life, and compatibility with renewable energy sources. This validation process extends to evaluating the system's grid integration capabilities, assessing its responsiveness to demand fluctuations, and ensuring safety in stationary energy storage applications such as grid-scale batteries and home energy storage systems. By rigorously assessing these aspects, manufacturers and developers prioritize reliability, efficiency, and safety in the utilization of energy storage technology, thereby contributing to the advancement and sustainability of renewable energy solutions [26].

For industrial equipment, validation is important, with a focus on ensuring reliability, durability, and optimal performance even in the most challenging operating environments. Testing procedures are designed to validate the battery's lifespan under heavy usage conditions, assess its resistance to shock and vibration, and confirm its compatibility with a wide range of industrial equipment including forklifts, robotics, and power tools. By rigorously evaluating these aspects, manufacturers of industrial equipment aim to deliver products that can withstand the demands of industrial operations, enhancing productivity, safety, and efficiency in various industrial settings [27].

8.4 Caution Regarding Golden Samples

In the battery industry, the concept of "golden samples" poses significant concerns in the battery qualification process. Golden samples refer to specially selected batteries provided by suppliers for testing, which are engineered to meet or exceed qualification standards. While these samples may pass rigorous testing, they may not be representative of the batteries produced in large quantities on the manufacturing line. This discrepancy raises critical concerns, discussed in this section, regarding the reliability, safety, and performance of batteries used in real-world applications.

Golden samples do not accurately reflect the variability and potential defects present in mass-produced batteries between lots. As a result, qualification tests conducted on golden samples may not capture the full spectrum of performance characteristics and failure modes exhibited by batteries in actual use scenarios. Qualification tests performed on golden samples may yield misleading results, giving the impression that the batteries are safe and reliable when, in fact, they may not meet the same standards once produced in larger quantities. This can create a false sense of security for manufacturers and end-users alike, leading to unforeseen safety hazards and operational failures. Golden samples typically undergo testing in controlled laboratory environments, where manufacturing variations and process inconsistencies are not adequately accounted for. Consequently, qualification tests may overlook critical factors that can impact battery performance and reliability during mass production. Using golden samples as the basis for qualification may compromise end-user safety, as the batteries deployed in actual products may exhibit different behaviors and failure modes than those observed during testing. This discrepancy increases the risk of safety incidents, including fires, explosions, and malfunctions, posing significant risks to consumers and stakeholders.

To mitigate the risks associated with golden samples in battery qualification, it is essential to adopt more robust and representative sampling protocols. This includes using a mass production line to build a large battery batch during the new product introduction (NPI) stage and conducting qualification tests on batteries randomly sampled from the NPI production, ensuring that the testing process accurately reflects real-world manufacturing conditions and variability. Random sampling helps ensure that the samples are representative of the overall population of batteries being manufactured.

It is critical to develop comprehensive testing criteria that encompass a wide range of performance parameters and failure modes relevant to the specific application of the batteries. This ensures that qualification testing adequately evaluates the performance and reliability of the samples under various conditions. Blind testing procedures need to be conducted where the testers are unaware of whether the samples being tested are qualification samples or production units. Importantly, blind testing helps eliminate biases and ensures that the testing process remains objective and impartial. Testers must periodically reevaluate the qualification process and adjust sampling procedures as needed based on feedback and performance data from production batches. Continuously improving the qualification process helps to ensure that the samples selected for testing remain representative of the production batches over time.

By implementing these strategies, manufacturers can enhance the integrity of their qualification process and ensure that the samples selected for testing accurately represent the performance and reliability of the batteries being produced. This, in turn, helps build confidence in the quality and safety of the batteries for various applications. The prevalence of golden samples in battery qualification presents a significant challenge for the industry, undermining the reliability, safety, and performance of batteries used in various applications. By addressing the concerns associated with golden samples and adopting more rigorous testing practices, manufacturers can enhance the integrity of the qualification process and ensure the delivery of safe and reliable battery products to consumers.

8.5 Analysis of Qualification Test Data

Interpreting qualification test data is essential for informed decision-making in battery development. Various approaches, including statistical analysis, trend examination, and comparison with predefined benchmarks, help extract meaningful insights from the data. Assessing battery performance based on qualification test results requires a systematic evaluation of multiple factors and parameters. In battery qualification testing, employing statistical techniques is imperative for determining appropriate sample sizes and establishing confidence levels.

Statistical analysis plays a pivotal role in ensuring that the samples selected for qualification yield statistically significant results that accurately represent the entire production batch. By employing statistical methods, such as hypothesis testing, confidence intervals, and power analysis, researchers can calculate the minimum sample size required to detect meaningful differences or deviations with a desired level of confidence. This ensures that the testing process is both efficient and effective, avoiding unnecessary costs and resources while still providing reliable insights into the quality and performance of the battery batch. Moreover, statistical techniques enable the establishment of confidence levels, allowing stakeholders to make informed decisions based on the reliability and robustness of the testing results. Ultimately, the integration of statistical analysis enhances the rigor and reliability of battery qualification testing, ensuring that products meet stringent quality standards and performance criteria.

Performance criteria should be aligned with the battery's intended application, encompassing metrics like capacity retention, cycle life, voltage stability, safety margins, and design life. Once the tests are run, the test data is compared against these specifications to ascertain that the battery meets the defined requirements. Consistency across tests and conditions will indicate the degree of stability and repeatability. Furthermore, sample sizes should be significant enough to assess normal variability and provide a confidence level.

The process begins with organizing the raw data gathered from the qualification tests in a structured manner. Next, the data must be meticulously cleaned by identifying and eliminating any outliers, errors, or anomalies that could potentially distort the analysis. Subsequently, the cleaned data needs to be converted into a suitable format for further analysis, such as spreadsheets or databases. Following this, pertinent performance metrics are calculated using the test data and specifications as guidelines. These metrics commonly include capacity, voltage, internal resistance, self-discharge rate, cycle life, and safety parameters such as temperature rise and response to abuse conditions.

In battery qualification studies, conducting thorough statistical analysis is paramount to unraveling the underlying trends, patterns, and variations present within the collected data. This analytical process involves several critical steps. Initially, efforts are directed toward identifying any discernible trends or patterns that may exist over time or across different conditions. Following this, measures of central tendency, such as the mean and median, are calculated to provide insights into the typical or average values exhibited by the data. Additionally, measures of dispersion, including standard deviation and range, are computed to assess the extent of variability or spread around the central tendency. These calculations offer valuable insights into the consistency and stability of battery performance. Furthermore, statistical tests, such as *t*-tests or analysis of variance (ANOVA), are employed to compare different data sets and ascertain if significant differences exist between them. By rigorously applying statistical methods, researchers can gain deeper insights into the performance characteristics of batteries, facilitating informed decision-making and driving improvements in battery design and quality assurance processes.

In battery qualification studies, the visualizations serve as powerful tools for understanding data distributions and identifying trends. Through the creation of visual representations such as histograms, box plots, scatter plots, and line charts, researchers can gain insights into the distributional characteristics of the data and observe any discernible patterns or trends. Moreover, employing color-coding or annotations enables the highlighting of data points that deviate from acceptable limits or specifications, facilitating quick identification of potential issues or outliers.

Subsequently, the calculated performance metrics are compared against the predefined specifications and acceptance criteria outlined in the test plan. Any observed deviations from these specifications are thoroughly investigated to assess their significance in terms of product performance and compliance. In instances where discrepancies or failures are noted during the qualification tests, a rigorous root cause analysis is conducted to identify the underlying factors contributing to these outcomes. Factors such as manufacturing variations, material properties, test conditions, and environmental influences are carefully considered during this investigation process. Finally, the analysis findings are summarized in a comprehensive report, providing stakeholders with a clear understanding of the observed outcomes, potential areas for improvement, and recommendations for further action.

8.6 Ongoing Reliability Test

Besides qualification tests to verify battery design, an ongoing reliability test (ORT) is important to assess the continued performance and reliability of battery cells and packs during mass production. Ongoing reliability testing includes continuous monitoring of key parameters such as temperature, voltage, and capacity. The collected data is analyzed to evaluate any deviations from the expected performance metrics, allowing for proactive identification of degradation or failure trends. The insights gained from ORT contribute to refining the design, improving manufacturing processes, and ensuring the continued reliability and safety of battery cells and packs throughout their operational lifespan.

8.6.1 Cell- and Pack-Level ORT

The preference for ORT at the cell level versus the pack level depends on the specific goals of the testing, the nature of the battery system, and the resources available. Considerations for each approach are described in this section.

Cell-level ORT provides valuable insights into individual battery cells, offering a comprehensive assessment of their performance. Specifically, it evaluates crucial metrics like capacity, voltage, internal resistance, and cycle life, providing detailed information about the characteristics of cells within a production batch. It is important to note that ORT is not a 100% screening process, but rather it offers representative insights into the performance of the batch. This allows for a better understanding of variations between production lots in terms of performance indices post-ORT. Moreover, this approach enables early detection of cell-specific issues or defects that could potentially impact the overall pack performance. This can lead to earlier corrections of manufacturing shifts or problems and avoid defective products getting into later stages in the supply chain.

In case inconsistent cells get into the market, the problems identified from cell-level ORT can help with proactive troubleshooting and maintenance. Additionally, cell ORT serves as a valuable tool for identifying cell-level defects or degradation mechanisms, evaluating the performance of new cell chemistries or designs, and monitoring changes in a cell's performance over production time. By leveraging the benefits of cell-level ORT, manufacturers can enhance the reliability, efficiency, and longevity of battery packs, ensuring optimal performance across various applications.

Pack-level ongoing reliability testing (ORT) offers several advantages in assessing the performance of battery systems. By testing the entire battery pack, including cell-to-cell interactions, thermal management, and safety features, pack-level ORT provides a comprehensive evaluation that reflects real-world operating conditions. This approach allows for the assessment of the overall reliability and performance of the battery system, ensuring that it meets the required standards and specifications. Moreover, pack-level ORT simplifies testing logistics and resource allocation by focusing on the final product rather than individual components, streamlining the testing process and facilitating efficient evaluation of the battery system. Overall, pack-level ORT serves as a valuable tool for ensuring the integrity and functionality of battery packs during mass production.

In practice, a combination of both cell-level and pack-level testing may be beneficial to ensure comprehensive reliability assessment and consistent quality. Cell-level ORT provides detailed insights into individual cell behavior and characteristics, while pack-level testing evaluates the integrated system performance under real-world conditions. The choice between cell-level and pack-level testing depends on the goals, resources, and risks of the testing program. However, starting testing early and detecting manufacturing issues as soon as possible are key for battery manufacturers.

8.6.2 Cell-Level ORT Guidelines

Tailored to the production capacity of the battery cell manufacturing facility, the individual ORT plan is designed to uphold high-quality standards while optimizing available resources. The process begins by determining the sample size for ongoing reliability testing, which is a calculation based on statistical methods such as the acceptance quality limit (AQL) to ensure a desired level of confidence in the manufactured battery cells' quality. The frequency of sampling is then established, dependent on production rates and the level of quality assurance sought, and it often involves testing battery cells representative from each production batch, shift, or time interval. Random selection of battery cells from the production line according to predetermined criteria ensures the sample's representativeness, enabling an accurate assessment of overall quality.

ORT may encompass measurements of key parameters like capacity, voltage, internal resistance, and visual inspections for defects or abnormalities, as well as normal and accelerated charge-discharge cycling testing, electrical testing, and environmental testing. Test results are compared against established acceptance criteria, which are usually obtained from qualification testing during the new product introduction (NPI) stage, to determine whether the battery cells meet the required quality standards. If the sample passes, the production batch proceeds for further processing or shipment. However, if issues arise, investigations or failure analyses are conducted, and corrective actions are initiated before resuming production, if the production was suspended. The ORT process is subject to continuous monitoring and evaluation, aimed at detecting manufacturing process shifts and inconsistent quality. Additionally, it may further identify opportunities for improvement through refining sampling techniques, adjusting acceptance criteria, or implementing new testing methodologies to enhance quality control and ensure the reliability of battery cells produced (Table 8.3).

Table 8.3 Cell-level battery ORT guidelines.

Test item	Test condition	Sample size
	Electrical Test	
External short-circuit test	*External short-circuit test at room temperature* The test method involves storing a rated charged cell at an ambient temperature of 25 °C. Each test sample undergoes short-circuiting by connecting the positive and negative terminals with a wire of less than 80 mΩ ± 20 mΩ resistance, with the duration of the test set to 1 hour. *External short-circuit test at high temperature* The test procedure involves storing a rated charged cell at an ambient temperature of 55 °C. Each test sample undergoes short-circuiting by connecting the positive and negative terminals with a wire of less than 80 mΩ ± 20 mΩ resistance, with the duration of the test set to 1 hour.	5–10

Table 8.3 (Continued)

Test item	Test condition	Sample size
Battery capacity and internal resistance test	The battery is discharged from the rated top charge voltage to 3.0 V at a rate of 1 °C, or optionally 0.4 °C, and then recharged at a rate of 0.4 °C to 1 °C. This cycle is repeated and monitored for 50 cycles, with the battery not allowed to return to its original state.	5–10
Battery cycling test	Battery cycling testing involves subjecting the battery to repeated charge and discharge cycles to evaluate its performance under various temperature conditions. In this test, the battery undergoes 100–500 charge–discharge cycles, depending on the application, while being exposed to temperatures of 0 °C, 25 °C, and 45 °C. By cycling the battery at different temperatures, its ability to maintain consistent performance and capacity across a range of operating conditions can be assessed. This testing helps determine how the battery behaves under real-world usage scenarios and provides valuable insights into its durability and reliability across different temperature environments.	5–10
ESD test	The electrostatic discharge (ESD) test is conducted on the battery pack in its shipment state to assess its resistance to electrostatic discharge events during transportation. The test involves subjecting the pack to both contact and air discharges at specified voltages. For contact discharge, the voltage is set at ±6 kV, and the test is performed 10 times. Similarly, for air discharge, the voltage is set at ±10 kV, and the test is repeated 10 times. By conducting these tests, the battery pack's ability to withstand electrostatic discharge incidents during shipment is evaluated, ensuring its integrity and safety during transit.	5–10
	Environmental test	
High-temperature storage test	Following a complete charge, the battery is stored in a thermostatic chamber at 75 °C for 72 hours. Subsequently, the battery is left at room temperature for a duration of 2 hours.	5–10
Low-temperature storage test	Following a complete charge, the battery is stored in a thermostatic chamber at −40 °C for 72 hours. Subsequently, the battery is left at room temperature for a duration of 2 hours.	5–10
High-temperature and high-humidity storage test	After being fully charged, the battery is placed in a thermostatic chamber at 60 °C and 90% humidity for 72 hours. Subsequently, the battery is transferred to room-temperature storage for a period of 2 hours.	5–10
Thermal shock test	Following a full charge, the battery undergoes 50 cycles of alternating exposure to −40 °C for 30 minutes and 75 °C for 30 minutes. The cell is then removed during the final −40 °C conditions to confirm dewfall. Subsequently, the battery is stored at room temperature for 2 hours.	5–10
	Mechanical test	
Crush test	The battery crush test involves applying a force of 10–15 kN to a rated charged cell, focusing on crushing the cell longitudinally.	5–10
Vibration test	Charge the battery pack to full capacity. Subject the battery pack to vibrations ranging from 7 to 18 Hz with a constant peak acceleration of 1 g. Apply vibrations with a constant amplitude of 0.8 mm and a peak acceleration increasing from 1 g to 8 g. Further, subject the battery pack to vibrations up to 200 Hz with a constant peak acceleration of 8 g. Repeat the entire vibration sequence for 15 minutes per set, completing a total of 12 sets over 3 hours. Vibrate the battery pack in three directions: X, Y, and Z. After completion of the test, allow the sample to rest for 2 hours before inspection.	5–10

8.6.3 Pack-Level ORT Guidelines

Battery pack-level ongoing reliability testing (ORT) guidelines are crucial for assessing the long-term performance, durability, and safety of battery packs throughout their life cycle. These guidelines typically involve a series of rigorous tests and evaluations conducted regularly on representative samples of battery packs to ensure their continued reliability and adherence to performance standards. The ORT guidelines may include various tests such as cycle life testing, thermal cycling, vibration testing, shock testing, and environmental testing to simulate real-world usage conditions and environmental factors that the battery packs may encounter. These tests help identify potential failure modes, degradation mechanisms, and performance issues that may arise over time, allowing manufacturers to make informed decisions about design improvements, materials selection, and manufacturing processes.

Additionally, ongoing reliability testing provides valuable data on the aging characteristics and degradation rates of battery packs, enabling predictive maintenance strategies and ensuring the long-term reliability and safety of battery-powered systems. Adhering to comprehensive ORT guidelines helps manufacturers mitigate risks, meet regulatory requirements, and deliver high-quality battery packs that meet the demanding requirements of various industries and applications (Table 8.4).

Table 8.4 Pack-level battery ORT guideline.

Test item	Test condition	Sample size
	Electrical test	
External short-circuit test	The test procedure involves storing a rated charged cell at an ambient temperature of 55 °C. Each test sample undergoes short-circuiting by connecting the positive and negative terminals with a wire of less than 80 mΩ ± 20 mΩ resistance, with the duration of the test set to 1 hour.	3–5
Battery capacity and internal resistance test	The battery is discharged from the rated top charge voltage to 3.0 V at a rate of 1C, or optionally 0.4C, and then recharged at a rate of 0.4 °C to 1 °C. This cycle is repeated and monitored for 50 cycles, with the battery not allowed to return to its original state.	3–5
ESD test	The electrostatic discharge (ESD) test is conducted on the battery pack in its shipment state to assess its resistance to electrostatic discharge events during transportation. The test involves subjecting the pack to both contact and air discharges at specified voltages. For contact discharge, the voltage is set at ±6 kV, and the test is performed 10 times. Similarly, for air discharge, the voltage is set at ±10 kV, and the test is repeated 10 times. By conducting these tests, the battery pack's ability to withstand electrostatic discharge incidents during shipment is evaluated, ensuring its integrity and safety during transit.	3–5

Table 8.4 (Continued)

Test item	Test condition	Sample size
	Environmental test	
High-temperature storage test	After a full charge, the battery is stored in the thermostatic chamber at 75 °C for 72 hours. Then, the battery is stored at room temperature for 2 hours.	3–5
Low-temperature storage test	After a full charge, the battery is stored in the thermostatic chamber at −40 °C for 72 hours. Then, the battery is stored at room temperature for 2 hours.	3–5
High-temperature and high-humidity storage test	After a full charge, the battery is stored in a thermostatic chamber of 60 °C and 90% humidity for 72 hours. Then, the battery is stored at room temperature for 2 hours.	3–5
Thermal shock test	After full charge, the battery is subjected to −40 °C for 30 minutes, then 75 °C for 30 minutes. It is carried out in 50 cycles. The cell is then removed during the final −40 °C conditions to confirm dewfall. Subsequently, the battery is stored at room temperature for 2 hours.	3–5
Liquid ingress test	As per IEC standard 60,529: The liquid ingress test classifies levels of protection against liquid exposure. IPX2 indicates suitability for indoor use, IPX4 for both indoor and outdoor applications, and IPX6 for outdoor use. Ratings of IPX7 and higher ensure complete protection against liquid ingress. The test assesses the device's functionality immediately after water exposure and again after 48 hours to ensure consistent performance under varying liquid exposure conditions.	3–5
	Mechanical test	
Safety-of-free-fall test	The battery is charged to full capacity at an ambient temperature of 20 °C ± 5 °C using a specified design method. Subsequently, at the same temperature, the battery is dropped six times from a height of 1 m onto a granite floor in random directions.	3–5
Vibration test	Charge the battery pack to full capacity. Subject the battery pack to vibrations ranging from 7 to 18 Hz with a constant peak acceleration of 1 g. Apply vibrations with a constant amplitude of 0.8 mm and a peak acceleration increasing from 1 g to 8 g. Further, subject the battery pack to vibrations up to 200 Hz with a constant peak acceleration of 8 g. Repeat the entire vibration sequence for 15 minutes per set, completing a total of 12 sets over 3 hours. Vibrate the battery pack in three directions: X, Y, and Z. After completion of the test, allow the sample to rest for 2 hours before inspection.	3–5

8.6.4 Statistical Testing for ORT

To assess whether the ongoing reliability testing results for the battery meet or exceed established criteria, statistical analysis is imperative because it allows for the comparison of means between two groups or conditions. This is fundamental in research and decision-making processes across various fields, including medicine, psychology, economics, and engineering. It is important to have a statistical analysis that is robust even with small sample sizes. This makes it particularly useful when dealing with limited data, which is common in many research studies. The results of the statistical analysis must be relatively easy to interpret, especially in comparison to more complex statistical methods. This makes it accessible to researchers and practitioners with varying levels of statistical expertise.

In general, the *t*-test plays a crucial role in hypothesis testing, parameter estimation, and decision-making processes, making it a cornerstone of statistical analysis in various fields of research and practice. The *t*-test is relatively easy to understand and implement, making it accessible to researchers and practitioners with varying levels of statistical expertise. Additionally, the *t*-test is robust even with small sample sizes, making it suitable for studies where sample sizes are limited. It can provide reliable results with as few as ten observations per group under certain conditions. Furthermore, the *t*-test comes in various forms, such as an independent samples *t*-test, paired samples *t*-test, and one-sample *t*-test, accommodating different study designs and research questions. The *t*-test allows for testing assumptions about the population, such as normality and homogeneity of variances, ensuring the validity of the statistical inference drawn from the data. With all of that said, the *t*-test is one of the most widely used and recognized statistical tests, making it easier to communicate and compare results across studies.

Typically, a *t*-test is employed to compare the means of two sets of data, in this case, the test results and the established criteria. By calculating the *t*-value and subsequently determining the p-value, it can be ascertained whether the difference between the two means is statistically significant or if it occurred by chance. If the p-value is less than a predetermined significance level (often denoted as α), it indicates that the results are significantly different from the established criteria. Conversely, if the p-value is greater than α, it suggests that there is no significant difference, implying that the ongoing reliability testing results are comparable to the established criteria. This rigorous statistical approach ensures that decisions regarding the reliability of the battery are grounded in empirical evidence, facilitating informed decision-making, and ensures adherence to quality standards.

Another method of statistical analysis is the sequential probability ratio test (SPRT), which is a method used for sequential hypothesis testing. This method is particularly useful when dealing with sequential data collection processes, where decisions need to be made based on accumulating evidence over time. The SPRT allows for the continuous monitoring of data, enabling early detection of significant changes or patterns while minimizing the number of observations needed. In the context of battery ongoing reliability tests, the SPRT can be applied to monitor the performance and reliability of batteries over time. Specifically, the SPRT allows for the continuous monitoring of key parameters or metrics related to battery performance, such as capacity degradation, voltage stability, or cycle life. Instead of waiting for a predetermined number of observations or a fixed testing duration, the SPRT updates decisions based on each new piece of data collected. By continuously analyzing incoming data, the SPRT can detect changes or deviations from expected behavior early in the testing process. This early detection is crucial for identifying potential issues or anomalies before they escalate, allowing for timely interventions or adjustments.

Traditional hypothesis tests often require a fixed sample size or testing duration, which can lead to inefficient resource allocation, especially if the data collection process is expensive or time-consuming. The SPRT dynamically adjusts the sample size based on the observed data, potentially reducing the overall testing effort while maintaining statistical rigor. Notably, the SPRT allows researchers to specify flexible decision criteria based on their specific requirements and priorities. For example, they can set different thresholds for detecting improvements versus deteriorations in battery performance, or adjust the decision boundaries based on acceptable error rates. The SPRT is adaptive, meaning it can update to changing conditions or trends in the data over time. This adaptability is particularly valuable in dynamic environments where battery performance may evolve or fluctuate, due to factors like usage patterns, environmental conditions, or aging effects.

8.7 Summary

The battery qualification process is one step in ensuring the design meets the performance, reliability, and safety requirements for the intended applications. The process consists of subjecting batteries to tests that simulate real-world life cycle operational and environmental conditions, over the full design life of the product. The tests should evaluate performance metrics, including capacity, voltage stability, temperature tolerance, and safety features, to ensure that batteries meet the design requirements. In some cases, it will be necessary to conduct accelerated tests to assess the life of the battery in a reasonable time. The test methods and results need to be documented to provide transparent records to ensure traceability and compliance.

To qualify batteries, many manufacturers have adopted standards-based qualification testing. However, a lack of correlation between field use and test conditions may lead to inadequate qualification. This can be especially true for manufacturers whose product lifetimes are long (e.g., more than a few years), in situations where environments or operation modes are harsher than normally expected (e.g., high- or low-temperature applications), or where the consequences of failures are severe (e.g., medical electronics). In other words, a one-size-fits-all qualification test will not be meaningful for all applications.

Battery qualification testing, while essential for assessing certain aspects of battery performance and safety, does not guarantee immunity against all potential failure modes or hazards. Despite companies conducting thorough qualification tests, incidents such as fires have occurred, indicating that these tests alone are not sufficient to prevent catastrophic failures. Qualification testing may overlook certain failure modes or environmental conditions that batteries could encounter in real-world usage scenarios. The qualification tests also do not guarantee that every product that is manufactured will be exactly like those tested (e.g., there will be variability due to the materials or manufacturing and assembly processes). Moreover, unforeseen interactions between components and external factors beyond the scope of qualification tests can contribute to safety hazards. Therefore, while battery qualification testing is valuable for evaluating specific parameters, it should be complemented with quality assurance processes, continuous monitoring, risk assessments, and proactive measures to ensure comprehensive reliability and safety.

References

1 Bashir, S.M.A., Zaidi, S.K.H., and Zaheer, M. (2015). Design and development of accelerated life tester for qualification of batteries. *Journal of Space Technology* V (1): 33–39.

2 Wang, Q., Mao, B., Stoliarov, S.I., and Sun, J. (2019). A review of lithium-ion battery failure mechanisms and fire prevention strategies. *Progress in Energy and Combustion Science* 73: 95–131.

3 Kaliaperumal, M. et al. (2021). Cause and mitigation of lithium-ion battery failure—a review. *Materials* 14 (19): 5676.

4 Bubbico, R., Greco, V., and Menale, C. (2018). Hazardous scenarios identification for Li-ion secondary batteries. *Safety Science* 108: 72–88.

5 Borujerd, S.V.N. et al. (2023). Fuzzy logic approach for failure analysis of Li-ion battery pack in electric vehicles. *Engineering Failure Analysis* 149: 107233.

6 Deng, Z., Xu, L., Liu, H. et al. (2023). Prognostics of battery capacity based on charging data and data-driven methods for on-road vehicles. *Appl. Energy* 339: 120954.

7 Omar, N., Bossche, P.V.D., Coosemans, T., and Mierlo, J.V. (2013). Peukert revisited - critical appraisal and need for modification for lithium-ion batteries. *Energies* 6: 5625–5564.

8 Deng, J., Bae, C., Denlinger, A., and Miller, T. (2020). Electric vehicles batteries: requirements and challenges. *Joule* 4 (3): 511–515.

9 Energy Assurance, Battery Cell, Module & Pack Testing - The Basics. Available online: Energy Assurance eBook.

10 Hendricks, C., Williard, N., Mathew, S., and Pecht, M. (2015). A failure modes, mechanisms, and effects analysis (FMMEA) of lithium-ion batteries. *Journal of Power Sources* 297: 113–120.

11 Kim, G.H., Pesaran, A., and Spotnitz, R. (2007). A three-dimensional thermal abuse model for lithium-ion cells. *Journal of Power Sources* 170: 476–489.

12 Maleki, H. and Howard, J.N. (2006). Effects of over discharge on performance and thermal stability of a Li-ion cell. *Journal of Power Sources* 160: 1395–1402.

13 Vetter, J. et al. (2005). Aging mechanisms in lithium-ion batteries. *Journal of Power Sources* 147: 269–281.

14 Monroe, C. and Newman, J. (2003). Dendrite growth in lithium/polymer systems: a propagation model for liquid electrolytes under galvanostatic conditions. *Journal of the Electrochemical Society* 150 (10): A1377–A1384.

15 Kirana, R.C., Purwanto, N.A., Azis, N.A. et al. (2023). Failure assessment in lithium-ion battery packs in electric vehicles using the failure modes and effects analysis (FMEA) approach. *Journal of Mechatronics, Electrical Power, and Vehicular Technology* 14: 94–104.

16 Al-Hallaj, S. et al. (1999). Thermal modelling and design considerations of lithium-ion batteries. *Journal of Power Sources*.

17 Zhang, Y. et al. (2011). Cycling degradation of an automotive $LiFePO_4$ lithium-ion battery. *Journal of Power Sources* 196: 1513–1520.

18 Sato, N. (2001). Thermal behavior analysis of lithium-ion batteries for electric and hybrid vehicle. *Journal of Power Sources* 99: 70–77.

19 Han, S. et al. (2014). A practical battery wear model for electric vehicle charging applications. *Applied Energy* 113: 1100–1108.

20 Karadeniz, H. et al. (2009). An integrated reliability-based design optimization of offshore towers. *Reliability Engineering and System Safety* 94: 1510–1516.

21 Zhang, C.W. et al. (2013). Reliability modeling and analysis for a novel design of modular converter system of wind turbines. *Reliability Engineering and System Safety* 111: 86–94.

22 Element, Energy Assurance, Cell Level vs. Pack Level Battery Testing — let's sort this out. https://www.element.com/nucleus/2022/ev-battery-testing-from-cell-to-module-to-pack.

23 Unico, The Fundamentals of Battery Pack and Module Testing. https://unicous.com/the-fundamentals-of-battery-pack-and-module-testing/.

24 Energy Assurance, Consumer Goods Battery Testing. https://www.energy-assurance.com/consumer-goods-battery-testing.
25 Energy Assurance, EV Battery Testing. https://www.energy-assurance.com/ev-battery-testing.
26 Energy Assurance, Medical Battery Testing. https://www.energy-assurance.com/medical-battery-testing.
27 Energy Assurance, Energy Storage Battery Testing. https://www.energy-assurance.com/energy-storage-battery-testing.

9

Quality Control in Li-Ion Battery Production: Best Practices and Challenges

Dulja Bamunusinghe, Thisali S. Rathnayake, Raveen Sanjaya De Silva, Logeeshan Velmanickam, and Rashed A. Islam

The production of lithium-ion batteries (LIBs) is a complex process that requires stringent quality control measures to ensure the safety, reliability, and performance of the final product. With these batteries powering an expanding range of devices and vehicles, from smartphones to electric cars, maintaining high quality standards is essential. Battery manufacturers must implement a comprehensive quality control framework that spans the entire production process. This begins with incoming quality control (IQC) to verify the integrity of raw materials, continues with monitoring of key manufacturing parameters during production, and concludes with rigorous outgoing quality control (OQC) before the batteries are released to the market. By following these practices, manufacturers can reduce risks, improve product consistency, and deliver LIBs that meet the stringent requirements of modern applications, all while prioritizing user safety and satisfaction [1].

9.1 Incoming Quality Control

In lithium-ion battery (LIB) production, rigorous IQC processes are indispensable for ensuring the safety, reliability, and performance of the final products. These processes encompass a series of critical steps to mitigate risks and uphold high standards.

Detection of metallic impurities in battery materials is paramount to prevent potential hazards such as internal short circuits and thermal runaway. Advanced X-ray inspection techniques, capable of identifying and analyzing impurities as small as 20 μm, can be deployed before materials enter production [2, 3]. This proactive screening ensures that only materials free from contaminants proceed further, enhancing both safety and manufacturing efficiency.

Ensuring the integrity and proper alignment of battery electrodes is crucial for minimizing defects and optimizing performance across different battery form factors (e.g., cylindrical, prismatic, and pouch). Specialized X-ray and imaging technologies can be employed for thorough inspection, ensuring precise electrode alignment and spacing between positive and negative electrodes [4]. This approach reduces the risk of internal faults that could compromise the battery function.

Beyond physical inspection, evaluating the electrochemical characteristics of incoming battery cells is essential. Techniques such as electrochemical impedance spectroscopy (EIS) provide detailed insights into parameters like internal resistance, capacity, and aging behavior [5, 6].

The Safety Challenges and Strategies of Using Lithium-Ion Batteries, First Edition. Edited by Michael G. Pecht.

By identifying cells with suboptimal performance early in the production process, manufacturers can take corrective actions to maintain consistent quality and reliability. In addition to detecting metallic impurities, the quality and composition of active materials, listed next, undergo rigorous scrutiny. Spectroscopic and chromatographic analyses ensure that these materials meet purity and performance criteria, contributing to enhanced battery longevity and consistent performance [7].

- Cathode materials: lithium, cobalt, nickel, manganese, aluminum
- Anode materials: graphite, silicon
- Electrolytes: lithium salts (e.g., lithium hexafluorophosphate), organic solvents
- Separators: polypropylene, polyethylene
- Current collectors: copper, aluminum

Safety testing protocols are important to ensure compliance with industry standards and regulations. These protocols encompass comprehensive tests for factors such as thermal stability, voltage tolerance, and cycle life under controlled conditions. Compliance with safety standards is paramount to mitigating risks associated with Li-ion battery (LIB) usage across diverse applications [8]. Implementing a comprehensive incoming QC program that integrates these techniques enables LIB manufacturers to detect and address potential defects early in the production chain. This proactive approach not only enhances product reliability but also fosters overall customer satisfaction and safety.

9.2 Process Control Measures

To ensure high-quality production, manufacturers must implement robust process control measures to guarantee the performance, safety, and reliability of LIBs. Any defects or inconsistencies in the production process can lead to reduced battery life, capacity, and overall performance, ultimately affecting the safety and reliability of the battery [9]. Process control measures, as shown in Table 9.1, can help to identify and mitigate potential defects, reduce variations, optimize the component design for efficient manufacturing, and enhance the manufacturing process by identifying critical factors affecting battery performance. By understanding and implementing these techniques, organizations can significantly improve their manufacturing operations and produce high-quality LIBs that meet the demands of the rapidly growing market while maintaining safety and reliability standards.

9.2.1 Core Process Control Techniques in Lithium-Ion Battery Production

Maintaining consistent quality and safety throughout LIB production is crucial. Several core process control techniques help manufacturers achieve this by identifying potential issues, optimizing designs for production and refining processes through experimentation. The core process control techniques discussed next are used in LIB production.

Process failure mode and effects analysis (PFMEA) serves as a proactive risk assessment tool, specifically designed to identify the causes and effects of failures within various stages of the battery production process. This focus ensures reliability and safety in the final product. A general failure mode and effects analysis (FMEA) workflow, according to IEC standard 60812 [13, 14], as illustrated in Figure 9.1, involves a structured process to identify and mitigate potential failures in a product or process. The workflow begins by defining the scope and objectives of the FMEA, followed by the identification of all possible failure modes and their causes. The severity,

Table 9.1 Quality control for steps and process parameters in a Li-ion battery manufacturing process.

Major manufacturing steps	Process parameters	Quality control steps
Raw material selection	• Material purity • Particle size distribution • Surface area	1. Chemical composition analysis 2. Impurity testing 3. Particle size measurement [10]
Electrode preparation (mixing, coating, calendaring, slitting, vacuum drying, tab welding, taping)	• Material properties • Slurry composition • Slurry viscosity • Mixing conditions (speed and time) • Coating equipment • Coating thickness • Drying temperature • Tab welding strength	1. Raw material inspection 2. Process parameter monitoring 3. Slurry homogeneity check 4. Coating uniformity inspection and electrode thickness measurement [11] 5. Pull-test strength checks for tabs 6. Visual inspection for weld spatter, burrs, and arcing
Cell assembly	• Electrode alignment, • Top insulator quality • Bottom insulator quality • Electrolyte-filling volume • Sealing pressure	1. Visual inspection for misalignment 2. Visual inspection of top and bottom insulators for defects (folds, cuts, and discoloration) 3. Electrolyte leakage test 4. Seal integrity check [10, 12]
Electrolyte filling and sealing	• Electrolyte properties • Filling equipment • Cell parameters • Environmental conditions	1. Electrolyte inspection 2. Filling process monitoring 3. visual inspection
Formation and aging	• Charge–discharge current • Voltage limits • Temperature • Formation time	1. Capacity measurement 2. Internal resistance test 3. Self-discharge rate check [12]
Final testing	• Charge–discharge cycles • Temperature range • Vibration frequency • Ongoing reliability tests	1. Capacity verification 2. Safety tests (overcharge, short circuit) 3. Cycle life assessment [12]

occurrence, and detection of each failure mode are then assessed using specific scales, typically ranging from one to ten. The risk priority number (RPN) is calculated by multiplying these rankings, with higher RPNs indicating higher risks that require urgent attention. The failure modes are then prioritized based on their RPN, and corrective actions are implemented to mitigate the risks. Finally, the PFMEA process and results are documented and maintained to ensure the ongoing reliability and safety of the product or process.

Real-world examples of PFMEA in action include identifying potential failure modes, which could include short circuits, thermal runaway, and leakage of LIBs. By proactively identifying the risks and assessing their severity, occurrence, and detectability, battery manufacturers can implement design changes, process controls, and inspection methods to mitigate the risks. This helps to ensure the safety and reliability of the batteries, which is critical in applications like electric vehicles (EVs) and consumer electronics. Companies like LG Chem, Samsung, and Panasonic use FMEA to identify potential failures in their LIB manufacturing processes [15].

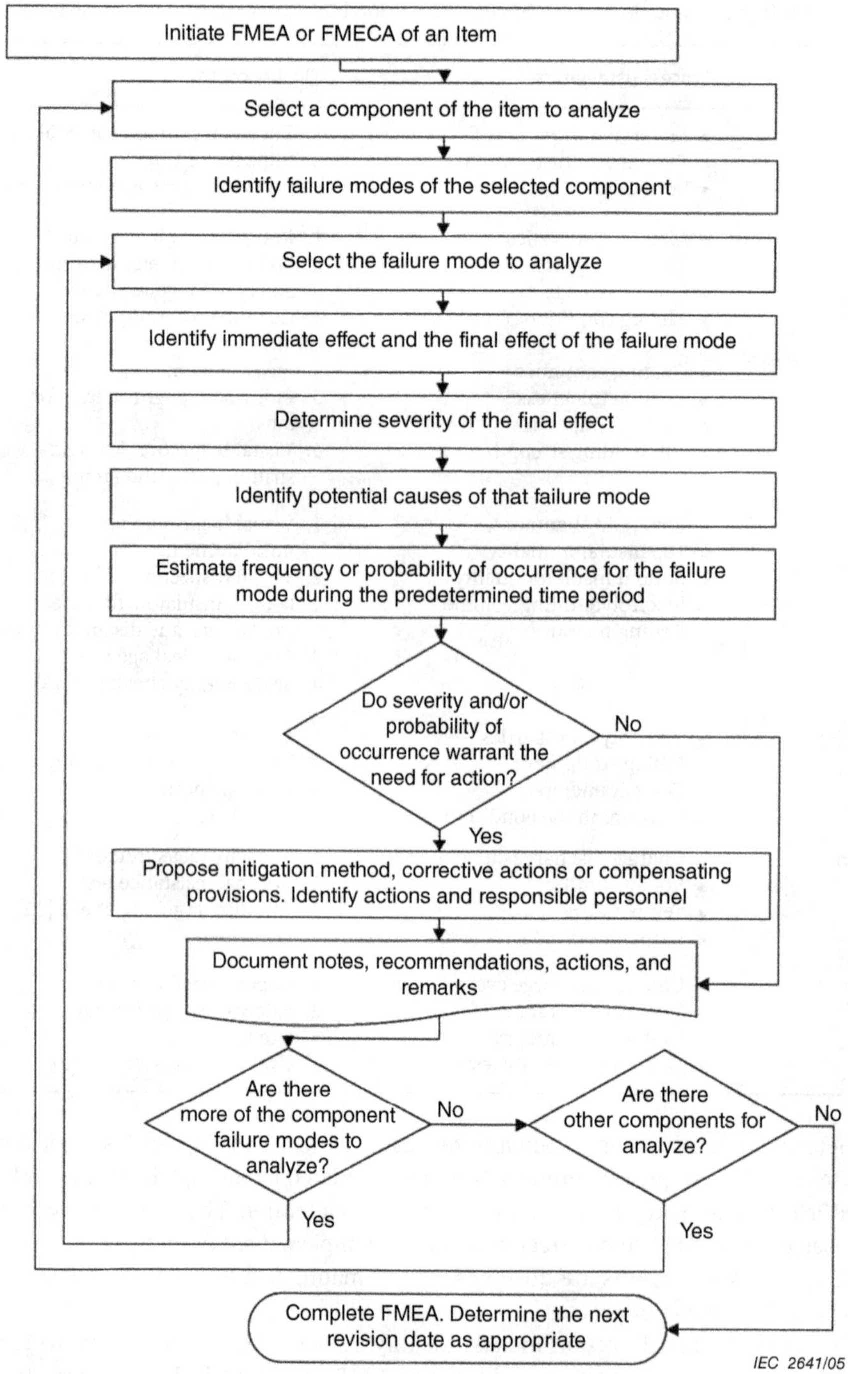

Figure 9.1 FMEA workflow according to IEC standard 60812. Source: [13, 14].

Six Sigma is a data-driven approach that minimizes defects and variations in processes, leading to near-perfect production of LIBs [16]. The two methodologies used in Six Sigma are (a) define, measure, analyze, improve, control (DMAIC), and (b) define, measure, analyze, design, verify (DMADV). While both methodologies follow a structured approach with specific steps, the focus of each is different [9, 16, 17]. DMAIC is used to improve existing processes, whereas DMADV is used to design new processes or products. DMAIC is a reactive process that focuses on the process, while DMADV is a proactive process that focuses on the customer. DMAIC requires a small team and a short- to medium-term time frame, but DMADV requires a large team and a long-term time frame. DMAIC is used when continuous improvement is required; however, DMADV is used when reengineering is required.

The DMAIC (Define, Measure, Analyze, Improve, Control) cycle, shown in Figure 9.2, identifies and eliminates root causes of problems and tackles existing problems. By defining the specific battery quality or safety issue (e.g., reduced capacity or overheating), DMAIC guides manufacturers through measuring current performance, analyzing root causes (e.g., material defects or process inefficiencies), implementing corrective actions (e.g., optimizing processes or stricter material control), and controlling the improvements for sustained quality.

DMADV, illustrated in Figure 9.3, focuses on designing new or improved processes. Here, the target specifications for batteries (e.g., capacity, safety features) and manufacturing processes (e.g., cost efficiency) are clearly defined. Key parameters influencing performance are measured and analyzed to identify optimization opportunities [18]. This knowledge is then used to design the

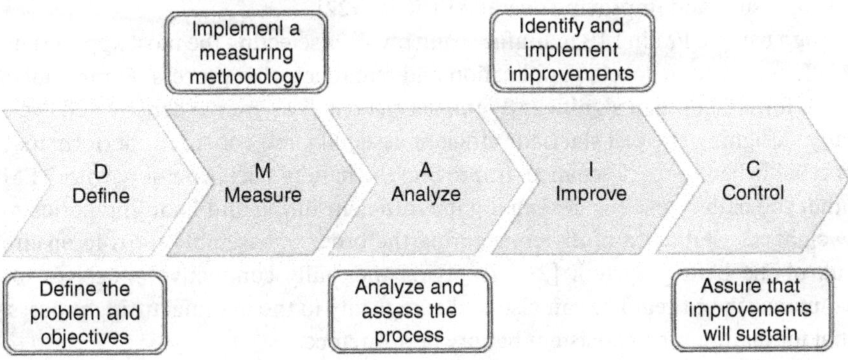

Figure 9.2 Lean Six Sigma (DMAIC) framework and methodology.

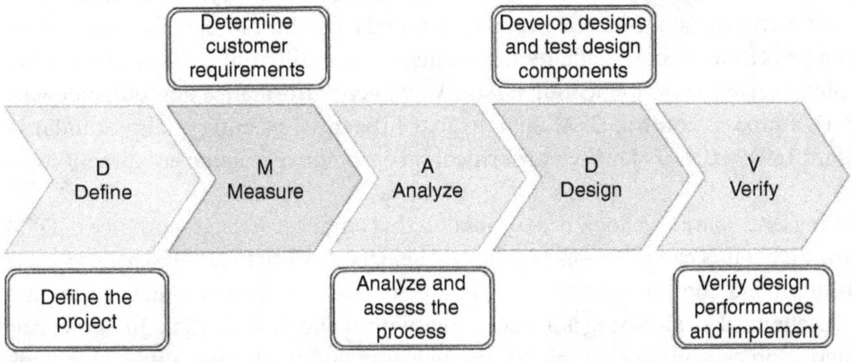

Figure 9.3 Lean Six Sigma (DMADV) framework and methodology.

new process or battery configuration, with thorough testing and validation to ensure it meets all requirements before mass production. By following DMADV, manufacturers can design efficient processes that produce high-quality and future-proof LIBs.

Six Sigma's process capability (Cp) and process capability index (Cpk) metrics are important statistical parameters for quality control. Cp, focusing on process variation, measures the capability of the process to produce within the specification limits, while Cpk measures the capability of the process to produce within the specification limits while accounting for the variability of the process. A Cp > 1.33 suggests the process is naturally capable. Cpk takes it a step further, and it considers both variation and centering of the process data. A Cpk > 1.33 indicates the process is centered within the limits, minimizing out-of-spec batteries. Low Cp and Cpk values signal quality issues, prompting investigation and improvement. High values indicate a well-controlled process, allowing for a focus on efficiency or tighter future specifications. By monitoring Cp and Cpk, manufacturers gain data-driven insights to ensure consistent, high-quality battery production [19].

At Northvolt Labs in Sweden, Six Sigma DMAIC was implemented to improve the quality assurance of LIBs in the cell assembly process [20]. The study identified key factors affecting technical cleanliness and particle contamination, and it provided recommendations to reduce these issues. These examples demonstrate the effectiveness of Six Sigma in improving quality and reducing defects in LIB manufacturing [16, 21].

The **Design for Manufacturability (DFM)** principles, outlined in Figure 9.4, optimize the manufacturing process of LIBs by integrating these principles into the design phase. Manufacturers can ensure that their products are designed with manufacturability in mind, reducing the risk of costly manufacturing issues and improving overall efficiency [22].

Optimizing cell design using DFM in LIB manufacturing involves selecting the most appropriate cell form factor and chemistry for a specific application and manufacturing process. For instance, choosing the right cell form factor can significantly impact the battery's energy density and cycle life [22]. Additionally, designing the cell stack for efficient assembly and consistent performance is critical to ensure reliable battery performance. Improving the battery pack assembly with DFM involves ensuring that the battery pack is designed with sufficient lifting and handling points to accommodate its weight, as well as carefully considering the order of assembly steps to ensure proper powering up of the battery system [23]. Applying thermally conductive adhesives for demanding thermal management regimes can also add complexity to the manufacturing process, but they are essential for maintaining consistent battery performance.

DFM also enhances manufacturing flexibility by enabling the development of flexible production facilities that can quickly adapt to changing design requirements and produce prototypes or small-volume runs. This flexibility is particularly important in the LIB industry, where rapid development and production are crucial for meeting market demands. Furthermore, DFM allows manufacturers to customize LIBs to meet specific requirements, such as different cathode chemistries or cell formats, which is critical for applications like EVs where performance and efficiency are paramount. In terms of process control, DFM ensures that LIBs meet exacting quality standards, such as IATF standard 16949:2016 [24], which are critical for automotive original equipment manufacturers (OEMs).

Axeon, Imagineering, and Sparkling Logic are companies that emphasize the importance of DFM in their LIB manufacturing processes. Axeon highlights the need for a detailed understanding of both the end application and chosen cell chemistry to ensure effective battery management systems (BMSs) and mechanical engineering for packaging within the vehicle [24]. Imagineering optimizes its production process by considering factors such as material selection, dimensional tolerances, and secondary processing like finishing [23]. Sparkling Logic helps leading custom battery

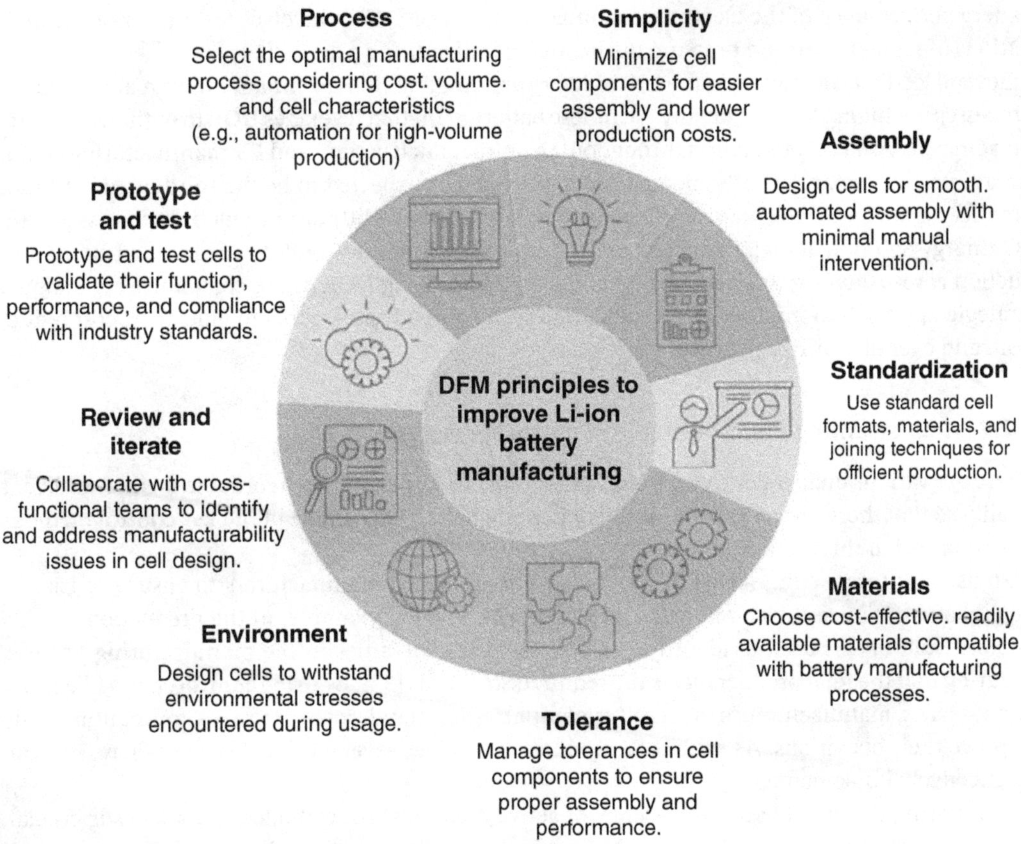

Figure 9.4 DFM principles to improve Li-ion battery manufacturing.

manufacturers expedite and optimize their design process by defining the relationship between materials, suppliers, and manufacturing processes to quickly produce a design decision based on automaker requirements [25].

Design of Experiments (DOE) helps manufacturers to systematically investigate various factors and their impact on the final product, which involves a structured methodology of planning, conducting, analyzing, and interpreting controlled experiments to identify key factors influencing the process. This approach is particularly useful in LIB manufacturing, where precise control over factors such as material types, processing conditions, and techniques is crucial for optimal performance, cost, and sustainability [26, 27].

The key advantage of DOE lies in its ability to optimize anode-free LIB development. DOE allows researchers to analyze how various factors, including cathode material (e.g., lithium iron phosphate (LFP), NMC), current collector material, electrolyte composition, and protective coatings, influence critical battery characteristics [28]. These characteristics encompass volumetric energy density, longevity (i.e., cycling performance), lithium deposition behavior, stability of the solid electrolyte interphase (SEI) layer, and dendrite suppression. By systematically studying these factors and their interactions, DOE facilitates the identification of the optimal combination of materials, processing conditions, and design choices for creating high-performing and long-lasting products [29]. Another critical aspect of battery performance is the electrode manufacturing process, and by using DOE, manufacturers can investigate the effects of different processing conditions on the battery performance of the electrode manufacturing process. This involves testing various combinations of temperature and pressure during the manufacturing process [29].

Several leading companies in the LIB industry are utilizing DOE to enhance their manufacturing processes, including BYD Company, a Chinese battery giant that leverages DOE to optimize production efficiency and cost reduction throughout their integrated battery and EV manufacturing chain. Arcadium Lithium, a recently merged industry leader, is expected to be the third largest lithium producer globally by 2027 and implements DOE to optimize lithium production processes [30]. LG Energy Solution, a major supplier to top automakers, employs DOE to streamline battery production across their expansive global operations. By adopting DOE, these companies are taking a strategic approach to improve LIB production, leading to advancements in efficiency, cost reduction, and overall industry progress.

9.2.2 Implementing Effective Quality Control Measures

There are various quality control measures that must be implemented in order to ensure sufficient quality so that there are no failures or safety issues when fielded. Some of the key control methods are explained in this section.

Statistical process control (SPC) is a powerful tool for LIB manufacturers to ensure consistent quality and optimize their production processes [28, 31]. For example, in the production of LIBs at Northvolt, SPC is used to identify and address potential issues in the manufacturing process, ensuring that the final product meets the required standards [32]. By using control charts like X-bar and R-charts, manufacturers can identify potential issues, standardize processes, and continuously improve their operations. As such, LIB production involves several critical stages where SPC can be effectively implemented.

Monitoring the quality and consistency of active materials like cathodes and anodes is crucial, and manufacturers can use X-bar charts to track the average particle size or specific capacity of materials across batches, ensuring they meet tight specifications [33]. Maintaining uniform coating thickness and density is essential for battery performance, and R-charts can help monitor the range of coating thicknesses, allowing manufacturers to quickly identify and address any variability [33, 34]. Precise control of parameters like electrode alignment and electrolyte-filling volume is critical during cell assembly, and X-bar charts can track the average values of these parameters, while R-charts monitor the range, ensuring consistent assembly. During the formation and aging processes, SPC can help track key performance indicators (KPIs) like capacity, resistance, and self-discharge, and manufacturers can use X-bar charts to monitor the average values and identify any shifts in performance [34].

In practice, SPC is applied to monitor manufacturing processes, such as the electrode coating process. By plotting the average thickness of each subgroup against the subgroup number, an X-bar chart can be constructed, as shown in Figure 9.5. The center line represents the overall average thickness, while the upper and lower control limits are calculated based on the average range of the subgroups. If any points fall outside these limits, it indicates that the process is out of control, and adjustments may be needed to maintain the desired coating thickness.

The R-chart is another type of visualization tool, as shown in Figure 9.5, and it monitors the range of coating thicknesses within each subgroup. It helps identify changes in process variability, which could be caused by factors like equipment wear or raw material changes. The center line represents the average range, and control limits are calculated accordingly. If the range exceeds the upper control limit, it suggests that the process variability has increased, and corrective actions need to be taken. By analyzing both the X-bar and R-charts, a battery manufacturer can ensure

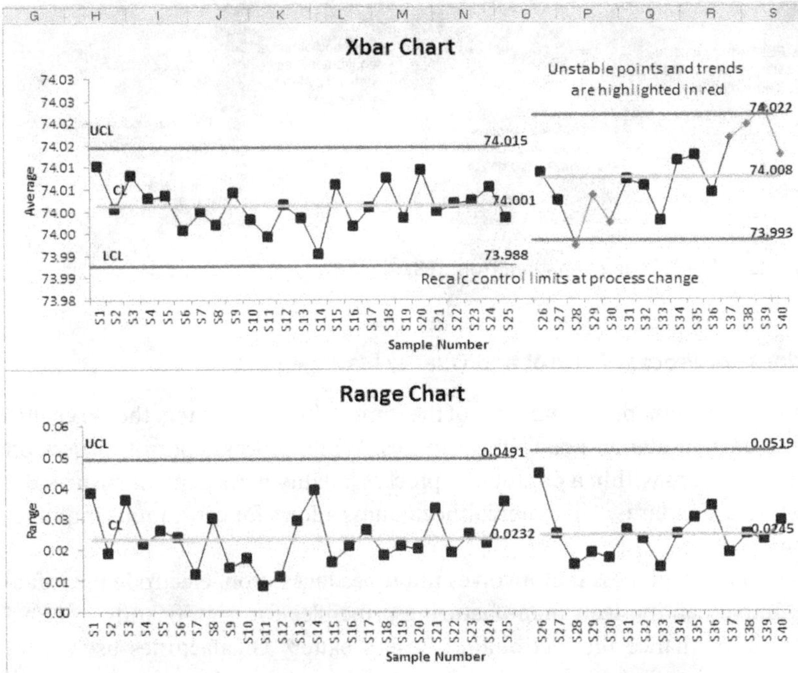

Figure 9.5 Sample X-bar chart and R-chart. Source: Adapted from Persson [35].

that their electrode coating process is stable, consistent, and operating within desired parameters. This helps to maintain high-quality electrodes, which is essential for producing reliable and high-performing LIBs.

The production part approval process (PPAP) is a critical process for ensuring that suppliers meet customer requirements and produce high-quality parts. PPAP guidelines provide a standardized procedure for completing this process, including detailed information, instructions, and sample documents [28, 36].

One of the key elements of PPAP is design documentation, which includes the technical drawings of batteries from both the supplier and customer, as well as the purchase order. This ensures that everything aligns with the customer's requirements and that the correct part is being delivered. The supplier must also provide evidence that the customer's engineering department has approved the sample production parts [36, 37]. Additionally, PPAP requires the supplier to conduct comprehensive risk assessments, such as design failure mode and effects analysis (DFMEA) and the previously described PFMEA. These assessments identify potential failures within the design and manufacturing processes, and they help to ensure that the process is stable and reliable.

The PPAP guidelines, shown in Figure 9.6, outline five submission levels, ranging from a simple part submission warrant (PSW) to a complete package with product samples and supporting data. The specific requirements are negotiated during the quoting process and depend on the complexity of the part and the customer's needs [39]. By following the PPAP guidelines, suppliers can ensure that their batteries meet customer specifications and quality requirements, reducing the risk of product recalls and customer dissatisfaction. However, PPAP can be complex and time-consuming, requiring significant resources and effort. Suppliers must plan, communicate effectively with customers, and regularly monitor and review the process to ensure its effectiveness and efficiency [36, 38, 39].

Figure 9.6 Submission levels of PPAP. Source: Adapted from [38].

9.2.3 Interconnectedness of Process Control and Quality Management

Process control and quality management are two sides of the same coin, working together to ensure the consistent production of high-quality goods and services. While process control focuses on monitoring and regulating the steps within a production process, quality management ensures the final output meets established standards. This interconnectedness allows for continuous improvement and the prevention of defects.

LIB manufacturing is a complex process that involves multiple stages, from electrode manufacturing to cell finishing. Implementing lean manufacturing principles can significantly improve efficiency, reduce waste, and enhance product quality. Tesla's battery Gigafactories use a lean approach to minimize waste and optimize production flow, thereby reducing inventory costs [40]. LG Chem also employs lean techniques to streamline its process, minimize transport waste, and improve overall efficiency [41]. By reducing unnecessary steps and optimizing material flow, these companies can produce higher-quality batteries while lowering costs.

In the cell finishing stage, lean manufacturing helps to optimize the formation and aging processes. With surging demand driven by European (EU) climate targets, manufacturers like Tesla are implementing lean principles to reduce waste and improve efficiency. By streamlining the formation and aging steps, they can produce batteries faster while using less space. This allows them to rapidly scale up production to meet the growing market need for LIBs. Furthermore, quality control is critical in lithium-ion manufacturing, and lean principles emphasize defect prevention. As an example, LG Chem has a lean approach to quality control, ensuring every battery meets rigorous performance, safety, and reliability standards. They use quality gates and process monitoring to catch issues early, which allows them to deliver consistently high-quality products that meet customer requirements.

Continuous improvement is a core tenet of lean. Battery manufacturers use tools like value stream mapping (VSM) and SPC to identify waste, streamline processes, and ensure consistent quality. VSM visualizes the production flow to pinpoint improvement opportunities, while SPC uses data to monitor processes, detect variations, and maintain tight control. By continuously enhancing its processes, manufacturers can drive ongoing efficiency gains [42, 43].

Real-world examples demonstrate the power of lean manufacturing in lithium-ion production. Tesla's Gigafactories, LG Chem's facilities, and European battery makers are all using lean principles to optimize production, reduce waste, and deliver high-quality batteries [44]. By implementing lean techniques like just-in-time inventory, process flow optimization, quality systems, and continuous improvement, lithium-ion manufacturers can significantly boost efficiency and competitiveness [45].

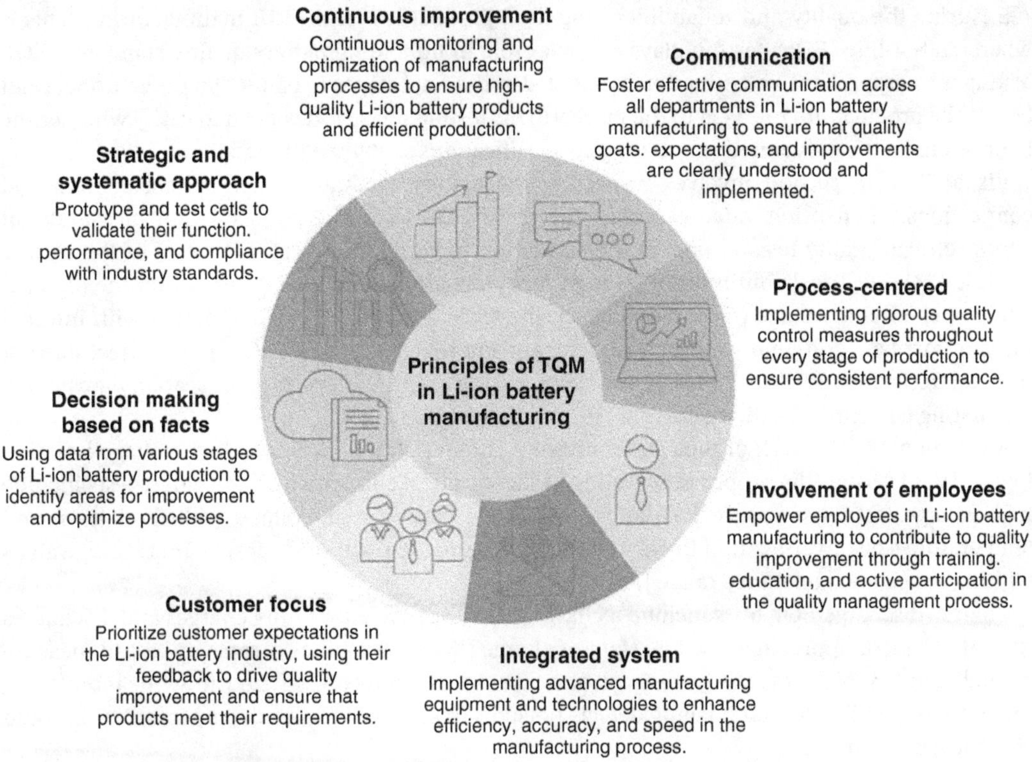

Figure 9.7 Principles of TQM in Li-ion battery manufacturing. Source: Adapted from [48].

Total Quality Management (TQM) is a company-wide approach that focuses on continuous improvement of quality across all departments. In LIB manufacturing, implementing TQM is essential for ensuring high-quality products while maintaining efficiency [46, 47].

One key aspect of TQM is the use of process control techniques like Six Sigma and SPC. These methods provide data-driven approaches to quality improvement, enabling manufacturers to identify and address defects or variations in the production process [47].

To achieve this, manufacturers must identify key parameters, shown in Figure 9.7, that directly impact the quality and performance of the batteries. Once these characteristics are defined, the manufacturer can implement SPC to monitor and control these parameters. For example, they might use control charts to track the mean and range values of certain production steps, allowing them to detect any deviations and promptly take corrective actions. In addition to process control, TQM also involves continuous improvement and innovation. LIB manufacturers can foster a culture of continuous improvement by encouraging employees to suggest new ideas and solutions. This can be achieved through regular feedback sessions, suggestion boxes, or even incentivizing employees for their contributions to quality improvement. By embracing a culture of continuous improvement, manufacturers can stay ahead of the competition and see that their products meet the evolving demands of the market. To ensure that employees are equipped to handle the latest technologies and manufacturing techniques, manufacturers must provide regular training sessions and workshops. This not only improves the quality of the products but also enhances the overall efficiency of the production process [48].

Ensuring the quality and reliability of the final product is vital in LIB manufacturing. This is where end-of-line (EOL) testing plays a crucial role in upholding battery quality standards. EOL testing is a comprehensive evaluation conducted at the final stages of battery manufacturing, right before the products are released to the market. This testing phase is designed to verify whether the batteries meet the predefined standards and specifications set by the manufacturer [49–52].

Visual EOL inspection involves inspecting the battery modules for proper installation and connections, often using video cameras. Electrical EOL tests include insulation measurement and electrical rigidity tests to ensure the insulation integrity and electrical safety of the battery module [53]. Electrical continuity EOL tests verify the electrical connections within the module, while open-circuit voltage (OCV) tests check the OCV of the cells and compare it with internal measurements. Additionally, alternating current internal resistance (ACIR) and direct current internal resistance (DCIR) pulse power tests evaluate the battery module's ability to handle alternating current (AC), direct current (DC), and pulse power [54, 55].

A functional EOL check of the cell monitoring circuit (CMC), BMS, and high-voltage isolation logic (HVIL) verifies the proper functioning of these critical components [56]. Safety tests include high-voltage tests to measure insulation resistance, battery high-voltage withstand tests, and temperature sensor verification to ensure that all internal temperature sensors are reading values within the specified tolerance range [56].

Charge–discharge EOL tests include a charge–discharge cycle that fully charges and discharges the battery to an approximately 30% state of charge (SOC), as well as a capacity test to measure the total amount of energy the battery can store. Leak tests apply pressurized air while monitoring for any leaks, while thermal imaging detects hotspots or areas of increased temperature that could indicate a problem [52].

Other EOL tests include cell voltage measurements to check that minimum, maximum, and delta voltages are reading properly and within tolerances. Pack voltage verification verifies that the BMS internal pack voltage measurement matches the pack output voltage. DC internal resistance measurement performs charge–discharge pulses and calculates the internal resistance of the battery pack. The charge–discharge pulse test ensures that battery charges and discharges and the BMS responds correctly to each cycle. Finally, battery management programming clears any diagnostic trouble codes (DTCs) generated during testing, updates cell voltage and temperature information, and reads/writes any data acquired during the test [53].

Companies like Tesla, LG Chem, and Panasonic employ EOL testing in their production chains to ensure the quality and reliability of their LIBs. For instance, Tesla's EOL testing includes visual inspections, electrical tests, and functional tests to verify the performance and safety of their battery modules. LG Chem's EOL testing involves a series of assessments, including insulation measurement, electrical continuity testing, and OCV control, to ensure the integrity and safety of their battery cells. Panasonic's EOL testing includes thermal imaging and leakage tests to detect any anomalies that might compromise the performance or safety of their battery modules. Also, it is notable that companies like Digatron from Germany and Unico from the United States focus on specific EOL testers for their products to ensure that batteries meet performance and safety standards throughout their intended lifespan, as in Figure 9.8 [57, 58].

By implementing EOL testing, LIB manufacturers can mitigate potential risks and uphold quality standards. EOL testing provides assurance that each battery leaving the manufacturing line meets rigorous quality and safety standards. This comprehensive evaluation ensures that the battery delivers the promised performance metrics, such as energy density, cycle life, and efficiency. Furthermore, EOL testing generates a wealth of data that manufacturers analyze to identify patterns, trends, and anomalies, helping in continuous process improvement and identifying potential areas for enhancement.

Figure 9.8 Battery EOL tester designed by Digatron and DSA. Source: [57].

9.3 Quality Gate Concept

LIB production undergoes crucial quality control checks to ensure the safety, performance, and reliability of the final product. Two primary methods used for this purpose are the quality gate concept and PFMEA. As mentioned earlier, PFMEA is a structured analytical tool designed to identify and evaluate potential failures in the manufacturing process, establish the impact of these failures, and prioritize action items to mitigate risks. On the other hand, the quality gate concept [59] involves strategically placed checkpoints to evaluate the quality of materials, components, and the overall manufacturing process, as illustrated in Figure 9.9. At each quality gate, batteries undergo a series of tests and inspections to ensure adherence to predefined standards and specifications. These checkpoints act as a prequality alert system, detecting and rectifying potential issues before they escalate. Thus, it minimizes defects and ensures that only high-quality batteries progress to subsequent production stages, thereby achieving high yields.

Figure 9.9 shows a quality gate concept, consisting of a prequality checking system, tests and inspection, data analytics, and a feedback mechanism. Specifically, the concept involves strategically placed checkpoints to evaluate the quality of materials, components, and the overall manufacturing process. At each quality gate, batteries undergo a series of tests and inspections to ensure adherence to predefined standards and specifications. These checkpoints act as a prequality alert system, detecting and rectifying potential issues before they escalate. Consequently, it minimizes defects and ensures that only high-quality batteries progress to subsequent production stages, thereby achieving high yields.

Figure 9.9 Simplified quality gate concept applicable to lithium-ion battery manufacturing.

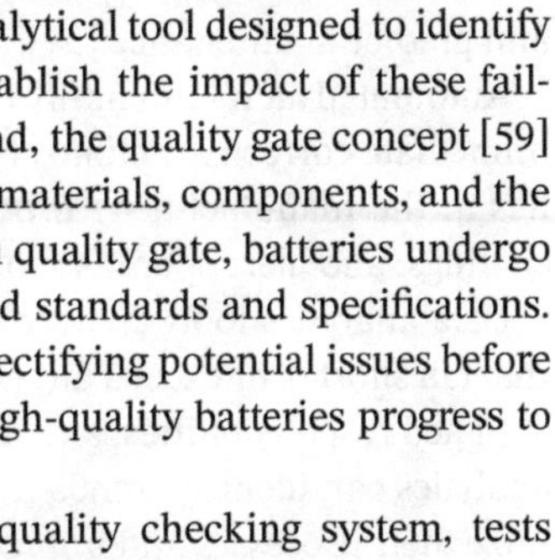

A key aspect of the quality gate concept is the incoming inspection of raw materials prior to their integration into the manufacturing process. This inspection ensures that only materials meeting specified standards are utilized, preempting potential issues downstream. Furthermore, the quality gate concept facilitates traceability and documentation throughout the production journey of each battery. By maintaining comprehensive records of materials, processes, and inspections, manufacturers can establish a transparent and accountable system that supports thorough quality analysis and effective problem-solving. In the event of postproduction issues, the detailed documentation enables swift identification of root causes and implementation of corrective measures, ensuring the integrity and reliability of the final product.

In an industry characterized by rapid innovation and evolving customer demands, real-time monitoring and feedback mechanisms within the quality gate concept are essential [59]. By leveraging an integrated approach that combines sensors, data analytics, and automated alerts within the production process, manufacturers can continuously monitor KPIs and address deviations from established standards promptly. This proactive methodology begins by identifying KPIs that align with the organization's operational goals and specific industry challenges, defining realistic and measurable targets and implementing robust data collection mechanisms.

Real-time monitoring involves the use of sensors and data analytics to track the production process and identify potential issues before they escalate. This includes monitoring parameters such as temperature, humidity, and pressure, as well as tracking the performance of critical equipment and machinery. By leveraging real-time data, manufacturers can detect anomalies, optimize processes, and predict maintenance [59].

Automated alerts can notify operators of any deviations from the established standards, enabling immediate corrective actions. For instance, in LIB production, automated alerts can detect anomalies in the manufacturing process, such as variations in cell thickness or inconsistent electrode coatings, and alert operators to take corrective action before the issue escalates [27].

Data analytics tools analyze collected data to uncover trends, patterns, and performance areas that fall short of or exceed targets. This process involves developing action plans with specific steps, assigned responsibilities, and timelines to address performance gaps effectively. For example, data analytics can identify trends in battery performance and quality, enabling manufacturers to optimize their processes and improve overall efficiency [60].

Regularly reviewing and updating KPIs ensures they remain aligned with the evolving goals and priorities of the organization. By integrating these technologies and practices, manufacturers can maintain consistent quality, meet customer expectations, and sustain a competitive advantage in the market. For instance, in LIB production, data analytics can help manufacturers optimize their processes to reduce scrap rates and improve overall efficiency, ultimately leading to higher-quality products and increased customer satisfaction [59].

Interactive communication channels and feedback loops between manufacturing and quality checkpoints foster a collaborative environment where insights from the quality control process indicate process deviations and product nonconformity. This interconnected approach allows for a holistic view of the production process and facilitates timely adjustments to optimize efficiency and quality. To effectively implement interactive communication and feedback loops in LIB production, manufacturers should establish clear communication channels between manufacturing and quality checkpoints [60]. This also involves employing feedback mechanisms to provide real-time insights on process deviations and product nonconformity. Subsequently, it feeds into monitoring and analyzing data from quality control checks to identify trends and patterns, and then adjusting processes to optimize efficiency and quality based on these insights [59]. Additionally, manufacturers need to regularly review and update their quality control procedures to ensure they remain aligned with evolving goals and priorities. By integrating interactive communication channels and feedback loops in LIB production, manufacturers can maintain consistent quality, meet customer expectations, and sustain a competitive advantage in the market.

Leveraging data analytics and predictive maintenance within the quality gate framework provides actionable insights into production trends and performance metrics [27]. This data-driven approach facilitates proactive decision-making and enables the identification of potential issues before they impact product quality or production efficiency. By harnessing the power of data, manufacturers can optimize processes, minimize downtime, and continuously enhance product quality, thus maintaining a competitive edge in the market.

Analyzing sensor data from condition-monitoring sensors allows manufacturers to assess the health and performance of critical equipment, enabling early detection of potential issues and timely interventions. This ensures machinery operates at peak efficiency and reduces the risk of costly downtime and defects in the production line [61]. Data analytics also plays a crucial role in process optimization by scrutinizing production data to identify bottlenecks and inefficiencies. For instance, predictive maintenance models detect periods of valve degradation, providing timely notifications for maintenance and reducing unplanned downtime. In supply chain management, data analytics offers visibility into demand patterns, supplier performance, and logistics operations, which facilitates optimized inventory management, improved delivery times, and reduced costs.

Data analytics support better decision-making by providing hard evidence and predictive analysis, eliminating guesswork and enhancing overall business performance. It identifies opportunities for cost reduction, streamlines supply chain operations, balances inventory to demand, and reduces utility waste. By implementing predictive maintenance and data analytics within the quality gate framework, manufacturers can continuously monitor and optimize their production processes, minimizing downtime, improving product quality, and enhancing overall operational efficiency. This approach ensures the delivery of high-quality products consistently and efficiently, which aids their competitive advantage.

Ultimately, the quality gate concept ensures that LIB production consistently meets regulatory requirements and adheres to industry best practices. By upholding rigorous quality standards and embracing continuous improvement initiatives, manufacturers can drive innovation, enhance competitiveness, and maintain customer satisfaction in an increasingly dynamic and demanding market landscape.

9.4 Screening Technologies for Batteries

With increasing demand for reliable and high-performance batteries, ensuring their quality has become crucial for manufacturers. In battery production, various screening technologies are employed to ensure the quality and reliability of batteries. These technologies include X-ray inspection, thermal imaging, EIS, acoustic emission testing, optical inspection, and ultrasonic testing. Specifically, they detect potential defects and issues affecting battery performance or safety. By using these methods, manufacturers can ensure battery reliability and performance, meeting application demands [62].

9.4.1 Optical Inspection

Optical inspection is a non-invasive technique in battery manufacturing, utilizing high-resolution cameras and advanced image processing to detect surface defects and irregularities in battery components like electrodes and separators. It identifies visible issues such as surface cracks, which can

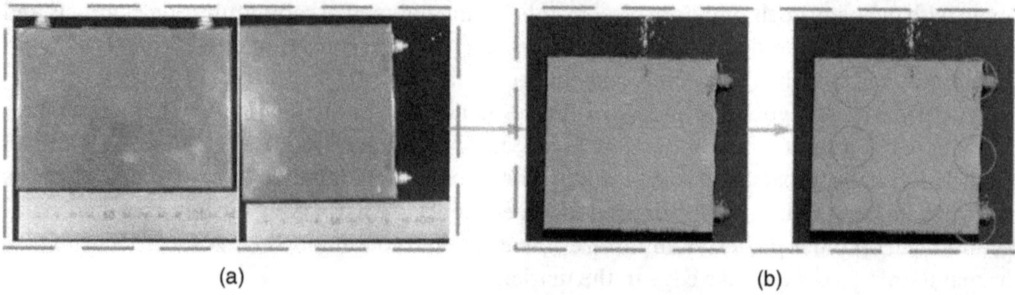

Figure 9.10 Surface defect detection using a depth camera. (a) The battery used for surface defect detection, and (b) The depth camera image of the battery and the defects detected (in gray circles). Source: [64]/Springer Nature/CC BY 4.0.

compromise battery integrity, and contamination, which can cause short circuits and operational failures. Optical inspection techniques include visual inspection, microscopy, and surface profilometry. Visual inspection identifies surface defects, scratches, and irregularities that may compromise the quality and performance of batteries, as shown in Figure 9.10 [64]. Microscopic analysis examines surface features at a microscopic scale to detect subtle defects and assess coating uniformity. Surface profilometry provides quantitative measurements of surface roughness and topography, offering insights into dimensional accuracy and consistency.

This technique verifies proper alignment of the anode and cathode, confirming the absence of scratches, voids, and particle agglomeration and ensuring a uniform appearance in terms of color and surface quality. Electrode edges are inspected to ensure they are straight, free of damage, and without burrs, foreign particles, or exposed current collectors, which could compromise the integrity of the cell.

For separators, optical inspection ensures they are free from discoloration, oxidation, wrinkles, scratches, and tears, as depicted in Figure 9.10(b). Maintaining separator quality is vital for preventing short circuits and ensuring reliable operation. Tabs undergo inspection to confirm clean-cut edges, proper tape placement, and the absence of weld spatter, all of which prevent interference with the separator and maintain mechanical integrity during assembly. By integrating optical inspection at every stage, including electrode fabrication, assembly, and postproduction quality control, manufacturers maintain high standards of quality and consistency, delivering reliable and safe battery products to the market.

9.4.2 Ultrasonic Testing

Ultrasonic testing utilizes high-frequency sound waves (>20 kHz) to penetrate the material and inspect internal defects such as voids, cracks, and bond quality [65]. This nondestructive testing method involves analyzing sound wave responses, as illustrated in Figure 9.11 [67].

In battery manufacturing, the evaluation of welded or bonded joints is critical for ensuring robust connections that are essential for the overall safety and longevity of battery systems [67, 68]. Ultrasonic testing is key to the quality assurance process, detecting potential weaknesses and thereby upholding structural integrity standards [69].

The use of ultrasonic testing in battery manufacturing can have significant cost implications, primarily due to its cost efficiency, ability to reduce waste, and enhancement of safety and reliability. Ultrasonic testing is described as a cost-efficient method that provides greater accuracy, safety, and performance for testing, managing, and reusing LIBs [70]. Recent technological advancements have led to increased automation in ultrasonic testing, making the inspection process more efficient

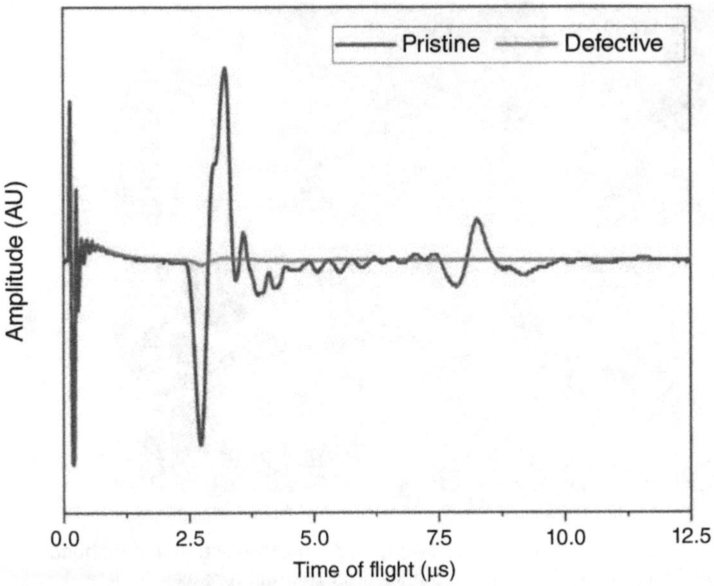

Figure 9.11 Comparison of ultrasonic ToF waveforms obtained in transmission mode through pristine and defective commercial cells. Source: Adapted from [66].

and accurate and reducing labor costs [67]. Additionally, the versatility of ultrasonic testing systems allows for flexibility in adapting to different battery sizes, geometries, and production requirements, potentially reducing capital expenditures [71].

9.4.3 X-Ray Inspection

X-ray inspection enables manufacturers to examine the internal structure and material distribution of batteries without causing harm, ensuring product safety and quality control. Additionally, X-ray inspection can detect internal defects such as impurities, distortion of positive and negative poles, short circuits, and fractures, which can lead to safety risks, reduced battery life, or even explosions [72].

X-rays penetrate the battery cell, and sensors capture the image, which advanced software reconstructs in 3D, as shown in Figure 9.12. One of the primary defects identified through X-ray inspection is internal short circuits, which can cause overheating and potential fires. These short circuits can result from manufacturing defects, such as misaligned components or impurities within the cell, or damage during use. Misalignment of internal components, detectable through X-ray inspection (illustrated in Figure 9.13), can lead to uneven wear and performance issues over time, often due to poor assembly practices or material deformation during the battery's life cycle. X-ray inspection allows for the examination of internal components, such as electrode alignment, and the evaluation of welds and solder joints in connectors and tabs, identifying weak or incomplete joints that could impact electrical conductivity. Additionally, it verifies the integrity of various components like the casing and terminals, detecting hidden defects such as cracks or voids.

During electrode and cell assembly, X-ray inspection ensures proper positioning and alignment, and before final packaging, it checks for any remaining defects or issues in the assembled cells, confirming that they are sealed correctly and free from internal flaws. Overall, X-ray inspection plays a vital role in maintaining the quality and reliability of LIBs by providing comprehensive and accurate assessments of their internal structure. By ensuring proper alignment and detecting

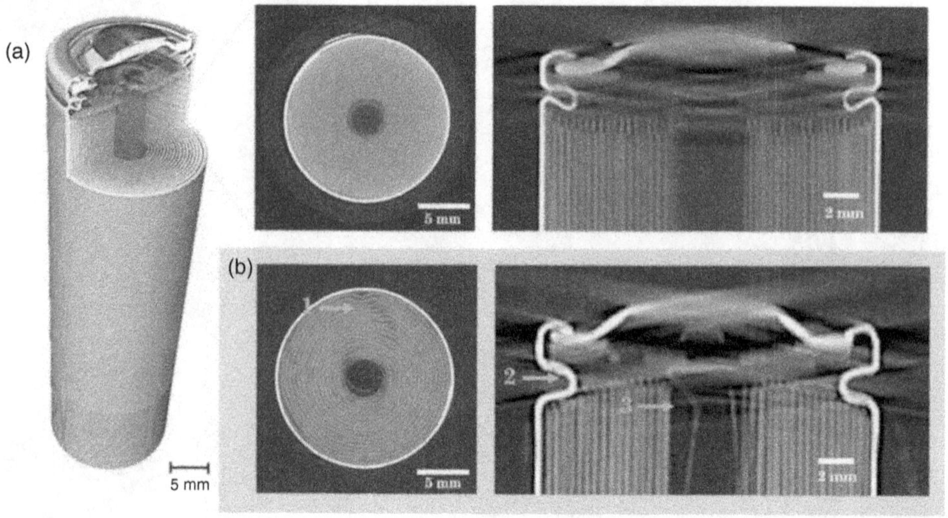

Figure 9.12 X-ray imaging results for a cylindrical battery, showing (a) 3D reconstruction and orthogonal slices in the XY and YZ planes, and (b) slices in the XY and YZ planes after thermal runaway. Source: [73]/IOP Publishing/CC BY 4.0.

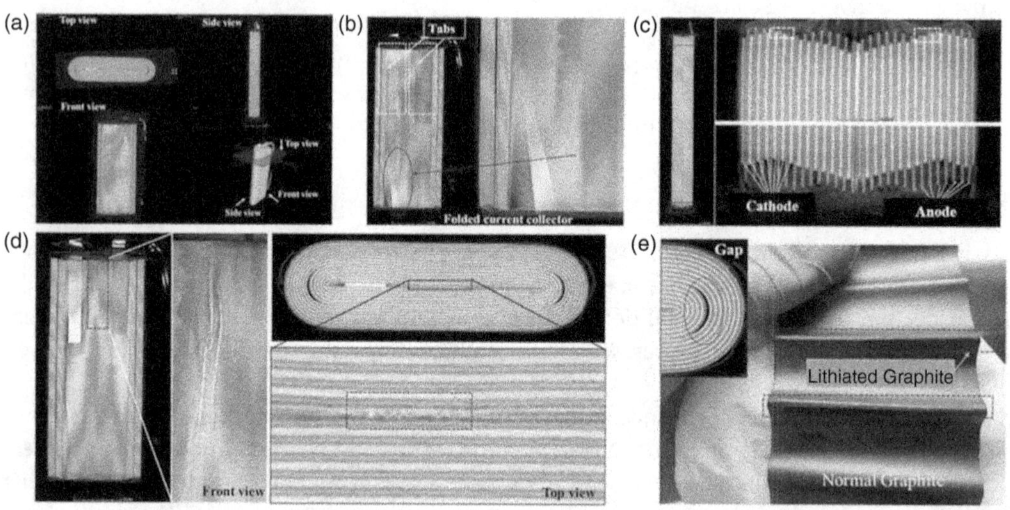

Figure 9.13 X-ray imaging system for identifying defects, showing (a) foreign particle inclusion, (b) cell assembly flaws, (c) electrode misalignment, (d) anode overhang, and (e) cracks or voids or air gap. Source: Kong et al. [74]/IOP Publishing/CC BY 4.0.

internal defects, X-ray inspection ultimately helps manufacturers produce safer and more reliable battery cells.

9.4.4 Thermal Imaging

Thermal imaging technology is a noncontact method that uses infrared cameras to detect temperature variations on battery surfaces, revealing potential issues such as internal short circuits, high resistance connections, or other defects (Figure 9.14).

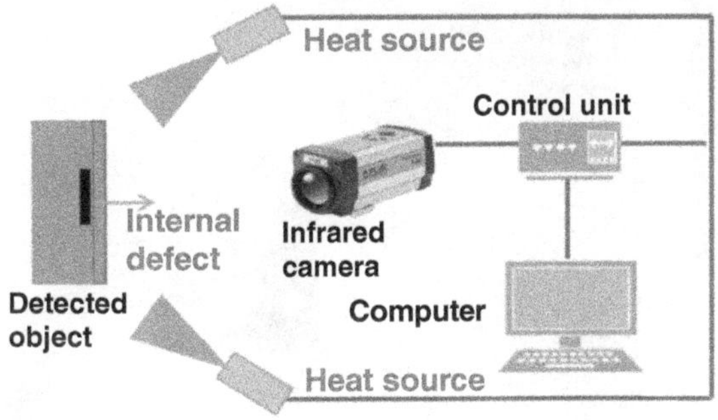

Figure 9.14 Using thermography for battery inspection. Source: Qu et al. [75]/MDPI/CC BY 4.0.

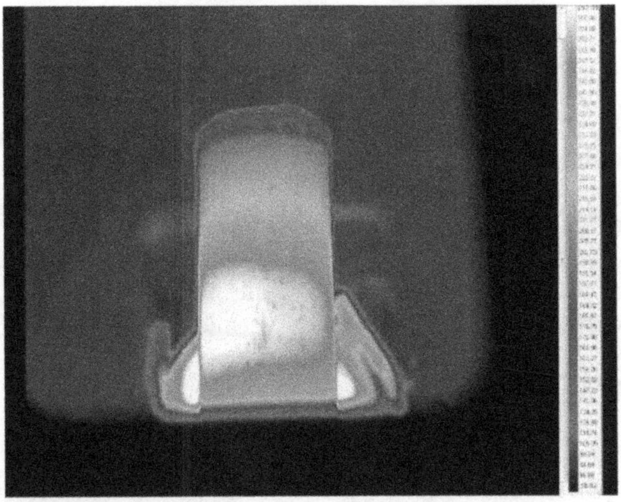

Figure 9.15 Thermal image of a battery immediately prior to thermal runaway. Source: [77]/Springer Nature/CC BY 4.0.

Elevated temperatures or hot spots can signal inefficiencies or risks of thermal runaway, which can lead to catastrophic failures [76]. These hot spots may be caused by internal short circuits, high resistance connections, or other defects that generate heat. In manufacturing, thermal imaging assesses weld quality and temperature distribution in battery packs, identifying defects like poor welds that can compromise performance and safety. During production, thermal imaging ensures the quality of welds and connections, by revealing inconsistent temperature patterns. Poor welds or faulty connections can be detected as they often create areas with abnormal heat distribution (Figures 9.15, 9.16).

9.4.5 Electrochemical Impedance Spectroscopy (EIS)

EIS measures the impedance of a battery cell over a range of frequencies, providing insights into the electrochemical processes occurring within the cell [79]. A small AC signal is applied to the battery cell, and the response is measured and analyzed across different frequencies. EIS is particularly useful for identifying defects related to the electrochemical performance of the battery. For instance, electrolyte degradation, which leads to reduced capacity and efficiency, can be detected through EIS. Electrolyte degradation can result from chemical reactions within the cell that consume or alter the electrolyte over time. Issues with the SEI layer, which can impact

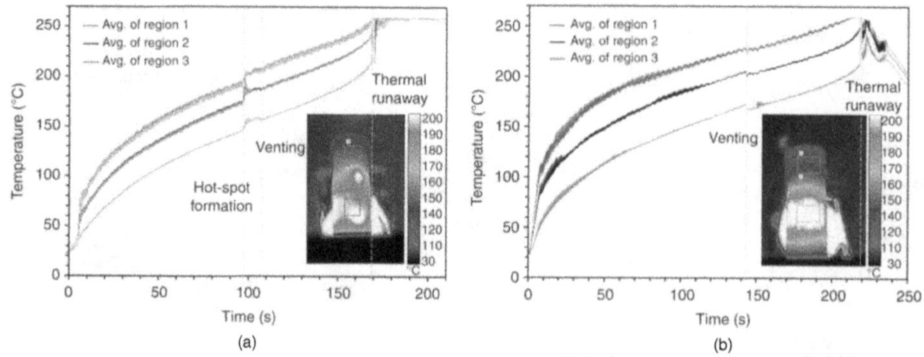

Figure 9.16 Mean surface temperature changes of three regions during a thermal abuse test, showing (a) cell 1, with the image taken at t = 97 s, and (b) cell 2, with the image taken at t = 80 s. Source: Finegan et al. [78]/Springer Nature/CC BY 4.0.

battery life and performance, can also be identified using this method [80]. The SEI layer forms on the anode during the initial cycles and plays a crucial role in stabilizing the interface between the electrode and the electrolyte. However, defects in the SEI layer can lead to increased resistance and capacity loss.

EIS enables the initial screening of batteries right after assembly, quickly identifying cells that deviate from specified impedance criteria and that may have potential issues [81]. EIS is also applied for batch testing, where it assesses samples from various production lots to uncover any anomalies or deviations that might indicate quality control problems. By monitoring impedance changes throughout the production process, EIS facilitates real-time process control and optimization, allowing for immediate corrective actions if impedance characteristics exhibit undesirable shifts. Additionally, EIS plays a key role in detecting internal defects, such as issues with electrode coatings and electrolyte distribution, ensuring that only the cells that meet stringent quality requirements proceed to the next stages of production and ultimately reach the market.

9.4.6 Acoustic Emission Testing

Acoustic emission is an effective technique for monitoring the mechanical integrity of battery cells, providing early warning signs of potential failures, and ensuring the structural soundness of the cells [82, 83]. Acoustic sensors capture sound waves emitted by the battery cell, and the data is analyzed to detect and locate defects. Additionally, crack propagation, or the growth of cracks within the battery structure, can be identified through acoustic emission monitoring. It is used in both laboratory settings and field applications to monitor battery health and detect signs of mechanical damage.

Acoustic emission monitoring is particularly valuable for large-scale battery systems, where early detection of defects can prevent costly failures and ensure safe operation. For example, acoustic emission data can show crack propagation patterns, highlighting areas within the battery structure that are at risk of failure. Areas of mechanical deformation, illustrated by acoustic emission data, indicate structural changes due to mechanical stress, affecting a battery's integrity. By continuously monitoring the mechanical state of battery cells, acoustic emission testing helps to maintain the safety and reliability of battery systems, providing essential data for preventive maintenance and early intervention.

9.5 Best Practices in Battery Quality Assurance

Battery quality assurance is a critical aspect of battery manufacturing and design. Importantly, it ensures that batteries meet the required performance specifications, safety standards, and reliability expectations. Battery quality assurance is a multifaceted process crucial to ensuring the performance, durability, and safety of batteries across diverse applications [84]. The most common quality issues in LIBs include manufacturing defects, cycle life issues, thermal runaway, and inconsistent performance of batteries. These common quality issues highlight the importance of rigorous quality control measures and advanced manufacturing practices in ensuring the safety, reliability, and performance of LIBs. By addressing these challenges, manufacturers can improve the overall quality of their products. This chapter provides a comprehensive overview of the essential best practices for ensuring high quality in LIB manufacturing.

Before analyzing physical manufacturing processes, companies frequently employ battery simulation and modeling tools to refine battery designs and forecast performance attributes [1]. This early-stage simulation serves as a pivotal step in the battery development process, enabling manufacturers to explore various electrode configurations, electrolyte compositions, and cell geometries. Simulation tools include finite element analysis (FEA), computational fluid dynamics (CFD), electrochemical modeling, system-level simulation, and multiscale and multiphysics modeling. FEA analyzes mechanical stress, thermal behavior, and electrochemical performance, while CFD simulates fluid flow and heat transfer. Electrochemical modeling focuses on reactions within the battery, predicting voltage, capacity, and efficiency. System-level simulation examines the overall battery system, including interactions between cells, modules, and BMSs, and multiscale and multiphysics modeling integrates various physical phenomena and scales for comprehensive understanding. Additionally, simulation enables the exploration of novel materials and manufacturing techniques, providing insights into their potential impact on battery performance and durability. With the exploration of novel materials, simulations aid in identifying performance bottlenecks and safety considerations early in the design process. This approach enhances performance, safety, efficiency, and cost-effectiveness in LIBs.

One of the best practices in this field is the implementation of comprehensive IQC for raw materials. The quality of raw materials directly impacts the performance, safety, and lifespan of LIBs since any impurities or inconsistencies can lead to defects and failures in the final product. Therefore, establishing strict specifications for raw materials and conducting thorough inspections upon arrival is vital. Advanced analytical techniques like X-ray diffraction (XRD), scanning electron microscopy (SEM), and inductively coupled plasma (ICP) analysis are used to verify the purity and composition of materials [60, 85].

Maintaining a controlled manufacturing environment is another critical best practice. Environmental factors such as temperature, humidity, and cleanliness can significantly affect the quality and reliability of battery components during production. To address this, cleanroom standards, like personnel and equipment protocols, a regular cleaning schedule, and other aspects, are implemented in critical areas of the manufacturing process to minimize contamination. Furthermore, dehumidifiers and climate control systems are used to maintain optimal temperature and humidity levels. Regular monitoring and recording of environmental conditions ensure they remain within specified limits, thus safeguarding the production process [86].

Implementing electrostatic discharge (ESD) protection on the manufacturing floor, infrastructure, and personnel is also essential. ESD can cause significant damage to battery components during production. Measures include grounding of equipment and personnel, using ESD-safe materials for work surfaces and tools, and maintaining appropriate humidity levels to reduce static buildup. Also, ESD training for employees and regular audits ensure ongoing compliance with ESD protection standards (EN 61340-5-1 or ANSI/ESD S20.20).

Precision in electrode fabrication is also essential for high-quality battery production, as the uniformity and consistency of electrodes are crucial for a battery's electrochemical performance and safety. Variations in thickness, composition, or coating can lead to imbalances and reduced efficiency. To achieve precision, automated coating machines and high-precision calendaring equipment are employed to produce uniform electrodes. In-line measurement systems continuously monitor electrode thickness and composition during fabrication, allowing for real-time adjustments based on feedback from these systems. For example, Panasonic uses automated coating and calendaring equipment to ensure uniformity and consistency in electrode production, which is critical for the performance of its batteries used in EVs [87].

Strict process control and monitoring throughout production are imperative for maintaining consistent quality, which can only be achieved if all stages of the manufacturing process are precisely controlled and monitored to detect and correct deviations in real time. Advanced process control systems (APCSs) integrated with manufacturing execution systems (MESs) are utilized to oversee production parameters. SPC methods identify and rectify variations, and real-time data analytics and machine learning algorithms predict and prevent potential issues before they impact quality [88]. Tesla, for instance, employs APCSs and real-time data analytics at its Gigafactories to ensure consistent quality and optimize production efficiency [89].

Comprehensive in-line and EOL testing are best practices that help to ensure high-quality batteries. Regular testing at different stages of production can identify defects early, ensuring that only high-quality batteries move forward in the production process. Automated optical inspection (AOI) systems check for surface defects and misalignments in real time, while EIS and cycle life testing provide an in-depth analysis of battery performance. The final EOL tests, including capacity testing, thermal runaway testing, and vibration testing, ensure the batteries meet all specifications and safety standards [90].

Effective nonconformance management procedures are essential for maintaining high quality, which include a robust system for documenting, tracking, and resolving nonconformances. Promptly identifying and addressing nonconformances is critical to prevent defective products from reaching the market. Accordingly, root cause analysis tools like fishbone diagrams and the 5 Whys Technique can help to identify the underlying causes of issues. Implementing corrective and preventive actions (CAPA) addresses nonconformances and prevents their recurrence. Regularly reviewing and updating quality control procedures based on nonconformance data ensures continuous improvement [91]. As an example, Samsung SDI implements robust nonconformance management procedures to maintain high quality and address any issues promptly [92].

Advanced employee training programs are crucial for maintaining high-quality standards and effectively operating complex manufacturing equipment, as skilled personnel are vital for ensuring the success of quality control processes. Comprehensive training programs that cover battery technology, quality control procedures, and safety protocols are crucial, while hands-on training sessions and certification programs for specialized roles ensure proficiency. Continuous learning through workshops, seminars, and online courses keeps employees updated on the latest industry developments and best practices [93].

The implementation of the quality gate concept in production, which was previously discussed, is another best practice. Quality gates act as checkpoints in the production process, ensuring that each stage meets predefined standards before proceeding to the next step. At each quality gate, specific inspections and tests verify that the components and processes meet the required standards.

Automated systems track and document the results of each quality gate, enabling traceability and accountability [94].

Effective supply chain management is crucial for securing high-quality materials, maintaining production continuity, and meeting customer demand in battery manufacturing. Companies must establish robust relationships with trusted suppliers, implement rigorous supplier qualification processes, and continuously monitor and audit supply chain operations to ensure compliance with quality and ethical standards. The PPAP plays a vital role in this by ensuring suppliers can consistently produce parts that meet specifications, thus reducing the risk of defects. Additionally, proactive risk management strategies, like dual sourcing and strategic inventory management, help mitigate supply chain disruptions and minimize production schedule impacts, ensuring a steady supply of high-quality battery components.

Leveraging data analytics and artificial intelligence (AI) for quality control is increasingly important. Advanced data analytics and AI provide deeper insights into the production process, identifying patterns and predicting potential issues that may not be apparent through traditional methods. Data analytics platforms collect and analyze data from various stages of the manufacturing process, while AI algorithms detect anomalies, predict failures, and optimize production parameters. Additionally, dashboards and real-time monitoring tools visualize KPIs and quality metrics, facilitating informed decision-making [95]. Specifically, BYD Company utilizes AI and data analytics to enhance its quality control processes, ensuring high standards and efficient production [96].

Establishing a culture of continuous improvement and feedback loops is essential for adapting to evolving industry standards, technological advancements, and customer expectations. Encouraging employees to identify and suggest improvements fosters a proactive approach to quality control. Regularly reviewing and analyzing production data, customer feedback, and industry trends helps to identify areas for enhancement. Implementing iterative improvements, through pilot projects and controlled trials before full-scale implementation, ensures that changes lead to measurable improvements in quality and efficiency. Feedback loops ensure that these changes are effective and sustainable [96].

9.6 Challenges and Pitfalls

A significant challenge in LIB production stems from managing the complex supply chain associated with sourcing raw materials and components. LIBs require critical materials such as lithium, cobalt, and nickel, sourced from diverse suppliers across multiple regions and countries. Consequently, supply chain disruptions, geopolitical tensions, and fluctuating material prices can impact production timelines and increase manufacturing costs, posing challenges for companies striving to deliver high-quality batteries on schedule and within budget [97, 98]. Ensuring a resilient and efficient supply chain is essential for maintaining production continuity and meeting customer demand. However, achieving supply chain resilience requires proactive risk management strategies, diversification of suppliers, and strategic stockpiling of critical materials. Collaborative partnerships with suppliers and transparent communication channels help to build resilience and agility within the supply chain, enabling companies to respond swiftly to disruptions and mitigate potential risks.

Increasingly, companies are focusing on ethical sourcing practices and environmental sustainability in their supply chain operations. Ensuring responsible sourcing of raw materials, including addressing concerns related to child labor and human rights abuses in cobalt mining regions, is essential for upholding corporate social responsibility commitments and maintaining stakeholder trust. These concerns must be addressed through transparent supply chain practices and certification programs.

9.6.1 Raw Material Quality

The quality of raw materials significantly impacts the performance and safety of LIBs. Variations in the purity and composition of materials for the cathode, anode, electrolytes, and separators can lead to inconsistencies in battery performance. Contaminants and impurities can cause unwanted chemical reactions, reducing battery efficiency and increasing the risk of thermal runaway. Ensuring consistent quality in raw materials requires rigorous supplier selection, comprehensive material testing, and maintaining strict quality standards throughout the supply chain.

One major challenge in this context is material variability. Variability in raw material quality can lead to significant issues during battery operation. For example, inconsistent particle size distribution in cathode materials can affect electrode packing density, leading to capacity loss and a reduced cycle life [99]. Such inconsistencies result in uneven electrical performance, increased internal resistance, and reduced overall efficiency of the battery. To overcome this challenge, manufacturers need to implement quality control measures at the supplier level.

This includes regular audits of suppliers, comprehensive material testing upon receipt, and maintaining detailed records of material properties. Advanced analytical techniques, such as ICP analysis and SEM, can be used to verify the purity and composition of raw materials. These methods allow for the detection of impurities and inconsistencies in particle size and distribution. By ensuring the consistency and purity of raw materials, manufacturers can produce more reliable and high-performing batteries, ultimately enhancing the safety and longevity of the battery cells [100]. Implementing these quality control measures helps to maintain a high standard throughout the supply chain, reducing the risk of defects and performance issues in the final product.

9.6.2 Electrode Manufacturing

The electrode manufacturing process involves coating, drying, calendaring, and slitting, each of which presents unique quality control challenges. Coating uniformity is crucial to ensure consistent electrochemical performance, since variations in coating thickness can lead to uneven current distribution, hotspots, and reduced battery efficiency. The drying process must be carefully controlled to prevent solvent residues, which can cause side reactions during battery operation. Calendaring, which compresses the electrode material to improve energy density, must be precisely controlled to avoid damaging the electrode structure.

Inconsistent coating thickness can result in areas of higher resistance within the electrode, leading to localized heating and potential safety hazards during battery operation. To address these coating uniformity issues, manufacturers can use automated, precision coating machines equipped with real-time monitoring systems. These systems can detect and correct variations in coating thickness, ensuring a uniform application across the entire electrode. Additionally, implementing rigorous process controls and regular calibration of equipment can help maintain consistency. Advanced techniques such as laser profilometry can be employed to measure and verify coating thickness, allowing for immediate adjustments and quality assurance [99]. By employing these strategies, manufacturers can enhance the reliability and performance of LIBs, mitigating risks associated with uneven electrode coatings.

9.6.3 Cell Assembly

During cell assembly, the electrodes are combined with the separator and electrolyte, and the cell is sealed. This process must be strictly controlled to prevent contamination and ensure proper alignment of components. Misalignment of electrodes can lead to internal short circuits, while contamination from particles or moisture can cause degradation of the electrolyte and reduced battery performance. Ensuring a clean and controlled assembly environment is essential to prevent these defects [101].

A misaligned electrode can create a short-circuit path, leading to rapid discharge and potential thermal runaway, and contaminants introduced during assembly can degrade battery components and performance. To mitigate these risks, manufacturers must establish cleanroom environments for cell assembly, equipped with advanced air filtration systems to minimize particulate contamination. Automated assembly systems with precise robotic handling can ensure the accurate alignment of electrodes and other components. Implementing in-line inspection systems, such as vision systems and laser alignment tools, can further ensure the proper placement of all parts. Additionally, regular training for assembly line workers and rigorous hygiene protocols can help maintain a contaminant-free environment [102].

9.6.4 Electrolyte Filling

The electrolyte-filling process involves injecting the electrolyte into the cell and ensuring it is evenly distributed. Any inconsistencies in electrolyte distribution can lead to areas of poor ionic conductivity, reducing battery capacity and efficiency. Additionally, improper sealing after filling the electrolyte can result in leaks, which not only degrade performance but also pose safety risks [103].

An unevenly distributed electrolyte can create regions of high impedance within the cell, leading to localized heating and reduced overall performance. To safeguard uniform electrolyte distribution, manufacturers can use precision dispensing equipment with real-time monitoring capabilities. These systems can accurately control the volume and placement of the electrolyte, ensuring even distribution throughout the cell. Vacuum filling techniques can also help remove air pockets and ensure complete wetting of the electrodes. Post-filling, cells should be subjected to rigorous leak testing, using methods such as helium mass spectrometry to detect any sealing issues [104].

9.6.5 Formation and Aging

Formation is the initial charging process where the SEI layer forms on the anode. This layer is critical for the battery's long-term stability and performance. The formation process must be carefully controlled to ensure a stable and uniform SEI layer. Aging tests, where the battery is cycled multiple times, help to identify early failures and ensure long-term reliability. However, these processes are time-consuming and can introduce variability if not properly controlled [105].

An unstable SEI layer can lead to continuous electrolyte decomposition, increasing internal resistance and reducing battery lifespan. To control SEI layer formation, manufacturers can use precise charge and discharge protocols during the formation stage, tailored to the specific battery chemistry. Advanced diagnostics, such as in situ impedance spectroscopy, can monitor the SEI layer's development in real time, allowing for adjustments to the formation process. Automated aging test systems with controlled environmental conditions can ensure consistent cycling and reliable identification of early failures [106].

9.6.6 Testing and Inspection

Comprehensive testing and inspection are essential at every stage of the manufacturing process to ensure the quality and reliability of LIBs. This includes electrical testing, mechanical inspection, and advanced techniques such as X-ray and ultrasonic imaging. However, implementing these tests can be challenging due to the high volume of batteries produced and the need for nondestructive testing methods that can detect subtle defects without damaging the cells [107].

To overcome the challenges associated with high-volume and nondestructive testing, manufacturers can invest in automated testing systems that integrate multiple nondestructive testing methods into the production line. Systems equipped with X-ray CT and ultrasonic testing can provide detailed insights into the internal structure and integrity of each cell without causing damage. Advanced software algorithms analyze the test data in real time, identifying any defects or anomalies. This integration of technology ensures that high-quality batteries are produced consistently, while maintaining the high throughput necessary for large-scale production [108]. By incorporating these advanced testing and inspection technologies, manufacturers can significantly enhance the quality control process, ensuring the delivery of reliable and safe LIBs to the market.

9.6.7 Ensuring Consistent Quality in High-Volume Manufacturing

One major challenge in LIB manufacturing is maintaining consistent quality in high-volume production. Variations in the manufacturing process can lead to defects and inconsistencies in battery performance, exacerbated by the need to maintain high throughput while minimizing defects and rework.

To address this, manufacturers can implement SPC, which uses statistical methods to monitor and control the manufacturing process. SPC involves using control charts to track key process parameters, such as electrode coating thickness, over time. This real-time monitoring allows manufacturers to detect variability and take corrective actions before defects occur, ensuring process stability and product quality. For example, X-bar and R-charts can help maintain consistent coating thickness, preventing defects caused by variations in this critical parameter.

Additionally, SPC supports continuous improvement by identifying areas for process optimization. By analyzing process data and identifying trends, manufacturers can implement changes that reduce variability and enhance overall performance. This proactive approach helps manufacturers to ensure that their high-volume production processes consistently produce high-quality LIBs, reducing the risk of defects and enhancing product reliability [109].

9.7 Key Components of a Quality Control Facility

Ensuring the quality control of LIB production significantly affects the cost and performance of the batteries [86]. A quality control program requires careful planning, precise infrastructure, specialized equipment, and skilled personnel. An earlier chapter describes the production steps for LIBs in more detail, but this section, along with Figure 9.17, summarizes the steps within the context of quality control.

The first step of manufacturing LIBs is the slurry preparation, which involves mixing the active material, binder, and conductive agent in specific mass ratios and dispersing various components to achieve a homogeneous mixture. This slurry serves as the foundation for the electrodes, providing the necessary materials for efficient energy storage and transfer within the battery. Measuring

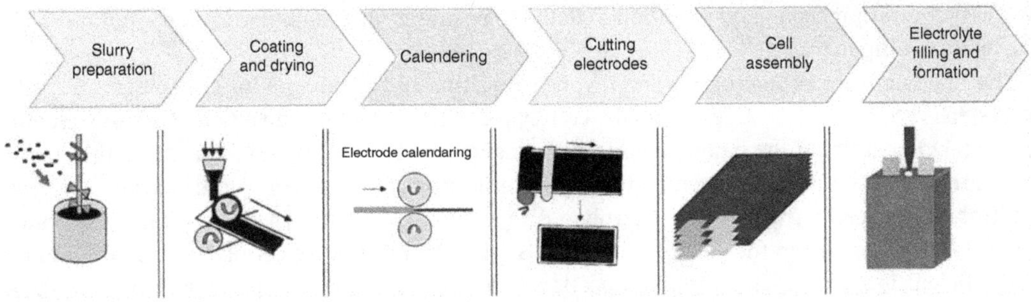

Figure 9.17 Manufacturing steps of Li-ion batteries. Source: [110]/Springer Nature.

the viscosity of the slurry helps to confirm that it is within the optimal range for proper coating and electrode formation. Further, thermal analysis techniques, such as differential scanning calorimetry (DSC) and thermogravimetric analysis (TGA), help to assess the thermal stability and decomposition behavior of the slurry components. This makes certain that the slurry does not degrade or decompose during the manufacturing process, which can negatively impact battery performance [111].

Subsequently, the electrodes are coated onto the current collectors during the coating and drying phase. This is accomplished through a tape-casting procedure, where the electrode slurry is precisely applied onto the current collector surface. Importantly, this coating process ensures uniform distribution of the active material and facilitates optimal electrochemical performance in the final battery assembly. Before progressing to the cell assembly stage, the porous electrodes undergo compression using large cylindrical machinery, a process known as calendaring.

The calendaring process in LIB manufacturing involves several critical quality control measures to ensure optimal electrochemical performance and mechanical stability. Temperature control is crucial, as it affects the plastic deformation of electrode materials, and proper regulation ensures malleability without compromising structural integrity [89]. Optimal speed control is also essential, since it influences processing time and energy consumption, while ensuring efficient processing and minimizing the risk of defects. Notably, the surface roughness and curvature of the rollers significantly impact the contact area and pressure distribution during calendaring, and proper roller geometry ensures uniform compaction and optimal porosity. Visual inspection is essential for detecting defects or irregularities in the calendared electrodes, confirming that they meet quality standards and are suitable for further processing. Microstructural analysis helps to understand the effects of calendaring on electrode microstructure, including particle rearrangement, compaction, and changes in pore size distribution. Electrochemical testing evaluates the performance of the calendared electrodes, including testing for capacity, rate capabilities, and cycling stability, while mechanical testing assesses their mechanical stability and durability with testing for tensile strength, elongation, and fracture resistance. By implementing these quality control measures, manufacturers can ensure the production of high-quality LIBs with optimal electrochemical performance and mechanical stability.

In the subsequent step of cutting electrodes, the electrodes are precisely cut or punched into strips of the desired shape. This precision-cutting process is essential for achieving uniform electrode dimensions and facilitating efficient assembly in the next stage of production. During cell assembly, the electrodes are carefully wound or stacked together with the separator, which prevents direct contact between the electrodes and facilitates ion transport within the battery. This assembly process requires precision and attention to detail to ensure proper alignment and configuration of the battery components.

The final step in LIB manufacturing involves injecting the electrolyte into the assembled cell. This step is crucial, as it ensures optimal electrolyte-electrode contact and efficient ion movement, ultimately maximizing the battery's performance and energy density [88]. To achieve this, quality control measures are implemented throughout this important process. These measures include electrolyte permeation, where the electrolyte is injected into the cell, ensuring it fully permeates and fills the pores within the electrodes, which is critical for optimal electrolyte-electrode contact and efficient ion movement. Visual inspection is performed to detect any defects or irregularities in the electrolyte injection process, ensuring proper distribution and no blockages or leaks in the cell.

Electrochemical testing evaluates the performance of the battery, including testing for capacity, rate capabilities, and cycling stability, which ensures optimal electrochemical performance. Mechanical testing assesses the mechanical stability and durability of the battery, including testing for tensile strength, elongation, and fracture resistance to ensure reliable performance. Finally, microstructural analysis is performed to understand the effects of electrolyte injection on the electrode microstructure, including particle rearrangement, compaction, and changes in pore size distribution, which influence electrochemical performance and mechanical stability. By implementing these quality control measures, manufacturers can ensure the production of high-quality LIBs that meet the demanding requirements of various applications, from consumer electronics to EVs and renewable energy storage systems.

To ensure the quality of LIBs, manufacturers must implement rigorous quality control measures [112]. One such measure is FMEA, which identifies potential failures and their effects on a battery's performance. Additionally, control charts are used to monitor and control the manufacturing process, ensuring that the battery meets the required standards. Furthermore, PPAP is utilized to ensure that batteries meet the customer's specifications and requirements. These quality control measures are crucial in ensuring the reliability and safety of LIBs.

9.7.1 Specialized Equipment

A quality management program for LIB production needs to have specialized equipment to perform various tests and measurements. Some of the specialized equipment include battery cell testers, thermal imaging cameras, and cycle life testers.

9.7.1.1 Battery Cell Testers

A battery cell tester works by analyzing the overall health of primary or secondary cell batteries through various parameters such as voltage, charge, impedance, ripple current, and others, ensuring they meet the required specifications [93]. One specific method used by battery cell testers is charge analysis, which measures a battery's ability to hold a charge and its capacity to be recharged. Another method, impedance analysis, measures the internal resistance of a battery, which affects its performance and efficiency. Ripple current analysis tests the battery's ability to handle the AC ripple current that is present in the charging circuit. These tests are typically performed using specialized instruments like the Fluke 500 Series Battery Analyzers and the Megger BITE3 Battery Impedance Testers, among others [94]. These instruments are designed to provide accurate and reliable results, enabling manufacturers to identify defects early in the production process and ensure the quality and reliability of LIBs across various applications.

9.7.1.2 Thermal Imaging Cameras

Thermal imaging cameras are used alongside battery cell testers to provide additional insights into battery performance and safety. These cameras capture thermal signatures and heat distribution within battery cells and packs, offering valuable information on thermal behavior during various operating modes. By detecting potential hotspots and thermal anomalies early on, thermal imaging cameras help to minimize safety risks such as overheating and thermal runaway. Additionally, they aid in optimizing battery pack design and thermal management strategies for enhanced performance and longevity. The cameras work by capturing thermal infrared (TIR) images of the battery cells and packs, which are then analyzed to identify areas with abnormal temperature variations. This analysis is often performed using advanced algorithms and techniques, such as saliency analysis [113], to differentiate between normal temperature variations and thermal anomalies.

9.7.1.3 Cycle Life Testers

A cycle life tester is an essential instrument for manufacturers aiming to predict battery lifespan, optimize design parameters, and ensure product reliability, particularly in demanding applications like EVs and renewable energy storage systems. By simulating real-world usage conditions, a cycle life tester provides critical insights into the number of charge and discharge cycles a battery can withstand. This information is invaluable for predicting a battery's lifespan, enabling manufacturers to optimize its design parameters to extend durability and performance.

In addition to lifespan prediction, cycle life testers play a pivotal role in optimizing design parameters. Manufacturers can test various battery formulations and operating conditions to identify the optimal parameters that yield the longest battery lifespan. This process allows for fine-tuning a battery's chemistry and construction to enhance its overall efficiency and longevity. By rigorously testing batteries to meet required standards for reliability and performance, manufacturers can guarantee that their products are safe and efficient, which is particularly important for applications such as EVs where battery reliability is paramount.

Moreover, a cycle life tester is instrumental in comparing different battery formulations. By evaluating the performance of various formulations under diverse operating conditions, manufacturers can determine the best formulation for specific applications. This comparative analysis helps in selecting the most suitable battery chemistry for a given use case, ensuring optimal performance and longevity. Additionally, cycle life testers are vital for testing BMSs. Confirming that BMSs are functioning correctly is critical for proper battery charging and discharging, as well as for preventing issues like overheating or overcharging.

A cycle life tester is utilized during various stages of battery development and production. During the development phase of new battery formulations, the tester ensures that they meet the necessary standards for reliability and performance. Before mass production, it serves as a final check to confirm that the batteries adhere to required reliability and performance standards. Furthermore, when comparing different battery formulations, a cycle life tester provides valuable data on their performance under various conditions. Lastly, particularly in high-demand applications such as EVs and renewable energy storage systems, a cycle life tester is an indispensable tool for verifying that batteries meet the strict standards required for these demanding environments. These testers (Figure 9.18) are used to simulate repeated charge and discharge cycles to assess a battery's durability and longevity over time.

9.7.2 Testing Tools

In addition to specialized equipment, a quality control facility needs to have a range of tools to conduct various tests and measurements [114].

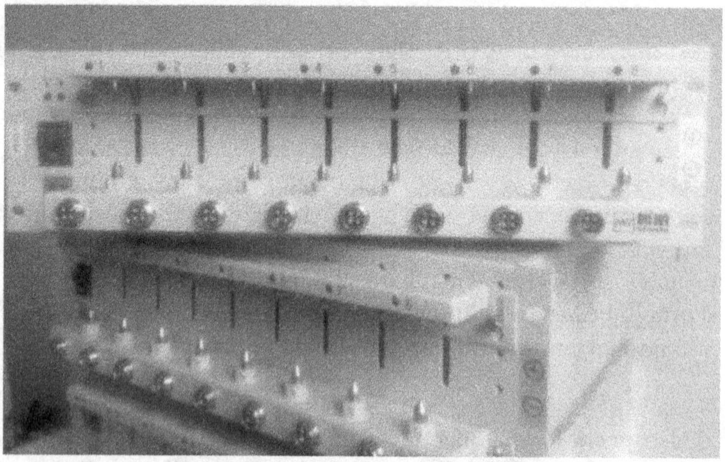

Figure 9.18 Cycle life tester. Source: [90].

Impedance analyzers offer crucial insights into battery health by measuring internal resistance or impedance, though interpreting these results requires careful attention to several factors. One important consideration is the impedance measurement method employed. Different analyzers may use different techniques, such as AC impedance at a specific frequency like 1 kHz, or methods outlined in standards like IEC 60896-21/22, which involves discharging the battery at two different currents and calculating the resistance. The chosen method can significantly impact the results, making it essential to compare values obtained using the same technique to ensure accuracy and reliability.

Establishing baseline impedance values is another critical step in assessing battery condition. The best time to establish baseline impedance for batteries is early in their life cycle, ideally right after manufacturing or after initial formation (conditioning) cycles. This allows for accurate monitoring of the battery's health over time. By comparing subsequent impedance readings to this baseline, it becomes possible to identify changes in a battery's condition over time. This approach helps to detect early signs of degradation and potential failures, which is vital for maintaining battery reliability and performance.

Interpreting impedance values also involves understanding impedance thresholds. There are no universal thresholds that define "good" or "bad" impedance values, but a significant increase from the baseline can indicate a decline in battery health. For instance, an increase of more than 20% in impedance compared to the baseline might warrant further investigation and potentially signal the need for battery maintenance or replacement.

However, relying solely on impedance measurements to predict battery performance or remaining useful life can be misleading. Other factors like the battery's age, usage history, and operating conditions also play a crucial role. Thus, impedance measurements should be considered alongside other tests (e.g., capacity tests) to gain a comprehensive understanding of a battery's condition.

Consistency in measurement is paramount when comparing impedance values. Measurements should be taken under the same conditions, such as at the same temperature, using the same points on the battery, and with the battery at the same SOC. Variations in these conditions can lead to inconsistencies and inaccurate results, which can misinform maintenance decisions and strategies.

By ensuring consistency in testing conditions, more reliable and meaningful comparisons can be made, enhancing the predictive power of impedance testing.

In summary, impedance analyzers provide valuable insights into battery condition, but interpreting these results requires a holistic approach. This includes considering the measurement method, establishing and comparing baseline values, understanding impedance thresholds, correlating impedance with overall performance, and ensuring consistency in measurement. Regular impedance testing, in conjunction with other battery assessments, can significantly aid in predicting battery failures and optimizing replacement strategies, ultimately enhancing battery reliability and performance.

In addition, voltage and current meters play a crucial role in monitoring the charging and discharging processes. They provide voltage and current levels, ensuring that a battery operates within safe thresholds. Any deviations from expected levels can alert operators to potential problems, allowing for prompt intervention to prevent damage or hazards.

By integrating impedance analyzers, voltage and current meters, and environmental testing chambers into their workflow, battery manufacturers can ensure the delivery of robust and dependable energy storage solutions tailored to meet the demands of various applications.

9.7.3 Skilled Personnel

A quality control facility for LIB production needs a skilled team trained in battery testing and quality control, proficient in using specialized equipment and tools. The facility must implement robust processes for documentation, nonconformance management, CAPA, and continuous improvement.

Effective documentation includes records of all tests, equipment calibration, and any deviations from standard procedures, ensuring accountability and consistency. A nonconformance management system should swiftly identify and document discrepancies, allowing prompt corrective actions to maintain quality standards. Corrective and preventive measures must address nonconformances and prevent recurrences, including process changes, personnel training, and equipment updates.

Continuous improvement is crucial, with the facility actively seeking feedback, analyzing data, and making changes to enhance battery quality and reliability. Integrating these components ensures the production of high-quality, reliable LIBs that meet market and industry standards.

9.8 Future Trends and Advancements in Battery Quality Control

Manufacturers are placing greater emphasis on innovative quality management techniques and advanced technologies to address the rising demands of the market. As battery applications expand into critical areas like EVs, renewable energy storage, and consumer electronics, the need for reliable, high-performance batteries has never been more essential. To meet these expectations, manufacturers are adopting cutting-edge tools such as automated testing systems, predictive analytics, and real-time monitoring solutions. These innovations not only enhance efficiency in production but also ensure that strict quality standards are upheld throughout the entire battery life cycle, from manufacturing to end-of-life. This proactive approach helps to prevent failures, improve performance, and extend the longevity of batteries, meeting both regulatory requirements and customer expectations.

9.8.1 Digitalization and Automation in Quality Control

Automation plays a vital role in enhancing consistency and efficiency in quality control processes for LIBs. In automated manufacturing applications, continuous monitoring of sensor data allows for early identification of potential problems before they lead to critical failures. For example, vibration sensors and thermal imaging can detect issues like spindle wear, overheating, and clogging in manufacturing equipment. Automated inspection of machine performance enables fine-tuned maintenance schedules and extends machinery lifespan. For instance, Contemporary Amperex Technology Co., Ltd. (CATL) has deployed advanced technologies, including 5G, flexible automation, and IoT, to achieve a defect rate of a single defect per billion for its cells, which is the top in the industry [115]. Similarly, LG Energy Solution, Ltd. has put in place a quality management system across all facilities certified with standards such as IATF 16949:2016 Automotive Quality Management System and ISO 9001:2015 Quality Management System, ensuring high quality and reliability for all products and processes [116].

Automation is key to LIB manufacturing and quality control processes. For example, hi-pot tests, which are conducted during the separator manufacturing process, can be automated to ensure insulation integrity and safety [117]. Contamination-level control during the electrolyte solution manufacturing process can also be automated to maintain the purity of electrolytes, reducing the risk of performance issues or safety hazards.

Sartorius, a renowned provider of high-precision analytical instruments, offers a suite of tools for quality control in LIB production. Their instruments are designed to provide accurate measurements and robust control over manufacturing stages, ensuring high-quality battery production [118]. This focus on precision and automation is crucial for producing LIBs for EVs, which require higher-quality batteries and controls than smaller devices. Other notable companies in the LIB industry that prioritize quality control and automation in their manufacturing processes include A123 Systems, LLC; AESC Group; LG Chem, Ltd.; Panasonic Corporation; Samsung SDI Co., Ltd.; and Toshiba Corporation [119].

Additionally, automation is used to monitor contamination levels in the electrolyte solution manufacturing process, allowing manufacturers to maintain the purity of electrolytes and mitigate the risk of performance issues or safety hazards [117]. Another real-world example of automated quality control in LIB production is the use of computer vision systems at Northvolt's facilities in Sweden and Poland to inspect battery cells for defects [89]. Additionally, the implementation of Six Sigma methodologies at Northvolt Labs in Vasteras, Sweden, reduces nonconformities during the cell assembly process [89]. Automated quality control steps in LIB manufacturing offer significant benefits that enhance the overall quality control process, as illustrated in Figure 9.19.

First, automation ensures consistency by conducting quality control tests uniformly and accurately across all batteries, minimizing the risk of human error and variability. Second, automated processes are more efficient than manual testing, enabling a quicker assessment of battery quality, which is crucial in large-scale production settings. Additionally, automated systems provide real-time monitoring of quality control tests, allowing for immediate detection of issues or deviations from specifications to facilitate prompt corrective actions [120]. Moreover, automation leads to cost savings by reducing the need for manual labor in repetitive testing processes while maintaining high levels of accuracy. Finally, automated quality control steps enhance safety by identifying defects and deviations in battery components that could pose safety risks if left undetected, ensuring the production of safer battery products for consumers [121].

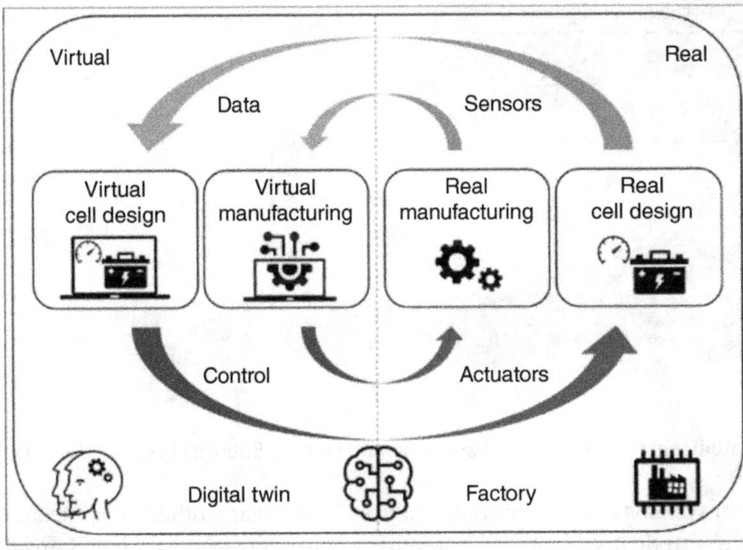

Figure 9.19 Digital twin concept for the battery manufacturing process. Source: Hidalgo et al. [28]/with permission of Elsevier.

9.8.2 Artificial Intelligence (AI), Predictive Maintenance, and Real-Time Monitoring in Quality Control

Data analytics extract valuable insights from real-time sensor data in battery manufacturing. AI algorithms analyze historical and current data, including electrochemical tests and CT scans, to identify defects and deviations. Machine learning processes large data volumes to detect subtle defects, while computer vision technologies, like CT scans, identify structural defects and misalignments. Additionally, predictive maintenance using AI can forecast potential failures, enabling proactive quality control. Natural language processing (NLP) algorithms analyze textual data related to battery specifications and quality control processes to extract insights for improving manufacturing quality. Real-world companies implementing these technologies include Contemporary Amperex Technology Co., Ltd. (CATL) [122], Grinntech Motors & Services Pvt. Ltd. [123], Battrixx [124], and Guoxuan High-Tech Co. Ltd. [125], among others.

These AI technologies play an important role in enhancing the efficiency, accuracy, and safety of LIB manufacturing processes by enabling real-time defect detection, comprehensive data analysis, and proactive quality control measures (Figure 9.20). One key area where AI excels is in the analysis of CT scans [28]. Specifically, AI algorithms enhance CT scan interpretation, improving the detection of structural defects, electrode misalignments, coating deformities, and other issues affecting battery performance and safety. The automation of CT image analysis through AI enables more efficient and precise defect detection compared to manual methods [127].

AI-powered quality control systems play a crucial role in LIB manufacturing. Automated defect detection algorithms analyze data from CT scans and other imaging techniques to automatically identify defects in battery components, such as electrode misalignments, coating deformities, and structural issues [127, 128]. Predictive maintenance AI models utilize historical data to predict potential failures or deviations in battery quality, enabling proactive maintenance measures to prevent issues before they occur [95]. Data analysis and integration AI technologies are used to comprehensively assess battery quality by analyzing and integrating data from various quality control tests, including electrochemical tests and material characterization [127, 128].

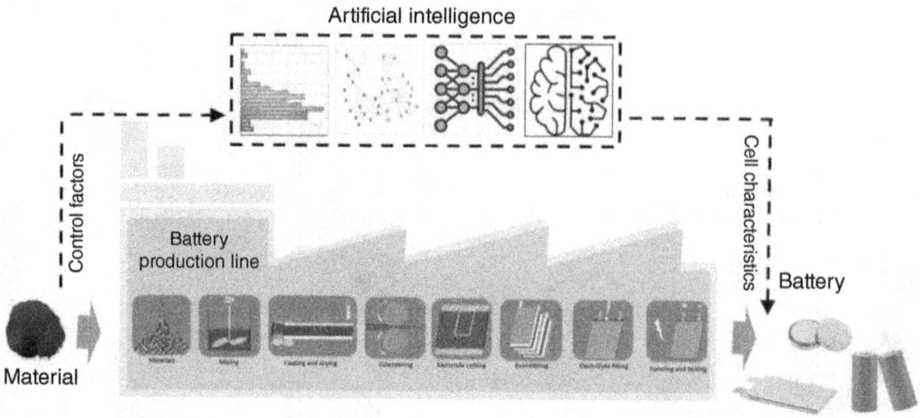

Figure 9.20 Role of artificial intelligence in lithium-ion battery manufacturing. Source: [126]/AZoNetwork.

AI-powered computer vision systems are employed to analyze CT scans and other imaging data to identify structural defects, electrode discrepancies, and other quality concerns that may impact battery performance and safety [127, 128]. Also, in LIB manufacturing, they enhance defect detection by accurately identifying structural defects, electrode discrepancies, and other quality issues, ensuring early detection and prevention. Importantly, these systems improve quality control through detailed analysis of CT scans and imaging data, assessing battery assembly, electrode alignment, and material composition for a comprehensive quality evaluation [121, 127, 129]. These systems increase efficiency by automating inspections, reducing manual labor, and speeding up defect detection, which enhances production efficiency and cuts quality control costs. AI provides real-time monitoring and analysis, allowing manufacturers to quickly identify and correct deviations from quality standards. Additionally, AI predicts potential failures based on historical data, enabling proactive maintenance to improve battery reliability and longevity.

9.8.3 Optimization and Quality Control in the Supply Chain Management

Collaboration among manufacturers, suppliers, and customers is vital for maintaining quality standards across the supply chain. This includes clear communication of expectations, timelines, and quality standards, fostering trust and respect. Quality control measures must be applied at every production stage, from raw material sourcing to final assembly, ensuring product specifications are met. Standards and certifications, such as ISO 9001 and IATF 16949, are crucial for ensuring consistency, reliability, and safety in the battery industry [130].

Integrating and consolidating the LIB supply chain has both benefits and risks. These strategies help improve efficiency, lower costs, and boost environmental performance by streamlining operations across different stages of production. By connecting activities from raw material sourcing to product delivery, companies can achieve better economic and environmental outcomes. However, these changes also come with potential supply chain challenges. Vertical integration can also mitigate supply risks for refiners, and boost returns for extractive companies, by ensuring reliable market access in cell fabrication [131, 132].

As an example, the UK government's battery strategy aims to establish a strong domestic battery industry, covering lithium mining, processing, and manufacturing. It includes investments in domestic lithium mining, processing, and production to enhance processing, procurement, and workforce development. The strategy also focuses on building local supply chains to support transportation electrification and reduce reliance on foreign lithium [132–134]. Local supply chains are another example, highlighting the importance of local refining for materials such as cathodes, which need to be protected from moisture during transit. Local refining is essential to ensure the longevity of the battery pack. The UK domestic battery industry is forecast to be worth GBP 5 billion (US $6.3 billion) by 2025, and demand for lithium-ion cells is increasing dramatically across various industries, including vehicle electrification [133, 134].

Companies like Samsung SDI, LG Energy Solution, and others are also implementing innovative strategies to optimize their supply chains. Samsung SDI, for instance, has 15 years of experience in LIB mass production and has developed advanced internal processes such as its manufacturing executing system (MES) to ensure strict and consistent quality control. The company is also investing in solid-state batteries, aiming to start mass production in 2027 [135]. LG Energy Solution, on the other hand, is implementing responsible raw material supply chain management, examining its suppliers' management systems through supply chain assessments, and conducting additional due diligence based on the findings. The company is also securing partnerships with major suppliers to ensure stable procurement of key raw materials, thereby delivering advanced battery technologies at competitive prices [136, 137].

Blockchain technology offers several applications in the LIB supply chain, enhancing transparency, traceability, and efficiency [138]. It enables tracking and traceability, establishing a secure database to monitor LIBs from extraction to recycling, ensuring authenticity, reducing counterfeiting, and enhancing transparency. Smart contracts automate supply chain processes, cutting costs, minimizing disputes, and boosting efficiency. Also, blockchain facilitates recycling and a circular economy, streamlining end-of-life battery collection, sorting, and recycling to reduce waste and promote sustainability [139]. It aids in supply chain risk management, monitoring risks like geopolitics and environmental concerns for a sustainable LIB supply chain. By integrating with data analytics, blockchain supports data-driven decision-making, optimizing processes, cutting costs, and enhancing performance [140, 141]. Importantly, successful blockchain implementation in the LIB supply chain necessitates collaboration among stakeholders like manufacturers, suppliers, recyclers, and regulators.

9.8.4 Advanced Material Testing and Inspection Methods

Component-level quality control is essential for detecting defective components before assembly [142, 143]. The choice of material in LIBs significantly affects their performance and safety. Different materials exhibit varying properties that directly influence battery performance metrics like energy density, cycle life, and safety. For instance, lithium nickel-manganese-cobalt oxide (NMC) batteries are known for their higher energy density and improved stability compared to lithium cobalt oxide (LCO) batteries, making them a preferred choice for many applications [144]. By opting for materials like NMC over LCO, manufacturers can enhance the overall quality and reliability of LIBs, ensuring better performance and longevity.

When selecting materials for LIBs, it is important to consider safety, electrochemical stability, cost, manufacturability, and recyclability, in addition to energy density and cycle life. The materials should be safe to handle, with minimal risk of thermal runaway. This is especially crucial for cathode and anode materials. Companies like Toray Industries, Inc. and Asahi Kasei Corporation are renowned for their high-quality battery separators, enhancing the safety and reliability of LIBs [145].

The materials need to have good electrochemical stability, with a low tendency to undergo side reactions or degrade over time. This is important for maintaining the performance and safety of the battery. Companies like Mitsubishi Chemical Group and Shenzhen Capchem are leading manufacturers of liquid electrolytes, which are critical for maintaining the stability of LIBs [146].

Additionally, materials should be cost-effective and readily available, as this can impact the overall cost of the battery and its manufacturing process. For example, ENTEK International LLC is a major supplier of battery separators, which are essential components in LIBs, and their cost-effectiveness is a significant factor in the battery manufacturing process [145]. The materials must be easy to process and form into the required shape and size, impacting manufacturing efficiency and cost. Companies like Freudenberg SE and Hollingsworth & Vose excel in producing battery separators, critical for LIB manufacturability [147]. The materials should be recyclable or have a low impact to reduce the environmental footprint of LIBs and support a circular economy. Notably, companies like Ampcera Inc. and Ube Industries are developing solid-state electrolytes to enhance LIB recyclability [146].

9.9 Summary

IQC, process control, OQC, and EOL testing measures are crucial for ensuring the quality, reliability, and consistency of LIB production. These measures include comprehensive monitoring of electrode fabrication, electrolyte filling, stringent quality control during cell assembly, and the adoption of advanced technologies. Such technologies include automated inspection systems, robotic assembly, and AI to boost efficiency and reduce errors, ensuring the production of high-quality batteries suitable for various applications.

Key components like battery modules, welding processes, postwelding testing, and EOL testing significantly impact manufacturing outcomes. Screening technologies are essential for enhancing battery performance, safety, and reliability by enabling defect detection, thermal dynamics assessment, and identification of mechanical stress-induced signals. Addressing challenges in LIB production, such as maintaining manufacturing consistency, optimizing supply chain management, and emphasizing safety and quality control through robust protocols and ethical sourcing practices, requires APCSs, strategic supply chain strategies, and stringent safety measures.

Balancing quality control, process optimization, regulatory compliance, and technological innovation is fundamental to the evolution of the LIB industry. Overcoming challenges, adopting forward-looking strategies, and prioritizing safety and quality propel manufacturers to sustainable energy solutions, drive progress, and meet market demands, with a commitment to delivering high-quality, reliable LIBs essential for future energy storage.

References

1 Liu, W., Placke, T., and Chau, K.T. (2022). Overview of batteries and battery management for electric vehicles. *Energy Reports* 8: 4058–4084.

2 Smith, J. (2022). Advanced X-ray inspection techniques for detecting metallic impurities in lithium-ion battery materials. *Journal of Battery Science* 10 (2): 45–58.

3 Brown, A. (2023). Detection and analysis of metallic impurities as small as 20 micrometer in lithium-ion battery materials. *Materials for Energy* 5: 112–125.

4 Johnson, B. (2020). Specialized X-ray techniques for electrode structure inspection in lithium-ion batteries. *Journal of Power Sources* 200: 210–225.

5 Garcia, C. (2021). Application of electrochemical impedance spectroscopy (EIS) in battery quality control. *Electrochimica Acta* 350: 78–92.

6 Patel, D. (2019). Insights into electrochemical characterization methods for lithium-ion batteries. *Journal of Electroanalytical Chemistry* 180: 335–348.

7 Lee, S. (2021). Material quality and composition analysis of cathode materials in lithium-ion batteries. *Journal of Materials Chemistry A* 25: 890–905.

8 IEC Standard (2020). *Safety testing standards for lithium-ion batteries.* International Electrotechnical Commission.

9 Reinhart, G., Kurfer, J., Westermeier, M., and Zeilinger, T. (2014). Integrated product and process model for production system design and quality assurance for EV battery cells. *Advanced Materials Research* 907: 365–378.

10 ThermoFisher SCIENTIFIC, Battery research and manufacturing — battery quality control, https://www.thermofisher.com/us/en/home/materials-science/battery-research/applications/quality-control.html.

11 Ricky, L. (2023). Ensuring battery quality: the role of stringent processes in quality control. https://www.linkedin.com/pulse/title-ensuring-battery-quality-role-stringent-processes-ricky-luo/ (accessed 8 July 2024).

12 Spectroscopy. Lithium-ion battery manufacturing and quality control Raman spectroscopy: an analytical technique of choice. https://www.spectroscopyonline.com/view/lithium-ion-battery-manufacturing-and-quality-control-raman-spectroscopy-an-analytical-technique-of-choice (accessed 8 July 2024).

13 Held, M. and Brönnimann, R. (2016). Safe cell, safe battery? Battery fire investigation using FMEA, FTA and practical experiments. *Microelectronics Reliability* 64: 705–710.

14 IEC Standard 60812 (2006). *Analysis techniques for system reliability-procedure for failure mode and effects analysis (FMEA).* International Electrotechnical Commission.

15 Kirana, R.C., Purwanto, N.A., Azis, N.A. et al. (2023). Failure assessment in lithium-ion battery packs in electric vehicles using the failure modes and effects analysis (FMEA) approach. *Journal of Mechatronics, Electrical Power, and Vehicular Technology* 14 (1): 94–104.

16 Mostafaee, M. (2021). Six sigma for quality assurance of lithium-ion batteries in the cell assembly process: a DMAIC field study at Northvolt, Technology and Arts.

17 Laurent, P.-G., Olivier, M., Marc, B., and Fabrice, F. (2019). Batteries for satellites constellation, using lean manufacturing for space industry. *2019 European Space Power Conference (ESPC)*, IEEE, 1–6.

18 Purushothaman, K. and Ahmad, R. (2022). Integration of six sigma methodology of DMADV steps with QFD, DFMEA and TRIZ applications for image-based automated inspection system development: a case study. *International Journal of Lean Six Sigma* 13 (6): 1239–1276.

19 International Journal of Innovative Science and Research Technology. International journal of innovative science and research technology. ISSN no:-2456-2165.

20 Nguyen, P. and Sahlberg, A. (2020). Measurement system analysis: for battery production using DMAIC at Northvolt AB. PhD dissertation. Luleå University of Technology, Department of Business Administration, Technology and Social Sciences.

21 TWI. What is design for manufacturing (DFM)? https://www.twi-global.com/technical-knowledge/faqs/faq-what-is-design-for-manufacture-dfm (accessed 6 June 2024).

22 Sparkling Logic (2022). Design for manufacturing in automotive. https://www.sparklinglogic.com/decision-management-in-manufacturing-expediting-and-optimizing-design-for-manufacturing-and-assembly-dfma-in-automotive/ (accessed 6 June 2024).

23 Johnson, J. (2012). Battery power online. https://www.batterypoweronline.com/markets/manufacturing-materials/the-challenges-of-manufacturing-lithium-ion-batteries-for-the-electric-vehicle-industry/ (accessed 6 June 2024).

24 Behind the Work. A guide to design-for-manufacturing (DFM) principles. Imagineering, Inc. https://www.pcbnet.com/blog/a-guide-to-design-for-manufacturing-dfm-principles/ (Accessed 9 August 2025).

25 (2024). How to conduct a design of experiments (DOE). Quality InfoCenter. https://www.qualitymag.com/articles/97774-how-to-conduct-a-design-of-experiments-doe (accessed 6 June 2024).

26 MoreSteam. Design of experiments. https://www.moresteam.com/toolbox/design-of-experiments (accessed 6 June 2024).

27 Örüm Aydin, A., Zajonz, F., Günther, T. et al. (2023). Lithium-ion battery manufacturing: industrial view on processing challenges, possible solutions and recent advances. *Batteries* 9 (11): 555.

28 Hidalgo, M.F.V., Apachitei, G., Dogaru, D. et al. (2023). Design of experiments for optimizing the calendering process in Li-ion battery manufacturing. *Journal of Power Sources* 573: 233091.

29 Hessing, T. and Ramana, P.V. (2014). X bar R control charts. https://sixsigmastudyguide.com/x-bar-r-control-charts/ (accessed 7 June 2024).

30 Edison Group (2024). Arcadium Lithium — When one plus one equals three. https://www.edisongroup.com/research/when-one-plus-one-equals-three/33478/ (accessed 26 May 2024).

31 Trydegård, M. and Blide, M. (2020). Suggestions for implementation of statistical process control in lithium-ion battery processes: a field study at Northvolt. PhD dissertation. Luleå University of Technology, Department of Business Administration, Technology and Social Sciences.

32 Avram, D. X-bar and R-chart: is there a difference and how are they used. https://www.alisqi.com/en/blog/x-bar-and-r-chart-is-there-a-difference-and-how-are-they-used (accessed 7 June 2024).

33 Boyd Blog (2019). SPC: what are control charts. https://www.boydcorp.com/blog/spc-what-are-control-charts.html (accessed 7 June 2024).

34 X-bar R chart excel. QIMacros. https://www.qimacros.com/control-chart/x-bar-r-chart-excel.png (accessed 7 June 2024).

35 Persson, T. (2023). What is PPAP, why is it important, and what are the challenges. https://www.chaintraced.com/posts/what-is-ppap-why-is-it-important-and-what-are-the-challenges (accessed 7 June 2024).

36 PPAP (2016). https://quality-one.com/ppap/ (accessed 7 June 2024).

37 PPAP graphic levels. WISCONSIN. https://www.wisconsinmetalparts.com/images/PPAP_Graphic_Levels-01_big.png (accessed 7 June 2024).

38 PLEX. Guide to production part approval process (PPAP). https://www.plex.com/knowledge-articles/mes/guide-production-part-approval-process-ppap (accessed 7 June 2024).

39 Amper. 9 real-life lean manufacturing examples. https://www.amper.xyz/post/lean-manufacturing-examples (accessed 7 June 2024).

40 Cooke, P. (2020). Gigafactory logistics in space and time: Tesla's fourth Gigafactory and its rivals. *Sustainability* 12 (5): 2040.

41 LG Chem (2019). LG Chemical Sustainability Report. Innovative Sustainability: deliver advanced, innovative, and sustainable solutions for our environment and society. https://www.lgchem.com/upload/file/sustainabilityreports/2019_LGChem_Sustainability_Report_ENG.pdf (accessed 26 May 2025).

42 Management and Strategy Institute (2023). Define statistical process control. https://www.msicertified.com/statistical-process-control/ (accessed 7 June 2024).

43 Slideshare (2010). International battery: applying lean manufacturing processes in the production of large format lithium-ion batteries. https://www.slideshare.net/slideshow/applying-lean-manufacturing-processes-in-the-production-of-large-format-lithium-ion-batteries/4703553 (accessed 7 June 2024).

44 Markgraf, B. (2013). Purpose of lean manufacturing. https://smallbusiness.chron.com/purpose-lean-manufacturing-69380.html (accessed 7 June 2024).

45 Abdulaziz, M., Abdurakhimov, F., Maksumov, S., and Kadamov, A. (2020). Quality Management practices of Samsung Electronics and Apple. https://www.researchgate.net/publication/347948245_Quality_Management_practices_of_Samsung_Electronics_and_Apple?channel=doi&linkId=5fe9befaa6fdccdcb80d2d81&showFulltext=true (accessed 7 June 2024).

46 Rajkhowa, V., Loganathan, M.K., and Pahwa, M.S. (2023). Total quality management (TQM) for clean, affordable and reliable energy in research and development lab – a case-study from north east India. *2023 IEEE Silchar Subsection Conference (SILCON)*, IEEE, 1–5.

47 Alawag, A.M., Alaloul, W.S., Liew, M.S. et al. (2023). The role of the total-quality-management (TQM) drivers in overcoming the challenges of implementing TQM in industrialized-building-system (IBS) projects in Malaysia: experts' perspectives. *Sustainability* 15 (8): 6607.

48 Battery University (2017). Battery test methods. https://batteryuniversity.com/article/battery-test-methods (accessed 8 June 2024).

49 Nichols, C. (2023). The crucial role of managing end-of-life Li-ion battery sustainability. https://www.idtechex.com/en/research-article/the-crucial-role-of-managing-end-of-life-li-ion-battery-sustainability/30302 (accessed 8 June 2024).

50 Reza, M.S., Mannan, M., Mansor, M. et al. (2024). Recent advancement of remaining useful life prediction of lithium-ion battery in electric vehicle applications: a review of modelling mechanisms, network configurations, factors, and outstanding issues. *Energy Reports* 11: 4824–4848.

51 Robben, T., Offermanns, C., Heimes, H., and Kampker, A. (2024). Advancements in battery cell finalization: insights from an expert survey and prospects for process optimization. *World Electric Vehicle Journal* 15 (5): 219.

52 A&D Company. End-of-line and in-line testing. https://aanddtech.com/applications/eol-testing/ (accessed 8 June 2024).

53 Yan, H. (2020). Battery internal resistance: ACIR, DCIR and RSS. https://www.linkedin.com/pulse/battery-internal-resistance-acir-dcir-rss-hongwei-yan/ (accessed 8 June 2024).

54 Suffiyan, M. (2022). Internal resistance: DCIR and ACIR. https://www.batterydesign.net/dcir-acir/ (accessed 8 June 2024).

55 ELE Times (2024). Dürr and Rohde & Schwarz collaborate on ADAS/AD functional testing for EOL and PTI. https://www.eletimes.com/durr-and-rohde-schwarz-collaborate-on-adas-ad-functional-testing-for-eol-and-pti (accessed 8 June 2024).

56 Cross, E. (2021). Digatron & DSA launch end-of-line (EOL) tester for lithium battery production. https://batterysystemsexpo.com/digatron-dsa-launch-end-of-line-eol-tester-for-lithium-battery-production/ (accessed 8 June 2024).

57 Unico LLC (2024). Unico introduces powerful, flexible and cost-effective cell formation and test channels for EV testing. https://www.globenewswire.com/en/news-release/2024/03/12/2844394/0/en/Unico-Introduces-Powerful-Flexible-and-Cost-Effective-Cell-Formation-and-Test-Channels-for-EV-Testing.html (accessed 8 June 2024).

58 Schnell, J. and Reinhart, G. (2016). Quality management for battery production: a quality gate concept. *Procedia CIRP* 57: 568–573.

59 Lee, C.K.M., Cao, Y., and Ng, K.H. (2017). *Big Data Analytics for Predictive Maintenance Strategies*, Supply Chain Management in the Big Data Era, 50–74. IGI Global.

60 Sai, V.C., Shcherbakov, M.V., and Tran, V.P. (2019). Data-driven framework for predictive maintenance in industry 4.0 concept. In: *Communications in Computer and Information Science* (ed. A.G. Kravets et al.), 344–358. Cham: Springer Nature.

61 Semco Infratech Pvt Ltd. (2023). Battery testing methods and techniques: different industries and their approaches. https://www.linkedin.com/pulse/battery-testing-methods-techniques-different-industries-approaches-9uurf/ (accessed 4 April 2024).

62 E-Mobility Engineering (2024). Battery inspection. https://www.emobility-engineering.com/battery-inspection/ (accessed 4 April 2024).

63 SAE (2022). Ultrasound battery testing. https://www.sae.org/news/2022/07/ultrasound-battery-testing (accessed 15 May 2024).

64 Liu, X., Wu, L., Guo, X. et al. (2023). A novel approach for surface defect detection of lithium battery based on improved K-nearest neighbor and Euclidean clustering segmentation. *The International Journal of Advanced Manufacturing Technology* 127 (1-2): 971–985.

65 Szewieczek, A. (2023). Non-destructive testing of battery pouches with imaging ultrasonic techniques. https://www.ndt.net/article/ecndt2023/presentation/ECNDT2023_PRESENTATION_153.pdf (accessed 15 May 2024).

66 Super User. Battery inspection. https://xarion.com/en/applications/non-destructive-testing/battery-inspection (accessed 15 May 2024).

67 Robinson, J.B., Owen, R.E., Kok, M.D.R. et al. (2020). Identifying defects in Li-ion cells using ultrasound acoustic measurements. *Journal of the Electrochemical Society* 167 (12): 120530.

68 Nature Research Custom Media By P. Peering into batteries for a rechargeable future. https://www.nature.com/articles/d42473-022-00333-5 (accessed 15 May 2024).

69 Sun, H., Muralidharan, N., Amin, R. et al. (2022). Ultrasonic nondestructive diagnosis of lithium-ion batteries with multiple frequencies. *Journal of Power Sources* 549: 232091.

70 Balz, I., Raad, E.A., Rosenthal, E. et al. (2020). Process monitoring of ultrasonic metal welding of battery tabs using external sensor data. *Journal of Advanced Joining Processes* 1: 100005.

71 Masuch, S., Gümbel, P., Kaden, N., and Dröder, K. (2022). Applications and development of X-ray inspection techniques in battery cell production. *Processes (Basel)* 11 (1): 10.

72 MSK-TE907-LD. X-ray imaging system for cylindrical, coin and pouch battery inspection. https://www.mtixtl.com/NFXBatteryInspection--MSK-TE907-LD.aspx (accessed 4 April 2024).

73 Patel, D., Robinson, J.B., Ball, S. et al. (2020). Thermal runaway of a Li-ion battery studied by combined ARC and multi-length scale X-ray CT. *Journal of the Electrochemical Society* 167 (9): –090511. https://doi.org/10.1149/1945-7111/ab7fb6.

74 Kong, L., Aalund, R., Alipour, M. et al. (2022). Evaluating the manufacturing quality of lithium ion pouch batteries. *Journal of the Electrochemical Society* 169 (4): 040541. https://doi.org/10.1149/1945-7111/ac6539.

75 Qu, Z., Jiang, P., and Zhang, W. (2020). Development and application of infrared thermography non-destructive testing techniques. *Sensors* 20 (14): 3851. https://doi.org/10.3390/s20143851.

76 Tomaszewska, A., Chu, Z., Feng, X. et al. (2019). Lithium-ion battery fast charging: a review. *eTransportation* 1: 100011.

77 Wikimedia Commons (2015). File:In-operando-high-speed-tomography-of-lithium-ion-batteries-during-thermal-runaway-ncomms7924-s2.ogv. Wikimedia.org. https://commons.wikimedia

.org/wiki/File:In-operando-high-speed-tomography-of-lithium-ion-batteries-during-thermal-runaway-ncomms7924-s2.ogv (accessed 06 August 2024).

78 Finegan, D., Scheel, M., Robinson, J. et al. (2015). *In-operando* high-speed tomography of lithium-ion batteries during thermal runaway. *Nature Communications* 6: 6924. https://doi.org/10.1038/ncomms7924.

79 Savca A, Azizighalehsari R and Venugopal P. Application of electrochemical impedance spectroscopy (EIS) on module level Li-ion batteries for echelon utilization. https://essay.utwente.nl/92423/1/Savca-EE-EEMCS.pdf (accessed 15 May 2024).

80 Qu, D., Wang, G., Kafle, J. et al. (2018). Electrochemical impedance and its applications in energy-storage systems. *Small Methods* 2 (8).

81 Meddings, N., Heinrich, M., Overney, F. et al. (2020). Application of electrochemical impedance spectroscopy to commercial Li-ion cells: a review. *Journal of Power Sources* 480: 228742. https://doi.org/10.1016/j.jpowsour.2020.228742.

82 Komagata, S., Kuwata, N., Baskaran, R. et al. (2010). Detection of degradation of lithium-ion batteries with acoustic emission technique. *ECS Transactions* 25 (33): 163–167.

83 ZEISS. Quality assurance for batteries. https://www.zeiss.com/metrology/solutions/emobilitysolutions/battery.html (accessed 4 April 2024).

84 Vergori, E., Mocera, F., and Soma, A. (2018). Battery modelling and simulation using a programmable testing equipment. *Computers* 7 (2): 20.

85 Jaguemont, J. and Bardé, F. (2023). A critical review of lithium-ion battery safety testing and standards. *Applied Thermal Engineering* 231: 121014.

86 Smekens, J., Gopalakrishnan, R., Van den Steen, N. et al. (2016). Influence of electrode density on the performance of Li-ion batteries: experimental and simulation results. *Energies* 9 (2): 104.

87 De Silva, R.S., Silva, H.P., Gunathilake, S. et al. (2023). Optimal BESS placement and sizing for frequency control in islanded microgrids using non-linear shepherd battery model. *2023 IEEE 2nd Industrial Electronics Society Annual On-Line Conference (ONCON)*, 1–6. IEEE.

88 Elharony, A. (2024). Inside Tesla's gigafactory: the future of sustainable manufacturing.

89 Chen, Y., Kang, Y., Zhao, Y. et al. (2021). A review of lithium-ion battery safety concerns: the issues, strategies, and testing standards. *Journal of Energy Chemistry* 59: 83–99.

90 Katari, J.S., Sneha, J.R., Vinayaka, K.U., and Rana, K. (2020). A concise review of different standards for performance testing of lithium-ion batteries for electric vehicle applications. *2020 IEEE International Conference on Power Systems Technology (POWERCON)*. IEEE.

91 Samsung Electronics. Samsung announces new and enhanced quality assurance measures to improve product safety.

92 Reis, A. (2023). Lithium-ion cell manufacturing workforce training.

93 Instrumart. Battery testers. https://www.instrumart.com/categories/6156/battery-testers (accessed 3 May 2024).

94 Sledz, A. and Heipke, C. (2021). Thermal anomaly detection based on saliency analysis from multimodal imaging sources. *ISPRS Annals of the Photogrammetry, Remote Sensing and Spatial Information Sciences* V-1-2021: 55–64.

95 Lipu, M.S.H., Miah, M.S., and Jamal, T. (2023). Artificial intelligence approaches for advanced battery management system in electric vehicle applications: a statistical analysis towards future research opportunities. *Vehicles* 6: 22–70.

96 Mayyas, A., Steward, D., and Mann, M. (2019). The case for recycling: overview and challenges in the material supply chain for automotive Li-ion batteries. *Sustainable Materials and Technologies* 19: e00087.

97 Fleischmann, J., Hanicke, M., Horetsky, E. et al. (2023). Battery 2030: resilient, sustainable, and circular. https://www.mckinsey.com/industries/automotive-and-assembly/our-insights/battery-2030-resilient-sustainable-and-circular (accessed 6 April 2024).

98 Zhang, Z., Sun, B., and Wang, Q. (2019). Advanced techniques for battery testing and diagnosis. *Journal of Power Sources* 412: 346–356.

99 Smith, K.J. and Bazant, M.Z. (2021). Electrochemical impedance spectroscopy in lithium-ion batteries. *Annual Review of Materials Research* 51: 497–523.

100 Goodenough, J.B. and Park, K.-S. (2013). ChemInform abstract: the Li-ion rechargeable battery: a perspective. *ChemInform* 44 (20): 1167–1176.

101 Manthiram, A. (2020). A reflection on lithium-ion battery cathode chemistry. *Nature Communications* 11: 1550.

102 Mauger, A., Julien, C.M., Armand, M. et al. (2020). (Invited) building better batteries in the solid state. *ECS Meeting Abstracts* MA2020-01 (2): 244–244.

103 Andre, D., Kim, S.-J., Lamp, P. et al. (2015). Future generations of cathode materials: an automotive industry perspective. *Journal of Materials Chemistry A, Materials for Energy and Sustainability* 3 (13): 6709–6732.

104 Arora, P., White, R.E., and Doyle, M. (1998). ChemInform abstract: capacity fade mechanisms and side reactions in lithium-ion batteries. *ChemInform* 29 (47): 3647–3667.

105 Winter, M., Besenhard, J.O., Spahr, M.E., and Novak, P. (1998). ChemInform abstract: insertion electrode materials for rechargeable lithium batteries. *ChemInform* 29 (37): 725–763.

106 Ramadass, P., Haran, B., White, R., and Popov, B.N. (2002). Capacity fade of Sony 18650 cells cycled at elevated temperatures: Part I. Cycling performance. *Journal of Power Sources* 112 (2): 606–613.

107 Zhou, X., Wang, X., Yuan, Y. et al. (2024). Fast capacity estimation for lithium-ion batteries based on XGBoost and electrochemical impedance spectroscopy at various state of charge and temperature. *Automotive Innovation* 7 (3): 473–491.

108 Montgomery, D.C. (2023). *Introduction to Statistical Quality Control*, 8e. Nashville, TN: Wiley.

109 CTCN. Batteries for electric cars: challenges, opportunities, and the outlook to 2020. https://www.ctc-n.org/resources/batteries-electric-cars-challenges-opportunities-and-outlook-2020 (accessed 2 April 2024).

110 TOB. The preparation process of lithium-ion battery anode electrode slurry. https://www.tobmachine.com/blog/the-preparation-process-of-lithium-ion-battery-anode-electrode-slurry_b88 (accessed 5 June 2024).

111 Falenius, E. Unlocking power: the calendering step in battery manufacturing. https://www.fomtechnologies.com/insights-blog/battery/calendering-step-in-battery-manufacturing (accessed 5 June 2024).

112 Keysight. How to test EV battery cells. https://www.keysight.com/us/en/solutions/ev-battery-cell-test.html (accessed 3 May 2024).

113 https://batterycycler.com/li-ion-battery-cycler-capacity-test-cycle-life-test-charge-and-discharge-test/ (accessed 2 April 2024).

114 Maisch, M. (2023). CATL: staying on top of the battery game. https://www.pv-magazine.com/2023/12/01/catl-staying-on-top-of-the-battery-game/ (accessed 28 April 2024).

115 Lgensol. LG energy solution. https://www.lgensol.com/en/product-quality-safety (accessed 28 April 2024).

116 Gupta, A.K. (2022). How to setup lithium-ion battery (battery assembly). https://www.linkedin.com/pulse/how-setup-lithium-ion-battery-assembly-ajjay-kumar-gupta/ (accessed 28 April 2024).

117 Sartorius. Rechargeable battery research, manufacturing and recycling. https://www.sartorius.com/en/applications/applied-industries/battery-manufacturing (accessed 8 June 2024).

118 imarc. Top 15 lithium-ion battery manufacturers. https://www.imarcgroup.com/top-lithium-ion-battery-manufacturers (accessed 8 June 2024).

119 https://www.diva-portal.org/smash/get/diva2:1573372/FULLTEXT01.pdf (accessed 28 April 2024).

120 Hioki. Lithium-ion battery pack testing. https://www.hioki.com/us-en/news/Lithium-Ion-Battery-Pack-Testing (accessed 2 April 2024).

121 CATL (2023). CATL's Liyang plant recognized as lighthouse factory by World Economic Forum. https://www.catl.com/en/news/6176.html (accessed 28 April 2024).

122 Grinntech. https://www.grinntech.com/ (accessed 28 April 2024).

123 Battrixx. Environmental and economic benefits of using lithium-ion batteries for electric two-wheelers. https://www.battrixx.com/environmental-and-economic-benefits-of-using-lithium-ion-batteries-for-electric-two-wheelers.php (accessed 28 April 2024).

124 Evlithium. Guoxuan high-tech 50GWh power battery project started. https://www.evlithium.com/Blog/guoxuan-battery.html (accessed 28 April 2024).

125 FST. FST blog. https://www.foundation.org.uk/Blog/2023/What-is-the-Role-of-Artificial-Intelligence-in-Lit (accessed 4 April 2024).

126 AZOM (2023). Driving the production of lithium-ion batteries through quality control and analysis. https://www.azom.com/article.aspx?ArticleID=22326 (accessed 28 April 2024).

127 Vu, T.-T.-H., Chang, T.-W., and Kim, H. (2024). Enhancing quality control in battery component manufacturing: deep learning-based approaches for defect detection on microfasteners. *Systems* 12 (1): 24.

128 Wiley Industry News. Intelligent quality assurance for batteries. https://www.wileyindustrynews.com/en/news/intelligent-quality-assurance-batteries (accessed 28 April 2024).

129 Niri, A.J., Poelzer, G.A., Zhang, S.E. et al. (2024). Sustainability challenges throughout the electric vehicle battery value chain. *Renewable and Sustainable Energy Reviews* 191: 114176.

130 Malik, A.T.A. Strategies for enhancing the circularity of lithium-ion batteries. https://www.diva-portal.org/smash/get/diva2:1768491/FULLTEXT01.pdf (accessed 28 April 2024).

131 Chen, J. (2020). Vertical integration strategy among EV OEMs. https://www.linkedin.com/pulse/vertical-integration-strategy-among-ev-oems-john-chen/ (accessed 28 April 2024).

132 Williams, M. (2023). UK government reveals battery strategy to support EV production. https://www.automotivelogistics.media/battery-supply-chain/uk-government-reveals-battery-strategy-to-support-ev-production/44961.article (accessed 28 April 2024).

133 Harrison, D. (2022). Strategies for a European battery supply chain. https://www.automotivemanufacturingsolutions.com/ev-battery-production/strategies-for-a-european-battery-supply-chain/42660.article (accessed 28 April 2024).

134 Samsung SDI. Automotive battery – innovation. https://www.samsungsdi.com/automotive-battery/innovation.html (accessed 28 April 2024).

135 Lgensol (2023). Responsible raw material supply chain management at LG energy solution. https://inside.lgensol.com/en/2023/07/responsible-raw-material-supply-chain-management-at-lg-energy-solution/ (accessed 28 April 2024).

136 LG Energy Solution. https://www.lgensol.com/en/esg-supplychain-management (accessed 28 April 2024).

137 Cheng, S. (2023). What makes a transparent Li-ion battery supply-chain? https://www.cellcycle.co.uk/what-makes-a-transparent-li-ion-battery-supply-chain/ (accessed 28 April 2024).

138 Grant Thornton (2024). Improving EV battery reverse supply chains with blockchain. https://www.grantthornton.in/insights/blogs/improving-ev-battery-reverse-supply-chains-with-blockchain/ (accessed 28 April 2024), Feb. 2024.

139 Moors, C. (2023). Battery-makers turn to blockchain as sourcing rules get stricter – Circular CEO. https://www.spglobal.com/marketintelligence/en/news-insights/latest-news-headlines/battery-makers-turn-to-blockchain-as-sourcing-rules-get-stricter-8211-circulor-ceo-74661967 (accessed 28 April 2024).

140 Robinson, E. (2018). Blockchain could be the answer for lithium-ion battery supply chain issues. https://www.picoanalytics.co.uk/insights/blog-post-title-four-b2kgz (accessed 28 April 2024).

141 Bird, R., Baum, Z.J., Yu, X., and Ma, J. (2022). The regulatory environment for lithium-ion battery recycling. *ACS Energy Letters* 7 (2): 736–740.

142 Ruiz, V., Pfrang, A., Kriston, A. et al. (2018). A review of international abuse testing standards and regulations for lithium ion batteries in electric and hybrid electric vehicles. *Renewable and Sustainable Energy Reviews* 81: 1427–1452.

143 Saaid, F.I., Kasim, M.F., Winie, T. et al. (2024). Ni-rich lithium nickel manganese cobalt oxide cathode materials: a review on the synthesis methods and their electrochemical performances. *Heliyon* 10 (1): e23968.

144 News of Market Research Industry (2023). Top ten companies in battery separators market. https://www.linkedin.com/pulse/top-ten-companies-battery-separators/ (accessed 28 April 2024).

145 MSE Supplies (2018). Innovations in electrolytes – market research of liquid and solid electrolyte for lithium batteries. https://www.msesupplies.com/blogs/news/innovations-in-electrolytes-market-research-of-liquid-and-solid-electrolyte-for-lithium-batteries (accessed 28 April 2024).

146 MarketsandMarkets. Battery separators market. https://www.marketsandmarkets.com/Market-Reports/battery-separators-market-194866297.html (accessed 28 April 2024).

147 Freudenberg Safety Separators Resolve Common Causes for Lithium-ion battery failure. Freudenberg-pm.com. https://www.freudenberg-pm.com/Press/2016/2016_09_06_Battery_Show. (accessed 9 August 2025).

10

Battery Supply Chain: Quality, Risks and Audits

Yulun Zhang, Kianoush Naeli, Virendra Jadhav, and Sanjay Tiku

The global supply chain for lithium batteries is complex, to begin with. Additional constraints related to meeting ethical and regulatory requirements, while ensuring the quality and safety of the end products, create further unique challenges. Recent industry trends related to traceability, sustainability, socio-political supply chain vulnerabilities, and counterfeiting have enhanced the risks in material procurement, manufacturing, and assembly. Rapidly growing competing uses for critical metals required for battery manufacturing, and the shifting focus by major battery suppliers toward EVs and BESSs, are other significant factors that have been stressing the supply chain for batteries.

The complex global supply chain for lithium batteries used in consumer devices and electric vehicles brings with it unique challenges. These challenges are centered around determining whether the materials and processes used to manufacture them meet established ethical and regulatory requirements while ensuring the quality and safety of the end products into which the batteries go. Risks occur at three distinct levels of the supply chain: the cell manufacturers, the pack manufacturers, and the original equipment manufacturers (OEMs). Strict adherence to quality practices throughout the battery supply chain is needed to ensure safety in end-product applications, along with audits to validate that the quality assurance processes are being implemented properly.

10.1 Introduction

Growth in the use of batteries, besides direct consumer electronics applications, has spurred further investments and innovation. Specifically, the expectation is that the cost per kWh of lithium battery capacity will go down to around $70/kWh in 2030, compared to about $230/kWh in 2018, which is almost a 70% reduction within a decade or so [1]. Presently, the cost has already been reduced by around 40% to around $140/kWh [2]. However, recent trends related to traceability [3, 4], sustainability [5, 6], and avoiding supply chain vulnerabilities, including supply disruptions [7] and counterfeiting [8], have made the supply chain of batteries even more complex, with enhanced risks in material procurement, manufacturing, and assembly.

Traceability refers to knowing exactly where the materials and components of a system are coming from and thereby avoiding potentially questionable sources. In a circular economy context

The Safety Challenges and Strategies of Using Lithium-Ion Batteries, First Edition. Edited by Michael G. Pecht.

and with the development of associated business models around sustainability considerations, traceability also involves tracking materials through their life cycle. Key traceability characteristics that are required to enable circularity are material provenance, uniqueness identifiers, information trustworthiness, interchangeability, interoperability, reconfigurability, and constituent elements [3]. Emerging technologies like blockchain and nested tokens are being proposed to track batteries through their life cycle. This allows tracking battery material handoffs and transactions and easy retrieval of information about the raw materials contained within them [4].

Sustainability is associated with reducing the carbon footprint of sourcing, manufacturing, assembly, and transportation through recycling and reuse of materials and parts. This is also driven by the paucity, and sometimes uncertainty, of the availability of raw materials for future production, thereby making reuse, repurposing, and recycling of materials the only viable option. For example, it is estimated that 90% recovery of lithium from lithium-ion battery recycling can alleviate the supply risk of lithium for future anticipated growth [5].

One vulnerability for the battery supply chain is driven by rapidly growing competing uses for critical metals (e.g., nickel, cobalt, or lithium) required for battery manufacturing in automotive, aerospace, biotechnology, and other applications [9]. Another significant factor that has been stressing the supply chain for batteries used in commercial electronics products is the shift by major battery suppliers toward manufacturing standardized cells used in electric vehicles (EVs) or battery energy storage systems (BESSs). Consequently, manufacturers are transitioning their manufacturing capacity to build batteries for these larger-volume, high-margin sectors. As a result, the challenges for lithium-ion batteries used in electric vehicles are different. These, however, do influence the supply chain for batteries used in other applications because of technology-innovation curves, scale, and cost [10–12].

This chapter covers manufacturing and quality risks at the cell, pack, and device designer or original equipment manufacturer (OEM) levels. Risk mitigation options are provided to ensure quality and safety. Then, auditing approach and methodologies are discussed for these different entities in the supply chain.

10.2 Quality Assurance: A Tool for Risk Mitigation for Battery Safety

For batteries, like many safety-critical components or systems, the words "quality" and "safety" can be considered synonymous, particularly when they go into mass production. Every action or activity to improve quality ensures safety, and any quality misses impair safety. Accordingly, for all cell manufacturers, pack manufacturers, and even device makers or OEMs, creation, implementation, and adherence to a quality plan is imperative to avoid any safety concerns in the field. This is only useful if the fundamental requirements for designing safe cells, packs, safety mechanisms, and device integration protocols are built-in and strictly adhered to during the new product introduction (NPI) phase for any battery system.

A quality plan for batteries includes quality practices, assembly, handling and manufacturing protocols, documentation, monitoring schemas, control procedures, remediation, and corrective action processes used during product development and after launch to provide customers with a high-quality, safe product. This section provides a broad overview of a quality system that must exist throughout the supply chain, without going into the associated regulatory and compliance requirements. In its most general form, quality is the totality of the characteristics of a product that affects its ability to satisfy the stated and implied needs of customers. In a practical sense,

Table 10.1 Five stages of quality maturity.

Stage		Definition
1	Uncertainty	We do not know why we have problems with quality.
2	Awakening	Is it necessary to always have problems with quality?
3	Enlightenment	Through management commitment and quality improvement, we are identifying and resolving our problems.
4	Wisdom	Defect prevention is a routine part of our operation.
5	Certainty	We know why we do not have problems with quality.

Source: Adapted from Crosby [13].

for designers and engineers developing the product, quality is the ability of a product to meet the workmanship and performance criteria set forth under the product specifications.

The quality of any product, including a battery, has three distinct aspects to it. First, the design must be optimized to ensure that performance attributes that are highly desirable to the immediate customer are satisfied. Second, the costs involved in product development must be minimized – the goal for any business is the affordable realization of intended attributes. After all, a battery product that is cost-prohibitive will find no buyers. Third, the physical or software design implemented through design, manufacturing, or coding must exhibit low variability in performance. Based on planning, execution efficiency, and quality performance, an organization goes through five stages of quality maturity, as shown in Table 10.1. Following this table, any organization must aspire to reach "certainty" in the planning, execution, and maturity of their quality processes [13].

Customer-facing issues cause financial losses, loss of brand reputation, and customers turning to competing alternatives. In the case of batteries, there can also be safety implications affecting human lives. Direct financial costs are related to shipping, reworking, scrapping, and servicing defective batteries, if feasible. Indirect financial costs are connected to loss of business opportunity, lower market share, and potential liability in the case of an accident involving humans or pets. The intent of a quality plan is to minimize these losses and ensure that the quality of the outgoing batteries is achieved through cooperation, know-how, upfront monitoring control and prevention, and reducing the reliance on end-of-line detection.

A quality plan is a systematic way of guaranteeing that organized activities are conducted the way they are planned. Furthermore, a quality plan identifies and ties together all quality-related activities. The plan must include a schedule of activities, allocate resources and responsibilities, and specify decision criteria for altering plans. Consequently, documentation is a key part of quality planning. A quality plan is not a notional concept that everyone agrees to, but a guiding document that everyone adheres to.

A quality plan requires that workmanship criteria be defined. Specification documents must exist to elucidate product hardware and software features. Released drawings need to be prepared for all manufactured or sourced mechanical or electromechanical parts and subassemblies. Drawings should include clear definitions of materials, dimensional tolerances, surface finish definitions, and change control history. Because of this, a process for change control and approval must exist. Gerber files, firmware versions, and software releases need to have a release process and identifiers.

A quality plan has six fundamental structural aspects: metrics (what and why), metrology (how), supply chain management (where), data analysis (intervention), training (skills), and feedback and audit (improvement). It is a continuous process extending from preliminary product design to end-of-life, and these aspects are described in detail in subsequent sections. A quality plan identifies these items:

- What should be measured
- Why, how, and where it should be measured
- How the data is monitored and analyzed
- What the process is for response if the monitored metrics go out of acceptable bounds
- How field performance is tracked
- How quality plans are improved
- How the quality of products is tracked based on different inputs
- Assurance that all pertinent activities are implemented and done correctly the first time

10.2.1 Metrics

Metrics correspond to what should be measured and why those measurements are relevant to the immediate battery customer or the eventual product user. Attributes of product value for battery customers and corresponding quality metrics can be functional (capacity, charge rates, discharge rates, etc.), spatial (size, weight, packaging, etc.), or temporal (performance stability, safety, reliability, etc.). Other considerations for metrics include information about potential extreme corner cases of how the end product may be used, the criticality of the application, cost and schedule limitations, business constraints like yearly volumes, information about competitive products, and field data from previous generation products.

Every metric identified must be accompanied by a definition of acceptable limits or bounds that define what is acceptable and what is unacceptable. However, limits should not be based on what can be achieved, what gives an acceptable yield, or what people can generally agree on. Instead, limits must be directly related to the customer experience for the feature involved and, more importantly, safety considerations that provide sufficient margins to prevent unexpected exposure to dangers.

10.2.2 Metrology

Metrology corresponds to how quality metrics are measured and how repeatable and reproducible those measurements are, and it can consist of direct measurement or a test. Measuring anything involves equipment, fixtures, procedures, test scripts, consumables, and personnel, and every quality metric must have documented requirements for those six elements. A clear definition of "goodness" and rationale thereof (from the definition of metrics in Section 10.2.1) is required for any quality inspection or test, whether destructive or nondestructive, failure or time-terminated, and in-line or offline. Quality inspections or tests may be conducted according to industry standards or based on a customized procedure for metrics like thickness measurement, coating uniformity, or porosity analysis. Testing procedures may be generic (i.e., common for all batteries), or the tests may be customized for specific cases.

Metrics and elements of metrology used for the evaluation of goodness must be the same throughout the supply chain. Every piece of test equipment used anywhere in the supply chain must go through regular calibration procedures and repeatability and reproducibility studies to verify the stability and precision of the measurements. Fixtures must be replicated from the same design and similarity authenticated, and procedures must be similarly interpreted and implemented at all sites. Additionally, test scripts must be kept current at all test locations, and personnel must be trained to execute the plans reliably.

10.2.3 Supply Chain Management

Once quality metrics are defined, reliable and repeatable equipment is verified, and protocols for measuring the metrics exist, it must be clearly identified where in the supply chain it makes sense to measure them. After all, a product is only as good as its worst-performing part. Potential quality deviations must be identified wherever they occur in the supply chain to prevent possible defective components, enable faster fixes, and reduce risk exposure. It is imperative to select the right supplier, have a direct insight into the supplier's quality practices, and have the ability to improve processes through the exchange of technological expertise. Accordingly, supplier selection must be based on their ability to deliver quality components and their demonstrated ability to control their own supply chain. An audit of suppliers' procurement and quality processes can help provide the required information for such decisions.

It is important that a supplier quality plan exists for each supplier. This plan must include definitions of all in-process and assembly controls, a list of all quality metrics that will be measured and reported, and the corresponding sampling rates. Additionally, the plan must have clear guidelines for handling, storage, shipping control, component traceability, change notices, and end-of-life or obsolescence warnings.

10.2.4 Data Analysis

Data analysis corresponds to how data streams from various parts of the supply chain are viewed, analyzed, and responded to, which includes both online and offline data. Data streams are only valuable when analyzed and when there are identified triggers to initiate proactive intervention to fix problems before they manifest as bigger quality issues further down the supply chain. Appropriate statistical methods and interpretation must be used to clearly define the risk.

Data visualization—converting raw data into actionable information in a visually appealing representation—is a precursor for active monitoring and quick responses to fix quality issues as soon as they occur. Furthermore, artificial intelligence and machine learning methods can be used to track implementation, identify patterns, flag inconsistencies, and initiate action. For example, training a model to correlate critical battery in-line measurements, including slurry viscosity, anode and cathode thickness, electrolyte volume, or other metrics, with eventual battery capacity can allow the manufacturer to fine-tune these metrics to create cells with predictable and repeatable capacities.

10.2.5 Training

Training is key to achieving the safety of the end product through a quality mindset ingrained in people. Every person involved in battery design, material sourcing, supplier management, or manufacturing should have a clear understanding of quality plans, metrics, and goals, and have enough expertise in the methods required to achieve those goals. Training enhances the specialized skills and knowledge of people, so they can perform their roles effectively.

Managers must appreciate the importance of training and enumerate the specific training needs of their team. Specifically, training can consist of in-class training, mentoring, web-based training, guided self-study, or a formal on-the-job training program. Managers must identify the right medium for every individual and piece of training content, while employees must be trained in quality concepts and general statistical methods. Training is particularly needed for select employees in specific areas like fault isolation, failure analysis, and root cause analysis, among others.

10.2.6 Feedback and Audit

Feedback provides the ability to identify opportunities for improvement and deliver better results. Indicators for any quality issues, safety concerns, and customer feedback must be monitored through all means possible. Notably, the most valuable quality input is to track and collect manufacturing in-process data, test data, and data on any defective returns. This set of data relates to identifying the root causes of quality deviations that can be corrected for current and future improvement. All anomalies or failures must be analyzed to isolate and identify root causes. Documented records for each anomaly must include the date and lot code of the component, the failure point (production testing, quality testing, or field), the return date, the failure site, an identified root cause, and recommendations for preventing the anomaly in existing and future products.

Internal or external reviews/audits of quality practices, metrics, metrology, and supplier management help to ensure that planned quality activities are implemented as intended, so the product fulfills the specified quality requirements. Benchmarking to compare practices against known best-in-class industry leaders can also be used to study internal or external practices, finding ways to improve facilities, equipment, and methodologies, and producing better results.

10.3 Cell Manufacturing and Quality Risks

The cell manufacturing process involves several intricate and sequential steps, such as the mixing of active materials, applying the electrode coating on current carrier metal foils, and calendaring the coated electrodes to improve their density and conductivity, among others. The final step involves testing the cells to ensure that they meet the desired performance and quality metrics.

The two primary risks for cell manufacturing are the continuity of supply (i.e., the availability of raw materials) and inadequate or ineffective quality control plans impacting the durability or safety of the cells. The risk associated with the supply of materials is a pivotal element of the overall battery supply chain, given that the battery cell is the principal cost component [2]. The intricate supply chain for battery cells includes a variety of materials, including lithium, nickel, cobalt, and other essential materials for the cathode, anode, separators, and electrolytes. For example, cobalt is a crucial element in the cathodes of lithium-ion batteries, valued for its ability to enhance energy density and ensure thermal stability, and the Democratic Republic of Congo (DRC) is known for having the largest cobalt reserves in the world. However, the mining of cobalt in DRC has been associated with significant ethical and environmental concerns [14]. As highlighted before, the quality and safety of batteries can be synonymous, considering the potential consequences. Manufacturing variations beyond acceptable specifications, lack of procedures for evaluating in-process quality, incorrect equipment settings, uncalibrated metrology equipment, and a manufacturing organization not immersed in a quality culture can create significant risks for the end product delivered.

10.3.1 Risk Mitigation Practices for Cell Manufacturing

An expedient way to deal with issues relating to continuity of supply is to find alternate sources or materials. For example, the battery industry has been exploring alternatives to cobalt, such as cobalt-free, high-nickel, layered oxide cathodes, although practical application remains challenging due to intrinsic issues like thermal instability and phase transitions during cycling [15]. There is also a growing interest in adopting recycled cobalt to alleviate the environmental impact and lessen the reliance on mining operations [16]. Market dynamics also play a significant role, as increased costs due to demand from the electric vehicle sector and supply manipulation (controlling supply to affect prices) affect the supply chain. The recent oversupply and subsequent price collapse of lithium highlights the volatility and risk associated with lithium supply [17], and similar situations can impact the supply of other manufacturing elements as well.

Risk mitigation at the cell level starts with identifying the design and processing risks by conducting design failure modes and effects analysis (DFMEA) and process failure modes and effects analysis (PFMEA), respectively. Understanding these analyses, the best practices for quality control at the cell level involve a comprehensive approach to maintaining the quality of lithium-ion batteries throughout their lifecycle. This includes practices to ensure that all raw materials and components used in the manufacturing process meet the required quality standards, statistical process control (SPC) is utilized to monitor and control the key process parameters, a robust quality management system is implemented to ensure continuous improvement, and regular inspections and audits are conducted to ensure that all processes are being followed correctly.

Due to the sequential nature of battery manufacturing, starting with the highest possible quality materials is critical. Incoming raw materials like lithium cobalt oxide (LCO) need to meet high criteria for parameters like lithium composition, amount of moisture, particle size distribution, and presence of impurities. The metal foils used as current carriers must meet the requirements to withstand the pressure applied during calendaring and tension applied during the winding process. Even the pouch material must survive the stresses applied during the forming process. Similar metrics must be defined and met for anode materials, cathode materials, electrolytes, metal casings, and other raw materials used in cell manufacturing. Suppliers need to define critical limits on strength-related parameters in such a way that once in mass production, sufficient margins are available for normal manufacturing variations.

Battery cell metrics designated as "critical to quality" (CTQ) and "critical to safety" (CTS) are crucial for in-line quality control, detailing the parameters for various processes involved in the manufacturing of battery cells. These include sampling plans, control parameters, and quality metrics for various components such as cathode, anode, electrolyte, current collectors, separators, tabs, and other parts. Of course, all of these build toward the end goal of maintaining the high quality and reliability of battery cells.

Outgoing quality control (OQC) for battery cells is an essential quality assurance measure, ensuring that all cells meet the required standards prior to distribution. Additionally, the final quality control (FQC) and out-of-box audit (OBA) are critical inspections that utilize a well-defined sampling plan and switching rules to maintain consistent quality across all products.

Line audits are integral in battery cell manufacturing to guarantee the integrity and safety of the batteries. They encompass a comprehensive review of both online and offline operations, such as storage, quality inspection stations, warehouse management, and material processing. Also, the audits ensure that each pivotal phase in cell manufacturing adheres to stringent standards. Key items to review during an audit are storage facilities for incoming materials and final products, handling practices in storage areas and the production line, systems for inventory and data management, and testing areas. Manufacturing and quality teams need to review the operations end-to-end and identify specific best practices and prohibited actions in certain areas. This includes monitoring parameters on machines being used (e.g., the line pressure on pneumatic pick-and-place tools) and specific rules for the production line crew (e.g., no food on the manufacturing floor). The FIFO method, standing for "first in, first out," is a key principle in inventory management, particularly pertinent to battery stocks where prolonged storage can lead to irreversible capacity degradation. By prioritizing the usage of the oldest inventory, FIFO mitigates the risks associated with the long-term storage of batteries.

Ongoing reliability testing (ORT) is an integral part of the manufacturing process, designed to ensure that the quality of battery cells remains consistent with their initial specifications over time. ORT serves to monitor any deviations from the baseline of qualified products, and it maintains the capability to address any anomalies in manufacturing tools, methods, or operations. At the battery cell level, ORT typically encompasses mechanical assembly, environmental storage, and capacity testing. It must be made mandatory during both the ramp-up and sustaining phases of production. Crucially, all ORT failures need to be reported using standard data reporting procedures, and a thorough failure analysis must be conducted to understand and rectify any issues. The list of cell ORT items may include tests for cycle life, mechanical, environmental, and storage. It is important to note that ORT should not be used as a catch-all mechanism at the end of the production line, since it is not a substitute for process and material controls, but a mere validation of them.

10.4 Pack Manufacturing and Quality Risks

Once the cells are manufactured, they go on to pack manufacturing. At the pack manufacturing line, the cells are first connected in series, in parallel, or in a hybrid configuration, as discussed in Chapter 5. Subsequently, the final battery pack is manufactured by connecting the cells to the protection circuit module (PCM), which houses battery control electronics. Then, any additional mechanical support structures like frames or pressure-sensitive adhesives (PSAs) are added. If the pack manufacturing line is at a different location, additional time must be planned for shipping and added to the lead time for pack availability.

Usually, most of the engineering effort is spent during the new product introduction phase (NPI) of product development to improve and optimize the design. However, a pack not only needs the appropriate design but also requires well-controlled quality in manufacturing to meet the design and performance specifications during mass production. Notably, this is not a challenge unique to lithium-ion batteries. What makes the quality control for a battery even more critical is the safety concerns about the malfunction of the pack, as the energy stored in a lithium-ion pack can be on the order of a hand grenade [18]. Battery safety incidents can harm or kill users, affect the business, and tarnish the brand, as described earlier.

10.4.1 Risk Mitigation Practices: Pack

As in the case of a cell, the best practices for risk mitigation of a pack involve a systematic approach to a safe design, followed by scrupulously adhering to manufacturing and quality control actions. It starts with identifying the pack-level design and manufacturing risks by conducting DFMEA and PFMEA. A cross-functional team must conduct these assessments to keep it as inclusive as possible of all relevant viewpoints. During manufacturing, line operators or other quality personnel may change over time; if the manufacturing or assembly procedures are not well-documented, or not correctly implemented, they may be subtly altered during personnel changes. Although there must be emphasis placed on documentation, in practice, some critical points may get blurred over time. While process control and monitoring can reduce such occurrences, a periodic line audit is an effective way to find deviations and misses in the operating procedures.

By conducting FMEA, a list of processing parameters that are designated as "critical to safety" (CTS) and "critical to quality" (CTQ) must be identified. The advantage of placing CTS and CTQ parameters under more scrutiny is a faster reaction time for preventing an out-of-spec product from being distributed in the field and into the hands of the customer. The CTQ and CTS parameters are a subset of parameters that are needed for production control. However, to ensure that the manufactured packs meet specifications and expectations, a more comprehensive list of parameters must be constantly monitored through the production line, and warnings must be flagged when the control limits are not met. This extended list of parameters can be obtained from FMEA, especially by incorporating the learnings from previous pack products.

Battery pack CTS lists should include overcurrent cutoff response time, overvoltage and undervoltage cutoff response times, over- and under-temperature cutoff response times, and clearance between the cell and critical edges in the pack mechanical assembly. Additionally, the list should include pack capacity, weld joint strength, fuel gauge reporting accuracy of the relative state of charge (RSOC) and state of health (SOH), and pack RSOC at the end of the manufacturing line. Finally, the list should also include the bank-balancing efficiency (for multibank packs), pack dimensions (length, width, thickness) compared to the required tolerances, pack power line resistance, and mechanical strength of the enclosure.

Supply chain communication management is key for pack quality. Relevant risks identified through DFMEA and PFMEA must be communicated with the suppliers in every tier and modified as needed based on their feedback. Importantly, suppliers must be made fully aware of their responsibility for the quality control processes, with proper financial accountability included in contracts to ensure compliance. Suppliers must also have a functioning failure reporting, analyses, and corrective action system (FRACAS), or any equivalent failure isolation, failure analysis, and corrective action system. Similarly, they also must have an efficient channel for reporting quality data throughout their supply chain. A key aspect is accountability in storing and producing documentation, when asked, for quality data and any other in-line inspection data (tables, images, videos, etc.), in case the data must be reviewed in the future for root cause analyses or audit actions.

After completion of production and before leaving the line, the pack should go through a final quality check of its form, fit, and function, which is known as outgoing quality control (OQC). The pack supplier and the lower-tier suppliers for components inside the pack must record and provide the OQC data to the upper tiers in the supply chain. In a similar sense, the same parameters, or a subset of them, should be checked for incoming packs in the host device manufacturing line as part of the incoming quality control (IQC). Ideally, there should be no difference between the OQC and IQC data, except for the packs that are aging in the storage or shipping channel. The pack supplier, too, must conduct IQC for its lower-tier suppliers, such as its cell suppliers.

Reliability testing is an essential part of pack qualification during the NPI phase. Based on the results, the pack will be assessed to meet the reliability requirement. To make sure that there has been no deviation in the mass production of packs, a limited number of samples must be frequently put in a set of reliability tests. The difference between ORT and CTQ is that the latter only monitors the pack attributes at the time of manufacturing (time-0), but the former verifies that the aging and failure rates of packs in the field meet the expectations set in NPI. Consequently, the list of ORT items can include the pack capacity degradation assessment with cycling, pack capacity degradation assessment with storage, and pack mechanical degradation by usage or environmental conditions.

Storage conditions are especially important for batteries. This is because the battery is an active component with ongoing chemical reactions inside, and it decays over time. A stored battery, even if completely isolated and inactive, can discharge over time [19], and sometimes reach the undervoltage limit, which leads to permanent pack failure. This process, named self-discharge, is the main factor in limiting the battery's shelf life. Notably, batteries stored at a higher temperature or higher RSOC can have faster capacity degradation [19–21]. Also, dissimilar metals in the battery can experience galvanic corrosion in the presence of high humidity [22]. For these reasons, the storage temperature and humidity of battery packs must be controlled. Controlling the amount of outgoing RSOC (i.e., the amount of charge in the pack at the end of its manufacturing line) is a crucial factor too. However, finding the correct outgoing RSOC is a fine balance. To reduce the impact of aging, it is recommended to store the packs at an RSOC lower than the full charge capacity [23]. On the other hand, reducing the outgoing RSOC will result in a shorter shelf life, because of the battery's self-discharge. Even if the battery does not fail due to its self-discharge, its RSOC level in the host device may not be high enough to give an acceptable first impression to the customer who has just activated the device. To add to the complexity, regulatory requirements may set limits on a battery's outgoing RSOC (e.g., for air transportation) [24].

Concerns about battery storage also pertain to its shipping. However, unlike the designated storage room, the environmental conditions in the shipping channel may be unstable. As an example, solar loading can raise the temperature inside a cargo container; the humidity in the air, when accompanied by temperature fluctuations, can result in moisture condensation [25]. Subsequently, moisture can create a leakage path around the components that are experiencing a high voltage difference on their terminals (e.g., the power switches). This is a particular problem for batteries because they usually have a component under a high voltage delta by virtue of its cells' charge, unless the cells have reached a deep discharge stage. However, at this stage, the battery pack will be marked as failed by BMS regardless. Finally, to ensure that the battery pack has not been damaged, or undergone a degradation in the inventory storage or through the shipping channel, it is highly recommended to conduct IQC at any stage when the pack is assembled to the host device (e.g., the device manufacturing line, the repair centers, etc.).

10.5 OEM Device Integration and Quality Risks

The time between the definition of initial requirements for a lithium-ion battery and its final integration into a mass-produced device can take anywhere from months to years. Additionally, manufacturing a lithium-ion battery is a lengthy process. Manufacturing lead times can vary significantly from a few weeks to several months based on the complexity of the battery design, technology node choice, scale of production, and specific manufacturing processes used.

When an OEM chooses to customize the battery used in their next device, these lead times get stretched. For any custom cell, the typical development process consists of an NPI phase and a mass production (MP) phase. Battery samples are required during the very early stages of the NPI phase to ensure sufficient time to validate the performance, quality, and reliability metrics before production starts. As a result of the large upfront investments required for custom batteries, battery manufacturers have a specific minimum order quantity (MOQ) for a certain price. For lower volumes, the cell price per unit goes up. Accordingly, OEMs need to weigh the benefits of using a commercial-off-the-shelf (COTS) battery or a previously qualified battery versus a new battery design, taking lead times and cost implications into consideration.

For device integration, in-line transfer protocols, handling, assembly procedures, and storage and shipping conditions are of high importance, given the delicate nature of the pouch cells or packs. It is critical to ensure the best manufacturing and quality controls in the factory where the batteries are handled. This applies to the final assembly site as well as repair centers. To avoid variations and errors in manual assembly, it is often necessary to use automated assembly tools to install batteries.

Sometimes, it may be necessary to expose the incomplete assembly to higher temperatures after installing the battery, to address another operation down the manufacturing line. In this situation, it is important to understand the temperature limits for the safe and reliable operation of the battery and ensure that the assembly line does not exceed the allowable process window.

When an OEM qualifies a battery, aspects of safety are accounted for both in terms of battery technology and integration of the battery into the device. However, when designs keep changing, monitoring safety can add significant overhead to product development, especially in terms of time spent. Also, with frequently changing designs and supply chains, it would be nearly impossible to keep building on lessons learned, and mistakes of the past may be repeated in the future.

10.5.1 Risk Mitigation Practices: Device Integration

Batteries are integrated into devices by attaching them to the device enclosures using adhesives or mechanical methods. When the battery is attached to the device enclosure using PSA, it is possible to use automated pick-and-place tools and semiautomatic processes to remove lines from the PSA, followed by a fully automated PSA activation process. In this case, visual inspection (automated or manual) is still necessary, as the PSA used may have different degrees of adhesion to its liner, and instead of lying flat on the battery or enclosure, the PSA may fold. When pressure is applied to a battery with a folded PSA under it, a dent is created on the pouch cell, which can cause damage to the internal structure and consequently lead to premature failure of the battery.

When the battery pack is to be assembled mechanically, a semiautomatic process can be used to pick and place the battery in the device, followed by a manual process for screwing down the frame. In this case, it is necessary to ensure that screws are tightened with appropriate torque, since insufficient torque can lead to a loose screw inside the device, which can poke the pouch cells and lead to leakage. Conversely, overly tightened screws can damage the frame or the chassis, thus failing to couple the battery and enclosure. While full automation in the factory is possible, it may not always be the best choice. This is especially true when the design is meant to enable easy repairs by hand. In this case, performing the original assembly by hand may be best.

Despite the best assembly practices, it is necessary to ensure that quality control and quality inspections are implemented. These can be automated or manual visual inspections, looking for any dents or interference, or more involved inspections, involving X-ray or magnetic pulls, to look for any floating screws or debris in the vicinity of the battery. If specific clearances between cells and nearby components are deemed critical, it may be necessary to make measurements to ensure that the critical parameters are within acceptable limits. As the space within devices is very tight, some level of denting of batteries during manufacturing will occur despite the best practices. To address this, additional visual inspections for the battery should be ensured before closing the device. Finally, once the device is fully assembled, it needs to meet the temperature and humidity requirements acceptable for safe and reliable operations of the battery.

Given the fluidity of designs, certain safety aspects must be standardized. This involves first-level protection in terms of TCOs at cell suppliers, up to more advanced features at the pack level involving monitoring of current, voltage, and temperatures. Each of these measures must be verified at the OEM assembly site.

Batteries have a unique behavior of degradation with aging that is much more pronounced compared to other components used in a typical laptop, phone, EV, or any other device. This leads to specific requirements for shipping and storage of batteries and devices with batteries in them. Shipping of devices with batteries is regulated with UN3481 and other regulations, corresponding to lithium-ion batteries packed with equipment [26, 27], and general compliance is required to ensure the safe shipping of devices. However, the overall state of charge of the battery contained inside a device is not regulated; it is a parameter left up to the OEMs.

For some portable devices shipped in a cardboard packaging box, the ability of the customer to push the power button and start using the device immediately without having to charge provides a great out-of-the-box experience. To achieve this experience, devices must be shipped with some level of charge from the factory. Sometimes, a mechanical barrier is placed between terminals to prevent discharge, as in the case of remote controllers or cordless landline phones. When shipped and ready, the customer can pull off a tab to complete the circuit and start using the device. In some devices, like a typical laptop, the mechanical barrier may not be feasible, and a similar objective can be achieved by putting the battery in ship mode if the OEM allows it. This, however, requires the device to be plugged in for a short while to unlock the gas gauge. If this is not acceptable, the battery needs to be charged sufficiently to ensure that there is still sufficient charge, after accounting for leakage during shipping and warehouse storage, ensuring the intended experience. This can be highly variable, based on the logistics of the downstream supply chain.

One factor to balance against is the impact of a high state of charge on the permanent capacity loss of the battery. It should be noted that the higher the state of charge during long storage, the greater the permanent loss of capacity [23]. As such, this characteristic must be characterized at the battery and device levels, and a balance between providing a great customer experience at the time of purchase must be weighed against the risk of providing the customer with a lower initial battery capacity.

Batteries have a definite useful life, both in terms of the number of cycles and calendar age. If the battery cannot be replaced, then a perfectly working device may need to be disposed of when the battery stops working effectively. Given this, the solution is to make the battery in the device replaceable. Several regulations exist the world require OEMs to support battery replaceability. This, however, can solve the problem only if the battery to be replaced is available in the market a few years after the product launch. Given the constant churn in technology development in the battery industry, battery suppliers have little incentive to maintain older technology products for a long time. This is where it becomes important to have standard form factors that third-party battery manufacturers can produce.

Recently, an additional factor impacting the battery industry in consumer electronics is the arrival of EVs. The volume demand in the EV market is much larger than that in consumer electronics. As such, a cell or pack manufacturer starting with consumer electronics products may aspire to enter the EV market and subsequently exit the consumer electronics market [28]. This creates additional churn in the industry and puts greater responsibility on OEMs to ensure the availability of spare batteries, repair capability, and repair capacity.

10.6 Auditing Considerations

An audit is a verification activity involving the inspection and examination of the battery manufacturing process and the implementation of the quality system therein. Specifically, an audit can be focused on three distinct aspects: process, product, or system. ISO standard 19011:2018 provides guidelines for auditing management systems and defines the audit process as such:

> Systematic, independent and documented process for obtaining audit evidence [records, statements of fact or other information which are relevant and verifiable] and evaluating it objectively to determine the extent to which the audit criteria [a set of policies, procedures or requirements] are fulfilled [29].

The primary focus of an audit is on the manufacturing line, but it can also include the entire design and support activities, such as design capabilities, documentation, regulatory and environmental compliance, and testing and failure analysis capabilities, among many other aspects. Moreover, an audit can be applied to the entire organization of the battery supplier, or only focused on a section of it. Additionally, an audit can be a follow-up on a corrective action.

A battery supplier audit can be conducted directly by the customer or by a third party. Also, battery suppliers can conduct their own internal audits and share the results with suppliers at a regular cadence. During an audit of Li-ion battery suppliers, several common findings will correspond to nonconformity of critical manufacturing attributes, defects, process issues, data management deficiencies, or failure processing protocols. Nonconformity to specifications of the anode and cathode coatings, cell thickness, capacity, input resistance, and others may show up unless these parameters are monitored effectively through in-line control measures or an OQC process. Furthermore, defects such as the presence of external particles on the anode, cathode, separator layers, and other components may show up unless material handling and cleanliness protocols are strictly adhered to. Process issues, such as improper jelly roll winding or cell tab welding, may show up if the supplier has insufficient experience or expertise in battery manufacturing, which can include improper equipment settings or lack of proper equipment maintenance and calibration protocols.

An audit can examine the supplier's technical expertise by probing their reasons for actions taken in the manufacturing line. Moreover, an audit can identify production condition variations that have not been documented or approved since the NPI phase. Data management deficiencies in manufacturing data acquisition and retention can be identified in an audit. An audit can examine if the supplier maintains an appropriate record of production (e.g., in charts or pictures) and if there are parts of the manufacturing line that are missing. The incompetence of ancillary services like yield fall-out analysis, fault isolation, and failure analysis can be revealed in an audit too. A lack of ability to properly conduct failure analyses, reporting, and reliability and safety testing can be revealed in an effective audit.

Among the specific types of audits, a process audit is used to verify that the manufacturing process is conforming to the required limits. It assesses a procedure or method against required instructions or standards to evaluate how effective the instructions are and if the standards are met. A process audit can assess if suppliers follow the required specifications in cathode and anode material preparation, cell production steps, and pack assembly steps. Also, a process audit may include inspecting the resources (e.g., machinery, materials, and personnel) used to convert incoming materials to cells and packs. Additionally, a process audit can involve inspecting the manufacturing line environment and the methods utilized by the supplier to ensure production quality (e.g., operating procedures, instructions, and operator training) and the methods for evaluating process performance. A production process stability audit also includes temperature control and humidity control during battery assembly, as well as the consistency of the process. Finally, a process audit can be about evaluating the efficiency of process controls put in place by battery suppliers.

Another type of audit is a product audit, which is about how well the final product in each tier (e.g., cathode or anode material, cell, pack, etc.) meets the functional and quality requirements. It includes an audit of the quality of battery materials and components. It evaluates the sources and quality of anode and cathode materials, as well as the quality parameters of other components, such as separators and electrolytes. Testing and inspection parameters can be audited to evaluate the battery performance testing methods, reasonableness of test case design, and the final inspection process of the product. In fact, this type of audit also acts as a verification of supplier OQC.

Finally, a system audit is an audit of the quality management system. It determines how well different responsible functions (e.g., the teams of design, quality, program management, and supply chain assurance) are defined in the cell and pack supplier organizations. Safety measures and personnel training can be audited to evaluate safety at the production site, including staff training and emergency plans, to reduce the risk of fire and other accidents. Furthermore, quality systems can be evaluated to ensure that the production processes, safety testing procedures, and installation into end-use environments comply with international and local regulations, standards, and environmental protection requirements. Some examples of end-use applications include electric or hybrid vehicles [30], industrial applications [31–33], energy storage systems [34, 35], or light electrical rail [36].

10.6.1 Audit Process

The design of the audit process is a pivotal element in ensuring the quality and compliance of the end product delivered through a supply chain. It consists of a serialized set of activities to comprehensively assess various aspects of the process, the product, and the system at any entity of the supply chain. It starts with planning and preparation, wherein the audit team must define the scope, objectives, and plan of the audit. Considering the specificity of lithium batteries, this includes aspects such as battery assembly, material management, production processes, testing, and compliance. Importantly, comprehensive analyses of literature, technical standards, and previous audit reports related to lithium battery production must be conducted to provide a basis for subsequent field audits.

The next step is an actual field investigation in which inspections are conducted of production lines, production processes, equipment, personnel training, and safety measures. It is the most critical step in the audit used for all essential data gathering and helps create a mapping between documentation and reality. This data gathering helps assess the stability and compliance of the production process through observation of actual operations. The audit team must conduct an in-depth analysis by collecting production data, such as battery performance, process parameters, and quality control data, with special attention to key parameters that may affect quality and safety. This can lead to the discovery and identification of problems, especially those that may lead to safety risks such as fires. Notably, problem identification requires expertise and an in-depth understanding of the lithium battery production process.

The final step involves drafting a report and summarizing findings, issues, and recommendations. The report needs to include an assessment of key parameters, as well as proposals for improvement measures. The findings should be presented in an organized manner, with supporting data and evidence, highlighting areas of inefficiencies, noncompliance, and areas for improvement. Recommendations must be specific, actionable, measurable, and most importantly achievable. The report should also include a plan for implementing the recommendations, including timelines and responsible entities [37].

In lithium battery production, supplier assessment criteria are a pivotal component of the audit process, crucial for ensuring quality, safety, and compliance throughout the battery manufacturing process. This entails conducting regular assessments of suppliers that provide battery materials, components, and key technologies. Auditors must ensure that these suppliers adhere to international and industry standards, maintain stable quality systems, and possess reliable delivery capabilities. Additionally, the assessment criteria should encompass various aspects of the supplier's financial status, quality management system, environmental practices, social responsibility, and more, to guarantee the reliability and compliance of the entire supply chain.

The development of evaluation criteria for lithium battery suppliers necessitates a comprehensive consideration of multiple facets of the battery production chain, ensuring the selected suppliers meet the requirements of high-quality battery production. This includes assessing multiple aspects of battery materials, battery assembly, and production processes. Also, the audit must define a set of quantitative indicators such as the product defect rate, production efficiency, and environmental protection index, among others, to quantify a supplier's performance across various aspects. Qualifying thresholds for each indicator must be established in advance to determine whether assessment criteria are met, aiding in establishing clear criteria and reducing subjective judgment. These assessment criteria must be dynamically adjusted over time to adapt to industry development changes and supplier performance, keeping the assessment system current and forward-looking.

The primary parameters of the evaluation criteria encompass the quality system, technical strengths, safety standards, and social and environmental accountability. An audit must evaluate a supplier's quality management system to ensure it aligns with international standards (e.g., ISO 9001) and meets specific requirements of the battery industry. The focus must be on a supplier's quality assurance measures in material procurement, production process control, and product testing. Furthermore, an audit must evaluate the research and development (R&D) capability and technological capabilities of suppliers to ensure that they can consistently provide products that meet the latest industry standards and innovation requirements. This includes mastery and application of new materials and processes. Next, it must evaluate the supplier's safety management standards, particularly concerning hazardous chemicals and battery assembly processes involved in battery production, to ensure employee safety and a secure production environment.

Social and environmental governance needs to be evaluated to assess suppliers' social responsibility practices, encompassing aspects like employee welfare, labor rights, and community involvement, upholding ethical standards and social responsibility throughout the supply chain. This also includes an evaluation of suppliers' environmental policies and practices to verify compliance with relevant environmental regulations and their commitment to reducing environmental impact. Crucially, an audit is useful for enhancing the sustainability of battery products.

10.6.2 Auditing Frequency

Auditing frequency refers to how often a system, process, or supply chain undergoes audits as part of regular monitoring. Determining the appropriate audit frequency involves considering factors such as business risks, the rate of system changes, monitoring requirements, and the costs involved in conducting an audit. By strategically arranging audit frequencies based on scientific principles, issues can be identified and addressed promptly, thereby reducing the likelihood of potential risks. Auditors must develop suitable audit plans and frequencies tailored to specific circumstances, maximizing system stability and operational effectiveness. In the lithium battery industry, it is common to conduct audits at least once a year. However, for areas with higher risks or products with sensitive applications, more frequent audits may be necessary, such as every six months or quarterly. Nonetheless, the determination of audit frequency should be made on a case-by-case basis, considering the company's unique situation, product usage, and market demands.

Some factors that can influence auditing frequency, enhancing operational resilience and compliance, include the pace of technological innovation, battery application (consumer devices, EVs, or BESS), regulatory compliance, and sustainability considerations. Lithium battery technology and energy storage techniques undergo constant innovation and rapid product iterations; thus, it is advisable to enhance monitoring of technological advancements and changes to stay abreast of industry developments. Also, the industry is highly influenced by fluctuations in market demand and supply chain dynamics. Regular audits aid in understanding market shifts and supply chain risks, facilitating timely adjustments in production and procurement strategies.

Considering the widespread use of lithium batteries and energy storage products in critical sectors like electric vehicles and renewable energy, it is even more important to ensure high levels of safety and performance. Consequently, battery supply chains for these sectors may necessitate more frequent audits to establish stability in high-risk scenarios. Audit frequency should align with changes in regulatory compliance to ensure manufacturing processes meet the latest requirements mandated by the regulations and standards that govern particular industries. Notably, the emphasis on energy efficiency and environmental sustainability has intensified across several sectors of the economy. Audits should prioritize these aspects to ensure production processes align with eco-friendly and green manufacturing principles.

The significance of auditing lies not only in onetime assessments but also in establishing continuous monitoring mechanisms. Continuous monitoring and periodic audits must complement each other and should constitute an integral component of the quality management system. A single audit can offer insights into a specific point in time, yet system risks evolve with changes in business operations and the external environment. Conversely, continuous monitoring enables audits to promptly identify emerging risks and issues, empowering organizations with enhanced risk management capabilities and decision-making support. While onetime audits can serve as regular, in-depth assessments for identifying potential issues and ensuring compliance, their limitations lie in their static representation of a moment in time. Given the unique challenges posed by lithium batteries, including electrochemical reactions and temperature control complexities, onetime audits may struggle to capture evolving risks across the entire production cycle. With that said, continuous monitoring compensates for these limitations by establishing real-time data collection, monitoring, and feedback mechanisms to ensure ongoing scrutiny of the lithium battery production process. This benefits the timely detection of unexpected events, technological advancements, and market shifts impacting quality and safety.

10.7 Key Steps in Battery Selection

Selecting the right battery cells and packs is crucial for any application. The process involves a multistep evaluation that considers the intended usage environment, performance targets, desired lifespan, supplier capabilities, battery reliability and safety, environmental and sustainability considerations, and various business constraints around cost and time-to-market. There will always be trade-offs between these factors, and a balanced approach is needed to reach the optimal decision. However, prioritizing safety and reliability is still paramount to mitigate potential risks. Specific key steps involved in battery selection include battery requirement assessments, supplier assessments and audits, qualification and certification verification, safety and authenticity evaluation, social and environmental impact assessment, and warranty scrutiny. After battery requirements are established and potential suppliers identified, the rest of the steps can be conducted sequentially or in parallel and should involve evaluations of more than one supplier.

The first step in selecting a battery is to define its requirements, which lays the groundwork for the entire battery sourcing selection process. The OEM needs to define the specific needs of the application for both the battery cell and pack. Specifically, requirements include the intended use and life cycle application requirements, performance requirements, and service life. The intended use of the battery significantly impacts the selection process, whether it is for a portable electronic device, a medical instrument, an electric vehicle, an energy storage system, or something else entirely. Life cycle application requirements include the operational and nonoperational temperature ranges, temperature extremes, humidity levels, and potential for shock or vibration that the battery will encounter. Understanding the application will guide size, weight, and durability requirements. Additionally, performance requirements need to be defined in terms of the capacity, voltage range, discharge current, charging current and time, and power output needed from the battery pack. This will influence the number and configuration of cells required. The target service life requirement concerns the desired operational lifespan for the battery, which includes both cycle life and calendar life, which affects the choice of cell chemistry and impact factors like depth of discharge and cycle life.

Once the battery requirements are established, the potential suppliers need to be identified. Of course, there are two main suppliers to consider: the cell manufacturer and the pack manufacturer. In some cases, these may be the same company. Cell companies manufacture cells and may market them to pack manufacturers or directly to final product manufacturers. Pack houses assemble battery packs using premanufactured cells sourced from other cell manufacturers (i.e., upstream suppliers). They focus on pack design, configuration, and integration based on customer specifications. When working with a pack house, the product company may have the option to specify the desired cell source or allow recommended cells based on the pack house's expertise and supplier relationships. This decision will depend on the product company's access to cell information and their specific needs.

The choice between these options depends on the product company's specific needs and priorities. Pack houses often offer greater flexibility in pack design and customization, while cell and pack manufacturers may provide greater control over cell quality and consistency. Regardless of whether pack houses or cell and pack manufacturers are chosen, an evaluation of the manufacturers and their supply chain is crucial. This would include an evaluation of their track record, depth of experience in the field, and feedback from other customers to gauge their reliability and commitment to quality. Some suppliers may be favored because they are committed to environmentally friendly practices throughout the battery's life cycle, encompassing responsible material sourcing, manufacturing processes, disposal, and recycling initiatives. Additionally, the financial stability and reputation of the supplier should be assessed, recognizing that a financially sound supplier is more likely to provide consistent support and fulfill warranty obligations.

Auditing the supplier selection or evaluating conformance to existing industry standards can be used as a decision tool as well. Indeed, audits can be conducted to enhance trust. "Quality, health & safety and environmental" (QHSE) audits can help assess the integration and management of systems to enable risk management and sustainability. An ISO 45001 audit can help evaluate systems managing occupational health and safety, which is important for handling hazardous materials and preventing workplace accidents. Quality management system audits like ISO 9001 can help to evaluate a supplier's manufacturing processes, alignment with global quality standards, and areas for improvement. Manufacturing process audits can examine the efficiency, safety, and quality of manufacturing operations to identify issues impacting product quality and safety.

Environmental compliance audits like ISO 14001 can focus on responsible environmental management and regulatory compliance. Supply chain security compliance audits can help evaluate security measures in the supply chain, which is essential for mitigating theft, tampering, and other risks. An evaluation of the reliability capability of the battery manufacturer can be conducted, for example, according to the IEEE Standard for Organizational Reliability Capability (IEEE 1642). For some specialized applications, audits like an ISO 13485 audit are used for LIB manufacturers for medical devices, ensuring compliance with specific medical device quality requirements. In the context of negotiating bulk or long-term contracts, the geopolitical conditions and trade regulations of the manufacturer's originating country may also need to be assessed. If feasible, evaluations or reviews from independent third-party organizations specializing in battery technology assessments can be used to gain an unbiased perspective on the suppliers.

Once potential suppliers have been shortlisted, the focus shifts to evaluating their ability to provide documentation for qualification and certification. A supplier must provide a detailed datasheet with all of the required information, such as details on charging and discharging current limits at various temperatures or documents showing that the battery products meet pertinent industry standards, safety regulations, and compliance requirements. This includes verifying that the battery products have undergone rigorous testing and hold certifications from organizations such as Underwriters Laboratories (UL), International Electrotechnical Commission (IEC), Conformité Européenne (CE), Restriction of Hazardous Substances (RoHS), and others required for certain regions.

Suppliers must be evaluated for their ability to conduct customer-specified testing protocols during the product development phase. In general, it is wise for the customers to conduct their own performance testing before embedding a battery into a final product, even if the battery has some form of certification. This is especially important if the battery manufacturer is new to that particular business or if the batteries received were in storage for a long time after the cells were made. Typically, the manufacturer's testing protocols are outlined in the manufacturer's specifications. It is crucial to ensure they encompass assessments for performance under temperature, C-rate, and voltage ranges and their impact on battery health. If the OEM directly sources the battery cells and packs from the original manufacturers, then the OEM may define their own performance and qualification tests beyond the battery manufacturers specifications.

Safety considerations include identifying potential hazards associated with the application and selecting cells and pack designs that meet relevant safety standards. Safety testing may also be conducted to evaluate the design of the battery products, considering the robustness of the casing, incorporation of fail-safe mechanisms, effectiveness of thermal management systems, provisions for overcharge and overdischarge protection, and mechanisms for preventing short circuits. When a pack contains a battery management system (BMS), its functionality and reliability should also be evaluated. If, for some reason, the batteries can only be sourced from a third-party distributor, warehouse, or retailer, then the cells must be authenticated. The purpose is to determine whether the received cells are genuine and unused. This is especially of concern when purchasing battery packs directly from third-party sources with limited information on the cells used for manufacturing the packs. Authentication can be performed through a certificate provided by a trustworthy organization or through structural comparative analysis against a known authentic cell using CT scanning or any other nondestructive methods, or if needed, by using other destructive analysis methods.

The social and environmental impacts of battery material sourcing and manufacturing are important during the selection process. Prioritizing batteries with zero conflict minerals and a high percentage of recycled materials, to minimize environmental footprint and support a circular economy, may not be just a choice but may sometimes be required by regulation. Some EU regulations already mandate or incentivize using recycled materials in batteries and ensure that batteries have responsible end-of-life management plans to prevent environmental hazards from improper disposal. While it is tempting to choose cheaper cells and packs, balancing the cost with quality and safety features is crucial. Investing a bit more can often lead to better performance, better warranties, and enhanced safety. It is important to assess not only the upfront cost of the battery but also its total life cycle cost, which includes shipping, maintenance, charging, and replacement.

A review of the warranty terms provided by the manufacturer is needed, recognizing that the warranty coverage indicates the manufacturer's confidence in their product's reliability. Importantly, close attention must be paid to the duration of coverage, terms for repair or replacement, and any limitations or exclusions. Comprehensive warranty coverage can signal confidence. A supplier's ability to service warranty returns needs to be looked into, revealing the availability of technical support, customer service, spares planning, and other aspects.

10.8 Summary

The global supply chain for lithium batteries is complex, to begin with. Additional constraints related to meeting ethical and regulatory requirements, while ensuring the quality and safety of the end products, create further unique challenges. Recent industry trends related to traceability, sustainability, socio-political supply chain vulnerabilities, and counterfeiting have enhanced the risks in material procurement, manufacturing, and assembly. Rapidly growing competing uses for critical metals required for battery manufacturing, and the shifting focus by major battery suppliers toward EVs and BESSs, are other significant factors that have been stressing the supply chain for batteries.

For safety-critical systems like batteries, the words quality and safety are synonymous. For all cell manufacturers, pack manufacturers, and device makers integrating batteries into devices, strict implementation of a quality plan is imperative to mitigate field safety concerns. A quality plan for batteries must include quality practices, protocols, documentation, monitoring schemas, control procedures, remediation, and corrective action processes. It has six fundamental aspects: metrics, metrology, supply chain, data analysis, training, and feedback. Additionally, a quality plan must identify the metrics to be measured and how and where they are measured, how the data is monitored and analyzed, the process for response if anomalies are observed, and a plan to track field performance.

The two primary risks for cell manufacturing are continuity of supply or the availability of raw materials, and inadequate or ineffective quality control plans impacting the durability or safety of the cells. An expedient way to deal with continuity of supply issues is to find alternate materials, alternate sources of materials, and the ability to recycle and reuse materials. Quality control at the cell level involves a comprehensive approach to maintain quality throughout their life cycle. The necessary practices include statistical process control (SPC) to monitor and control the key process parameters and defining cell critical to quality (CTQ) and critical to safety (CTS) metrics for in-line quality control. Additionally, practices include outgoing quality control (OQC) as an essential quality assurance measure, line audits, and ongoing reliability testing (ORT) to ensure that the quality of battery cells remains consistent with their initial specifications over time.

Risk mitigation approaches to safe battery pack designs start with identifying the design and processing risks by conducting design failure modes and effects analysis (DFMEA) and production failure modes and effects analysis (PFMEA). By conducting FMEAs, a list of CTQ and CTS processing parameters must be identified. The advantage of placing these parameters under more scrutiny is to ensure a faster reaction time for isolating out-of-spec packs. For a pack, some of these parameters are overcurrent cutoff response time, overvoltage and undervoltage cutoff response times, over- and under-temperature cutoff response times, pack dimensions and clearances, pack capacity, and weld joint strength, among others. Storage conditions are particularly important for batteries, since their capacity decays over time due to self-discharge, which, if prolonged, can lead to permanent failure. To reduce the impact of aging, it is recommended to store the packs at a charge level smaller than the full charge capacity.

Lead times for lithium-ion batteries for device makers or OEMs can vary significantly from a few weeks to several months based on the complexity of the battery design, technology node, scale of production, and manufacturing processes used. Customization stretches these lead times further. For device integration, in-line transfer protocols, handling, assembly procedures, and storage and shipping conditions are of high importance, given the delicate nature of the pouch cells or packs. Controlled and repeatable handling and assembly procedures are imperative for the integration of batteries into devices, whether they are attached to the enclosures using adhesives or other mechanical methods.

It is necessary to ensure quality control through automated methods, manual visual inspections, or more involved inspections involving X-ray or magnetic pulls, to look for any floating screws, debris, or other defects in the vicinity of the battery. Batteries have specific requirements for shipping and storage due to the unique behavior of degradation with aging, even at room temperatures. Also, batteries have a specified useful life, both in terms of the number of cycles and calendar age. One solution for limited battery life is to make them replaceable, and several regulations exist or are in the stages of approval to make this mandatory.

An audit is a verification activity aimed at an examination of the battery manufacturing process and the implementation of its quality system by the supplier. Even though the primary focus of an audit is on the manufacturing line, it also must include an assessment of suppliers' design capabilities, documentation, risk assessment procedures, prototyping processes, data acquisition and retention, regulatory and environmental compliance, testing and failure analysis capabilities, resource management, and interactions with lower- and higher-tier suppliers. An audit can be conducted by the customer or a third party, and it must focus on auditing a process, product, or system. A process audit evaluates the conformance of the manufacturing process to the required limits. A product audit determines whether the final product meets the functional and quality requirements. A system audit is an audit of the quality management system.

Factors to be considered while determining an appropriate audit frequency are business risks, the rate of system changes, and monitoring requirements. In the lithium battery industry, it is generally recommended to conduct audits at least once a year. However, for areas with higher risks or products with more sensitive applications, more frequent audits may be necessary, such as every six months or quarterly. Besides a onetime snapshot assessment, an audit must also establish continuous monitoring mechanisms. Continuous monitoring and periodic audits must complement each other and should constitute an integral component of the quality management system.

Selecting a battery for a specific application is a multistep evaluation process that must account for intended use, performance targets, desired life span, supplier capabilities, environmental and sustainability considerations, and business constraints around the cost and time-to-market. Prioritizing safety and reliability is still paramount to mitigate potential risks. Key steps involved in battery selection are establishing battery requirements, identifying potential suppliers, conducting qualification and certification verification, evaluating product safety and authenticity, assessing social and environmental impacts, and scrutinizing warranty terms. After requirements are established and suppliers identified, the rest of the steps can be conducted concurrently or sequentially, as feasible.

References

1 Sarkar, E.M., Sarkar, T., and Bharadwaj, M.D. (2018). Lithium-ion battery supply chain: enabling national electric vehicle and renewables targets. *Current Science* 114 (12): 2453–2458.

2 Bloomberg, Lithium-Ion Battery Pack Prices Hit Record Low of $139/kWh [Online]. https://about.bnef.com/blog/lithium-ion-battery-pack-prices-hit-record-low-of-139-kwh/ (accessed 14 September 2024).

3 Agrawal, T.K., Angelis, J., Thakur, J.R. et al. (2021). Enabling circularity of electric vehicle batteries - the need for appropriate traceability. *IEEE International Conference on Technology Management, Operations and Decisions (ICTMOD)*, Marrakech, Morocco.

4 Perez, A., Gallo, P., and Sanseverino, E.R. (2023). Blockchain and nested tokens for tracking, reusing, and recycling batteries. *2023 Asia Meeting on Environment and Electrical Engineering (EEE-AM)*, Hanoi, Vietnam.

5 Krishna Mohan, T.V., Pratap, B., Amit, R.K., and Venugopal, S. (2021). A road map to sustainable mobility: Analyzing the dynamics of lithium-ion battery recycling. *2021 IEEE Transportation*, New Delhi, India.

6 Zhang, Y., Nguyen, R.T., and Liaw, B. (2021). Status and gap in rechargeable lithium battery supply chain: Importance of quantitative failure analysis. *Proceedings of the IEEE* 109 (6): 1029–1038.

7 Ren, H., Du, J., Wang, C. et al. (2024). Multiplex characteristics and vulnerability assessment of the global electric vehicle lithium-ion battery supply chain network: a focal firm-specific analysis. *IEEE Systems Journal* 18 (1): 339–350.

8 Kong, L., Das, D., and Pecht, M.G. (2022). The distribution and detection issues of counterfeit lithium-ion batteries. *Energies* 15: 3798.

9 Office of Energy Efficiency & Renewable Energy (2021). Battery Critical Materials Supply Chain Challenges and Opportunities: Results of the 2020 Request for Information (RFI) and Workshop, 20 December 2021 [Online]. https://www.energy.gov/eere/amo/articles/battery-critical-materials-supply-chain-challenges-and-opportunities-results-2020 (Accessed 24 March 2024).

10 Sutopo, W., Maryanie, D.I., Purwanto, A., and Nizam, M. (2013). A comparative value chains analysis of battery technologies for electric vehicles. *2013 Joint International Conference on Rural Information & Communication Technology and Electric-Vehicle Technology (rICT & ICeV-T)*, Bandung, Indonesia.

11 Radomski, D., Jezierski, K., Nelson, T., and Kibler, R. (2013). NextEnergy Center's highlights on global lithium ion battery supply chain dynamics and domestic opportunities. *2013 World Electric Vehicle Symposium and Exhibition (EVS27)*, Barcelona, Spain.

12 Shyam, H.S., Shuaib, I., and Pandey, S.P. (2023). Exploring the supply chain challenges in the area of lithium-ion batteries used in electric vehicles. *Macromolecular Symposia*, Jan 2023.

13 Crosby, P.B. (1979). *Quality Is Free: The Art of Making Quality Certain*. New York: McGraw-Hill.

14 Banza, C.L.N., Nawrot, T.S., Haufroid, V. et al. (2009). High human exposure to cobalt and other metals in Katanga, a mining area of the Democratic Republic of Congo. *Environmental Research* 109 (6): 745–752.

15 Kim, Y., Seong, W.M., and Manthiram, A. (2021). Cobalt-free, high-nickel layered oxide cathodes for lithium-ion batteries: progress, challenges, and perspectives. *Energy Storage Materials* 34: 250–259.

16 Apple Inc., Apple will use 100 percent recycled cobalt in batteries by 2025, April 2023 [Online]. https://www.apple.com/newsroom/2023/04/apple-will-use-100-percent-recycled-cobalt-in-batteries-by-2025/ (accessed 21 April 2024).

17 Burton, M., Lee, A., Attwood, J., and Li, Y.Y. (2024). Lithium market struggles to recover after epic boom and bust, Bloomberg, 9 April 2024 [Online]. https://www.bloomberg.com/news/articles/2024-04-09/lithium-prices-market-faces-hard-truths-after-epic-boom-and-bust (accessed 23 April 2024).

18 Barsukov, Y. and Qian, J. (2013). *Battery Power Management for Portable Devices*. Norwood, MA: Artech.

19 Li, R., Bao, L., Chen, L. et al. (2023). Accelerated aging of lithium-ion batteries: bridging battery aging analysis and operational lifetime prediction. *Science Bulletin* 68 (23): 3055–3079.

20 Spotnitz, R. (2003). Simulation of capacity fade in lithium-ion batteries. *Journal of Power Sources* 113 (1): 72–80.

21 Karger, A., Schmitt, J., Kirst, C. et al. (2023). Mechanistic calendar aging model for lithium-ion batteries. *Journal of Power Sources* 578: 233208.

22 Zwicker, M., Moghadam, M., Zhang, W., and Nielsen, C. (2020). Automotive battery pack manufacturing – a review of battery to tab joining. *Journal of Advanced Joining Processes* 1: 100017.

23 Han, X., Lu, L., Zheng, Y. et al. (2019). A review on the key issues of the lithium ion battery degradation among the whole life cycle. *eTransportation* 1: 100005.

24 IATA (2020). 2020 Lithium Battery Guidance Document: Transport of Lithium Metal and Lithium Ion Batteries, IATA.

25 Nikolayev, V.S., Beysens, D., Gioda, A. et al. (1996). Water recovery from dew. *Journal of Hydrology* 182: 19–35.

26 PHMSA (2021). Lithium Battery Guide for Shippers, September 2021 [Online]. https://www.phmsa.dot.gov/sites/phmsa.dot.gov/files/2021-09/Lithium-Battery-Guide.pdf.

27 IATA (2024). IATA, 30 April 2024 [Online]. https://www.iata.org/en/publications/store/lithium-battery-shipping-regulations/.

28 T. J. News. Panasonic's pivot from home appliances to EV batteries fraught with risk, 15 June 2021 [Online]. https://japannews.yomiuri.co.jp/business/companies/20210615-53202/ (accessed 28 June 2024).

29 International Standards Organization (2018). ISO 19011:2018, Guidelines for auditing management systems, Geneva: Switzerland.

30 Ruiz, V., Pfrang, A., Kriston, A. et al. (2018). A review of international abuse testing standards and regulations for lithium ion batteries in electric and hybrid electric vehicles. *Renewable and Sustainable Energy Reviews* 81: 1427–1452.

31 Secondary cells and batteries containing alkaline or other non-acid electrolytes – Safety requirements for secondary lithium cells and batteries, for use in industrial applications, Standards Publication BSI, EN 62619, 2017.

32 Secondary lithium cells and batteries for use in industrial applications - Part 2: Tests and requirements of safety, Japanese Industrial Standards Committee (JISC), JISC 87512-2, 2019.

33 Secondary cells and batteries containing alkaline or other non-acid electrolytes – Safety requirements for secondary lithium cells and batteries, for use in industrial applications, International Electrotechnical Commission, IEC 62619, 2017.

34 Secondary cells and batteries containing alkaline or other non-acid electrolytes – Safety requirements for secondary lithium cells and batteries for use in electrical energy storage systems, International Electrotechnical Commission, IEC 63056, 2020.

35 Energy storage systems and equipment, Joint Canada - United States National Standard, NSI/CAN/UL-9540, 2016.

36 Batteries for Use in Stationary, Vehicle Auxiliary Power and Light Electric Rail (LER) Applications, Joint Canada - United States National Standard, Standard for Safety ANSI/CAN/UL-1973, 2018.

37 The Institute of Internal Auditors, Audit Report Writing Toolkit [Online]. https://www.theiia.org/globalassets/site/auditing-report-writing-toolkit.pdf (accessed 14 September 2024).

11

Storage of Lithium-Ion Batteries

Haibo Huo, Gifty Pamela Afun, Manoj Kumar Lohana, and Sahithi Maddipatla

Storage refers to the period when the battery cell, pack, or system is held in a nonoperational no-load power dissipation state (no-load situation) or is completely electrically disconnected (storage situation). As lithium batteries grow in use globally, proper lithium-ion battery (LIB) storage is critical for maintaining optimum battery performance and reducing the risk of fire or explosion.

11.1 Introduction

Lithium-ion batteries (LIBs) have become the power supply of choice in mobile electronics, vehicles, and electric storage due to their high energy density and long lifespan over other battery technologies [1–4]. With the increasing demand for LIBs on the market, the scale and number of battery storage warehouses where stacks of batteries are placed for future use, are also increasing significantly. This storage is referred to as daily warehousing (DW). Another rapidly growing industry where LIBs play a crucial role is the energy storage system (ESS) sector, where batteries may be only used for energy backup and cycled infrequently.

Several safety issues can arise when a battery is stored without an electrical connection or with no load for extended periods. Batteries can experience degradation, leading to capacity loss and increased internal resistance. Over time, this can cause swelling, leakage, or even short circuits when reconnected. Additionally, if the battery is stored at a high state of charge or in high-temperature conditions, the risk of thermal runaway (TR) increases [5].

The California-headquartered Electric Power Research Institute (EPRI) database reveals that 89 energy storage incidents occurred globally from 2011 to 2024 [6]. The most reported energy storage-related fire incidents occurred in South Korea (39) and the United States (24). Since 2017, more than 30 fire accidents involving energy storage stations in China have caused heavy casualties or property losses [7]. The April 2021 explosion of a lithium battery station in Beijing, resulting in at least two deaths, is the worst accident in China's battery storage sector in recent years [8].

Effective risk assessment and mitigation for LIB storage requires an approach that combines design, operational, administrative, and emergency response strategies. The internationally influential safety standards for LIB ESSs include Underwriters Laboratories (UL) standard 1973, Batteries for Use in Stationary and Motive Auxiliary Power Applications [9], International Electrotechnical Commission (IEC) standard 62619, Safety Requirements for Secondary Lithium Cells and Batteries for Use in Industrial Applications [10], and IEEE standard 1679.1-2017, Guide for

The Safety Challenges and Strategies of Using Lithium-Ion Batteries, First Edition. Edited by Michael G. Pecht.

the Characterization and Evaluation of Lithium-Based Batteries in Stationary Applications [11]. Europe, China, Japan, South Korea, Australia, and other countries have referenced or compiled their applicable domestic standards according to those standards. The United States places great emphasis on energy storage safety, and the development of its regulations is relatively advanced and well-established. Specifically, in the United States and Canada, ESSs need to comply with UL standard 9540, ESSs and Equipment [12]. The UL 9540A [13] Test Method for Evaluating Thermal Runaway Fire Propagation in Battery Energy Storage Systems (BESSs) helps identify potential hazards and vulnerabilities in ESSs. Additionally, the National Fire Protection Association (NFPA) standard 855 for the installation of stationary ESSs [14] can be followed to control the fire risk of ESSs.

TAS/NZS standard 5139, jointly developed by Australia and New Zealand, specifies requirements for general installation and safety requirements for BESSs [15, 16]. Chinese standard GB/T 36276-2023, LIBs for Electric Energy Storage [17], describes the corresponding test methods and stipulates standards for coding, normal working environment, inspection, marking, packaging, transportation, and storage. KS C IEC standard 62933-5-2 [18] is South Korea's specific guideline for electrical ESSs, focusing on safety and performance.

11.2 Incidents During Lithium-Ion Battery Storage and Analysis

Despite the implementation of various regulations and standards, safety incidents are still occurring at an increasing rate. For example, BESS and 10 DW incidents occurred between 2019 and 2023 [5], with several selected battery storage fire incidents listed in Table 11.1.

Accidents involving LIB fires are often linked to inadequate storage conditions, such as poor environmental management, including insufficient ventilation, improper temperature control, and high humidity levels. Consequently, these conditions can lead to the accumulation of heat and gases, increasing the risk of fire. However, it is important to note that inadequate storage conditions may not directly cause internal short circuits (ISC), a significant safety hazard associated with LIBs [20–22]. ISCs are typically caused by internal defects, manufacturing flaws, or mechanical damage rather than external storage conditions. Nevertheless, inadequate storage environments can

Table 11.1 Selected global battery fire incidents during storage.

Mode	Date	Location	Description
BESS	May 15, 2024	San Diego, CA, USA	An energy storage facility caught fire and spread to others.
BESS	September 26, 2023	Bouldercombe, Queensland, Australia	A Tesla battery caught fire and exploded.
BESS	September 18, 2023	Valley Center, CA, USA	One battery storage unit caught fire.
BESS	August 22, 2023	Barban, Saucats, France	A container of LIBs burned.
BESS	August 19, 2023	Zhuhai, Xiangzhou District, China	A container of LIBs burned and exploded.
DW	January 16, 2023	Rouen, France	A fire spread through a warehouse containing thousands of LIBs.

Source: Adapted from [19].

exacerbate the consequences of an ISC. For example, high temperatures or poor ventilation can accelerate the onset of thermal runaway if an ISC occurs, leading to a fire or explosion. Other factors that can increase the risk of fire in storage areas include storage temperature, state of charge (SOC), and the length of the storage period.

Many countries have adopted or aligned their domestic standards with international guidelines, such as UL 1973 and IEC 62619, to ensure the safety and reliability of ESS. Additionally, in recent years, various nations have introduced their national standards related to ESSs to address specific safety concerns and regulatory requirements. Despite this, the safety of lithium battery energy storage power stations is still prominent. Understanding the warning signs, guidelines, and measures to prevent LIB fires and ensure safety in all devices using LIBs is crucial. This also shows that the standard lithium battery energy storage specifications are far from perfect.

Considering the wide range of incidents related to lithium battery storage, several critical best practices may decrease the hazards and improve the overall safety of the environment. Of course, safety also includes emergency preparedness, which comprises in-depth responsive plans and specific training for personnel on lithium battery hazards. The observance and enforcement of safety standards, such as UL 9540A, are important for evaluating thermal runaway in batteries. Consequently, a more secure operational environment is created by rigorously applying and enforcing the national and international safety standards regarding the storage of lithium batteries.

11.3 Safety Tests for Storage of Lithium-Ion Batteries

LIBs must meet specific testing standards prior to operation to ensure safety during both storage and transportation. UN standard 38.3 [23] and IEC standard 62281 [24] specifically address the safety of LIBs in the context of transportation.

11.3.1 UN Standard 38.3

UN standard 38.3 is part of the United Nations Recommendations on the Transport of Dangerous Goods, found in the Manual of Tests and Criteria. It is specifically designed to ensure the safety of lithium metal and lithium-ion cells and batteries. Importantly, it applies to all lithium batteries, whether they are stored alone or installed in equipment.

UN 38.3 includes eight rigorous tests to ensure that batteries can withstand the conditions encountered during transportation, with the specific test procedures for lithium-ion cells and batteries summarized in Table 11.2. In general, test procedures depend on whether the item is a cell or battery pack. All cell types shall be subjected to tests T.1 through T.6 and T.8. All rechargeable battery pack types, including those composed of previously tested cells, shall be subjected to tests T.1 through T.5 and T.7. In addition, rechargeable single cell batteries with overcharge protection shall be subjected to test T.7. A cell as a component of a battery that is not transported separately from the battery only needs to be tested according to tests T.6 and T.8. A component cell that is transported separately from the battery shall be subjected to tests T.1 through T.6 and T.8 [23].

UN 38.3 typically requires 16 samples for testing. Before 2019, shippers were only required to confirm that batteries had passed the UN 38.3 tests. However, since January 21, 2022, lithium cell and battery manufacturers and distributors must have test summaries available. The updated UN guidelines incorporate a consolidated summary of LIB testing, and after test completion, a report summary is mandatory.

Table 11.2 UN tests T.1 to T.8 for lithium-ion cells and batteries prior to being stored.

Test step	Test type	Specific procedure
Test T.1	Altitude simulation	Test cells and batteries are stored at a pressure of 11.6 kPa or less for at least 6 hour at ambient temperature (20 °C ± 5 °C).
Test T.2	Thermal	Rapid thermal cycling between high (72 °C ± 2 °C) and low (−40 °C ± 2 °C) storage temperatures; stored for at least 6 hour at the test temperature; time interval between high and low test temperature change of less than 30 min; 10 cycles.
Test T.3	Vibration	Vibration is a sinusoidal waveform with a logarithmic sweep between 7 Hz (1 g_n peak acceleration) and 200 Hz (8 g_n/2 g_n peak acceleration) and back to 7 Hz in 15 min; 12 cycles; 3 mutually perpendicular mounting positions.
Test T.4	Shock	Subjected to a half-sine shock (150 g_n/50 g_n peak acceleration) and pulse duration (6 ms/11 ms), 3 shocks cycling in the positive and negative directions for each of 3 mutually perpendicular mounting positions (total of 18 shocks).
Test T.5	External short circuit	Short circuit with a total external resistance of less than 0.1 Ω at (57 °C ± 4 °C) for a 1 hour duration.
Test T.6	Impact	A 15.8 mm diameter bar is placed across the sample cell center, and a 9.1 kg ± 0.1 kg mass is dropped from a height of 61 cm ± 2.5 cm onto the sample.
Test T.7	Overcharge	The overcharging test should be conducted for 24 hours with a charge current (twice the manufacturer's recommended maximum) and minimum test voltage. The minimum test voltage is defined in two categories: (a) when the recommended charge voltage ≤18 V, the lesser of 22 V or 2 times the maximum charge voltage, or (b) when the recommended charge voltage >18 V, 1.2 times the maximum charge voltage.
Test T.8	Forced discharge	Each cell is forced discharged by connecting it in series with a 12 V DC power supply at an initial current equal to the maximum discharge current specified by the manufacturer.

11.3.2 IEC Standard 62281

IEC 62281 [24] is an international standard developed by the IEC. It specifies the safety requirements for primary and secondary (rechargeable) lithium cells and batteries to ensure their safety during transport, other than for recycling or disposal. It is part of a broader set of IEC standards that cover various aspects of battery safety, including manufacturing, usage, and disposal. While both UN 38.3 and IEC 62281 aim to ensure the safety of lithium batteries, they differ in the number of test samples required, specific test conditions, and overall focus. UN 38.3 is more transportation-centric, with stringent packaging and handling requirements, whereas IEC 62281 is part of a comprehensive set of standards addressing multiple aspects of battery safety. For example, the thermal test in IEC 62281 has different temperature ranges and durations compared to UN 38.3. Additionally, IEC 62281 generally requires 33 sealed and 11 open samples for testing, while UN 38.3 requires 16 samples.

These regulations and standards are crucial for ensuring LIBs' safe production, handling, storage, and disposal. Compliance helps to mitigate risks associated with fire, explosion, and environmental hazards. While the UN 38.3 standard for the transportation of LIBs is widely accepted globally, the storage regulations for these batteries can vary significantly across countries. Literature searches show that the United States, Japan, South Korea, and China have particularly stringent regulations, with additional safety tests and requirements beyond international standards.

11.4 Regulations and Standards for Daily Warehousing and Battery Energy Storage Systems

The efficient use of BESSs is essential for effectively managing energy supply and demand, integrating renewable energy sources, and enhancing grid stability. Optimizing warehousing processes is critical for maximizing storage efficiency and ensuring safe storage. As such, the safe and effective operation of these systems necessitates comprehensive regulations and standards that address safety, performance, environmental impact, and interoperability. The safety of a BESS is a paramount concern due to the potential risks of thermal runaway, fires, and explosions.

With the accelerating deployment of these systems, it is increasingly critical to establish comprehensive and harmonized international regulations. These regulations will be essential in supporting the safe, efficient, and sustainable operation of energy storage technologies across diverse applications and industries. These regulations and standards are crucial for ensuring the safe production, handling, storage, and disposal of LIBs. IEC develops international standards for all electrical, electronic, and related technologies, including ESSs. Key IEC standards for ESSs include IEC 62933-5-1 [25] and IEC 62933-5-2 [26], which outline safety requirements for grid-integrated ESSs. The NFPA develops codes and standards to minimize the risk of fire and other hazards, and NFPA 855 [14] is a key standard for the installation of stationary ESSs. UL develops safety standards for a wide range of products, including ESSs, and important UL standards for ESSs include UL 1973, UL 9540 [11], and UL 9540A [13].

The IEC has developed several standards, such as IEC 62619 [10], which specify safety requirements for secondary lithium cells and batteries for industrial applications and address crucial aspects like thermal stability, overcharge protection, and mechanical integrity to prevent incidents like fires and explosions. The IEC 62933 series provides guidelines for safety in grid-integrated ESSs, covering aspects from system design to operation and maintenance. IEC 62133 [27] is one of the most important international standards for rechargeable battery safety, particularly for lithium-ion and nickel-metal hydride batteries, and it outlines requirements and tests for ensuring the safety of batteries used in portable applications. ISO/IEC standard 17025 [28] outlines general requirements for the competence of testing and calibration laboratories. It ensures that energy storage devices undergo accurate and consistent testing for performance metrics such as energy density, cycle life, and charge-discharge efficiency.

Efficient warehousing is fundamental for enabling the management of BESSs and ensuring safety. With warehouses embracing advanced technologies like automation and energy storage, regulatory frameworks must adapt to address new challenges and guarantee safety and efficiency. Consequently, warehousing standards strongly emphasize safety and fire protection to proactively prevent accidents and prioritize the well-being of workers. The NFPA sets forth standards like NFPA 855, providing guidelines for installing ESSs; NFPA 13, which provides guidelines for installing sprinkler systems; and NFPA 72, which covers fire alarm systems. These standards are crucial for mitigating fire risks, particularly in warehouses that store flammable or hazardous materials. Additionally, BESSs and DWs are subject to a complex and ever-changing regulatory environment. Different countries and regions have created their legislation, in addition to international norms, to handle certain local needs and difficulties.

11.5 Lithium-Ion Battery Storage in the United States

Battery storage systems are a relatively new resource that is now experiencing rapid growth in adoption. In 2015 alone, the global installed capacity of such systems increased by almost 117%, from 166 MW to 361 MW, and this base further increased to 568 MW in 2016 (DOE Global Energy Storage Database, 2017a). In the United States, BESSs are essential for improving grid dependability and facilitating the integration of renewable energy. These systems offer advantages such as peak reduction, load balancing, and emergency power, thereby contributing to the electrical grid's stability. Also, the widespread use of lithium-ion technology in BESS is fueled by its high energy density, extended life span, and decreasing costs.

The deployment and operation of these systems are regulated by safety standards established by organizations like UL and NFPA to ensure safety and effectiveness. Furthermore, federal and state incentives, such as tax credits and grants, encourage the growth and application of lithium battery energy storage in various industries [29]. For example, the UL 9540-2020 product standard is the key product safety listing for stationary ESSs. The current standard is the second edition and requires installation via reference by one of the two model fire codes used in the United States: International Fire Code (IFC) and NFPA 1 Fire Code [30]. The IFC is used or adopted in 41 US states, the District of Columbia, New York City, Guam, and Puerto Rico [4].

California, Oregon, Massachusetts, and New York are the only states with energy storage mandates in the United States. Given California's 1.35 GW energy storage goal, California Independent System Operator (CAISO) has the largest capacity of planned energy storage projects compared to other states [30]. California Fire Code Chapter 12 provides specific requirements for ESS installations in California, while the New York State Energy Research and Development Authority (NYSERDA) Energy Storage Guidebook offers guidance on permitting and interconnection processes for ESSs in New York State. For local building and fire codes, many jurisdictions have adopted or modified versions of the IFC and International Building Code (IBC).

New York State and New York City have actively addressed issues with LIBs. In March 2023, New York City Mayor Eric Adams signed five bills into law after a series of fires were linked to LIBs. The bills aim to prevent fires, provide information to educate the public on fire risks, and set safety and manufacturing standards. Further, on March 20, 2023, the NYC Council passed Introduction 663-A, which prohibits the sale, lease, or rental of powered mobility devices, such as e-bikes and e-scooters, and the storage batteries for these devices, that do not meet UL 2849, 2272, and 2271 standards [31]. State Senator Iwen Chu introduced four bills aimed at LIBs in New York City in November 2023, which would require all New York City storage shops to have fire suppression measures and for all batteries to be tested and approved by UL. Most recently, in January 2024, New York State Governor Kathy Hochul announced that she would propose a state-wide ban on the sale of uncertified LIBs, additional training for local fire departments on LIB fires, and expanded online public service safety announcements [31].

11.5.1 US Battery Storage Specifications

The regulatory framework for LIB storage systems in the United States is intricate, involving a combination of federal, state, and local regulations. The US Department of Energy (DOE) and its

Figure 11.1 Exterior and interior photos of LIB storage containers. Source: [33]/U.S. Chemical Storage.

Office of Electric Delivery and Energy Reliability encourage the development of energy storage technologies through its Energy Storage Systems Program (ESSP) [32]. The program conducts fundamental research and system analysis and provides monetary and knowledge-based support for demonstrating and deploying multiple energy storage technologies. Notably, special storage precautions protect workers and the environment. For example, US Chemical Storage manufactures prefabricated lithium battery storage buildings designed specifically for storing batteries, as shown in Figure 11.1. Optional conditions and equipment for LIB storage and warehousing include power outlets, fire suppression systems, LED lights, sensors, and alarms, workbenches (for buildings with charging rooms), temperature and climate controls, multitiered shelving or cubbies, fiberglass flooring, and chemical-resistant paint.

11.5.2 US Daily Warehousing

DW for battery storage systems involves the routine handling, storage, and management of batteries in a warehouse setting. According to the existing literature, there are no specific regulations for DW concerning battery storage systems. However, general regulations like NFPA 855 apply to any facility that stores LIBs regularly.

In the United States, storing LIBs in warehouses requires adherence to specific safety and operational guidelines. These standards ensure the safe handling and storage of batteries, mitigating risks such as thermal runaway, fire, and degradation. For temperature control, it is advised to maintain between 15 °C and 25 °C [34, 35]. This range helps prevent thermal degradation and extends the batteries' lifespan. Additionally, controlling humidity levels is vital to prevent corrosion and other moisture-related issues, with a recommended relative humidity of 50% for battery storage [36]. Proper ventilation and dehumidification systems should be in place to maintain these conditions. Storing batteries at an optimal SOC helps minimize stress and degradation, and it is recommended to store batteries at 25–30% SOC, which balances performance and longevity [37]. Regular checks should be conducted to ensure that the SOC remains within this range.

Proper segregation of batteries is essential to prevent cross-contamination and reduce risks, so new, used, and damaged batteries should be stored separately. Damaged batteries must be isolated immediately in designated containers to manage potential hazards. Fire safety measures are critical in battery storage; therefore, warehouses should use fire-resistant storage cabinets and maintain safe distances from flammable materials. Implementing fire suppression systems and ensuring easy access to emergency equipment are also essential for safety compliance. Clear and accurate labeling of battery storage areas and individual batteries is essential for safe handling and inventory management. Labels should include the batteries' type, condition, and SOC.

11.5.3 US Battery Energy Storage System (BESS)

Varying levels of LIB management policies exist across US states, but federal regulations require state standards to be either identical or more stringent. Thus, certain states, such as California, have taken the initiative to implement stricter regulations regarding LIB management. California has passed regulations relevant to the proper disposal of LIBs, regulating activities such as dismantling, transportation, storage, disassembly, installation into ESSs, hazardous waste treatment, and export [38].

UL 9540 is a safety standard for ESS and equipment; it does not include a specific propagation test for ESS because it references UL 1973 mandatory battery requirements. In the 2nd edition of UL 9540, UL 9540A is referenced to evaluate thermal runaway propagation and characterize resulting fire and explosion hazards, if any, for BESSs. Consequently, a BESS appropriately designed and evaluated today will have the UL 1973 propagation test conducted on the battery system, and if necessary, it will be subjected to UL 9540A large-scale fire testing. The UL 9540A test indicates that once propagation is initiated within a module, the installation shall be provided with sufficient protection to safely contain it. It states that ESS enclosures that can be fully entered by persons, such as walk-in units, shall be designed to automatically provide mechanical ventilation using 100% outdoor air when anyone is working within the enclosure. The minimum amount of ventilation air shall be 5.1 $L/s/m^2$ (1 $cf/min/ft^2$). In addition, lighting shall be provided in enclosed working spaces associated with the ESS. The lighting source shall provide at least 100 lux and be controlled only manually.

Besides the batteries, the standards also state that inverters and other equipment connected to the output terminals of the battery or other storage devices of the ESS shall be able to safely withstand a potential short-circuit current from the storage device. The inverter and other equipment's input short-circuit current shall be equal to or greater than the battery's rated output short-circuit current or other energy storage mechanism. Compliance can be determined by evaluating the output short-circuit ratings of the energy storage mechanism (e.g., for the battery) as determined by testing in UL 1973. For equipment capable of providing an output short-circuit current, the rating shall reflect the protection provided by any internal branch circuit protectors, including electronic circuit regulation, such as limitation of peak currents, inrush currents, and time. An energy storage mechanism relying on a branch circuit protector installed on the output of the energy storage mechanism shall provide the maximum short-circuit current from that device. Furthermore, ESSs with integral fire detection or fire suppression systems shall be provided with instructions for installing, commissioning, maintaining, and testing these systems per applicable installation codes and standards. NFPA 1 and IFC provide information on applicable requirements and codes for fire suppression systems.

NFPA 855 specifies that the installation of stationary ESSs shall be permitted on buildings that do not obstruct fire department rooftop operations when approved. ESSs comprised of units with a maximum stored energy greater than 20 kWh and associated equipment that are located on rooftops and not enclosed by building construction shall comply with the following: the roofing materials under and within 5 ft (1.5 m) horizontally from an ESS or associated equipment shall comply with one of the following: (a) be noncombustible or (b) shall have a Class A rating when tested in accordance with ASTM E108 or UL 790. ESSs and associated equipment shall be located from the roof's edge at a distance equal to at least the height of the system, equipment, or component but not less than 5 ft (1.5 m). Installations on rooftops over 75 ft (23 m) in height above grade shall be permitted when approved by the authorities having jurisdiction (AHJ). The ESS shall be a minimum of 10 ft (3 m) from the fire service access point on the rooftop.

UL 9540A or an equivalent test should evaluate the fire characteristics of the composition of gases generated at the cell, module, and unit. Additionally, it should evaluate the installation levels for ESS undergoing thermal runaways, such as what might occur due to a fault, physical damage, or exposure hazard. The evaluation of the fire characteristics during fire vent testing at the unit level and installation testing should document whether the fire event propagates to the neighboring ESS units. Also, it needs to include radiant heat flux measurements at enclosing wall surfaces and at various distances from the ESS being tested at the unit level. The fire and explosion testing data are intended to be used by manufacturers, system designers, and AHJs to determine if the required separation distance for an ESS installation can be reduced.

IEEE 1547 [39] is a standard for interconnection and interoperability of ESSs with the electricity grid, and it is applied with the understanding that the distributed energy resource is a 60 Hz source. Figure 11.2 shows the IEEE 1547 standards used in the United States.

ANSI/NEMA C18 [40] is a safety standard that defines performance requirements for portable, rechargeable cells and batteries to ensure their safe operation under normal use and reasonably foreseeable misuse. The dimensions of batteries shall remain within specified dimensional tolerances under conditions of storage and discharge specified herein. The ambient temperature surrounding the battery shall be 21 °C ± 2 °C. During short periods, the temperature may deviate from these limits without exceeding 21 °C ± 5 °C. The relative humidity shall be between 35% and 65% unless otherwise specified [41].

For IFC 2018, the energy system covers a wide range of ESSs with mandatory battery and storage space specifications. Notably, permits are to be obtained before the installation and operation

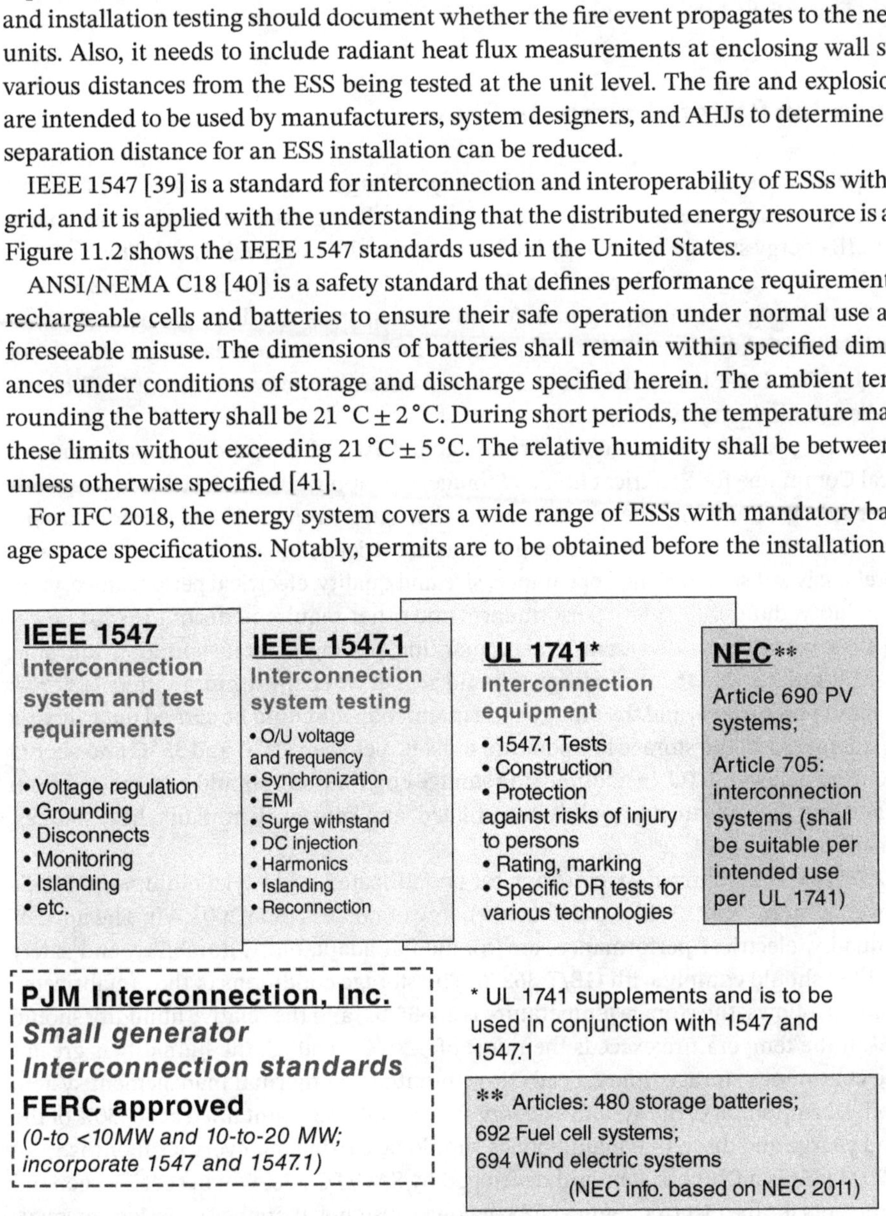

Figure 11.2 IEEE 1547 standards used in the United States. Source: [39]/National Renewable Energy Laboratory/Public Domain.

of a BESS. The minimum allowable threshold quantity is 20 kWh, and the maximum allowable quantity of lithium batteries in a storage space is 600 kWh. Systems shall not be located in areas 75 ft (23 m) above for vehicle access or with a floor level over 30 ft (9.1 m) below the finished floor. Storage batteries shall be listed according to UL 1973 and UL 9540. The ventilation system shall be designed to limit the maximum concentration of flammable gases to 25% of the flammability unit, with continuous ventilation provided at a rate of 1 cf/min/ft^2 of floor area.

Safe storage implements a BMS for real-time monitoring and management, incorporates fire suppression systems, manages and dissipates heat effectively, follows installation zoning and setback requirements, and establishes clear emergency response and maintenance protocols.

11.6 Lithium-Ion Battery Storage in China

LIB energy storage power stations have entered the stage of large-scale commercial applications in China. However, in this process, many shortcomings restricting the healthy and sustainable development of LIB energy storage technology have been gradually exposed [42, 43]. The National Electric Power Energy Storage Standardization Technical Committee (SAC/TC 550) is mainly responsible for formulating and managing national and industry standards in the field of power energy storage in China. Additionally, it is responsible for the International Organization for Standardization IEC/TC 120 Electrical ESSs and other standardization work in the field of test certification bodies and industry associations.

GB/T36276-2023 [17] is a Chinese standard formulated by SAC/TC 550, the National Standardization Technical Committee for Electrical Energy Storage, and applicable to the test requirements of LIBs for power storage, including the test of battery cells, battery modules, and battery clusters. Compared to the 2018 version, the technical requirements are stricter, and the testing requirements are more comprehensive. It specifies the appearance, size and quality, electrical performance, environmental adaptability, durability, safety performance, and other requirements of LIBs for power storage, as well as the normal working environment, marking, packaging, transportation, and storage. This standard stipulates that these conditions should be met when the lithium battery is stored: the initial SOC should be 30–50%, and the energy state maintenance should be carried out every six months during storage. Also, the storage temperature must be between 20 °C and 35 °C and should not exceed 50 °C or be below −30 °C. In addition, the storage environment should be protected from direct sunlight and rain, kept clean, dry, and well-ventilated, and kept away from fire, heat sources, and corrosive media.

GB/T 44026-2024 [44] is a technical specification for prefabricated cabin-type lithium-ion BESSs with a rated power of more than 100 kW and a rated energy of not less than 200 kWh. The appearance, size and quality, electrical performance, environmental adaptability, durability, and safety performance of LIBs should comply with GB/T 36276. The storage conditions of the prefabricated storage system are as follows: the storage temperature is 20–35 °C, and the relative humidity should be less than 95%. If the temperature exceeds the range of −30 °C to 50 °C, the humidity is greater than 95%, or the continuous storage time exceeds three months, the thermal management system should be started. During storage, the system's energy state should be maintained at 30–50% of the rated energy, and charge and discharge maintenance should be carried out every six months.

GB/T 36558-2023 [45] is a Chinese standard developed by SAC/TC 550, which specifies general technical requirements for the electrochemical ESS of power systems. It applies to designing, manufacturing, testing, operating, maintaining, and repairing electrochemical ESSs with LIBs, water electrolysis, and fuel cells as energy storage carriers in power systems. Moreover, the electrical performance, durability, safety performance, labeling, packaging, transportation, and storage of the electrochemical ESS's LIBs shall meet the GB/T 36276 requirements.

For energy storage, the standards UL 1973-2022 [9], UL 9540A-2019 [13], and GB/T 36276-2023 all require a TR test for the battery system. In UL 9540A, TR test methodologies include short circuits, external heating, overcharge, and nail penetration. In GB/T 36276-2023 and UL 1973-2022, the requirement for test compliance is that the test sample should have no fire, no explosion, and no rupture outside the explosion-proof valve. GB/T 36276-2023 requires the surface temperature of a single battery cell to be greater than 90 °C when it undergoes thermal runaway throughout its entire lifecycle. It also states that there should be no fire, explosion, or TR trigger after the TR of any single battery cell in the battery module. Electrical, mechanical, and thermal abuse are more likely to occur in the BESS system, so electrical, mechanical, environmental, and thermal safety performance tests must be conducted before lithium batteries are stored.

The commonly used test methods for TR include overcharging, overdischarging, short-circuiting, crushing, dropping, and vibration performance tests. GB/T 36276-2023 also requires additional environmental performance testing requirements, including a salt spray test, alternating humidity-heat test, high-altitude insulation test, and high-altitude pressure resistance test.

11.7 Lithium-Ion Battery Storage in South Korea

In South Korea, ESSs play a crucial role in the country's renewable energy strategy. South Korea uses these systems to improve grid stability, manage peak loads, and facilitate the integration of renewable energy sources. There is a strong emphasis on safety improvements due to past incidents, leading to the implementation of stringent regulations and standards.

South Korea can produce a fair amount of renewable energy under the right conditions, especially in some regions like the islands of Jeju and Honam. These regions often produce more power than they need, resulting in grid instability and thus mandating power output restrictions. Large-scale ESS facilities are designed to manage this excess supply by storing extra generated power for use during periods of high demand. The government actively promotes ESS deployment through incentives and policies to enhance energy efficiency and reduce dependence on fossil fuels. For example, systems below 600 kW are restricted to indoor installations [46] and require mandatory installation of safety devices such as surge protection, monitoring, and black box systems and propose specific guidelines for temperature, humidity, and dust control in battery storage environments. Another requirement is that indoor systems' SOC is limited to 70%, while outdoor systems are limited to 80% [46].

The Occupational Safety and Health Agency (OSHA) ensures a safe work environment for US workers, setting standards, providing training, and conducting inspections [47]. Many other countries base their occupational safety regulations on OSHA's standards. OSHA requires employers to train employees, maintain safety data sheets (SDSs), and appropriately label hazardous substances like LIBs. The South Korea Occupational Safety and Health Agency (KOSHA) sets safety standards and guidelines that serve as benchmarks for organizations to create safe working conditions in South Korea; therefore, by establishing clear guidelines, KOSHA helps organizations to assess risks, implement control measures, and maintain compliance with battery storage safety requirements [48]. Korean Industrial Standards (KS) also offer specific requirements for battery storage, ensuring alignment with global safety practices. Battery storage systems must comply with Korean Certification (KC) standards [49], which involve rigorous testing and certification to ensure safety and reliability.

South Korea's regulatory framework for battery storage systems is comprehensive, focusing on safety, environmental sustainability, and industry growth. Indeed, this demonstrates close alignment with international standards while incorporating specific national requirements to address local conditions and safety concerns. The adoption of international standards like NFPA (855, 110, 111), IEC (62133, 62619), and UL (1973, 9540A) ensures compatibility and safety, while national standards and certifications provide additional layers of safety and reliability. Enhanced fire safety measures and environmental regulations further demonstrate South Korea's commitment to safe and sustainable energy storage solutions.

In South Korea, lithium storage systems undergo DW following strict guidelines to ensure safety, efficiency, and durability. These practices are based on international standards and specific national regulations tested and tailored to the unique needs of the South Korean market.

Within these guidelines, batteries are categorized and stored separately based on their condition: new, used, and damaged. This helps prevent cross-contamination and reduces safety risks. Specific areas within the warehouse are designated for different types of batteries, with clear signage and labeling for easy identification and handling. Nonconductive materials are used for shelving and storage bins to prevent accidental short circuits. Batteries, especially those damaged or at risk, are immediately isolated and stored in fire-resistant containers to mitigate fire hazards. According to Korean codes and standards, it is mandatory to follow local regulations and industry best practices for the safe disposal of damaged LIBs.

Warehouses are equipped with temperature monitoring systems to ensure that the environment remains within the recommended storage temperature for LIBs of between 15 °C and 25 °C (59 °F to 77 °F). This helps prevent thermal degradation and ensures long-term performance, and any deviations are promptly addressed to avoid damage to the batteries. Advanced dehumidification systems are often employed in warehouses to control moisture levels, especially during the humid summer months. Specifically, the relative humidity should be maintained between 45% and 75% to prevent moisture-related issues such as corrosion. Batteries should be stored at a 40–60% state of charge to minimize stress and prolong battery life, with the state of charge regularly monitored and adjusted as needed. This includes periodic recharging for batteries in storage for extended periods.

Warehouses are expected to have advanced fire suppression systems, including sprinklers and gas-based extinguishing systems, designed to handle lithium battery fires. Comprehensive emergency response plans are put in place, with regular drills conducted to ensure staff are prepared to handle potential incidents. The Ministry of Trade, Industry, and Energy has implemented policies to enhance safety standards in response to multiple fire incidents involving ESSs. These include stricter verification processes for critical components such as batteries and power conversion systems (PCSs), improved insulation systems, and enhanced fire safety procedures and standards [50]. South Korea has enforced regulations and standards to oversee battery storage systems' safety, installation, and operation, particularly in response to several fire incidents involving ESSs. ESSs must comply with the KC standards, which involve rigorous testing and certification to ensure safety and reliability. This includes standards like KC 62133:2018 [51] for electrical appliances.

Based on the international standard IEC 62619, the Korean Agency for Technology and Standards (KATS) issued the industrial large-capacity lithium battery safety standard KS C IEC 62619 (i.e., KC 62619) [52]. This standard proposed the safety requirements for secondary lithium cells and batteries used in industrial applications. The scope of application includes portable outdoor energy storage power supplies that are within the scope of more than 500 Wh and less than 300 kWh. It adds requirements for battery system design, system lock requirements, and thermal propagation test procedures for thermal runaway triggered by laser. Furthermore, the IEC standard requires R (R is 1 or more) for the number of samples, while in the new version of KC 62619, three samples are required for each battery cell test, and one sample is required for the battery system. The new version of KC 62619 has additions that refine the evaluation method for battery systems below 5 kWh. Additionally, the overall accuracy of controlled or measured values, relative to the specified or actual parameters, shall be within these tolerances: ±0.5% for voltage, ±1% for current, ±2 °C for temperature, ±0.1% for time, ±1% for mass, and ±1% for dimensions. Batteries are kept at 25 °C ± 5 °C, at a constant discharge current of 0.2 A, and the SOC should be kept at 50%.

KS C IEC 62933-5-2 [18] is the specific guideline for electrical ESSs, focusing on safety and performance. A BESS designed and constructed to provide a specified level of reliability and durability shall include not only the levels of safety as a design feature of the overall system but also the subsystem safety level, which is necessary to achieve the specified level. At the subsystem level, all integrated electrochemical energy storage subsystems shall comply with appropriate safety standards (e.g., IEC 62477-1 and IEC 62619). The location of a BESS can range from single domestic situations to commercial and industrial applications and up to utility-scale systems; risks need to be assessed accordingly. The selection of chemistry for the electrochemical accumulation subsystem of the BESS can depend on the environment, performance characteristics, and associated costs and benefits. Temperature, voltage, and SOC are to be obtained from the battery manufacturer.

KC safety certifications consist of type testing and initial factory inspection, and after obtaining the certificate, a factory inspection is required every two years. Additionally, a KC safety certification holder must be a local company in South Korea. After the battery cell obtains the KC 62619 safety certification for the entire ESS, the system also needs to have safety confirmation. The system's production factory then applies for safety confirmation, and after passing the test, the KC 62619 safety confirmation certificate will be issued.

Based on ISO 45001, the KOSHA sets out regulations and guidelines to ensure workplace safety and health, including specific regulations for handling and storing hazardous materials like LIBs. Batteries must be stored in conditions that prevent damage and minimize hazards, which includes controlling temperature and humidity and preventing exposure to direct sunlight or heat sources. Employers must ensure a safe working environment by identifying potential hazards and implementing appropriate control measures. Importantly, adequate ventilation in storage areas is necessary to prevent the accumulation of hazardous gases that could be emitted from batteries. Also, reporting serious incidents or accidents involving batteries to KOSHA is mandatory for investigation and follow-up.

The NFPA 855 standard lists the maximum aggregate capacity for lithium battery storage at 20 KWh, and the maximum stored energy should be 600 kWh. The storage system must be located 100 ft away for outdoor installations, and 50 kWh groups are spaced 3 ft from groups & walls; other arrangements are based on large-scale fire tests. Exhaust ventilation should be installed at 1 $ft^3/min/ft^2$ to keep flammable gases under 25%.

The NFPA 110 [53] standard aims to achieve maximum system reliability. Specifically, NFPA 111 [54] is the standard for stored electrical energy emergency and standby power systems. This standard saves costs by extending the life of the storage system through proper care and maintenance. It requires that the electrical conversion equipment (ECE) shall be tested and certified by the manufacturer at the full-rated load for the specified class. Also, it lists that the output of an ECE shall be DC or AC and of the voltages, waveform, and frequency rated for the load. Moreover, energy converters shall be designed to operate without failure over the expected environmental temperature range. For indoor units, this range shall be 10 °C to 40 °C (50 °F to 104 °F). For outdoor units, or units in areas having no temperature control, the range shall be −34 °C to 50 °C (−30 °F to 122 °F). In addition, the ECE shall be designed to function in an environment with a relative humidity that can vary from 5% to 95% and is noncondensing.

South Korean companies involved in producing, storing, and using LIBs must conduct regular safety inspections and audits to ensure adherence to standards and provide comprehensive training programs for employees on safe handling, storage, and emergency procedures. Additionally, the companies need to maintain detailed records of battery storage conditions, incidents, and maintenance activities and report to regulatory bodies as required.

11.8 Recommendations for Safe Storage Practices

Safety is crucial in the storage facility management of lithium batteries. Utilizing the previously described regulations and instructions and recent research, this recommendation set can guarantee the safe operation of lithium battery storage facilities. Segregated storage, avoiding stacking, monitoring systems, periodic audits, and providing comprehensive personnel training is crucial for optimal LIB storage that ensures safety.

11.8.1 Segregation and Separation Requirements

Segregation and separation are vital requirements for the safe storage of LIBs to prevent hazardous situations. Batteries should be stored in dedicated, fire-resistant areas separate from other flammable materials and substances to mitigate the risk of fire propagation. Different types of batteries should be segregated within storage facilities based on their chemistry and condition—new, used, and damaged batteries should be stored in distinct sections to prevent cross-contamination and reduce the risk of incidents. Damaged or defective batteries must be isolated immediately and stored in specially designated, clearly marked containers to handle potential hazards such as leakage or thermal runaway.

The spatial separation between battery units is underscored in safety regulations by NFPA 855. This requisite encompasses individual battery cells, packs, and products incorporating these components. Specifically, NFPA 855 delineates explicit separation distances to avert fire spread across diverse battery technologies and installation magnitudes. The IFC 2018 prescribes a fire-resistant barrier rated for at least one hour to segregate battery systems exceeding specific energy capacities from the remainder of a building's contents. Chiefly, the energy capacity thresholds that trigger this requirement are 20 kWh for flow and LIBs and 70 kWh for all types of lead-acid batteries [55]. This protocol is designed to diminish the likelihood of fire migration from the battery system to other building sectors.

The IFC stipulates minimal spacing prerequisites for outdoor battery placements, considering the system's energy capacity and the presence of fire suppression mechanisms. UL 9540A complements these stipulations by detailing preventive measures against thermal runaway [56]. IEC 62619 provides an auxiliary perspective focusing on the prudent storage and management of batteries, thus contributing to an overarching strategy for battery safety [9]. Despite the guidance from standards like NFPA 855, IFC 2018, UL 9540A, and IEC 62619, the complications of assuring battery safety necessitate a multidimensional strategy.

11.8.2 Ventilation and Temperature Control Measures

Ventilation prevents the accumulation of gases such as hydrogen, which may explode in the presence of an ignition factor likely to be released from batteries in case of malfunction. Subsequently, ventilation helps to maintain safe temperature levels in the environment [57]. NFPA, IEC, and

OSHA have issued requirements for minimizing flammable gases, such as hydrogen and ensuring suitable temperatures to prevent overheating and subsequent industrial explosions [58]. Mechanical ventilation systems that enable the reduction of the concentration of hazardous gases and clime technologies that maintain batteries within suitable temperature ranges are recommended.

Ventilation and temperature control measures are prioritized since they allow control of the byproducts of battery operation and avoid potential fires and explosions caused by this factor. Indeed, adequate cooling of overheated batteries may significantly reduce the risk of subsequent explosions [59]. These safety measures can be implemented using heating, ventilation, and air conditioning (HVAC) tailored for storage areas, gas detection systems, and thermal management systems. It was found that minimizing the temperature effect improves the operational stability of metal-air batteries and other types [60, 61]. This method guarantees that locations with battery storage are less prone to severe hazards, such as devastating fires, explosions, or gas poisoning. By properly ensuring good storage working conditions, an organization can prevent harm to personnel and destruction to the facility.

11.8.3 Fire Detection and Suppression Systems

The continuous advancement of battery technologies, particularly LIBs, and the associated risks demand implementing critical fire safety measures such as fire detection and suppression systems [62]. Standards such as NFPA 70 (National Electrical Code), NFPA 1 (Fire Code), NFPA 72 (National Fire Alarm and Signaling Code), UL 9540 (Standard for ESSs and Equipment), IEC 62619, and IEC 62485-3, guide the design, selection, and maintenance of fire protection systems in a facility with LIBs.

The key areas of consideration include guidelines for safe electrical installations to prevent them from malfunctioning and comprehensive fire safety provisions spanning from the need for fire-resistant materials to keeping the escape routes clear. Also to be considered are detailed guidelines for fire alarm systems for early detection and response and protective measures aimed at minimizing dangers associated with ESSs, such as thermal runaway. Additionally, precautions specify that LIBs should be stored and charged at the temperature at which they are stored. Facilities must monitor such standards and any updates and interpretations to remain safe, given the instability and risks that battery technologies face. Routine risk assessments and compliance with best practices are also necessary to keep the premises safe.

11.8.4 Emergency Response Planning and Personnel Training

An effective emergency response can only be achieved if it is based on detailed planning and well-trained professionals [63]. Successive incidents, like the 2019 Arizona utility battery fire and explosion incident, South Korea's multiple fires of ESSs between 2017 and 2018, and the 2020 Liverpool battery fire, underscore the need for a well-coordinated response team. Training programs should cover the dangers of lithium batteries, including the potential for thermal runaway and the use of fire suppression tools and skills. Consequently, the simulation drills and continuing education ensure the team's ability to act quickly and efficiently to minimize the outcomes of an emergency.

11.8.5 Monitoring and Inspection Protocols

Monitoring and inspection at regular intervals are key to safe lithium battery storage facilities [64]. These include periodic checks to ensure that all safety systems are in good working order and that batteries are stored according to guidelines (e.g., degraded, damaged, or being repaired). With industry best practices in reserve for monitoring, such as BMSs, which can provide real-time information about the status and operation of each cell and its health, it is possible to identify issues before they become serious [65]. Facilities must also protect personnel and property from the risks associated with lithium battery storage by prioritizing segregation and separation, ventilation and temperature control, fire detection and suppression, emergency preparedness, and regular monitoring and inspection.

11.9 Summary

LIBs are widely used across multiple industries for their high energy density. However, they require stringent storage guidelines to ensure safety and longevity. Key standards and regulations include those issued by the UN, UL, IEEE, IEC, IFC, and NFPA.

UN 38.3 is the main testing standard that LIBs must pass prior to storage, and it is recommended that companies follow UN test criteria for cells, packs, and batteries for safe storage. Depending on the regions in which the BESS will be installed, companies should check both international and local regulations, such as IEC 62619, UL 9540, UL 9540A, and FPA 855, for safe storage requirements based on their unique storage requirements.

There is a need to upgrade existing safety storage standards in line with battery developments. With the escalation of BESS capacity, the risk of fire and potential hazards increases, increasing demand for the system's reliability and safety. Companies should be encouraged to proactively participate in international standards organizations to ensure standards are properly implemented.

Safe storage practices for LIBs are essential to prevent hazardous situations. The key recommendations include the following:

- Every time a battery is not used actively (e.g., for a few days), it should be placed in the storage area to avoid being damaged and unsafe.
- When not using a LiPo/LIB pack, it should be stored at 60–70% of the pack's rated capacity. Lithium-ion cells must never be stored fully charged; it is suggested to store them within a voltage of around 3.8 V. Most of the chargers have a "storage mode" that will either charge or discharge the cell to the proper storage voltage. Experts recommend placing the cells in storage mode after every run, as this will help the battery to lengthen its usable life span.
- Remove LIBs from a device before storing it.
- Use a LIB fireproof safety bag or other fireproof container when storing batteries. Always follow manufacturer recommendations on fireproof bags for details on correctly using them. Do not buy cheap fireproof bags since they might not be effective.
- Cell terminals must be protected by electrical insulating material.
- Store batteries in a dry and well-ventilated place at room temperature or lower. While batteries can be used safely between −20 °C and 60 °C (−4 °F to 140 °F), it is strongly suggested to avoid storing them at a temperature that is close to the upper or lower limits. Storing batteries in a refrigerator may create internal condensation when the battery is brought to room temperature, and they may become dangerous when operated.
- It is best to have a reserved area *only* for LIB storage. It has to be a cool and dry place, away from heat sources. The area should be free from any materials that can catch fire, such as wood tables, carpets, or gasoline containers. The ideal surface for storing LIBs is concrete, metal, ceramic, or any nonflammable material.

- Batteries can be stored in a metal cabinet, such as a chemical storage cabinet, as long as the batteries are not touching each other. It is recommended to have a fire detector in place within the storage area.
- Never leave batteries unattended where someone can damage them.
- Have a class ABC or CO_2 fire extinguisher near the storage area.

References

1 MIT (2019). Lithium battery safety guidance. https://ehs.mit.edu/wp-content/uploads/2019/09/Lithium_Battery_Safety_Guidance.pdf (accessed 3 July 2024).

2 Lisbona, D. and Snee, T. (2011). A review of hazards associated with primary lithium and lithium-ion batteries. *Process Safety and Environmental Protection* 89: 434–442.

3 Feng, X.N., Ren, D.S., He, X., and Ouyang, M.G. (2020). Mitigating thermal runaway of lithium-ion batteries. *Joule* 4 (4): 743–770.

4 TÜV Rheinland (2021). Ensuring energy storage safety and accelerate to build a zero-carbon future. https://www.tuv.com/content-media-files/master-content/services/products/p06-solar/solar-download-page/tuv-rheinland-global-ess-brochure-en-20211008.pdf (accessed 3 July 2024).

5 Xu, C. and Liu, E. (2024). Hazards during transport and storage. In: *Encyclopedia of Electrochemical Power Sources*. Elsevier Science.

6 Tamarindo. Where do most major energy storage fires occur?https://tamarindo.global/articles/where-do-most-major-energy-storage-fires-occur/ (accessed 6 July 2024).

7 Battery News (2022). Causes and countermeasures of accidents in energy storage power stations. Tycorun Energy (takomabattery.com) (accessed 5 July 2024).

8 Energy Iceberg Analysis (2021). Storage: battery and EV "China's battery storage after the explosion. https://energyiceberg.com/chinas-battery-storage-after-the-explosion/ (accessed 5 July 2024).

9 UL 1973 (2022). *Batteries for use in stationary and motive auxiliary power applications*. UL.

10 IEC 62619 (2022). *Secondary cells and batteries containing alkaline or other non-acid electrolytes-safety requirements for secondary lithium cells and batteries, for use in industrial applications*. International Electrotechnical Commission. https://cdn.standards.iteh.ai/samples/102658/921553c48a634ba292a91f563b8c7aec/IEC-62619-2022.pdf (accessed 10 July 2024).

11 IEEE Std 1679.1-2017 (2018). *IEEE guide for the characterization and evaluation of lithium-based batteries in stationary applications*, 1–47. https://doi.org/10.1109/ IEEESTD. 2018.8262521.

12 UL 9540 (2020). *Energy storage systems and equipment*. UL.

13 UL 9540A (2019). *Test method for evaluating thermal runaway fire propagation in battery energy storage systems*. UL.

14 NFPA 855 (2024). *Standard for the installation of stationary energy storage systems*. National Fire Protection Association. https://www.nfpa.org/codes-and-standards/nfpa-855-standard-development/855 (accessed 10 July 2024).

15 AS/NZS 5139:2019. *Electrical installations – Safety of battery systems for use with power conversion equipment*. Australian/New Zealand Standard. (standardsau.com) (accessed 5 July 2024).

16 AS/NZS 5139:2019 (2021). *Electrical installations – Safety of battery systems for use with power conversion equipment*. Battery Energy Storage System Installation, RequirementsElectrical Regulatory Authorities Council (ERAC). https://www.erac.gov.au/wp-content/uploads/2021/03/Battery_Energy_Storage_System_02Feb2021.pdf (accessed 5 July 2024).

17 GB/T 36276-2023 (2023). *Lithium ion battery for electrical energy storage*. National Standard of the People's Republic of China. https://std.samr.gov.cn/gb/search/gbDetailed?id=0DF2F72AE3AF403DE06397BE0A0A87C4 (accessed 10 July 2024).

18 KS. KS C IEC 62933-5-2. https://www.kssn.net/search/stddetail.do?itemNo=K0010101237 (accessed 11 July 2024).

19 EPRI. BESS failure incident database. https://storagewiki.epri.com/index.php/BESS_Failure_Incident_Database (accessed 4 July 2024).

20 ABC News (2023). Tesla battery on fire at bouldercombe energy storage site. Genex Confirms. https://www.abc.net.au/news/2023-09-27/tesla-battery-fire-at-queensland-renewable-energy-project/102905302 (accessed 4 July 2024).

21 Valley Center. Fire burns tonight at TerraGen battery storage. https://www.valleycenter.com/articles/fire-burns-tonight-at-terragen-battery-storage/ (accessed 4 July 2024).

22 RTL. VIDÉO – Rouen: 100 pompiers envoyés pour éteindre l'incendie d'un site industriel. https://www.rtl.fr/actu/justice-faits-divers/video-rouen-100-pompiers-envoyes-pour-eteindre-l-incendie-d-un-site-industriel-7900225800 (accessed 5 July 2024).

23 United Nations (2019). *Recommendations on the Transport of Dangerous Goods, Manual of Tests and Criteria*, 7e, 420–433. New York, NY; Geneva, Switzerland: United Nations Publications.

24 IEC 62881 (2019). *Safety of primary and secondary lithium cells and batteries during transport*. International Electrotechnical Commission. https://cdn.standards.iteh.ai/samples/101582/d03a929a7d5d4bb9bfe0bc1d5959c21f/IEC-62281-2019.pdf (accessed 10 July 2024).

25 IEC 62933-5-1 (2017). *Electrical Energy Storage (ESS) Systems-Part 5-1: safety considerations for grid-integrated EES systems-General specification*. International Electrotechnical Commission. https://cdn.standards.iteh.ai/samples/21836/47dcab8441bc495ebb5bf97afb2c6e97/IEC-TS-62933-5-1-2017.pdf (accessed 10 July 2024).

26 IEC 62933-5-2 (2020). *Electrical Energy Storage (ESS) Systems-Part 5-2: Safety requirements for grid-integrated EES systems-electrochemical-based systems*. International Electrotechnical Commission. https://cdn.standards.iteh.ai/samples/23472/ec3faf47fdd7409882de8f51eeebb68f/IEC-62933-5-2-2020.pdf (accessed 10 July 2024).

27 IEC 62133-2 (2017). *Secondary cells and batteries containing alkaline or other non-acid electrolytes-safety requirements for portable sealed secondary cells, and for batteries made from them, for use in portable applications-Part 2: Lithium systems*. International Electrotechnical Commission. https://cdn.standards.iteh.ai/samples/21479/fc1425401c494ac88927998737150b9b/IEC-62133-2-2017.pdf (accessed 10 July 2024).

28 ISO/IEC 17025. *General requirements for the competence of testing and calibration laboratories*. ISO. https://www.iso.org/files/live/sites/isoorg/files/store/en/PUB100424.pdf (accessed 10 July 2024).

29 Malloci, I. (2024). Lithium battery regulations and standards in the US: an overview. https://www.compliancegate.com/lithium-battery-safety-standards-united-states/ (accessed 10 July 2024).

30 Sakti, A., Botterud, A., and O'Sullivan, F. (2018). Review of wholesale markets and regulations for advanced energy storage services in the United States: current status and path forward. *Energy Policy* 120: 569–579.

31 Foley & Lardner LLP (2024). A spark in regulation: lithium-ion batteries. https://www.foley.com/insights/publications/2024/01/spark-regulation-lithium-ion-batteries/ (accessed 10 July 2024).

32 U.S. Department of Energy, Office of Electricity (OE)-Energy Storage Division (2021). The ESS Mission. https://www.sandia.gov/ess/ (accessed 10 July 2024).

33 U.S. Chemical Storage. Lithium ion battery storage buildings. https://www.uschemicalstorage.com/chemical-storage-applications/lithium-battery/ (accessed 10 July 2024).

34 U.S. Chemical Storage (2020). Lithium ion battery storage requirements. https://www.uschemicalstorage.com/lithium-ion-battery-storage-requirements/ (accessed 10 July 2024).

35 RENOGY (2024). How to store lithium batteries safely: a complete guide. https://www.renogy.com/blog/how-to-store-lithium-batteries-other-battery-maintenance-tips/ (accessed 10 August 2024).

36 AceOnGroup (2023). How to correctly store lithium-ion batteries. https://www.aceongroup.com/news/how-to-correctly-store-lithium-ion-batteries/ (accessed 10 July 2024).

37 Battery University (2021). BU-702: How to store batteries. https://batteryuniversity.com/article/bu-702-how-to-store-batteries (accessed 10 July 2024).

38 Meegoda, J., Malladi, S., and Zayas, I. (2022). End-of-life management of electric vehicle lithium-ion batteries in the United States. *Clean Technologies* 4 (4): 1162–1174.

39 IEEE 1547 and 2030 (2015). *Standards for distributed energy resources interconnection and interoperability with the electricity grid.* National Renewable Energy Laboratory. https://www.nrel.gov/docs/fy15osti/63157.pdf (accessed 10 July 2024).

40 ANSI C18.1M (2001). *Part 1-2001.* American National Standards Institute. https://www.batterybob.com/Battery-Reference/ansi_c18_1part1.pdf (accessed 10 July 2024).

41 Vartanian, C., Paiss, M., Viswanathan, V. et al. (2021). Review of codes and standards for energy storage systems. *Current Sustainable/Renewable Energy Reports* 8 (3): 138–148.

42 Energy Storage Committee of China Energy Research Society, Zhongguancun Energy Storage Industry Technology Alliance (2023). White paper: energy storage white paper 2023.

43 Wang, C., Zhang, J., Huang, T.F. et al. (2024). Analysis of standardization development of lithium-ion batteries for electrical power storage. *Standard Science* 4: 68–75.

44 GB/T 44026-2024 (2024). *Technical specification for prefabricated cabin type lithium-ion BESS.* National standard of the People's Republic of China. https://std.samr.gov.cn/gb/search/gbDetailed?id=19B8508AED464286E06397BE0A0A616D (accessed 11 July 2024).

45 GB/T 36558-2023 (2023). General technical requirements for electrochemical energy storage system of power system. National standard of the People's Republic of China. http://c.gb688.cn/bzgk/gb/showGb?type=online&hcno=CFC9F95B159C4ADA8945379C00D042A6 (accessed 11 July 2024).

46 Kim, Y. (2024). Li-ion Battery of BESS safety in Korea. https://www.esmap.org/sites/default/files/ESP/Lithium%20ion%20Battery%20Safety%20in%20Korea.pdf (accessed 10 July 2024).

47 International Safety Consultants (2024). Health and Safety Laws in Different Countries. https://www.internationalsafetyconsultants.com/health-and-safety-laws-in-different-countries/ (accessed 15 July 2024)

48 International Safety Consultants. Korea Occupational Safety and Health Agency (KOSHA). https://www.internationalsafetyconsultants.com/korea-occupational-safety-and-health-agency-kosha/ (accessed 15 July 2024)

49 UFine Blog (2024). What is the KC Certification of Lithium Battery?https://www.ufinebattery.com/blog/what-is-the-kc-certification-of-lithium-battery/ (accessed 10 August 2024).

50 InfoLink (2024). Fires raise concern over energy storage battery safety in South Korea. https://www.infolink-group.com/energy-article/Fires-raise-concern-over-energy-storage-battery-safety-in-South-Korea (accessed 11 July 2024).

51 Korean Agency for Technology and Standards (2015). Technical regulations for electrical and telecommunication products and components.

52 KS C IEC 62619 (2023). Secondary cells and batteries containing alkaline or other non-acid electrolytes safety requirements for secondary lithium cells and batteries, for use in industrial applications.

53 NFPA 110 (2019). Standard for emergency and standby power systems. National Fire Protection Association.

54 NFPA 111 (2024). National Fire Protection Association. https://www.nfpa.org/codes-and-standards/nfpa-111-standard-development/111 (accessed 11 July 2024).

55 International Code Council (2018). 2018 International Fire Code. http://books.google.ie/books?id=U2IOzgEACAAJ&dq=2018+International+Fire+Code&hl=&cd=2&source=gbs_api (accessed 26 March 2024).

56 UL 9540A (2024). *Test method*. UL. https://www.ul.com/services/ul-9540a-test-method (accessed on 26 March 2024).

57 MIT EHS (2024). Lithium ion battery safety. https://ehs.mit.edu/lab-research-program/lithium-ion-battery-safety/ (accessed 26 March 2024).

58 Occupational Safety and Health Administration (2024). Ventilation – hazard recognition. https://www.osha.gov/ventilation/hazards (accessed 26 March 2024).

59 Liu, B., Su, Y., Deng, Q. et al. (2023). Reducing lithium-ion battery thermal runaway risk based on an integrated cooling strategy for electric vehicles. *International Journal of Heat and Mass Transfer* 216: 124594.

60 Lencwe, M.J., Chowdhury, S.D., Mahlangu, S. et al. (2021). An efficient HVAC network control for safety enhancement of a typical uninterrupted power supply battery storage room. *Energies* 14 (16): 5155.

61 Asmare, M., Zegeye, M., and Ketema, A. (2024). Advancement of electrically rechargeable metal-air batteries for future mobility. *Energy Reports* 11: 1199–1211.

62 Yang, Y., Wang, R., Shen, Z. et al. (2023). Towards a safer lithium-ion batteries: a critical review on cause, characteristics, warning and disposal strategy for thermal runaway. *Advances in Applied Energy* 11: 100146.

63 Rhodes, A. (2024). Four critical elements of a battery storage emergency response plan. https://blog.fluenceenergy.com/four-critical-elements-battery-storage-emergency-response-plan (accessed 26 March 2024).

64 Luo, R. (2024). Lithium-ion battery safety: guidelines for handling and storage. https://www.linkedin.com/pulse/lithium-ion-battery-safety-guidelines-handling-storage-ricky-luo-xqggc/ (accessed 26 March 2024).

65 Pistoia, G. (2009). Vehicle applications. In: *Battery Operated Devices and Systems*, 321–378. Elsevier Science.

12

The Transportation of Lithium-Ion Batteries

Dinithi Senarath, Prabhashi Amanda Andrahennadi, Nipun Iranga Wijesekara, Logeeshan Velmanickam, Niles Perera, Haibo Huo, and Gifty Pamela Afun

The transportation of lithium-ion batteries poses unique challenges due to their potential for fire and explosion. International standards, including UN 38.3, IATA Dangerous Goods Regulations (DGR), and the International Maritime Dangerous Goods (IMDG) Code, provide safety regulations that encompass how the cells are charged, advanced packaging of cells and products, tracking technologies, and environmental conditions.

12.1 Introduction

From the acquisition of raw materials to their integration into products, the entire transportation process for batteries entails a sophisticated and well-coordinated worldwide logistics operation. Logistically, the handling and storage of batteries require temperature-controlled environments and careful management to prevent physical damage. Despite the advantages of lithium-ion batteries (LIBs), their transportation comes with considerable challenges, with one of the primary hurdles being regulatory compliance, as different countries and regions have varying classifications and labeling requirements for hazardous materials. LIBs are also highly prone to thermal runaway, a chain reaction that can result in overheating, fires, and even explosions, if batteries are damaged or improperly packaged. This multifaceted array of risks necessitates a comprehensive approach to ensure the safe and efficient movement of LIBs through supply chains.

12.1.1 Environmental Factors That Affect Battery Performance During Transportation

Environmental factors, including temperature, humidity, vibration, atmospheric pressure, and electrical stress, affect the performance, durability, and safety of batteries during transit and storage. Preserving battery integrity, maximizing performance, and guaranteeing safety hinge on understanding the impacts of these aspects.

Temperature is one of the main factors that affect battery performance. As such, in high-temperature environments, batteries can become thermally unstable and age faster. In addition, elevated temperatures can heighten the likelihood of thermal runaway, resulting in failures like fires or explosions. Once a lithium cell or battery ignites and catches fire, it may cause any nearby cell or battery to overheat and catch fire. These fires produce toxic and irritating fumes

The Safety Challenges and Strategies of Using Lithium-Ion Batteries, First Edition. Edited by Michael G. Pecht.

and are difficult to put out with normal fire extinguishers. Therefore, special fire extinguishers are recommended; lithium metal batteries require a Class D fire extinguisher, whereas LIBs require a foam extinguisher with CO_2, ABC dry chemical, powdered graphite, copper powder, or soda (sodium carbonate) [1].

Conversely, low temperatures can reduce lithium ions mobility, which might diminish the accessible capacity and amplify internal resistance. For batteries to uphold their integrity and perform optimally, it is imperative to maintain them within the ideal temperature range [2]. Storing LIBs at room temperature (20–25 °C) is optimal, as batteries stored in this range exhibit minimal capacity fading and maintain their performance over extended periods.

Humidity adds another layer of complexity. Moisture can have a negative impact on battery components, causing them to deteriorate and ultimately limiting their lifespan and performance. Elevated levels of humidity can result in moisture infiltration, which can cause battery component corrosion and hinder electrical functionality. Preventing degradation requires maintaining moisture control during transit and storage [3].

Vibrations and shocks can cause physical damage to batteries, resulting in internal short circuits, which poses yet another hazard. During transit, ensuring sufficient cushioning and packaging is essential to minimize these risks [4]. Vibrations and shocks reduce the battery's efficiency and pose significant safety risks, including the possibility of fires or explosions. Pressure variations, particularly during air transportation, might impact the battery's physical strength, perhaps resulting in leaks or breaches in the casing [5].

Electrical stress is another factor that may lead to electrolyte depletion, reduced battery efficiency, and an increased risk of thermal incidents. The existence of stressors may lead to the deterioration of battery components and electrolytes, leading to a reduction in the overall lifespan and energy storage capability of the battery. Implementing monitoring systems during the storage process can effectively mitigate such concerns [6].

12.1.2 Effects of Environmental Factors on Battery Performance During Transportation

Lithium-ion batteries dominate as the most common energy storage technology in products due to their exceptional energy density and robustness. However, UN Category 9 categorizes these items as dangerous due to their susceptibility to thermal and electrical instability, chemical dangers, electrical risks, and mechanical hazards when exposed to certain unregulated environmental conditions or mishandled during transportation [1]. It is important to investigate real-world examples of how environmental factors have affected batteries.

There have been incidents involving lithium batteries catching fire on board aircraft transporting airmail, including a UPS DC-8 on the ground in Philadelphia on February 7, 2006, a UPS Boeing 747 flight from Dubai to Cologne on September 3, 2010, and an Asiana Boeing 747 flight from Seoul to Shanghai on July 27, 2011 [7]. These types of fires are usually attributed to the batteries short-circuiting, this is why specialized packaging tailored to these products (such as ZARGES aluminum cases) is needed. Unprotected cells can cause an electrical short by touching, with propagation then causing a chain reaction that can release huge amounts of energy.

In December 2015, the leading US airlines; American, Alaska, Delta, Hawaiian, JetBlue, Southwest, and United Airlines implemented a prohibition on hoverboards on passenger flights. Additionally, the US Postal Service ceased air transportation of hoverboards due to the potential risk of fires. While e-cigarettes are not prohibited from being shipped, between August 2009 and January 2017, 44 out of the 214 reported incidents of e-cigarette explosions took place during transportation, in storage, or under unclear conditions [1]. Additionally, the Federal Aviation Administration

(FAA) has prohibited bringing the Samsung Galaxy Note 7 on board any aircraft since October 14, 2016, due to numerous burn incidents [1].

The risks associated with the transportation and storage of these batteries, such as thermal runaway and chemical instabilities, highlight the importance of strong safety precautions. It is essential to adhere to safety regulations, implement best practices, and employ appropriate storage solutions to effectively mitigate these risks and ensure the safe handling of LIBs throughout their life cycle.

12.2 Regulations and Standards (and Specifically UN 38.3)

International, national, and regional governments and other authorities have developed many regulations for the transportation of lithium batteries and battery-powered devices by different modes (e.g., air, road, rail, and sea). These regulations are helpful for ensuring that shippers transport these items safely within their country or internationally. National regulations are often similar to those defined in the international standards for specific transportation modes, but there are minor differences in compliance [8]. Considering the risks during the transportation of lithium batteries, norms, regulations, and standards should be consistent, easy to understand, and aggregable across borders. Hence, safety guidelines cannot be ignored.

UN Model Regulations, also known as the Orange Book, were originally developed by the UN Economic and Social Council's Committee of Experts on the Transport of Dangerous Goods. Specifically, the Orange Book presents a comprehensive framework for the safe transport of LIBs across different transport modes. This classification of LIBs as Class 9 dangerous goods underlines the diverse nature of hazardous materials. The UN Model Regulations have strict rules for testing and packaging to reduce risks like short circuits, overheating, and potential fires. The regulations also present an information guide for the transport of several kinds of lithium batteries, including new and undamaged batteries, low production runs, preproduction prototype batteries, disposable or recyclable lithium batteries, and damaged or defective lithium batteries.

A UN number is the four-digit number assigned by the United Nations Committee of Experts on the Transport of Dangerous Goods and by the Globally Harmonized System of Classification and Labeling of Chemicals. UN numbers, listed in Table 12.1, are for the identification of an article, substance, or a particular group of articles or substances [9].

One of the most significant sets of regulations governing the transportation of LIBs is the standard found in the United Nations Manual of Tests and Criteria, Part III, subsection 38 (UN 38.3). This section delves into the essence of UN 38.3 regulations, their importance, specific testing requirements, and a comparison with other relevant standards. The UN 38.3 regulations, established by the UN Committee of Experts on the Transport of Dangerous goods, address the classification, testing, and safe transportation of LIBs. The regulations serve as a comprehensive framework to assess the risks associated with battery transportation, particularly concerning issues like short circuits, thermal runaway, and fire hazards [10]. UN 38.3 outlines stringent testing protocols aimed at assessing the safety of batteries at various levels: cell, pack, and product. These tests are designed to simulate real-world scenarios and assess the batteries' performance under diverse conditions, ensuring that they meet minimum safety requirements for transportation and storage [11]. UN 38.3 outlines eight tests to ensure batteries can withstand transportation stress, specifically altitude simulation, thermal tests, vibration, shock, external short circuit, impact, overcharge, and forced discharge.

Table 12.1 UN numbers and their corresponding proper shipping names for lithium batteries.

UN number	Proper shipping name
UN 3090	Lithium metal batteries (including lithium alloy batteries)
UN 3091	Lithium metal batteries contained in equipment (including lithium alloy batteries)
UN 3091	Lithium metal batteries packed with equipment (including lithium alloy batteries)
UN 3480	Li-ion batteries (including Li-ion polymer batteries)
UN 3481	Li-ion batteries contained in equipment (including Li-ion polymer batteries)
UN 3481	Li-ion batteries packed with equipment (including Li-ion polymer batteries)

The UN 38.3 and IEC 62281 standards both address the safety of lithium batteries, particularly in the context of transportation, but they have some key differences in their requirements and focus areas. The testing of LIBs before transportation is essential for several reasons: safety, regulatory compliance, and consumer confidence.

12.2.1 Specific Testing and Compliance Requirements

At the cell level, UN 38.3 mandates tests such as altitude simulation, thermal cycling, external short circuit, and forced discharge to evaluate a battery's response to extreme conditions. These tests assess a battery's stability, thermal management capabilities, and resilience to abuse [12]. Moving up to the pack level, additional tests focus on assessing the pack's mechanical integrity, electrical performance, and resistance to environmental factors such as temperature fluctuations and vibrations. Tests like vibration, shock, and overcharge ensure that battery packs remain safe and functional under normal operating conditions and foreseeable abuse scenarios [13]. At the product level, UN 38.3 requires manufacturers to conduct comprehensive tests on the final assembled product, considering factors like mechanical robustness, electrical safety, and compliance with labeling and documentation requirements. These tests aim to verify that the assembled product meets all regulatory standards and is safe for transportation and use [14].

The UN 38.3 standard is specifically designed to ensure the safety of lithium metal and lithium-ion cells and batteries. Importantly, it applies to all lithium batteries, whether they are stored alone or installed in equipment.

UN 38.3 includes eight rigorous tests to ensure that batteries can withstand the conditions encountered during transportation, with the specific test procedures summarized in Table 12.2. In general, test procedures depend on whether the item is a cell or battery pack, and all cell types must be subjected to tests T.1 through T.6 and T.8. All rechargeable battery pack types, including those composed of previously tested cells, need to be subjected to tests T.1 through T.5 and T.7. In addition, rechargeable single-cell batteries with overcharge protection should be subjected to test T.7. A cell as a component of a battery that is not transported separately from the battery only needs to be tested according to tests T.6 and T.8. A component cell that is transported separately from the battery must be subjected to tests T.1 through T.6 and T.8 [15].

Table 12.2 UN tests T.1 to T.8 for lithium-ion cells and batteries.

Test step	Test type	Specific procedure
Test T.1	Altitude simulation	Test cells and batteries are stored at a pressure of 11.6 kPa or less for at least 6 h at ambient temperature (20 °C ± 5 °C)
Test T.2	Thermal	Rapid thermal cycling between high (72 °C ± 2 °C) and low (−40 °C ± 2 °C) storage temperatures; stored for at least 6 h at the test temperature; time interval between high and low test temperature change of less than 30 min; 10 cycles
Test T.3	Vibration	Vibration is a sinusoidal waveform with a logarithmic sweep between 7 Hz (1 g_n peak acceleration) and 200 Hz (8 g_n/2 g_n peak acceleration) and back to 7 Hz in 15 min; 12 cycles; 3 mutually perpendicular mounting positions
Test T.4	Shock	Subjected to a half-sine shock (150 g_n/50 g_n peak acceleration) and pulse duration (6 ms/11 ms); 3 shocks cycling in the positive and negative directions for each of 3 mutually perpendicular mounting positions (total of 18 shocks)
Test T.5	External short circuit	Short circuit with a total external resistance of less than 0.1 Ω at (57 °C ± 4 °C) for a 1 h duration
Test T.6	Impact	A 15.8 mm diameter bar placed across the sample cell center, and a 9.1 kg ± 0.1 kg mass is dropped from a height of 61 cm ± 2.5 cm onto the sample
Test T.7	Overcharge	Overcharging test should be conducted for 24 h with charge current (twice the manufacturer's recommended maximum) and minimum test voltage. The minimum test voltage is defined in two categories: (a) when the recommended charge voltage ≤18 V, the lesser of 22 V or 2 times the maximum charge voltage, or (b) when the recommended charge voltage >18 V, 1.2 times the maximum charge voltage
Test T.8	Forced discharge	Each cell is forced discharged by connecting it in series with a 12 V DC power supply at an initial current equal to the maximum discharge current specified by the manufacturer

The number of test samples required for UN 38.3 is generally higher compared to IEC 62281. Typically, UN 38.3 requires 16 samples for testing, and IEC 62281 requires a drop test that is not listed in the tests for UN 38.3. A comparison of the two standards is shown in Table 12.3.

IEC 62133 also specifies requirements and tests for the safe operation of portable sealed secondary cells and batteries. The details of UN 38.3 and IEC 62133 are compared in Table 12.4:

Prior to 2019, shippers were only required to confirm that batteries had passed the UN 38.3 test. However, since January 21, 2022, lithium cell and battery manufacturers and distributors must have test summaries available. The updated UN guidelines incorporate a consolidated summary of LIB testing and following test completion, a report summary is mandatory.

12.2.2 Cell-Level Tests and Concerns in Battery Transportation and Storage

At the cell level, batteries face a multitude of degradation mechanisms and safety hazards, presenting considerable challenges during transportation and storage. Among the primary concerns is thermal runaway, where uncontrollable temperature spikes occur due to internal short circuits,

Table 12.3 Comparison of UN 38.3 and IEC 62218 tests for lithium-ion cells and batteries.

Test	UN 38.3	IEC 62281
External short circuit	Yes	Yes
Abnormal charging	Yes	Yes
Forced discharge	Yes	Yes
Impact	Yes	Yes
Shock	Yes	Yes
Vibration	Yes	Yes
Thermal cycling	Yes	Yes
Altitude simulation	Yes	Yes
Drop	No	Yes

Source: Adapted from Yang [16].

overcharging, or external factors, potentially leading to catastrophic failures, fires, and explosions, which pose serious risks during transportation and storage [17]. Additionally, internal short circuits, stemming from defects or damage to battery components, can induce localized heating and thermal runaway within the cell, with factors like manufacturing defects, mechanical stress, or contamination heightening the risk, particularly in high-energy-density battery chemistries [18]. Moreover, exposure to extreme temperatures or mechanical stress can compromise the integrity of the cell, resulting in leakage and chemical reactions among battery components. This not only endangers safety, but it also accelerates degradation mechanisms such as electrode corrosion and capacity loss [19]. Over time, repeated charge–discharge cycles and exposure to environmental stressors can lead to capacity fading in battery cells, diminishing their ability to store and deliver energy effectively, thus making it a significant concern for long-term storage and transportation [20].

Addressing concerns at the cell level is essential for ensuring safety and preventing potential hazards during battery transportation and storage [1]. Cell-level concerns include cell chemistry, design, manufacturing, and aging. Different cell chemistries have varying levels of safety risks, with LIBs having a higher risk of thermal runaway than nickel-based batteries. The cell design, such as the shape, size, and structure, can impact its safety. Defects or inconsistencies in the manufacturing process can lead to safety issues, and regular testing and monitoring of battery cells are essential to detect and mitigate aging risks.

The testing protocols for battery transportation and storage at the cell level include necessary equipment and testing for safety, performance, quality, and certification. These tests ensure that battery cells can deliver the required power and energy, meet the required safety and quality standards, and withstand various conditions and hazards during transportation and storage.

The key points related to testing protocols and requirements for battery transportation and storage encompass several major areas [1]. Ensuring safety through testing is a top priority, which includes assessing abuse, transportation, and environmental factors to guarantee the battery cells can withstand diverse conditions without any negative effects. Performance testing evaluates capacity, power, and cycling performance over a battery's lifetime. Quality testing verifies dimensional accuracy, electrical properties, and adherence to standards. Certification testing ensures compliance with safety and quality standards such as Underwriters Laboratories (UL), IEC, and UN testing. Additionally, necessary testing equipment like battery testers and environmental

Table 12.4 Detailed comparison of UN 38.3 and IEC 62218 tests for lithium-ion cells and batteries.

Test	UN 38.3	IEC 62133
Altitude simulation	Test cells and batteries are stored at a pressure of 11.6 kPa or less for at least 6 h at ambient temperature (20 °C ± 5 °C)	Test cells and batteries are stored at a pressure of 11.6 kPa or less for at least 6 h at ambient temperature (20 °C ± 5 °C)
Thermal test	Rapid thermal cycling between high (72 °C ± 2 °C) and low (−40 °C ± 2 °C) storage temperatures; stored for at least 6 h at the test temperature; time interval between high and low test temperature change of less than 30 min; 10 cycles	Test cells are heated in an oven at a rate of 5 °C/min ± 2 °C/min to a temperature of 130 °C ± 2 °C; cells remain at this temperature for 30 min
Vibration	Vibration is a sinusoidal waveform with a logarithmic sweep between 7 Hz (1 g_n peak acceleration) and 200 Hz (8 g_n/2 g_n peak acceleration) and back to 7 Hz in 15 min; 12 cycles; 3 mutually perpendicular mounting positions	Full charge cells or batteries are tested with an amplitude of 0.76 mm; frequency varies at the rate of 1 Hz/min from 10 Hz to 55 Hz and back to 10 Hz, traversed in 90 min ± 5 min for each direction; tested in 3 mutually perpendicular mounting positions
Shock	Subjected to a half-sine shock (150 g_n/50 g_n peak acceleration) and pulse duration (6 ms/11 ms); 3 shocks cycling in the positive and negative directions for each of 3 mutually perpendicular mounting positions (total of 18 shocks)	Fully charged cell or battery subjected to a total of 3 shocks of equal magnitude; shocks are applied in each of 3 mutually perpendicular mounting positions; acceleration averages 75 g within the initial 3 ms, with a peak between 125 g and 175 g
External short circuit	Short circuit with a total external resistance of less than 0.1 Ω at (57 °C ± 4 °C) for a 1 h duration	Short circuit by connecting the positive and negative terminals, with a resistance of 80 Mω ± 20 Mω
Impact/Crush	A 15.8 mm diameter bar placed across the sample cell center and a 9.1 kg ± 0.1 kg mass is dropped from a height of 61 cm ± 2.5 cm onto the sample	Crush test: Each fully charged cell is crushed between two flat surfaces; crushing force is applied by a hydraulic ram exerting a force of 13 kN ± 1 kN. Failure occurs when there is an abrupt voltage drop of one-third of the original voltage, or 10% of deformation has occurred compared to the initial dimension, when the force is released
Overcharge	Overcharging test should be conducted for 24 h with charge current (twice the manufacturer's recommended maximum) and minimum test voltage. The minimum test voltage is defined in two categories: (a) when the recommended charge voltage ≤18 V, the lesser of 22 V or 2 times the maximum charge voltage, or (b) when the recommended charge voltage >18 V, 1.2 times the maximum charge voltage	A discharge cell or battery is subjected to a high-rate charge of 2.5 times the recommended charging current, for a time that produces a 250% charge input (250% of rated capacity)
Forced discharge	Each cell is forced discharged by connecting it in series with a 12 V DC power supply at an initial current equal to the maximum discharge current specified by the manufacturer	A discharge cell is subjected to a reverse charge at 1 A for 90 min

Source: Adapted from Yang [16].

chambers are detailed to facilitate such rigorous testing. Collectively, these requirements ensure that battery cells meet safety, quality, and performance benchmarks, enabling them to withstand transportation and storage challenges effectively.

12.2.3 Pack-Level Tests and Concerns in Battery Transportation and Storage

Battery packs, which typically consist of multiple cells arranged in series or parallel configurations, present unique challenges and considerations due to their larger size and complexity compared to individual cells [21]. Battery packs face several challenges during transportation and storage, primarily related to thermal management and mechanical integrity. During transportation, battery packs are exposed to various environmental conditions, including temperature fluctuations, which have distinct impacts on single cells and battery packs.

In single cells, fluctuations can compromise performance and safety, resulting in reduced capacity, heightened internal resistance, and the risk of thermal runaway. Conversely, for battery packs, temperature changes affect collective performance and safety, potentially inducing uneven cell degradation, imbalanced performance across the pack, and an elevated likelihood of thermal runaway spreading throughout the entire assembly. High temperatures, for instance, can accelerate chemical reactions within the battery cells, leading to increased degradation and potentially triggering thermal runaway. By comparison, low temperatures can hinder battery performance, reducing energy output and potentially causing irreversible damage to the cells. To mitigate these risks, it is imperative to implement effective thermal management systems [22].

Maintaining the structural integrity of battery packs is necessary for their safe transportation and storage, as they may encounter mechanical stresses like vibrations, impacts, and compression forces during handling and transport. These mechanical stresses can potentially compromise the structural integrity of the pack, leading to physical damage or deformation. Vibrations, resulting from vehicle or machinery movement, can induce mechanical oscillations within the pack, potentially leading to cell displacement or internal short circuits if not adequately dampened or absorbed [23]. Similarly, sudden shocks during loading and unloading procedures can exert significant forces on the pack, risking physical damage to the cells or compromising enclosure integrity, which could result in catastrophic failures. Compression forces encountered during transportation present a distinct challenge, which may result in the deformation of the pack enclosure. This deformation, in turn, carries the potential for cell rupture and exacerbation of thermal management concerns. Such deformations can hinder airflow within the pack, consequently increasing the risk of thermal runaway [21].

Addressing these challenges requires strong design and construction methods, which involve using durable materials and reinforcing pack enclosures structurally. Additionally, protective features like shock-absorbing mounts, cushioning materials, and structural bracing can help mitigate the impact of mechanical stresses [24].

Rigorous testing of battery packs under simulated transportation and storage conditions is mandated to ensure they meet safety and performance requirements. By adhering to robust design principles and conducting comprehensive testing, manufacturers can enhance the mechanical integrity of battery packs, thereby minimizing the risk of damage or failure during transportation and storage [24].

To ensure the safe transportation and storage of battery packs and comply with regulations, rigorous testing procedures and standards have been established. These procedures are designed to evaluate pack performance, safety, and reliability under simulated transportation and storage conditions [25]. Testing procedures for battery packs typically encompass a comprehensive set of environmental tests. Accordingly, these tests are conducted to assess the pack's ability to withstand various environmental stressors and operational conditions encountered during transportation and storage [20].

At the pack level, mechanical tests assess the durability and safety of battery packs during transportation and storage. Battery packs undergo mechanical shock testing to simulate the impact they may experience during transportation. This involves subjecting the pack to sudden, high-intensity shocks to assess its ability to withstand rough handling without damage or compromising safety. Testing procedures adhere to standardized protocols to ensure consistent evaluation across different pack designs and manufacturers.

The drop test assesses a pack's resilience to impacts by dropping it from a specified height onto a hard surface. This test helps to determine if a pack can withstand accidental drops during transportation or handling without sustaining damage. Additionally, the immersion test examines a pack's ability to withstand exposure to liquids, such as water or other fluids, during transportation or storage. The pack is submerged in liquid for a specified duration, and its performance and integrity are evaluated to ensure that it remains sealed and operational [26]. Another test, the crash test, assesses a pack's ability to withstand compressive forces, such as those experienced during accidents or collisions. Controlled crushing forces are applied to a pack, and the structural integrity and ability to protect internal components are evaluated. The rollover test simulates the scenario of a vehicle rollover, where a pack may be subjected to various forces and orientations. By rolling the pack in different directions, manufacturers can assess its ability to maintain structural integrity and prevent damage to internal components. Equally important, the vibration test evaluates a pack's resistance to mechanical oscillations and vibrations that may occur during transportation or operation. The pack is subjected to controlled vibrations, simulating real-world conditions, and its response is assessed to ensure it can withstand prolonged exposure to vibration without failure.

The sample sizes and passing criteria listed in Table 12.5 are based on common industry standards such as UN 38.3, IEC 62133, SAE J2929, and UL 1642, but specific requirements may vary depending on the regulatory body and the type of battery pack being tested. Additional criteria might be imposed depending on the specific application and regulatory requirements for different regions or industries.

At the pack level, environmental tests assess the resilience and safety of battery packs under various conditions. The thermal cycling test assesses a pack's performance under temperature variations encountered during transportation and storage. Packs are subjected to temperature cycling to simulate temperature fluctuations experienced in different thermal conditions. This test provides valuable insights into the pack's performance, safety, and resilience, aiding in identifying weaknesses, ensuring compliance with safety standards, and enhancing overall reliability for transportation and storage applications.

Testing procedures for battery packs are guided by industry standards and regulatory requirements established by organizations such as the International Electrotechnical Commission (IEC), the International Organization for Standardization (ISO), and various national regulatory agencies. These standards provide a framework for manufacturers to conduct comprehensive testing and ensure the safety, reliability, and performance of battery packs intended for transportation and storage applications [1] (Table 12.6).

UN 38.3 defines testing protocols, performance criteria, and acceptance criteria for battery packs, focusing on transportation testing requirements to ensure safe handling during transit. The UN Manual of Tests and Criteria outlines the UN classification schemes for hazardous materials and provides the necessary test methods and procedures for correctly classifying materials intended for transportation. Seen as an indispensable benchmark in transportation testing, this manual is obligatory for manufacturers who transport lithium batteries. It was developed in accordance with the

Table 12.5 Required sample size and passing criteria for tests at the pack level.

Test	Sample size	Passing criteria
Mechanical shock test	At least three battery packs	The battery pack must not exhibit any leakage, venting, rupture, or fire. The open-circuit voltage should stay within specified limits.
Drop test	At least three battery packs	The battery pack must not exhibit any leakage, venting, rupture, fire, or explosion. The open-circuit voltage should stay within specified limits.
Immersion test	At least three battery packs	The battery pack must remain sealed, with no leakage, venting, or rupture. It should maintain its integrity and functionality after the test.
Crash test	Typically, three to five battery packs	The battery pack must maintain structural integrity without causing short circuits, leakage, venting, or fire. The pack should protect its internal components from damage.
Rollover test	At least three battery packs	The battery pack should not exhibit leakage, venting, rupture, or fire. It should maintain its structural integrity and functionality.
Vibration test	At least three battery packs	The battery pack must not show any leakage, venting, rupture, or fire. It should continue to function correctly after the test, with the open-circuit voltage within specified limits.

Table 12.6 Common international battery standards.

Organization	Safety standard
IEC	IEC 62133: Secondary Cells and Batteries Containing Alkaline or Other Non-Acid Electrolytes—Safety Requirements for Portable Sealed Secondary Cells, and for Batteries Made from Them, for use in Portable Applications
	IEC 62281: Safety of Primary and Secondary Lithium Cells and Batteries During Transport
IEEE	IEEE 1625: Rechargeable Batteries for Multi-Cell Mobile Computing Devices
	IEEE 1725: Rechargeable Batteries for Cellular Telephones
SAE	SAE J2929: Electric and Hybrid Vehicle Propulsion Battery System Safety Standard Lithium-Based Rechargeable Cells
	SAE J2464: Electric and Hybrid Electric Vehicle Rechargeable Energy Storage System Safety and Abuse Testing
UL	UL 1642: Lithium Batteries
	UL 1973: Batteries for Use in Light Electric Rail (LER) Applications and Stationary Applications
	UL 2054: Household and Commercial Batteries
	UL 2580: Batteries for Use in Electric Vehicles
	UL 2271: Batteries for Use in Light Electric Vehicle Applications
	UL 2272: Electrical Systems for Self-Balancing Scooters
UN	UN 38.3: Transportation Testing for Lithium-Ion Batteries

Source: Adapted from Huo et al. [1].

UN Recommendations on the Transport of Dangerous Goods—Model Regulations and the Globally Harmonized System (GHS) of Classification and Labeling of Chemicals. Although the UN safety test allows self-certification, many manufacturers choose to involve third-party certified test labs because of possible liability issues [1].

In addition to industry standards, battery pack manufacturers must comply with regulatory requirements imposed by national and international bodies, focusing on safety, environmental sustainability, and consumer protection aspects [1]. Compliance with recognized standards is essential for manufacturers to market their battery packs, providing assurance to stakeholders and mitigating risks associated with product liability and reputational damage. Adherence to industry standards and regulatory requirements demonstrates a commitment to quality, safety, and environmental responsibility, enhancing customer trust and confidence in their products.

12.2.4 Product-Level Tests and Concerns in Battery Transportation and Storage

Battery-powered products encompass a wide range of devices, from consumer electronics to EVs. Ensuring the safe transportation and storage of these products involves addressing unique integration challenges and considerations. This section focuses on the specific concerns at the product level, examining integration challenges, testing requirements, and the economic implications of compliance with UN 38.3 regulations [1].

Battery-powered products are essential components in various industries; they serve diverse purposes, from powering consumer electronics to driving EVs and supporting renewable energy systems. However, ensuring their safe transportation involves more than just putting them in a box and sending them off. Factoring in compatibility with various transportation modes, such as air, sea, and road, is a key aspect to take into consideration [27].

Air transportation presents unique challenges due to the fire hazards associated with LIBs, which contain flammable electrolytes, making them susceptible to thermal runaway and fire incidents if mishandled. Strict regulations from organizations like the International Civil Aviation Organization (ICAO) and the International Air Transport Association (IATA) govern their air transportation, including packaging and labeling requirements aimed at minimizing the risk of fire during transit [28].

Sea transportation, while generally less restrictive than air transportation, still requires compliance with regulations such as the International Maritime Dangerous Goods (IMDG) Code. The IMDG Code contains information on terminology, packaging, labeling, markings, stowage, segregation, handling, and emergency response. For batteries, the code specifies criteria for packaging to prevent short circuits, mandates the use of appropriate labels to indicate hazardous contents, and requires detailed documentation to ensure proper handling and emergency response procedures are in place. These comprehensive guidelines are designed to minimize the risk of incidents during sea transport and ensure the safety of the vessel, crew, and environment [29]. Manufacturers must ensure their products meet these regulations and are designed to withstand the challenges of maritime shipping, including exposure to moisture, vibration, and fluctuating temperatures.

Surface transportation (road/rail) offers more flexibility but still necessitates adherence to national and international regulations. These may vary depending on the jurisdiction, but they often include requirements related to packaging, labeling, and vehicle transport conditions [1].

Product design is indispensable in addressing integration challenges during transportation and storage, involving thorough consideration of factors like weight distribution, impact resistance, and thermal management, to ensure product stability, durability, and safety. For example, evenly distributing weight throughout the product optimizes its center of gravity, minimizing the risk of instability during transportation. Incorporating impact-resistant materials and robust construction helps to mitigate damage from mechanical stresses and impacts. Effective thermal management solutions, such as heat sinks and insulation materials, maintain optimal operating temperatures within the product, safeguarding battery performance and safety during transit and storage [21].

The integration of batteries into products brings forth a host of considerations regarding storage conditions and shelf life. Preserving the battery's health and maintaining the product's integrity go beyond mere storage; they are integral to the process. Batteries, especially lithium-ion types commonly used in various devices, are quite sensitive to environmental factors like temperature and humidity. Any deviation from the recommended storage conditions can lead to detrimental effects such as degradation of battery performance or even safety hazards like overheating or leakage. To combat this, product designers need to establish precise parameters for storage that will safeguard both the battery's health and the longevity of the product itself [30].

To meet these requirements, designers often integrate a range of features into the product. This might include monitoring systems equipped with temperature sensors to keep tabs on internal conditions. If the temperature rises beyond safe limits, these systems can trigger cooling mechanisms or send alerts to prevent potential hazards. Similarly, controlling humidity levels within the product enclosure helps to prevent moisture-related damage to the batteries. In addition to monitoring, insulation materials and protective measures are often employed to shield batteries from external environmental factors. Importantly, insulation helps to keep internal temperatures stable, preventing thermal stress and degradation. Protective enclosures or impact-resistant materials provide a physical barrier against shocks and vibrations during storage and transportation, which safeguards the battery's integrity and performance [31].

Compliance with UN 38.3 regulations mandates rigorous testing of battery-powered products to assess their safety and performance characteristics during transportation and storage. These testing requirements are designed to ensure that products meet stringent regulatory standards and can withstand the challenges posed by various environmental conditions [20]. By assessing the structural integrity of battery-powered products and their packaging, mechanical testing ensures they are capable of withstanding the challenges posed by transportation and storage. This testing involves subjecting the product to various mechanical stresses and simulating real-world scenarios encountered during handling, transit, and storage [32].

Assessing a product's ability to withstand vibrations experienced during transportation, like those encountered during road, rail, or air transport, requires vibration testing. By subjecting the products to controlled vibrations across a range of frequencies and amplitudes, manufacturers can evaluate their structural integrity and functional performance under these conditions. As an example, shock testing evaluates a product's resilience to sudden impacts or drops that may occur during handling, loading, unloading, or in accidents. This testing helps to ensure that products can withstand such forces without sustaining damage or malfunction, thereby ensuring their integrity and safety during transportation. Impact testing further assesses a product's resistance to external impacts or collisions encountered during transportation. By subjecting the products to controlled impacts at specified velocities, manufacturers can evaluate their durability and implement measures to protect them from damage.

Testing for external short circuits is necessary to evaluate the safety of battery-powered products, especially in cases of short-circuit occurrences during transportation. For this test, a battery must remain intact without any disassembly, rupture, or fire. Additionally, the battery temperature must not exceed 170 °C during the test. Post-test observations must ensure no hazardous venting, leakage, or rupture within six hours after the test completion, and any emitted gases should not be toxic or harmful. These conditions ensure that the battery can safely handle a short-circuit situation without leading to dangerous outcomes such as fires or explosions, thereby mitigating risks during transportation. This testing helps to identify potential risks and ensures the implementation of safety mechanisms to mitigate them effectively. Overcharged and forced discharge testing evaluates the stability and performance of batteries under extreme charging or discharging conditions. By subjecting the batteries to controlled scenarios, manufacturers can assess their ability to withstand these conditions and ensure their safety during transportation [33].

Thermal testing is indispensable for battery-powered products due to the sensitivity of lithium-ion and other battery chemistries to temperature fluctuations. It evaluates the product's thermal management capabilities, ensuring it can withstand extreme temperature conditions encountered during transportation and storage, thereby guaranteeing safety, reliability, and regulatory compliance. To ensure the safe operation of battery-powered products across all environmental conditions, including both low-temperature and high-temperature scenarios, effective thermal management is paramount. Batteries are highly sensitive to temperature fluctuations, and extremes in temperature can significantly impact their performance, longevity, and safety.

For transportation in low-temperature conditions, batteries face challenges such as reduced capacity, increased internal resistance, and potential damage due to lithium plating. To mitigate these risks during transportation, effective thermal management solutions should be implemented. This may involve the use of advanced insulation materials within shipping containers to retain heat, active heating systems to maintain optimal temperatures during transit, and thermal blankets to shield batteries from extreme cold. Additionally, integrating thermal sensors and monitoring systems within the transport containers can provide immediate temperature feedback. This facilitates preemptive actions to prevent performance decline, capacity reduction, and safety risks linked to low temperatures.

Conversely, in high-temperature conditions, batteries are prone to accelerated chemical reactions, intensified heat generation, and increased risk of thermal runaway. Thermal management solutions for high temperatures may include heat sinks, fans, or phase-change materials to dissipate heat effectively. Similarly, thermal sensors and monitoring systems play a vital role in detecting and managing temperature spikes, allowing for timely intervention to prevent overheating and ensure the safety of battery-powered products. By addressing thermal management under both low- and high-temperature conditions, product designers can optimize the performance, reliability, and safety of battery-powered products throughout their operational life cycles, regardless of the environmental challenges they may encounter during transportation, storage, or usage [34].

Thermal testing protocols typically involve subjecting the product to temperature extremes and monitoring its response to these conditions. This includes conducting thermal cycling tests to simulate real-world temperature fluctuations during transportation and storage, exposing products to extreme temperatures, and performing thermal runaway tests to assess the effectiveness of safety mechanisms in preventing catastrophic failures. By ensuring safety and reliability with thorough thermal testing, manufacturers demonstrate their commitment to meeting regulatory standards and building trust among consumers and regulatory authorities [33]. In prioritizing regulatory compliance, manufacturers can mitigate risks, ensure the safety and reliability of their products, and safeguard their long-term business viability [35].

12.2.5 Analysis of Costs

The cost analysis for battery transportation and storage involves numerous factors [1]. Testing battery cells for safety and performance is necessary, involving a range of tests such as cell aging, thermal stability, and electrical abuse, all of which contribute to the overall expenditure.

Packaging costs are significant, as specialized materials and designs are required to meet safety regulations, particularly for LIBs. Transportation expenses vary depending on distance, mode, and safety requirements, with additional costs incurred for air shipping due to heightened safety measures. The costs of storage involve covering expenses for facility maintenance, regulating temperature and humidity, and ensuring safety through measures like fire suppression systems. Complying with UN 38.3 regulations imposes substantial economic costs on manufacturers of battery-powered products. These costs primarily stem from the rigorous testing and certification processes mandated by regulatory authorities. Testing and certification expenses encompass a wide range of costs, including the procurement of specialized equipment, materials required for testing, and skilled personnel to conduct the tests and interpret results accurately [36]. While specialized equipment for UN 38.3 testing, like temperature chambers and safety monitors, requires significant investment, outsourcing these tests to certified third-party labs can be a cost-effective alternative, especially for smaller companies [37].

The acquisition and maintenance of testing equipment can represent a significant upfront investment for manufacturers. Specialized instruments and facilities are often necessary to perform the battery safety tests required by UN 38.3, such as crush, impact, and thermal abuse tests. Additionally, ongoing calibration and maintenance of testing equipment is essential to ensure accurate and reliable results, further adding to the operational costs.

Procuring battery cells for testing, especially high-energy-density chemistries, and necessary materials like electrolytes contribute to expenses [38]. Accurate testing and interpretation of results require skilled personnel, which contributes to increased costs, particularly in regions where labor rates are high. Personnel costs associated with compliance efforts include salaries for qualified engineers, technicians, and safety specialists tasked with conducting tests, analyzing data, and preparing documentation for regulatory submissions. However, recruiting and retaining skilled personnel proficient in battery testing methodologies can be challenging, particularly in highly specialized fields such as battery safety engineering. Compliance with UN 38.3 may require third-party certification, audits, and ongoing quality assurance, further contributing to expenses. Regulatory compliance fees, training, and legal consultation also factor into the overall cost of ensuring compliance at the cell level [37].

Effectively managing the intricacies of cell-level considerations, testing procedures, and associated expenses is necessary to guarantee the secure transportation and storage of batteries in compliance with UN 38.3 regulations. By understanding the degradation mechanisms, safety hazards, and testing requirements at the cell level, companies can implement robust quality assurance processes and mitigate risks effectively. Prioritizing safety and following regulations builds trust, reliability, and sustainability in the battery industry over time. In addition to testing and certification expenses, achieving compliance with UN 38.3 may necessitate product re-designs and modifications to meet regulatory requirements. For instance, manufacturers might have to invest in specialized packaging solutions or include extra safety features to improve the product's durability during transportation and storage. These design changes can result in increased manufacturing costs, longer lead times, and potential disruptions to production schedules [25].

Despite the substantial financial investments required for compliance, the consequences of noncompliance far outweigh the costs of regulatory adherence. A product recall's legal liabilities and damage to brand reputation can have devastating effects on manufacturers, leading to financial losses and eroding consumer trust. Moreover, noncompliant products pose significant safety risks to end-users, potentially causing injury or property damage. Therefore, investing in compliance measures is not only a legal requirement but also a strategic imperative for manufacturers of battery-powered products.

12.3 Global Regulations Governing the Secure Transportation of Lithium-Ion Batteries

The UN model regulations serve as global guidelines for transporting hazardous materials, drawing insights from various organizations involved in crafting policies for secure and efficient cross-border transportation [1]. The UN Publication Recommendations on the Transport of Dangerous Goods Model Regulations and its counterpart Recommendations on the Transport of Dangerous Goods Manual of Tests and Criteria are the internationally recognized standards for testing, packaging, and shipping of dangerous goods.

UN offices worldwide contribute to the development of these regulations, which also encompass the International Air Transport Association Dangerous Goods Regulations (IATA DGR), the International Maritime Dangerous Goods (IMDG) Code, the European Agreement concerning International Carriage of Dangerous Goods by Road (ADR), and the European Regulation concerning the International Carriage of Dangerous Goods by Rail (RID). Together, these regulations aim to ensure the safe transportation of lithium batteries and minimize the risks associated with their shipment. The regulations surrounding the transportation of lithium batteries depend on various factors, including the type of battery, its capacity, and whether it is shipped alone, packed with, or installed in a device.

The International Civil Aviation Organization (ICAO) sets forth regulations governing the international shipment of hazardous substances by air. Building upon the UN/ICAO standards, the International Air Transport Association (IATA) incorporates specific airline and governmental prerequisites into their Dangerous Goods Regulations (DGR) publications. The United Nations Economic Commission for Europe (UNECE) sponsors the ADR to enhance the safety of international road transport of dangerous goods, including waste. Additionally, the Intergovernmental Organization for International Carriage by Rail (OTIF) formulates the RID regulations, which pertain to the international transport of hazardous materials by rail. Finally, the International Maritime Organization (IMO) acts as the regulatory authority for all dangerous goods shipments on the open seas [1].

Countries like the United States (DOT regulations), China (GB standards), and Japan (METI guidelines) have additional national regulations that must be followed. Regulatory requirements for storage, aside from UN test requirements, are based on standards from the International Electrotechnical Commission (IEC), Underwriters Laboratory (UL), the National Fire Protection Association (NFPA), and Occupational Safety and Health Administration (OSHA). Packaging requirements for LIBs necessitate using materials to contain leaks, prevent short circuits, and provide protection. Shippers must refer to the relevant transport regulations for detailed specifications, including the applicable Special Provisions for each UN number, outlined in Table 12.7 [39].

Proper labeling, handling instructions, and documentation are critical for compliance and safety. Importantly, adhering to regulations reduces transportation risks and promotes reliability. While the latest UN Model regulations, revision 23 [40], were published in June 2023, the ADR will not introduce these until January 2025. Also, there is a longstanding committee operating at the request of the International Civil Aviation Organisation (ICAO), and led by SAE, known as SAE G-27 Lithium Battery Packaging Performance [39]. This committee is tasked with the responsibility of developing packaging standards for the safe transport of lithium batteries on aircraft. Primarily, it addresses the concerns of lithium batteries that go into thermal runaway and the potential to reintroduce certain battery types as cargo on passenger aircraft. However, local authorities are responsible for the interpretation and implementation of the relevant regulations, and they can, at their discretion, make decisions differing from this guideline.

Table 12.7 UN numbers and their corresponding Packing Instructions for each mode of transport.

UN number	Battery type	Passenger & cargo aircraft (IATA)	Cargo aircraft only (IATA)	Road (ADR)	Sea (IMDG)
UN 3090	Lithium metal batteries	Forbidden	PI 968, Section IA & IB	PI 903	PI 903
UN 3091	Lithium metal batteries contained in equipment	PI 970, Section I & II	PI 970, Section I & II	PI 903	PI 903
UN 3091	Lithium metal batteries packed with equipment	PI 969, Section I & II	PI 969, Section I & II	PI 903	PI 903
UN 3480	Lithium-ion batteries	Forbidden	PI 965, Section IA & IB	PI 903	PI 903
UN 3481	Lithium-ion batteries contained in equipment	PI 967, Section I & II	PI 967, Section I & II	PI 903	PI 903
UN 3481	Lithium-ion batteries packed with equipment	PI 966, Section I & II	PI 966, Section I & II	PI 903	PI 903
UN 3090/ UN 3091/ UN 3480/ UN 3481	Prototype	Forbidden	Forbidden	PI 910	PI 910
UN 3090/ UN 3091/ UN 3480/ UN 3481	Damaged or defective	Forbidden	Forbidden	PI 908, PI 911	PI 908, PI 911
UN 3090/ UN 3091/ UN 3480/ UN 3481	Disposal or recycling	Forbidden	Forbidden	PI 909	PI 909

Source: Adapted from [39].

Several documents are required for shipping batteries internationally. These include a dangerous goods declaration (DGD), transport document, emergency response information, material safety data sheet (MSDS), pilot notification, packaging certifications, battery test summary, and exemption approvals.

12.3.1 Regulations for Transportation by Air

International regulations for the transportation of LIBs by air are primarily governed by the International Civil Aviation Organization (ICAO) and the International Air Transport Association (IATA). The IATA DGR sets detailed requirements for packaging, labeling, and transport documentation of LIBs, exceeding ICAO technical instructions. In compliance with IATA guidelines, lithium batteries can be transported by air if they meet UN test requirements for cells or batteries, ventilation, short-circuit prevention, reverse current flow prevention, and manufacturing standards. ICAO does not require the use of an MSDS, and test data is not part of the required documentation requirements when transporting lithium batteries. These regulations classify

LIBs under UN 3480 when shipped alone and UN 3481 when contained in or packed with equipment.

Due to the inherent risks associated with LIBs, strict safety measures are enforced. Packaging requirements are prioritized, mandating that batteries be isolated from conductive materials to prevent short circuits, and they must be packed in strong outer containers that can withstand impacts. LIBs must be labeled appropriately to indicate their hazardous nature, and shippers are responsible for ensuring compliance with all regulations, including documentation. Comprehensive documentation needs to accompany each shipment, ensuring all parties in the supply chain are aware of the contents and the associated risks.

In addition to these global standards, certain regions and countries may have supplementary regulations. For instance, the United States Department of Transportation (USDOT) has additional requirements for LIB shipments, including specific measures for damaged or defective batteries. Similarly, the European Aviation Safety Agency (EASA) enforces the ICAO and IATA regulations within Europe, but it may have additional guidelines to address regional safety concerns.

In accordance with the IATA-DGR requirements for air freight transit, LIBs (UN 3480) classified under Packing Group II must be labeled with ICAO Class 9 hazard labels and comply with Special Provisions A88, A99, A154, and A164, as well as Packing Instructions P965 to P970. Damaged, defective, or waste batteries are explicitly prohibited from air transport.

Lithium batteries generally require thorough testing in accordance with UN 38.3 requirements to verify compliance with safety protocols before they can be shipped. Nonetheless, there are specific exemptions to this rule, particularly for preproduction prototypes that have not yet undergone such testing. In general, preproduction prototypes of lithium batteries or cells that have not been tested to the requirements of UN 38.3 may be transported on cargo aircraft if approved by the appropriate authorities of the state of origin and the state of the operator. Additionally, Packing Instruction P910 of the Supplement to the Technical Instructions must be met. Waste lithium batteries and lithium batteries being shipped for recycling or disposal are forbidden from air transport unless approved by the appropriate national authority of the state of origin and the state of the operator [39].

Lithium-ion cells and batteries shipped by themselves, according to UN 3480 (not contained in or packed with equipment), are forbidden for transport as cargo on passenger aircraft. However, they can be shipped on cargo aircraft when packed in accordance with Packing Instruction P965. Lithium-ion cells and batteries (UN 3480) must be shipped at a state of charge (SOC) not exceeding 30% of their rated capacity. Conversely, cells and batteries at an SOC higher than 30% can only be shipped with the approval of the state of origin and the state of the operator [41] (Figure 12.1).

Lithium metal cells and batteries shipped by themselves, according to UN 3090 (not contained in or packed with equipment), are forbidden for transport as cargo on passenger aircraft. However, they can be shipped on cargo aircraft when packed in accordance with Packing Instruction P968. All types of lithium metal and LIBs (including UN 3090, UN 3091, UN 3480, and UN 3481) are forbidden from transport if they are identified by the manufacturer as being defective or damaged, which can potentially produce a dangerous risk of heat, fire, or short circuit [1]. This also applies to lithium cells or batteries installed inside equipment where the device has been recalled because of safety concerns about the cell or battery.

Effective January 1, 2025, in the 66th edition of the IATA Dangerous Goods Regulations (DGR) and the 12th edition of the Lithium Battery Shipping Regulations (LBSR), it will be recommended that, when transported by air, LIBs packed with equipment, LIBs contained in equipment, and vehicles powered by LIBs should have the batteries at a state of charge not exceeding 30% of their rated capacity [41]. These changes have been adopted by ICAO into the 2025–2026 edition of the Technical Instructions for the Safe Transport of Dangerous Goods by Air (Technical Instructions).

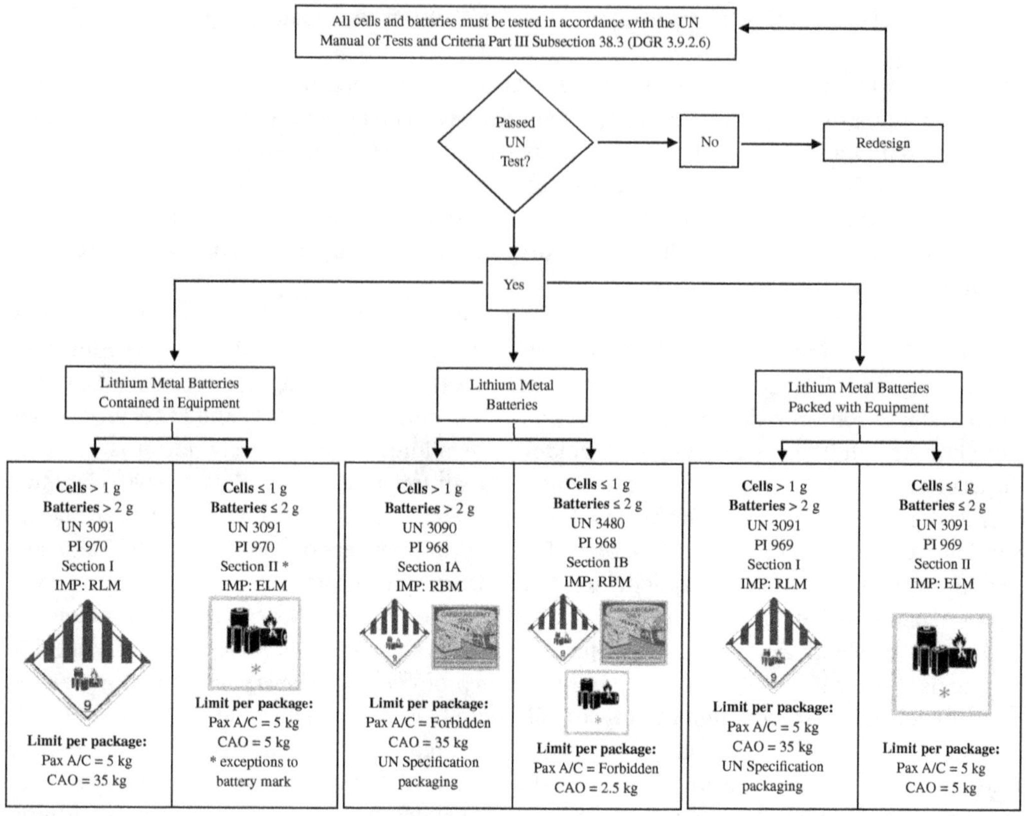

Figure 12.1 Classification flowchart for packing instructions of lithium metal batteries. Source: [41]/International Air Transport Association (IATA).

The objective of these changes is to reduce the potential risk posed by LIBs in air transport. With that in mind, it has been demonstrated that reducing the state of charge in a lithium-ion cell or battery reduces the potential for a lithium-ion cell to go into thermal runaway. Lithium-ion cells and batteries with a Watt-hour rating in excess of 2.7 Wh [41] must be transported at a state of charge not exceeding 30% of their rated capacity. Additionally, cells or batteries at a state of charge greater than 30% of their rated capacity must be transported in accordance with the provisions of Section I of PI 966 with the approval of the state of origin and the state of the operator under the written conditions established by those authorities (Figure 12.2).

12.3.2 Regulations for Transportation by Surface (Road/Rail/Sea)

When transporting LIBs by road, rail, or sea, it is important to adhere to a comprehensive set of international regulations that are put in place to prioritize safety and minimize the potential risks associated with these hazardous materials. The Inland Transport Committee (ITC) has been the highest policymaking body of the UNECE in the last 75 years in the field of transport [42]. Together with its subsidiary bodies, the ITC has provided a pan-European intergovernmental forum, where UNECE member countries come together to forge tools for economic cooperation, negotiate, and adopt international legal instruments for inland transport.

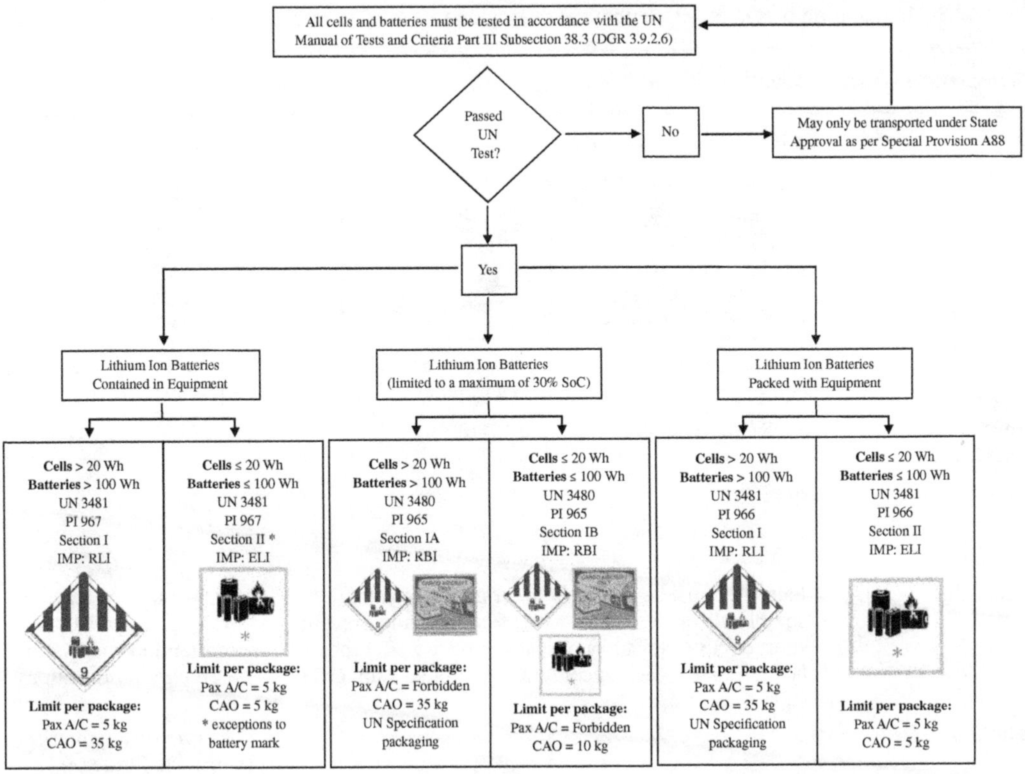

Figure 12.2 Classification flowchart for packing instructions of lithium-ion batteries. Source: [41]/International Air Transport Association (IATA).

The European Agreement concerning the International Carriage of Dangerous Goods by Road (ADR), regulated by the UNECE, provides a framework for ensuring the safe and secure transport of hazardous materials (dangerous goods) by road. Notably, it is the energy content and diverse conditions that classify which dangerous goods regulations must be taken into account for the transport of LIBs. Due to exemption regulations, simplified requirements apply to LIBs with a nominal energy up to a maximum of 100 Wh, whereas LIBs with a nominal energy of more than 100 Wh are always to be treated as fully regulated Class 9 Dangerous Goods [43] (Tables 12.8 and 12.9).

The ADR classifies LIBs as Class 9 dangerous goods and outlines stringent packaging, labeling, and documentation requirements. Cells and batteries, cells and batteries contained in equipment, or cells and batteries packed with equipment containing lithium in any form shall be assigned to UN Numbers 3090, 3091, 3480, or 3481 as appropriate. Each cell and battery incorporate a safety venting device to preclude a violent rupture under normal conditions of carriage, which is an effective means of preventing external short circuits. Additionally, battery-containing cells or a series of cells connected in parallel are equipped with effective means as necessary to prevent dangerous reverse current flow [44]. Packaging instructions under the ADR, such as P903, P908, and P909, specify the use of inner and outer packaging to prevent damage and short circuits and include provisions for the transport of damaged or defective batteries, highlighting the need for enhanced protective measures and clear identification.

According to ADR 2021, packages containing lithium cells, or batteries prepared in accordance with Special Provision 188, shall have a mark that indicates the UN number preceded by letters. For

Table 12.8 Lithium-ion batteries with a nominal energy up to a maximum of 100 Wh.

Transportation mode	Road/rail (ADR/RID), sea freight (IMDG Code)		
Nominal Energy	≤100 Wh (per battery)		
Description	 (a) Batteries without equipment icon	 (b) Batteries packed with equipment icon	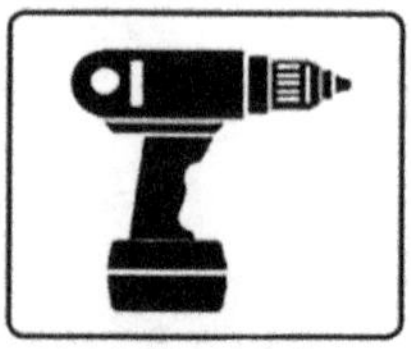 (c) Batteries contained in equipment icon
Special Provision/ Packing Instruction	ADR/RID SP 188, IMDG Code SP 188		
Max. Quantity	n/a		
Weight Limit	30 kg gross weight (per package)	n/a	
Packaging	Batteries must be placed in inner packagings that completely enclose them to protect them and prevent short circuits. Strong outer packaging, e.g., fiberboard box (Drop test passed: content shall not be damaged or shifted.		Strong outer packaging protection against unintentional activation short circuit protection
Marking	 (d) Lithium Battery Handling Mark – UN3480	 (e) Lithium Battery Handling Mark – UN3481	Lithium battery mark is similar to figure (e)  not applicable if no more than two batteries are installed and if there are no more than two packages in the consignment
Sea Freight Container—Marking	None		
Transport Document	n/a		n/a
Miscellaneous	Personnel shall be trained to commensurate with responsibilities		

Source: [43]/ZVEI.

Table 12.9 Lithium-ion batteries with a nominal energy of more than 100 Wh.

Transportation mode	**Road/rail (ADR/RID), sea freight (IMDG Code)**		
Nominal Energy	**>100 Wh (per battery)**		
Name and Description	See figure (a) from Table 12.8	See figure (b) from Table 12.8	See figure (c) from Table 12.8
Special Provision/ Packing Instruction	P903, LP903	SP 390, P903, LP903	
Max. Quantity	ADR 1.1.3.6: max. 333 kg (per transport unit, e.g., truck incl. trailer) If exceeded, further requirements for vehicle equipment and driver		
Weight Limit	n/a		
Packaging	Batteries must be placed in inner packagings that completely enclose the battery, batteries must be protected to prevent short circuits. Batteries must be secured against movement within the outer packaging. UN-approved packaging (Packaging Group II: e.g., UN/4G/Y30/…)		Strong outer packaging protection against unintentional activation short circuit protection
Marking	Hazard label No. 9A (10 cm × 10 cm)  (f) Class 9A Lithium ADR: UN 3480 IMDG Code: LITHIUM-ION BATTERIES UN 3480	Hazard label No. 9A (10 cm × 10 cm)  (g) Class 9A Lithium Battery Hazard Label – UN 3481 ADR: UN 3481 IMDG Code: LITHIUM-ION BATTERIES PACKED WITH EQUIPMENT UN 3481 or LITHIUM-ION BATTERIES CONTAINED IN EQUIPMENT UN 3481	
Sea Freight Container—Marking	Container Placards (min. 25 cm × 25 cm)  (h) Class 9 Placard		
Transport Document	UN 3480 LITHIUM-ION BATTERIES, 9, (E) Number of packages and packaging type (e.g., 1 Fiberboard box) weight (e.g., 5 kg) Shipper's & consignee's address Sea freight (IMDG Code): (language English) IMO-DANGEROUS GOODS DECLARATION (SOLAS 74, KAP, VII, REG 5, MARPOL 73/79)	UN 3481 LITHIUM-ION BATTERIES PACKED WITH EQUIPMENT, 9, (E) Number of packages and packaging type (e.g., 1 Fiberboard box) weight (e.g., 5 kg) Shipper's & consignee's address Sea freight (IMDG Code): (language English) IMO-DANGEROUS GOODS DECLARATION (SOLAS 74, KAP, VII, REG 5, MARPOL 73/79)	UN 3481 LITHIUM-ION BATTERIES CONTAINED IN EQUIPMENT, 9, (E) Number of packages and packaging type (e.g., 1 Fiberboard box) weight (e.g., 5 kg) Shipper's & consignee's address Sea freight (IMDG Code): (language English) IMO-DANGEROUS GOODS DECLARATION (SOLAS 74, KAP, VII, REG 5, MARPOL 73/79)

Source: [43]/ZVEI.

example, the mark would be UN 3090 for lithium metal cells or batteries, UN 3480 for lithium-ion cells or batteries, UN 3091 for lithium cells or batteries contained in equipment, and UN 3481 for lithium cells or batteries packed with equipment [45]. Special Provision 188 elaborates that, for a lithium metal or lithium alloy cell, the lithium content must not be more than 1 g, and for a lithium-ion cell, the Watt-hour rating must not be more than 20 Wh. It also dictates that for a lithium metal or lithium alloy battery, the aggregate lithium content is not more than 2 g, and for an LIB, the Watt-hour rating is not more than 100 Wh. LIBs subject to this provision need to be marked with the Watt-hour rating on the outside case [44].

The testing requirements in UN 38.3 do not apply to production runs, which consist of not more than 100 cells or batteries, or to preproduction prototypes of cells or batteries when these prototypes are carried for testing and packaged in accordance with Packing Instructions P910 or P905. The transport document must include the following statement: "Carriage in accordance with Special Provision 310." Damaged or defective cells and batteries, or cells and batteries contained in equipment, must be carried in accordance with Special Provision 376. Cells, batteries, or cells and batteries contained in equipment carried for disposal or recycling may be packaged in accordance with Special Provision 377 and Packing Instruction P909.

These criteria, when relevant to a specific application, may be considered to assess the performance of large packaging, according to ADR regulations:

- The assessment shall be done under a quality management system, allowing for the traceability of test results and reference data used.
- The list of hazards expected in case of thermal runaway for the battery type must be clearly identified and quantified in the condition it is carried (e.g., usage of inner packaging, state of charge, sufficient noncombustible, electrically nonconductive, and absorbent cushioning material, etc.). The reference list of possible hazards for lithium batteries can be used for this purpose, and the quantification of hazards shall rely on available scientific literature.
- The mitigating effects of the large packaging must be identified and characterized based on the nature of the protections provided and the properties of the construction materials.

Several technical characteristics and drawings are used in support of this assessment:

- Density ($kg \cdot m^{-3}$)
- Specific heat capacity ($J \cdot kg^{-1} \cdot K^{-1}$)
- Heating value ($kJ \cdot kg^{-1}$)
- Thermal conductivity ($W \cdot m^{-1} \cdot K^{-1}$)
- Melting temperature (K)
- Flammability temperature (K)
- Heat transfer coefficient of the outer packaging ($W \cdot m^{-2} \cdot K^{-1}$)

The provisions of ADR do not apply to companies carrying goods as an ancillary process to their main business activity (e.g., deliveries or returns from building sites or demonstration purposes), as per the Craftsman Regulation in ADR section 1.1.3.1c. Also, ADR rules do not apply to private individuals when the batteries are packaged for retail sale and if the transport is intended for their personal use.

Rail transport of LIBs in Europe is governed by the RID, which aligns closely with the ADR and IMDG Code. However, lithium batteries are not subject to the provisions of RID if they meet the requirements of Special Provision 188 of the ADR [46]. The RID mandates similar packaging, labeling, and documentation standards to ensure the safe carriage of LIBs by rail. This alignment helps maintain consistency across different modes of surface transport, facilitating smoother transitions between road, rail, and sea shipments.

RID is very particular in terms of personnel handling dangerous goods like LIBs. Hence, it is important that, with the degree of injury risk or exposure arising from an incident involving the carriage of dangerous goods, including loading and unloading, personnel should be trained in the hazards and dangers presented. The training provided must aim to make personnel aware of the safe handling and emergency response procedures. If an RID-contracting state considers that safety is not reduced, it may in its domestic legislation transfer the obligations to a specific participant. These derogations need to be communicated by the RID-contracting state to the Secretariat of OTIF, which will bring them to the attention of the other RID-contracting states [46].

The International Maritime Dangerous Goods (IMDG) Code is the international guideline for the safe shipment of dangerous goods by sea. It contains information on terminology, packaging, labeling, markings, stowage, segregation, handling, and emergency response, and it is intended for use not only by mariners but also by all those involved in industries and services connected with shipping. Of course, the safety approach for batteries transported by sea is different than by land. Specifically, the hull of a vessel gets slammed with waves at a very high frequency when the weather is rough. The shock and vibration effects can eventually lead to structural damage to the battery casing, potentially triggering a short circuit, and, above all, it takes much more time for emergency services to reach a burning ship [47].

The IMDG Code mandates the use of robust packaging materials and stipulates that batteries must pass the UN 38.3 tests, which assess their ability to withstand environmental and mechanical stressors such as temperature variations, vibration, and shock. Appropriate labeling and marking are required to inform handlers and emergency responders of the risks, including hazard labels and handling instructions. The transport of LIBs by sea freight, and its regulations, packaging instructions, and special provisions, are similar to ADR. LIB-related maritime accidents have intensified the IMDG regulations for the safe transport of unused and used LIBs for reuse and recycling in the near future. IMDG Code 5.2.2.2.2 uses the Special Provision 384 label for LIB dangerous goods packaging, and IMDG Code 5.2.1.10 uses the Special Provision 188 label for LIB nondangerous goods. Special Provision 377 is used to package batteries for disposal or recycling, with Packing Instruction P909 (disposal/recycle) and Packing Instruction P908-5 (battery with defects) (Table 12.10).

Table 12.10 IMDG Code packing guidelines for unused and used lithium-ion batteries.

Classification	Label and Mark for Lithium Battery packaging		
Special Provision SP 384	See figure (f) 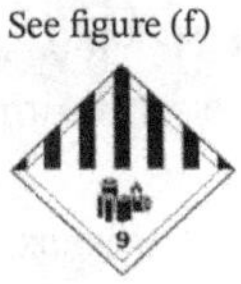	Lithium Battery Dangerous goods Label (No. 9A) -IMDG Code 5.2.2.2.2	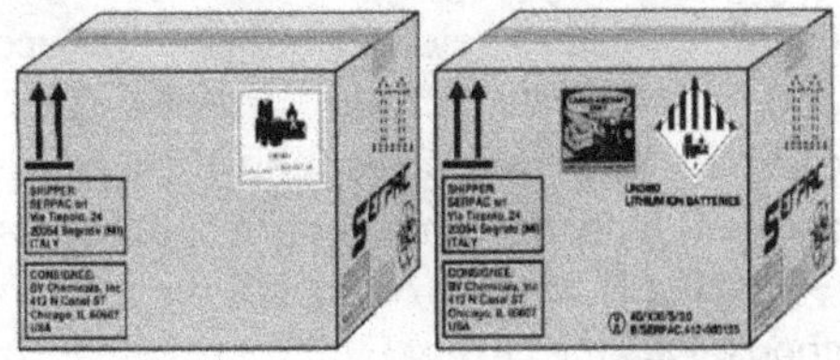 figure (h) Example: Packaging containing batteries
Special Provision SP 188	See figure (d) 	Lithium Battery Non-Dangerous goods Mark -IMDG Code 5.2.1.10	

Source: [48]/The Mock IMO Assembly.

Table 12.11 Proposed UN code and special provision for used lithium-ion batteries [48].

UN X123 Used Lithium Ion Batteries for Reuse & Recycling Purpose (IMDG Code, Amdt OO-OO)			
Proper Shipping Name (PSN)	**Used Lithium Ion Batteries for Reuse and Recycling Purpose**		
United Nations Number (Un No.)	**X123**	**Class**	**9**
Label & Placard	 Label (Class 9A)	 Label (Class 9A)	
	See figure (f) and figure (h)		
Special Provisions (SP)	Existing: 188, 230, 310, 348, 376, 377, 384, 387 (Li-Ion in general) **Newly proposed: XX1, XX2 (Used Li-Ion)**		
Stowage & Segregation	**Category A** if Special Provisions **XX1** and **XX2** are both satisfied; **Category D** if not		
Features of Used Batteries	1. The hazards of leakage, fire, and/or explosion exist. 2. Used Li-Ion batteries are potentially damages, with the probability depending on the causes of disposal. (e.g., accident, expiration of life-expectancy, leakage, etc.) 3. Dismantling used battery packs incorporates a complicate procedure, adding a challenge on safety inspection.		

IMDG, under the supervision of IMO, has proposed new UN Code (UN X123) provisions for the maritime transportation of used LIBs for reuse and recycling purposes. There are two new special provisions for used LIBs: XX1 and XX2 [48]. Special Provision XX1 elaborates that all vessels transporting used batteries for reuse and recycling shall obtain a document certification detailing the physical and electrochemical composition (e.g., cathode active material, anode active material, separator, and electrolyte) and the background behind the disposal (e.g., accident, expired life cycle, leakage) at least 14 days prior to the voyage. Special Provision XX2 states that all vessels transporting used batteries for reuse and recycling shall present a safety test result certification detailing electrical, thermal, mechanical, and chemical tests that are passed without any hazard identification, at least 7 days prior to the journey (Table 12.11).

The national/federal government of each member state shall issue the documentation or certification after the confirmation of test results by nationally/federally recognized testing laboratories unless a significant hazard exists. These new regulations are yet to be implemented to ensure the safe transport of LIBs at sea.

Detailed shipping documentation is also necessary to provide transparency and traceability, ensuring that all relevant information accompanies the shipment. According to the ADR, RID, and IMDG Code, lithium batteries (including UN 3090, UN 3091, UN 3480, and UN 3481) can be transported if they meet the provisions about battery UN tests, ventilation, short-circuit prevention, reverse current flow prevention, and manufacture.

In accordance with ADR/RID regulations for road transit, LIBs (UN 3480) are classified as Class 9 goods, Packing Group II, and are categorized under tunnel category E. It is imperative that these batteries are labeled with label 9 and listed as "Lithium-Ion Batteries" under the proper shipping name. Relevant ADR special provisions include 188, 230, 310, and 636. Moreover, the packaging must adhere to the guidelines outlined in documents P903, P903a, and P903b. In the event of damaged or defective batteries, it is essential to seek consultation with the national competent authority. As per the International Maritime Organization (IMO) regulations for sea freight transit, LIBs (UN 3480), which are classified under Class 9 with Packing Group II, must comply with the IMDG Code

Special Provisions 188, 230, and 310. They should be packed according to Packing Instruction P903, with EmS codes F-A and S-I, and stowed under Category A. Consultation with the national competent authority is mandatory for damaged or defective batteries.

When lithium batteries are carried for disposal or recycling, they must be designed or packed to prevent short circuits and the dangerous risk of heat. At the same time, they must be secured against excessive movement within the outer packaging, which must conform to the PG II performance level. Packages for these batteries must be marked "LITHIUM BATTERIES FOR DISPOSAL" or "LITHIUM BATTERIES FOR RECYCLING" [49]. Damaged or defective batteries may lead to short-circuiting or may catch fire through the release of stored energy or hazardous contents. Packages shall be marked "DAMAGED/DEFECTIVE LITHIUM-ION BATTERIES" or "DAMAGED/DEFECTIVE LITHIUM METAL BATTERIES," as applicable [46].

Proper packaging is one of the most critical measures that a shipper can consider to improve safety and prevent incidents. As such, ADR provisions have a condition that states a quality assurance system must be in place to ensure that the total amount of lithium cells and batteries per transport unit does not exceed 333 kg [49]. In accordance with the RID, lithium batteries are permitted for carriage as "express parcels." However, these parcels should not weigh more than 40 kg [50]. Also, the passage of a sea transport unit carrying lithium batteries is forbidden through category E tunnels if the gross mass of packages is less than 8 tons per transport unit marked in the IMDG Code [1].

12.4 Lithium Battery Transportation Regulations in Different Countries

The regulations for transporting LIBs vary across different countries [1]. In the United States, the Department of Transportation (DOT) has established regulations for transporting hazardous materials, including LIBs, under the Hazardous Materials Regulations (HMR). The International Air Transport Association (IATA) has also developed guidelines that are stricter than the DOT regulations for shipping LIBs.

In Europe, UN offices govern the transportation of LIBs. The UN does this via the International Air Transport Association Dangerous Goods Regulations (IATA DGR), the International Maritime Dangerous Goods (IMDG) Code, the European Agreement concerning International Carriage of Dangerous Goods by Road (ADR), and the European Regulation concerning the International Carriage of Dangerous Goods by Rail (RID). In China, the Civil Aviation Administration of China (CAAC) and the Ministry of Industry and Information Technology (MIIT) have established regulations for transporting LIBs.

12.4.1 Transportation Regulations in the United States

In order to ship LIB cells or packs in the United States, LIBs must pass the eight tests outlined in UN 38.3. In order to ship internationally, batteries must pass nine tests in IEC 62281, which are similar to the eight tests in UN 38.3 but with an extra drop test. UN 38.3 and IEC 62281 consist of the most fundamental tests a safe battery should pass, and many other regulatory bodies refer to these tests as a part of their passing criteria. The DOT defines shipping regulations for the United States under 49 CFR 173.185 [51]. However, compliance with 49 CFR does not guarantee acceptance by regulatory bodies outside the United States [52]. Hence, outside the United States (for transport to and from the United States) the UN Model Regulations are followed regarding all aspects of transporting batteries. A similarity between the two transport regulations is that the US regulations are also based on the size of the battery to be shipped. Conversely, there are distinct differences between 49 CFR and the international regulations due to the fact that most of the US regulations are based on UN 38.3 standards.

Part 49 of the Code of Federal Regulations (49 CFR Sections 100–185) of the US Hazardous Materials Regulations (HMR) governs the domestic transportation of lithium batteries for all modes of transport to, from, and within the United States by all modes [52]. It provides information on packaging, hazard communication (e.g., package marking, labeling, and shipping papers), and handling of batteries and battery-powered devices. The Hazardous Materials Regulations (HMR) specify different packaging instructions (PI), such as PI 965–967 for air transport and PI 903 for ground transport, with each detailing the necessary steps to properly secure batteries.

IEC 62619 is also an important standard in the United States because of how broadly it can be applied and how it dictates the system-level design parameters of large-format LIBs. This regulation deals with batteries on a system level and defers to other regulations, such as IEC 62133, for cell-level standards. For example, the BMS is pushed to operational regions where it should not function, and mitigating actions are observed (e.g., shutdown of the module or system). Propagation tests are also utilized to demonstrate the ability of module designs to contain full-fledged battery events. The drop tests consist of both battery cell-level drop testing and battery system-level drop testing. The American National Standards Institute (ANSI) C18 standard specifies tests and requirements for portable primary lithium cells and batteries, both the chemical systems and the types covered in ANSI C18.3 M, Part 1, to ensure their safe operation under normal use and reasonably foreseeable misuse.

The transportation of LIBs in the United States is governed by comprehensive internal regulations and best practices, primarily through the Department of Transportation's Hazardous Materials Regulations (HMR) outlined in 49 CFR. The US National Transportation Safety Board (NTSB) issued two safety recommendations: to "physically separate lithium batteries from other flammable hazardous materials stowed on cargo aircraft" and to establish "maximum loading density requirements" that restrict the quantities of lithium batteries and flammable hazardous materials [53]. The Pipeline and Hazardous Materials Administration (PHMSA) now has the authority to enforce stricter regulations for onboard fires caused by inadequate international transportation standards. In the United States, popular carriers like UPS, DHL, and FedEx comply with battery transportation regulations.

UPS transport regulations for lithium batteries are in accordance with the IATA DGR and 49 CFR, but with additional requirements for air shipping of lithium batteries. Compared with the IATA DGR and 49 CFR, small-size standalone lithium batteries (except lithium batteries transported according to PI 968) are not considered to be UPS dangerous goods. Lithium batteries for which the net quantity of each package exceeds limits, and standalone small-size lithium batteries transported according to PI 968, all need UPS dangerous goods contracts prior to being transported [54].

DHL Express provides international express delivery services for parcels and documents. To transport lithium batteries safely by air, DHL Express shippers must comply with the latest IATA DGR, in addition to its own norms based on the ICAO TI and the IATA DGR. According to the norms, standalone LIBs can still be transported by DHL on passenger aircraft after a separate account approval [55]. Lithium metal batteries packed in accordance with Section II of Packing Instruction 968 (loose lithium metal batteries) are forbidden in the DHL Time Definite Network. Additionally, LIBs packed in accordance with Packing Instruction 965 are forbidden on passenger aircraft.

FedEx has expanded its service by offering global freight-forwarding services. When shipping lithium batteries with FedEx's air service, all packages containing lithium batteries must strictly adhere to the ICAO TI, the IATA DGR, and 49 CFR, only with exceptions for a US territory or commonwealth such as Puerto Rico [56]. Shippers may use 49 CFR when prepared by air, as limited by the commodities FedEx accepts and prepared under these regulations, and the shipper is responsible for complying with all packing requirements, appropriate marking, labeling of the package, documentation, and compliance with all applicable local, state, and federal laws, regulations, ordinances, and rules. For standalone batteries, the state of charge (SOC) may not exceed 30% of the rated capacity, with a maximum of 10 kg net quantity per package, and it should be labeled for Cargo Aircraft Only (CAO) [57].

12.4.2 Transportation Regulations in China

China has become a particularly important country in the energy transition as the world's largest producer and transporter of LIBs. As a member of ICAO, China enforces these technical instructions rigorously. Lithium batteries transported by air in China must comply with quality management standards and pass the UN 38.3 tests. Chinese authorities, such as the Ministry of Transport (MOT) and the Civil Aviation Administration of China (CAAC), enforce these standards through detailed national regulations. The CAAC has developed comprehensive regulations for the air transport of lithium batteries, including the Administrative Regulations on the Transportation of Dangerous Goods by Civil Aviation. These regulations stipulate that LIBs must be packaged in a strong outer packaging that can withstand physical impacts and prevent short circuits. Additionally, inner packaging is required to separate individual batteries and prevent contact that could lead to a short circuit. Packaging must include appropriate hazard labels, such as the Class 9 miscellaneous dangerous goods label and the lithium battery handling label, along with the UN number and proper shipping name.

GB/T 31485 [58] and IEC 62660-3 [59] address risks associated with the use and transportation of power lithium batteries, and they introduce three additional safety tests compared to UN 38.3: drop, seawater immersion, and low-pressure tests. These additions aim to evaluate the transportation safety of power lithium batteries. As of January 1, 2020, all lithium batteries shipped by air or sea from Chinese ports must comply with updated UN 38.3 testing requirements. Additionally, a lithium battery test summary must be provided, which certifies that the battery has met these safety standards.

Shippers must obtain a safe transportation certificate, known as the Identification and Classification Report for Air/Sea Transport of Goods. This certificate, along with a material safety data sheet (MSDS), is mandatory for all shipments containing lithium batteries. The testing reports must be issued by laboratories accredited by the China National Accreditation Service for Conformity Assessment (CNAS), ensuring that the testing adheres to the latest standards. For ground transportation, the Provisions on the Supervision and Administration of Railway Transportation Safety of Dangerous Goods apply, which classify consumer-grade lithium batteries as Class 9 dangerous goods, under miscellaneous dangerous substances and articles. The List of Dangerous Goods in Railway (TB/T 30006-2022) classifies them under codes 91042, 91043, 91045, and 91046 [60]. However, specific exemptions exist for batteries with low lithium content or energy ratings, indicating that not all lithium batteries are subject to the same stringent regulations when transported by rail. The focus is on ensuring that batteries deemed unsafe, damaged, or designated for disposal are not transported under any circumstances [61].

Industry-specific standards such as MH/T 1052-2013 [62] and MH/T 1020-2013 [63] outline the requirements for testing and transporting lithium batteries by air. To ensure safety, China uses a third-party appraisal system for lithium battery transportation, where recognized organizations issue appraisal reports based on test results. Lithium battery shipments by air are subject to both Chinese (CCAR-276-R1) and international safety regulations (including ICAO TI and IATA DGR). The CAAC also mandates comprehensive documentation, including detailed shipping papers and a declaration of dangerous goods, certifying that the batteries comply with all applicable regulations. Additionally, special provisions are in place for the transport of damaged or defective batteries, which pose higher risks. The CAAC often restricts these batteries to cargo-only aircraft, further enhancing safety. The ICAO TI, in the 2015–2016 edition (DOC 9284-AN/905) with corrigendum No. 1, and CCAR-276-R1 under CAAC, issued provisions for carrying lithium metal batteries. The CAAC specifies that standalone lithium metal batteries and LIBs are forbidden for transport aboard passenger aircraft, unless the country of origin, native country, transit state, and/or destination country grants an exemption from the prohibition. However, lithium metal batteries can be transported on aircraft according to the specific provisions of DOC 9284-AN/905 [64].

For maritime transport, the IMDG Code is the primary regulatory framework. The China Maritime Safety Administration (MSA) ensures compliance with the IMDG Code, classifying LIBs as Class 9 dangerous goods and outlining specific packaging instructions. These instructions involve using packaging that can withstand the rigors of maritime transport, including exposure to moisture and mechanical stresses. Also, the MSA mandates the use of appropriate hazard labels and comprehensive shipping documentation, similar to other transport modes. Road transport of LIBs in China is regulated under national standards that align with the European Agreement concerning the ADR. Specifically, the regulations require robust packaging to prevent short circuits and physical damage, detailed labeling to inform handlers and emergency responders of the risks, and comprehensive documentation to ensure transparency and traceability throughout the transport process.

Chinese authorities conduct rigorous checks to ensure that all aspects of LIB transport, from packaging to documentation, adhere to the specified standards. This includes inspections at ports, airports, and other transportation hubs, where authorities verify that batteries are properly labeled, packaged, and documented. The enforcement of these regulations by Chinese authorities, coupled with mandatory training programs for personnel, ensures the safe and efficient movement of LIBs within and beyond China's borders.

Adhering to local regulations and implementing best practices in China is essential for maintaining the safety and compliance of transporting LIBs. Chinese transportation industry standards for ships carrying lithium batteries, in accordance with the current domestic laws and regulations of the MSA, follow the Provisions on the Safety Supervision and Administration of Dangerous Goods on Ships (Order No. 11 of the Ministry of Transport, 2018), which is consistent with UN 38.3 and the IMDG Code [65]. The technical requirements are based on the Marine Dangerous Cargo Container Safety Technical Requirements (GB 40163) and Port Operation Safety Requirements Part 3: Dangerous Cargo Container (GB 16994.3).

Due to the more stringent regulations of the IATA DGR for the transportation of lithium batteries, Chinese airlines must follow and enforce the rules rigorously. Additionally, China's airlines are subject to the requirements for UN tests, parameter limits, packaging, marking, labeling, and transport documents of the ICAO TI, the IATA DRG, and the CCAR-276-R1. In most cases, airlines permit up to two spare LIBs (more than 100 Wh, but not exceeding 160 Wh) [66]. Except for the above common provisions, each Chinese airline also has its own requirements (Table 12.12).

Table 12.12 Safety technical requirements for container-loaded vehicles.

Transportation category	Mandatory standard	Recommended standard	Integrity of the standard
Internal trade	Technical Requirements for Container Safety of Marine Dangerous Cargo Container (GB 40163-2021)	Technical Requirements for Transportation of Container, Commercial Vehicles (JT/T 1434-2022)	Vehicle loading scheme, transportation support requirements, binding system, fixing equipment requirements, binding system fixing method, safety clearance, and other contents are specified in detail in the recommended standards
External trade	Technical Requirements for Container Safety of Marine Dangerous Cargo Container (GB 40163-2021)	No provision	No detailed provision

Courier services in China handle LIB shipments with heightened safety protocols and regulatory compliance compared to other transportation methods. Courier services like DHL, FedEx, USPS, and UPS have established specific protocols for shipping LIBs that differ from standard freight practices.

To ship lithium batteries internationally, DHL requires full adherence to IATA rules and regulations on the transport of lithium batteries and UN Packing Instructions PI 965 to PI 970. Producers and successive distributors of batteries or cells need to provide a test summary that follows UN 38.3. USPS permits specific quantities of lithium batteries (when placed in the device they are to operate) to be shipped internationally if permitted by the destination country. However, they cannot ship to the postal operators in Germany and Italy, which do not accept items containing lithium batteries [67].

FedEx, which recently acquired TNT Express, requires full compliance with IATA and ICAO dangerous goods regulations before shipping lithium batteries. Also, packaging needs to comply with Section II IATA shipping requirements [67]. FedEx prohibits the transport of standalone lithium batteries packed separately, requiring that they be contained within equipment or packed with equipment.

UPS requires that all lithium batteries shipped by air are required to be treated as dangerous goods. The courier also requires shippers to consent to a UPS dangerous goods service agreement before shipping lithium batteries without equipment by air [67].

DHL mandates that lithium batteries adhere to the IATA DGR. This, of course, includes detailed packaging instructions and the necessity for batteries to have passed the UN 38.3 testing requirements.

Unlike other modes of transport, courier services often impose stricter restrictions on the shipment of LIBs. This contrasts with ground transport, where such restrictions may not be as stringent. Furthermore, courier services typically require shippers to have preapproved accounts, ensuring that only trained personnel handle the preparation and shipment of lithium batteries. Another significant difference is the documentation process. Courier services require comprehensive documentation for lithium battery shipments, including a dangerous goods declaration that details the type of cargo, associated hazards, and compliance with relevant regulations.

While international regulations, such as the IATA DGR and IMDG Code, consider environmental impacts, Chinese regulations often place a stronger emphasis on these aspects. Differences include more rigorous enforcement of state-of-charge limits, enhanced packaging requirements, detailed documentation, and stricter environmental considerations. These differences reflect China's commitment to maintaining high safety standards and ensuring the safe and efficient transport of LIBs within and beyond its borders. The Chinese government has implemented tighter controls on the chemical composition of LIBs, aiming to reduce environmental harm. China's approach includes additional reporting requirements under the Ministry of Ecology and Environment (MEE), which go beyond the basic international standards. These requirements ensure that manufacturers and shippers provide detailed information about the environmental impact of the batteries, facilitating better monitoring and control.

12.4.3 Transportation Regulations in Europe

Road transportation is the preferred method of transporting lithium batteries within Europe. Being geographically well-connected through its road systems, Europe has the advantage of using inexpensive road transportation that also offers less rigid transportation regulations compared to air and sea. The United Nations Economic Commission for Europe (UNECE) sponsors the European Agreement Concerning the International Carriage of Dangerous Goods by Road (ADR) to increase the safety of international transport of dangerous goods (including waste products) [68]. European Norms (EN) and regulations for Li-ion battery transportation mostly follow the international norms provided by UN regulations. Within Europe, the ADR has been signed by 46 countries, and it defines the transportation of dangerous goods similarly to UN class 9, following similar regulations for safety, pretesting, and packaging.

The ADR outlines thorough packaging, labeling, and documentation requirements aimed at minimizing risks during transit. It stipulates that LIBs must be enclosed in durable outer packaging capable of withstanding physical impacts and preventing short circuits. The RID, which governs the transport of dangerous goods by rail in Europe, mirrors ADR in many respects. The alignment between the ADR and RID ensures that LIBs can be transported seamlessly across road and rail networks within Europe, maintaining consistent safety standards.

The European Union has established regulations to improve the safety of transporting LIBs. Notably, EU Directive 2008/68/EC [68] consolidates the ADR, RID, and ADN into a single regulatory framework to ensure consistent high safety standards across all modes of inland transport within the EU. Additionally, the European Chemicals Agency (ECHA) imposes requirements under the Registration, Evaluation, Authorization and Restriction of Chemicals (REACH) regulation for transporting LIBs, focusing on their chemical composition and potential environmental impacts. Manufacturers and shippers must ensure that all components of LIBs comply with REACH requirements, thereby enhancing safety and environmental protection.

Besides the battery safety, pretransportation testing, and transportation regulations described in the EN and IEC standards, there are a few other regulations in practice for battery transportation in Europe. Lithium battery safety and transport regulations are largely covered by two EN standards: EN 50272 and EN 62281 [1]. These test requirements are comparable to the tests defined in UN 38.3; however, these standards do not cover any guidelines for the recycling or disposal of lithium batteries and cells. Extensive testing is also necessary according to IEC 62281 and UN 38.3 cell-level testing, IEC 62619 system-level testing, sensor failure mode testing, and functional safety testing (e.g., emergency stop functions, alarms and shutdowns, BMS temperature protection, and overvoltage protection and undervoltage protection).

Det Norske Veritas has developed strict guidelines for implementing LIBs on ships, which dictate the design principles for battery spaces and battery systems. Specifically, battery spaces require risk analyses, fire integrity management, fire control plans, fire extinguishing system documentation, fire detection arrangement, fire detection systems, ducting diagrams, ventilation systems, and gas detection systems.

Cargo fires and cargo flight accidents led to the banning of Li-ion battery transportation through air cargo. IAG, Emirates, and others do not accept booking of UN 3480 Li-ion battery cargo, although this restriction does not apply to batteries packed or contained in the equipment (under UN 3481 and UN 3091). Emirates has a strict policy for carriage of LIBs and polymer batteries, under PI 965. Importantly, these cases show that air transportation of LIBs is getting more difficult. Within the IMDG Code, requirements for lithium batteries are included in Special Provisions 38, 240, 312, and 363. The amendments of the maritime-related SP 900 series for safety matters are especially applicable to the transport of lithium batteries contained in vehicles, engines, and machinery. It is important to note that the testing requirements in UN 38.3 do not apply to production runs consisting of 100 or fewer cells and batteries or to preproduction.

For batteries ≤100 Wh, there is no limit on the quantity per package, and it must be marked with the UN number and labeled with Class 9 hazard labels [7]. For batteries >100 Wh, the maximum quantity is 333 kg per transport unit (e.g., a truck and its trailer) [49]. If exceeded, further requirements for the vehicle, equipment, and driver apply. A dangerous goods transport document is required for batteries >100 Wh, listing the UN number, proper shipping name, hazard class, packaging group, number and type of packages, total quantity, and emergency contact information [46]. For LIBs or devices with LIBs >100 Wh, a weight limit of 333 kg (battery weight) normally applies in connection with reduced requirements on transport devices (e.g., lorry equipment and driver's qualification). Damaged or defective cells or batteries, whether they have been identified as noncritical or critical, are forbidden for air transport (according to IATA DGR SP A154).

Europe has established specific regulations governing the transportation of LIBs, which are complemented by national legislation in individual EU member states in adherence to international standards. UN, ADR, ADN, IMGO, and IATA transport regulations are used in Europe, while, in practice, EN-IEC standards are also used for lithium battery transportation. Safety regulations during transportation, goods categories, and handling are defined by various authorities like the UN, EN, IATA, and IMO, and norms cover testing, safe parameter limits, and packaging. In addition to ADR regulations and special provisions, European countries have multilateral agreements (M285, M292, M294, M295, and M296) regarding various aspects of transportation and packaging of damaged, recyclable, prototype, and low-production-run lithium batteries [1].

The M285 agreement facilitates relaxed packing and sizing of lithium batteries when contained within equipment that guarantees safety during disposal or recycling. M292 deals with damaged or defective Lithium-ion and lithium metal batteries identified as nonconforming to the provisions of the UN Manual of Tests and Criteria, and according to SP 376, they may not be carried unless approved by the competent authority of an ADR-contracting party. The M294 agreement pertains to SP 310 for the preproduction of large LIBs exceeding 100 kg and not tested under UN subsection 38.3. However, a large preproduction battery can be transported in a strong, nonapproved package that meets the specified packaging requirements in the agreement.

M295 states that prototype or damaged cells/batteries contained in the equipment should be transported according to SP 376 and packaged according to the packing instructions P908 or LP904. Additionally, the transport document must include the phrase "carriage agreement under the terms of section 1.5.1 of the ADR (M295)". The M296 document defines the carriage of hybrid lithium batteries containing both primary lithium metal cells and rechargeable Li-ion cells. These batteries are not designed to be externally charged; rechargeable LIBs may only be charged from the primary lithium cells. The entire battery must undergo testing according to UN 38.3, and overcharging is precluded by design; there are weight and capacity constraints while transporting such a hybrid battery. The contents of lithium metal included in all batteries must not exceed 1 g, and the total capacity of lithium-ion cells is limited to 20 Wh.

DHL Global Forwarding UK is recognized for its operational excellence in this area, confirming compliance with safe packing, handling, and shipment of lithium batteries by air. It has become the first organization in the United Kingdom to be certified as an International Air Transport Association (IATA) Center of Excellence for Independent Validators Lithium Batteries (CEIV Li-batt) [69]. The certification covers six sites: Birmingham, Glasgow, London, Luton, Manchester, and Purfleet.

The CEIV Li-batt certification was introduced in response to concerns from within the aviation industry about a lack of awareness of the danger that lithium batteries pose, plus an increasing number of incidents of intentionally mis-declared and noncompliant lithium batteries in cargo shipments [69]. Importantly, it establishes baseline standards to improve the level of competency and quality management in the handling and carriage of lithium batteries throughout the supply chain.

Local authorities in Europe play a crucial role in interpreting and enforcing the ADR for LIBs. These regulations classify lithium batteries as dangerous goods, necessitating strict compliance with safety standards during transport. Consequently, local authorities are responsible for ensuring that shippers, carriers, and other stakeholders adhere to these regulations, which include specific packaging, labeling, and documentation requirements. Enforcement agencies, such as transport safety authorities and customs officials, conduct inspections and audits to ensure compliance with ADR regulations. However, the ADR allows for some flexibility, enabling authorities to adapt regulations to regional safety concerns or specific circumstances. When local regulations differ from ADR guidelines, authorities typically assess the nature of the discrepancy to determine the appropriate course of action. Thus, if local regulations impose stricter packaging or labeling requirements than those outlined in the ADR, local authorities may enforce these additional measures to enhance safety.

12.4.4 Transportation Regulations in South Korea

The transportation of LIBs in South Korea is subject to a comprehensive regulatory framework that prioritizes safety and compliance with international standards. Both South Korean and international regulations for the transport of LIBs are fundamentally based on the United Nations (UN) Model Regulations. South Korea adheres to these international standards through its national regulatory bodies like the Ministry of Land, Infrastructure and Transport (MOLIT), the Korea Maritime Safety Authority (KMSA), the Korea Agency for Technology and Standards (KATS), and the Korea Customs Service (KCS). There are notable differences in the implementation, enforcement, and specific requirements between South Korean regulations and those set by international bodies. South Korean regulations often provide more detailed guidance and enforcement measures to address local safety concerns and logistical challenges.

For air transport, the International Civil Aviation Organization (ICAO) Technical Instructions form the basis of South Korea's regulatory framework. These guidelines include strict controls on the state of charge (SOC) for LIBs, typically limiting them to 30% to reduce the risk of thermal runaway during flight. In addition to compliance with ICAO guidelines, MOLIT enforces additional documentation and packaging requirements. Shippers must provide detailed shipping papers, including a declaration of dangerous goods that certifies compliance with all relevant regulations.

The International Maritime Dangerous Goods (IMDG) Code governs the maritime transport of LIBs, and the Korea Maritime Safety Authority (KOMSA) ensures compliance with this code within South Korean ports and waters. KOMSA also pursues efficient domestic adoption and implementation of international treaties by responding to organizations such as the IMO and global cooperation projects to strengthen international maritime safety systems [70]. South Korean regulations often include additional national standards to address specific maritime conditions in the region. The KOMSA mandates enhanced packaging and labeling requirements to ensure that all packages are clearly marked and easily identifiable. Regular inspections at ports ensure that these standards are consistently met and that any noncompliance is addressed promptly.

For the transport of LIBs by road and rail, South Korea aligns with the ADR and the RID. These frameworks provide detailed requirements for packaging, labeling, and documentation to ensure safe transport. The Ministry of Land, Infrastructure and Transport (MOLIT) oversees the enforcement of these regulations, with additional national provisions tailored to South Korea's specific logistical and infrastructural challenges, which require enhanced securement of batteries during transport to prevent shifting and potential damage. Also, there are specific requirements for the transport of large quantities of LIBs, including route planning and the use of specialized transport vehicles to minimize risks.

KATS works under the Ministry of Trade, Industry and Energy (MOTIE) to regulate the testing, inspection, and certification processes. This ensures that only products meeting strict safety criteria can get KC certification, and the KC mark is a key prerequisite for lithium battery logistics in South Korea. Lithium batteries transported in South Korea should pass the KC 62133 test standard [71], which is harmonized with IEC 62133, and they should be identified by the KC mark. The KC 62133 standard was similar to the 1st edition of IEC 62133 and covered: (a) lithium secondary cells equal to or more than 400 Wh/L, (b) batteries assembled by the cells for use in portable devices, and (c) cells and batteries for use in portable devices with a navigation function. These standards were identical except the storage time for the temperature cycling test was only required to be for 24 hours after testing [72]. However, in September 2014, the new KC 62133 standard was established with added coverage for batteries with <400 Wh/L per volume. The new KC 62133 requires testing of all LIBs irrespective of their energy density value (Wh/L). Accordingly, the old standard was abolished in August 2015, and the requirement of batteries with <400 Wh/L per volume energy density became mandatory in the transportation process.

When a battery is already KC 62133 certified, some additional requirements for the battery system might still apply. A certified battery marked with the KC 62133 label does not mean that any shipping company accepts KC 62133 instead of UN 38.3, because the latter is a requirement directly from IATA for the safe transport of lithium batteries. However, in most cases, the UN 38.3 report would be provided by battery manufacturers prior to KC 62133 [73, 74].

Only certified lithium batteries with the KC mark can be transported in South Korea, and in-country testing is mandatory. Usually, sample batteries are required to be tested for performance and safety, which can be conducted by an authorized laboratory collaborating with a South Korean standard testing institute, such as the Korea Testing Laboratory [71]. KC certification tests for capacity, charging cycles, and durability are under South Korean standards, and the time required to obtain KC certification can vary depending on factors such as the battery design's complexity, the documentation's completeness, and the certification body's workload. Labeling requires specific details in Korean, including the product name, model name, designation, nominal voltage, rated capacity, manufacturer, customer service number, and country of manufacture.

Regular inspections and audits are conducted to verify that all aspects of LIB transport, from packaging to documentation, meet the specified standards. Penalties for noncompliance can be severe, reflecting the importance placed on safety and adherence to regulations.

The Civil Aviation Safety Authority of South Korea enforces its guidelines thoroughly. South Korean regulations may require more detailed shipping papers and a declaration of dangerous goods, that certifies compliance with all relevant regulations. Special provisions are in place for transporting damaged or defective batteries, which are often restricted to cargo-only aircraft to enhance safety. South Korea's Maritime Safety Authority ensures compliance with the IMDG Code but includes additional national standards to address specific maritime conditions. South Korean regulations may mandate enhanced packaging that can withstand the specific environmental conditions of Korean ports, which often experience extreme weather. The Korea Maritime Safety Authority's inspection protocols at ports can be rigorous, ensuring all packaging and labeling requirements are strictly met before allowing the batteries to be shipped. The ADR and RID provide the primary framework for road and rail transport of LIBs in Europe, so South Korea aligns its national standards for road and rail transport with these frameworks. However, regulations often go further in terms of enforcement and specific requirements for the transport of large quantities of LIBs. This level of detail, enforcement, and scrutiny can exceed the general international guidelines.

Different airlines in South Korea have their own norms for the transportation of lithium batteries. Most South Korean airlines, such as Korean Air, Air Busan, and Asiana [75–77], clearly regulate that up to five spare lithium batteries are allowed as carry-on only and must be protected to prevent short circuits. Each battery's energy cannot exceed 100 Wh; however, a maximum of two spare lithium batteries are permitted as a carry-on if they exceed 100 Wh, but not 160 Wh, only with the approval of the operator. Specifically, the two batteries of 100 Wh to 160 Wh each can be included with the previously mentioned five spare batteries. T'way Air permits up to four spare batteries that are less than 100 Wh as carry-on, and up to five batteries not exceeding 100 Wh mounted on the device are permitted as check-in baggage [78].

These rules are different from FAA regulations, which limit spare lithium batteries (more than 100 Wh, but not exceeding 160 Wh) to two per passenger. Even though there are no strict rules on specific battery products in South Korean airports, some products are still banned, such as portable chargers restricted as spare batteries in China and e-cigarettes in Taiwan [75–77]. However, lithium batteries within 160 Wh for consumer electronics and medical devices are allowed onboard.

Air Busan has more specific restrictions for battery-powered devices. For example, small devices equipped with lithium batteries are forbidden in either carry-on or check-in baggage. Yet, medical devices powered with lithium batteries, such as wheelchairs, are permitted with some restrictions. The batteries must be removed from the wheelchair and are only permitted in carry-on baggage with the approval of the operator. For devices operated by one battery, a maximum of two batteries with a rating of 300 Wh are permitted, whereas for devices with two batteries, a maximum of four batteries not exceeding 160 Wh are permitted [76].

Air cargo companies such as Korean Air accept shipments of lithium batteries according to UN 3480, 3090, 3481, and 3091 (PI 965 for lithium metal batteries only, PI 968 for LIBs only, and PI 966, 967, 969, and 970 for batteries with equipment), with SOC not more than 30% and a net weight limit range of 5 kg to 35 kg [79]. South Korean transport companies, including CJ Korea, Korea Post, Hanjin, Lotte, and KGB, covering air, road, and sea shipping, do not list specific prohibited items on their official websites. Products equipped with lithium batteries cannot be shipped by the South Korean post office EMS, but they can be shipped by FedEx and DHL with up to two products per box and a total of two boxes. Thus, products exceeding 100 watts (excluding electric bicycles) cannot be shipped [80]. South Korean Office Post does not mention specific requirements for the batteries if they are equipped in the appliance. However, for electric appliances equipped with batteries to be transported in South Korea, they must be affixed with a specific label (i.e., the KC certification).

While international regulations such as the IATA DGR and the IMDG Code consider environmental impacts, South Korean regulations often place a stronger emphasis on these aspects.

Notably, the South Korean government has implemented stricter controls on the chemical composition of LIBs, aiming to reduce environmental harm. Seoul Metropolitan Government and Provincial Environmental Regulations have additional requirements, such as the Korean Environment Corporation (KECO) Act on Resource Circulation of Electrical and Electronic Equipment, for packaging to reduce the environmental impact within the city limits, while other provinces in South Korea under the Provincial Environmental Regulations may also have their own regulations regarding the disposal and recycling of packaging materials [81, 82]. South Korea's Ministry of Environment implements national environmental laws [82], ensuring that manufacturers and shippers provide detailed information on the environmental impact of the batteries. Article 35 of the Constitution of the Republic of Korea firmly asserts the environmental rights of the citizens of South Korea and mandates the corresponding obligations of the State, laying the groundwork for the Framework Act on Environmental Policy. Manufacture of packing materials made of carton packs, metal cans, or plastic (synthetic resin foam excluded) whose previous year revenues are KRW 1 billion or higher, and importers whose previous year revenues are KRW 0.3 billion or higher under Extended Producer Responsibility (EPR) requirements are mandated to recycle a certain amount of waste from products or packaging materials [83]. The Act on Resource Circulation of Electrical and Electronic Equipment mandates that producers and importers of such equipment must collect and transfer damaged and defective batteries to a licensed recycling business [84].

Lithium-ion batteries are demonstrably safer for transport at a lower state of charge. They are considered more stable and less risky when partially charged, typically with the SOC around 30% or less [11]. This is reflected in international regulations that typically restrict the SOC for the transport of LIBs to a maximum of 30%. In Japan, the Ministry of Land, Infrastructure, Transport and Tourism (MLIT) and the Ministry of Economy, Trade and Industry (METI) have established regulations for transporting LIBs. In Canada, Transport Canada has established regulations for transporting LIBs under the Transportation of Dangerous Goods Regulations (TDGR). In Australia, the Civil Aviation Safety Authority (CASA) and the Australian Dangerous Goods Code (ADG Code) govern the transportation of LIBs. In India, the Directorate General of Civil Aviation (DGCA) and the Ministry of Shipping have established regulations for transporting LIBs [1].

12.5 Global Regulations on Lithium Battery Disposal

Lithium-ion batteries are classified as dangerous goods or hazardous materials under most international regulations due to their potential fire and explosion risks. This classification impacts how they must be handled, transported, and disposed of at end-of-life. Lithium battery disposal is heavily regulated worldwide due to hazardous waste concerns. For example, recycling is mandated in many regions, with specific recovery rates and recycled content requirements. Proper handling, labeling, and shipping procedures must be followed to comply with international dangerous goods regulations. In May 2023, the EPA released a guidance memorandum addressing the hazardous waste status of LIBs under the Resource Conservation and Recovery Act (RCRA) [85]. Specifically, international shipments of lithium batteries must comply with RCRA requirements for waste exports and imports, including prior notification and consent. However, entities that accumulate more than 5000 kg of universal waste at any time are subject to more stringent requirements [85].

In the United States, the EPA considers most LIBs to be hazardous waste when disposed of, due to ignitability and reactivity characteristics. The US Universal Waste Rule promotes safe recycling and disposal of lithium batteries, though some states have additional regulations. Lithium batteries can be managed as "universal waste" under the 40 CFR Part 273 streamlined regulations [85]. Generators must train employees, label containers, limit accumulation time, and comply with shipping requirements. As the EPA notes, most states have delegated authority to implement their own RCRA programs and can impose more stringent requirements. The contents of LIBs are likely to be subject to hazardous waste regulations, and, as noted by the National Renewable Energy Laboratory (NREL) [86], "Federal hazardous waste laws and regulations are the most stringent and govern the generation, handling, storage, treatment, domestic or international transport, and disposal of hazardous wastes." Accordingly, generators, recyclers, and others who manage LIBs at end-of-life should carefully evaluate potential state requirements [85].

Regulations on battery recycling, particularly in the EU, make recycling a must, but there are no clear regulations for LIB recycling in the United Kingdom [87]. The EU has a comprehensive regulatory framework for lithium battery recycling, setting specific collection rate targets and material recovery efficiency goals. EU Directive 2006/66/EC, also known as the EU Battery Directive, and its amending EU Regulation 2019/1020 [86] mandate that all waste lithium batteries be collected free of charge for end-users, regardless of chemistry or origin. The Battery Directive requires lithium batteries to be collected and recycled, with recovery rates of 50% for lithium by 2025 and 80% by 2030 [87]. It also mandates that all waste lithium batteries be collected free of charge for end-users, regardless of chemistry or origin. EU regulations include a requirement for battery passports that help inform consumers about the content, origin, and environmental impact of batteries. Production and reuse standards for batteries are established within the framework of newly proposed rules, and content standards mandate 12% recycled lithium in new batteries by 2035 [87].

Asian countries have differing levels of emphasis on and methods of implementation for LIB recycling, particularly China and Japan. Regulations in China, the largest market for electric cars, have been expanded to cover many aspects of production, collection, and recycling, resulting in a comprehensive used battery management system [87]. In addition, there are many laws in China that prevent the import of waste from other countries. In Japan, a nonprofit organization, the Japan Portable Rechargeable Battery Recycling Center (JBRC), manages Japanese battery recycling [86]. There is also an active program to reuse batteries for home emergency power [88].

12.6 Packaging and Safety Best Practices for Shipping Lithium-Ion Batteries

The lithium battery industry is prioritizing the improvement of safety through advancements in battery design, chemistry, and internal safety features, rather than focusing solely on enhanced packaging. It is crucial to develop safe battery technologies, as they will not only enhance safety during transportation but also throughout the entire life cycle, including storage, charging, and use. Both SAE G-27 and the UN Informal Working Group have inspired packaging companies to develop innovative packaging solutions with design features that enhance safety, minimize risk, and contain thermal effects [89]. Universities, government labs, and lithium battery manufacturers are all racing to find the next innovation that will make lithium batteries safer and less volatile.

A new focus on packaging performance has been initiated in recent years based on recommendations made by ICAO during three multidisciplinary meetings and by the United Nations Transport of Dangerous Goods Sub-Committee (UN TDGSC) [89]. ICAO members affirmed that a fire involving significant quantities of lithium batteries (UN 3090 and UN 3480) could exceed the fire suppression capability of the aircraft and could lead to a catastrophic failure of the airframe. They concluded that there is a need to develop performance-based standards on the principle that hazardous effects from the batteries would be contained within the package. Many regulators and safety experts believe [89] that the current requirements are not enough and more needs to be done to mandate appropriate packaging requirements for the transport of lithium batteries, particularly air transport. As a result of the concerns expressed by regulators, in 2016 the SAE formed the SAE G-27 Lithium Battery Packaging Performance, which is a technical committee in SAE's General Projects Systems Group, tasked with developing a minimum performance package standard for the safe shipment of lithium batteries as cargo on aircraft [89].

Adherence to key safety practices is essential for the secure shipment of batteries. When preparing batteries for shipping, it is imperative to assess the Watt-hour rating, which serves as an indicator of the battery's energy capacity. A heightened level of caution should be exercised with batteries featuring higher Watt-hour ratings. Additionally, the SOC necessitates scrutiny; accordingly, IATA regulations stipulate that the SOC should not exceed 30% for air transport, thereby mitigating the likelihood of thermal runaway. Training employees in proper battery handling is imperative, and implementation of no-smoking policies for battery shipments is essential. Also, it is vital to establish an emergency response plan for potential incidents such as leaks or fires, encompassing battery-specific response protocols and pertinent fire suppressants.

12.7 Summary

Batteries will be transported and stored numerous times in their life cycle, including in cell manufacturing, battery pack manufacturing, assembly of the product incorporating the battery, replacement of a defective battery, and recycling of a battery. Compliance with the UN Model Regulations, including test requirements, specific packing instructions, and limitations on state of charge and environmental conditions, is essential for implementing safe transportation practices.

UN 38.3 regulations involve rigorous testing requirements at the cell, pack, and product levels. At the cell level, tests evaluate stability, thermal management, and resilience to abuse. Pack-level tests focus on mechanical integrity, electrical performance, and resistance to environmental factors. Product-level tests verify mechanical robustness, electrical safety, and compliance with labeling requirements. Ensuring both effective thermal management and mechanical integrity are essential for the safe transportation and storage of batteries. Monitoring systems, cooling mechanisms, and insulation materials are utilized to maintain optimal operating temperatures, preventing performance degradation or safety hazards due to temperature fluctuations. Robust design and testing guarantee that battery packs endure transportation stresses such as vibrations, impacts, and compression forces without damage.

The number of regulations and the differences in regulations make the logistics of battery transport time-consuming and costly. There is definitely a need to consolidate norms and regulations; thus, safety testing requirements for battery shipments need to not only cover cells but also focus on the batteries and the products. Also, strict regulations that address infrastructure requirements at the transport terminals for storing battery consignments are necessary. It is recommended that battery manufacturers, distributors, and organizations involved with batteries in their supply chain follow UN test criteria for cells, packs, and damaged and low-production-run products that include batteries. However, in most cases, low-production-run and damaged cells are exempted from the test criteria. Companies should then check the appropriate UN, IATA, ADR, and IMO regulations for packaging, marking, and labeling requirements, because regulations often have their own unique requirements pertaining to weight, size, marking, and labeling per the mode of transportation.

Finally, there are two compliance resources that are helpful when transporting batteries. The first is "The Lithium Battery Guide for Shippers—A Compliance Tool for All Modes of Transportation." This guide was prepared to assist a shipper to safely package lithium cells and batteries for transport by all modes of transportation according to the latest regulatory requirements. Each distinct shipping guide in this document refers to the regulatory requirements for a specific lithium cell or battery type, configuration, and size [90].

The second compliance resource is the "Interactive Guide to Shipping Lithium Batteries" published by the FAA in January 2023. This easy-to-use guide provides awareness of the requirements for transporting lithium batteries in the ICAO 2023–2024 Edition of the Technical Instructions (Doc 9284) [91].

References

1 Huo, H., Xing, Y., Pecht, M. et al. (2017). Safety requirements for transportation of lithium batteries. *Energies* 10 (6): 793. https://doi.org/10.3390/en10060793.

2 Asef, P., Milan, M., Lapthorn, A., and Padmanaban, S. (2021). Future trends and aging analysis of battery energy storage systems for electric vehicles. *Sustainability* 13 (24): 13779. https://doi.org/10.3390/su132413779.

3 Casals, L.C., García, B.A., Aguesse, F., and Iturrondobeitia, A. (2015). Second life of electric vehicle batteries: Relation between materials degradation and environmental impact. *International Journal of Life Cycle Assessment* 22 (1): 82–93. https://doi.org/10.1007/s11367-015-0918-3.

4 Sharma, D.K. and Prabhakar, A. (2021). A review on air cooled and air centric hybrid thermal management techniques for Li-ion battery packs in electric vehicles. *Journal of Energy Storage* 41: 102885. https://doi.org/10.1016/j.est.2021.102885.

5 Yi, Z. and Bauer, P.H. (2017). Effects of environmental factors on electric vehicle energy consumption: a sensitivity analysis. *IET Electrical Systems in Transportation* 7 (1): 3–13. https://doi.org/10.1049/iet-est.2016.0011.

6 Fanoro, M., Božanić, M., and Sinha, S. (2022). A review of the impact of battery degradation on energy management systems with a special emphasis on electric vehicles. *Energies* 15 (16): 5889. https://doi.org/10.3390/en15165889.

7 Coombes, R. (2024). Shipping Lithium Batteries: Transporting lithium batteries by air, sea and road. Beginners guide to UN3480, UN3481 and IATA regulations, GWP Group, February 2024. https://www.gwp.co.uk/guides/un3480-regulations/.

8 Pipeline and Hazardous Materials Safety Administration (PHMSA), Shipping batteries safely by air [Online]. http://phmsa.dot.gov/staticfiles/PHMSA/DownloadableFiles/Files/SafeTravel/PHMSA_battery_guide.pdf (accessed 18 October 2015)

9 International Civil Aviation Organization (ICAO) (2011). *Annex 18, The Safe Transport of Dangerous Goods by Air*, 4e. Montreal, QC, Canada: ICAO.

10 United Nations (2024). Recommendations on the transport of dangerous goods: Manual of tests and criteria, April 2024. https://unece.org/fileadmin/DAM/trans/publications/ST-SG-AC10-11-Rev5-english.pdf.

11 International Air Transport Association (2024). 2021 Guidance Document - Battery Powered Cargo Tracking Devices/Data Loggers Revised for the 2021 Regulations - Revision 1, April 2024. https://www.iata.org/contentassets/05e6d8742b0047259bf3a700bc9d42b9/lithium-battery-guidance-document-2021-for-pharma-en.pdf.

12 European Commission (2024). Guidance document: Transport of lithium batteries by road - recommendations, April 2024. https://ec.europa.eu/transport/sites/transport/files/legislation/swd20190146.pdf.

13 International Electrotechnical Commission (2024). IEC 62133: Safety requirements for portable sealed secondary lithium cells, and for batteries made from them, for use in portable applications, April 2024. https://www.iec.ch/standards/62133.htm.

14 Federal Aviation Administration, Lithium battery safety resources. https://www.faa.gov/hazmat/resources/lithium\LY1\textbackslash_batteries.

15 United Nations (UN) (2019). *Recommendations on the Transport of Dangerous Goods, Manual of Tests and Criteria*, 7e, 420–433. New York, NY, USA; Geneva, Switzerland: United Nations Publications.

16 Yang, T. What's the different test procedures between UN 38.3 AND IEC 62213, LinkedIn. https://www.linkedin.com/pulse/whats-difference-test-produce-between-un-383-iec-62133-tim-yang/ (accessed 20 July 2024).

17 European Chemicals Agency (2017). Guidance on requirements for substances in articles version 4.0, June 2017. https://doi.org/10.2823/470616.

18 European Commission (2024). Mobility and transport - European Commission, April 2024. https://ec.europa.eu/transport/sites/transport/files/legislation/swd20190146.pdf.

19 Wang, Q., Ping, P., Zhao, X. et al. (2012). Thermal runaway caused fire and explosion of lithium-ion battery. *Journal of Power Sources* 208: 210–224. https://doi.org/10.1016/j.jpowsour.2012.02.038.

20 Ruiz, V., Pfrang, A., Kriston, A. et al. (2018). A review of international abuse testing standards and regulations for lithium-ion batteries in electric and hybrid electric vehicles. *Renewable and Sustainable Energy Reviews* 81: 1427–1452.

21 Lin, J., Liu, X., Li, S. et al. (2021). A review on recent progress, challenges and perspective of battery thermal management system. *International Journal of Heat and Mass Transfer* 167: 120834. https://doi.org/10.1016/j.ijheatmasstransfer.2020.120834.

22 Wu, W., Wang, S., Wu, W. et al. (2019). A critical review of battery thermal performance and liquid based battery thermal management. *Energy Conversion and Management* 182: 262–281. https://doi.org/10.1016/j.enconman.2018.12.051.

23 Arora, S. (2018). Selection of thermal management system for modular battery packs of electric vehicles: a review of existing and emerging technologies. *Journal of Power Sources* 400: 621–640. https://doi.org/10.1016/j.jpowsour.2018.08.020.

24 Arora, S., Shen, W., and Kapoor, A. (2016). Review of mechanical design and strategic placement technique of a robust battery pack for electric vehicles. *Renewable and Sustainable Energy Reviews* 60: 1319–1331. https://doi.org/10.1016/j.rser.2016.03.013.

25 Bell (2019). UN38.3 standard requirements for lithium battery production, December 14, 2019. https://belltestchamber.com/un38-3-standard-requirements-for-lithium-battery-production.html.

26 SAE International (2013). SAE J2929: 2013 Safety standard for electric and hybrid vehicle propulsion battery systems utilizing lithium-based rechargeable cells.

27 International Air Transport Association (2020). 2020 lithium battery guidance document transport of lithium metal and lithium-ion batteries revised for the 2020 regulations. https://www.iata.org/contentassets/05e6d8742b0047259bf3a700bc9d42b9/lithium-battery-guidance-document-2020.pdf.

28 International Air Transport Association, Lithium battery shipping regulations (LBSR). https://www.iata.org/en/publications/store/lithium-battery-shipping-regulations/.

29 International Maritime Organization (2014). *International Maritime Dangerous Goods (IMDG) Code (Volume I)*. Geneva, Switzerland: International Maritime Organization.

30 Zhang, J., Zhang, L., Sun, F., and Wang, Z. (2018). An overview on thermal safety issues of lithium-ion batteries for electric vehicle application. *IEEE Access* 6: 23848–23863. https://doi.org/10.1109/ACCESS.2018.2824838.

31 Friel, D.D. (2016). Electronic options for lithium-ion batteries, Elsevier eBooks, pp. 361–386, January 2016. https://doi.org/10.1016/b978-0-444-59513-3.00016-9.

32 Eurofins E&E North America (2023). UN/DOT 38.3 Transportation testing required for lithium battery safety during shipping, October 29, 2023. Retrieved April 3, 2024. https://metlabs.com/battery/un-38-3-transportation-testing-required-for-lithium-battery-safety-during-shipping/.

33 Remenick, B.M. (2009). Lithium battery transportation test requirements, 2009 IEEE Symposium on Product Compliance Engineering, Toronto, ON, Canada, pp. 1–12. https://doi.org/10.1109/PSES.2009.5356016.

34 Ma, S. et al. (2018). Temperature effect and thermal impact in lithium-ion batteries: a review. *Progress in Natural Science: Materials International* 28 (6): 653–666. https://doi.org/10.1016/j.pnsc.2018.11.002.

35 Brandt, K. and Garche, J. (2019). General battery safety considerations, Electrochemical Power Sources: Fundamentals, Systems, and Applications, pp. 1–19. https://doi.org/10.1016/b978-0-444-63777-2.00001-3

36 Chen, Y. et al. (2021). A review of lithium-ion battery safety concerns: the issues, strategies, and testing standards. *Journal of Energy Chemistry* 59: 83–99. https://doi.org/10.1016/j.jechem.2020.10.017.

37 Martinez, M., Soto, F.A., and Balbuena, P.B. (2015). Effect of the electrolyte composition on SEI reactions at Si anodes of Li-ion batteries. *Journal of Physical Chemistry* 119 (13): 7060–7068. https://doi.org/10.1021/acs.jpcc.5b01228.

38 Larcher, D. and Tarascon, J.-M. (2014). Towards greener and more sustainable batteries for electrical energy storage. *Nature Chemistry* 7 (1): 19–29. https://doi.org/10.1038/nchem.2085.

39 AIRSEA (2023). A Beginners Guide to Lithium Batteries Dangerous Goods Regulations: Packaging Guidance, AirSea Containers, October 16, 2023. https://www.airseadg.com/packaging-guidance/a-beginners-guide-to-lithium-batteries-dangerous-goods-regulations/.

40 United Nations (UN) (2023). Recommendations on the Transport of Dangerous Goods, Model Regulations, 23rd revised ed., United Nations Publications, New York, NY, USA; Geneva, Switzerland, 2023, ST/SG/AC.10/1/Rev.23(Vol.1). https://unece.org/sites/default/files/2023-08/ST-SG-AC10-1r23e_Vol1_WEB.pdf (accessed 25 July 2024).

41 IATA, Lithium Battery Guidance Document (2024): Transport of Lithium Metal and Lithium-ion Batteries Revised for the 2024 Regulations, ICAO Technical Instructions for the Safe Transport of Dangerous Goods by Air (Technical Instructions) and the 65th Edition (2024) of the IATA Dangerous Goods Regulations (DGR). https://www.iata.org/contentassets/05e6d8742b0047259bf3a700bc9d42b9/lithium-battery-guidance-document.pdf (accessed 25 July 2024).

42 United Nations Economic Commission for Europe, UNECE Inland Transport Committee - The Bureau, https://unece.org/about-bureau (accessed 25 July 25 2024).

43 Fachverband Batterien (2020). Shipping Lithium-ion Batteries and Lithium-ion Batteries in/with equipment: Implementation of Dangerous Goods Transport Regulations, ZVEI information leaflet No. 36 Edition December 2020. https://www.zvei.org/fileadmin/user_upload/Presse_und_Medien/Bilder_Allgemein/user_upload/Versand-von-Lithium-Ionen-Batterien-2021-EN_TD210104.pdf (accessed 25 July 2024).

44 Transport Regulations for Road and Rail Transport, ADR 2021 – Overview. https://tadiranbatteries.de/wp-content/uploads/2021/05/Transport_Regulations_Road_and_Rail_Transport.pdf (accessed 25 July 2024).

45 Economic Commission for Europe Inland Transport Committee, Agreement Concerning the International Carriage of Dangerous Goods by Road Volume II, 2021, ECE/TRANS/3009(Vol. II), United Nations, New York and Geneva, 2020. https://unece.org/sites/default/files/2021-01/ADR2021_Vol2e.pdf (accessed 25 July 2024).

46 Convention concerning International Carriage by Rail (COTIF) (2019). Appendix C – Regulations concerning the International Carriage of Dangerous Goods by Rail (RID). https://otif.org/fileadmin/new/3-Reference-Text/3B-RID/RID%202019%20E.pdf (accessed 25 July 2024).

47 LithiumSafe, Battery Fire Safety Marine: the Rising Risk of Lithium Batteries [Online]. http://www.lithiumsafe.com/battery-fire-safetymarine (accessed 16 August 2024).

48 Team ReLiB, IMDG Code for Lithium-ion Battery Reverse Logistics, The Mock IMO Assembly-IMO. https://www.imo-contest.org/images/award/2020/03_2020_LoveReLiB.pdf (accessed 25 July 2024).

49 Economic Commission for Europe Inland Transport Committee (2022). Agreement Concerning the International Carriage of Dangerous Goods by Road Volume I, 2022, ECE/TRANS/326 (Vol. I), United Nations, New York and Geneva. https://unece.org/sites/default/files/2023-01/ADR2023_Vol1e.pdf (accessed 25 July 2024).

50 Convention concerning International Carriage by Rail (COTIF) (2023). Appendix C – Regulations concerning the International Carriage of Dangerous Goods by Rail (RID). https://otif.org/fileadmin/new/3-Reference-Text/3B-RID/RID_2023_e_23_January_2023.pdf (accessed 25 July 2024).

51 Top 11 Lithium-ion Battery Regulations, Li-ion Tamer. https://liiontamer.com/wp-content/uploads/Top-11-Lithium-Ion-Battery-Regulations.pdf.

52 Pipeline and Hazardous Materials Safety Administration (PHMSA), Electronic code of federal regulations, Title 49-Part 173-Subpart A-173.1-Purpose and Scope [Online]. https://www.ecfr.gov/current/title-49/subtitle-B/chapter-I/subchapter-C/part-173/subpart-A/section-173.1 (accessed 3 August 2024).

53 US safety agency NTSB adds to warnings on lithium batteries in freighters, Aircargo News, 10 February 2016 [Online]. http://www.aircargonews.net/news/region/america/single-view/news/us-safety-agency-ntsb-adds-to-warnings-on-lithium-batteries-in-freighters.html (accessed 3 August 2024).

54 United Parcel Service (UPS), International lithium battery regulations [Online]. https://www.ups.com/media/news/en/pack_batteries.pdf (accessed 3 August 2024).

55 DHL, Shopping goods with or containing lithium batteries [Online]. https://mydhl.express.dhl/yt/en/help-center/international-shipping/lithium-batteries.html (accessed 3 August 2024)

56 FedEx, 2024 Service Guide [Online]. https://www.fedex.com/content/dam/fedex/us-united-states/services/Service_Guide_2024_preview.pdf (accessed 3 August 2024).

57 FedEx, FedEx Logistics-Lithium Batteries and Cells Shipping Guide [Online]. https://www.fedex.com/content/dam/fedex/us-united-states/logistics/Lithium-Battery-Shipping-Tool.pdf (accessed 3 August 2024).

58 National Technical Committee for Automobile Standardization (2015). Safety requirements and test methods for power batteries for electric vehicles, GB/T 31485-2015[S], Beijing: China Standards Press.

59 International Electrotechnical Commission (2016). Secondary lithium-ion batteries for electric road vehicle propulsion Part 3: Safety requirements, IEC 62660-3:2016[S], Geneva: International Electrotechnical Commission.

60 Ji, S. (2023). China Releases Guide on Railway Transportation of Consumer-grade Lithium Batteries, CHEMLINKED. https://chemical.chemlinked.com/news/chemical-news/china-releases-guide-on-railway-transportation-of-consumer-grade-lithium-batteries (accessed 15 July 2024).

61 Yingzhi (2019). China's New Regulations for Lithium Battery Shipping, TianHeng Consulting Inc. https://www.tianhengconsulting.com/china-new-regulations-for-lithium-battery-shipping/ (accessed 15 July 2024).

62 Civil Aviation Administration of China (2013). MH/T 1052-2013: Test Specification for Lithium Batteries in Air Transportation.

63 Civil Aviation Administration of China (2013). MH/T 1020-2013: Specification for Lithium Batteries in Air Transportation.

64 Civil Aviation Administration of China (CAAC), Notice on the safety management of the air transport of lithium metal battery cargo, 26 February 2015 [Online]. http://www.caac.gov.cn/XXGK/XXGK/ZFGW/201605/t20160518_37441.html (accessed on August 5, 2024).

65 https://www.baidu.com/link?url=q3uCIdRj5lQng5b4cSVi7sfh2glmHcnAgQICVtcwSk8TjLSu3sFJw2UFD_Sq4RBMO9Ensv7asrhOtvCRzbIG9kGyTzHRvEKn6YK-8eFjZpwRy8PQwFhfbOC0f52_zTxU&wd=&eqid=c9b52401009f77820000000566ab2ac6.

66 Civil Aviation Administration of China (CAAC), Air Travel Guide-Lithium battery carrying regulations [Online]. http://www.caac.gov.cn/CXCK/HLZN/201512/t20151214_15866.html (accessed 31 July 2024).

67 Shipping Lithium Batteries from China: The Ultimate FAQ Guide, China Freight News, 2022. https://www.tj-chinafreight.com/shipping-lithium-batteries-from-china-the-ultimate-faq-guide/ (accessed 31 July 2024).

68 Europe Transportation of Dangerous Goods-ADR, ATIC. https://www.atic-ts.com/european-transportation-of-dangerous-goods-adr/ (accessed 31 July 2024).

69 DHL Global Forwarding (2022). DHL Global Forwarding is the first organization to achieve IATA CEIV Li-Batt Certification in the United Kingdom. https://group.dhl.com/en/media-relations/press-releases/2022/dhl-global-forwarding-is-the-first-organization-to-achieve-iata-ceiv-li-batt-certification-in-the-united-kingdom.html (accessed 1 August 2024).

70 KOMARINE (2024). KOMSA. https://www.komarine.com/en/companies/komsa/ (accessed 2 August 2024).

71 Henry (2024). What is the KC Certification of Lithium Battery? Ufine Blog, June 7, 2024. https://www.ufinebattery.com/blog/what-is-the-kc-certification-of-lithium-battery/ (accessed 2 August 2024).

72 Florence, L. IEC 62133 2nd Edition: All You Need to Know about the New Standard, Battery Power Online. https://www.batterypoweronline.com/news/iec-62133-edition-2-0-archived-battery-webinar/ (accessed 10 August 2024).

73 Millsaps, C. Second Edition of IEC 62133, Battery Power- Energy Assurance. https://www.batterypoweronline.com/markets/manufacturing-materials/second-edition-of-iec-62133-the-standard-for-secondary-cells-and-batteries-containing-alkaline-or-other-non-acid-electrolytes-is-in-its-final-review-cycle/ (accessed 10 August 2024).

74 LG Chem, Global Certification for Small Batteries. https://www.prba.org/wp-content/uploads/15-Small-Battery-Certification.pdf (accessed 10 August 2024).

75 Korean Air, Baggage Services - Restricted Items. https://www.koreanair.com/contents/plan-your-travel/baggage/restricted-item (accessed 10 August 2024).

76 Air Busan, Restricted Items to Be Transported. https://en.airbusan.com/content/common/service/baggage/restriction (accessed 10 August 2024).

77 Asiana Airlines, Restricted Transport Items. https://www.flyasiana.com/C/US/EN/contents/restricted-transport-items (accessed 10 August 2024).

78 T'way Aviation (2017). Prohibited items on board: restricted items on board the plane-restricted carry-on products for liquids and gels-100ML or less (based on container size regardless of residual amount), October 12, 2017. https://m.blog.naver.com/sun2ace/221115199479 (accessed 10 August 2024).

79 Korean Air Cargo, Notice for 2022 yr Lithium Battery Regulations. https://cargo.koreanair.com/en/cargo-news/notice-2022-yr-lithium-battery-regulations (accessed 10 August 2024)

80 Allthatsales Blog (2023). Can I ship a product with batteries from Korea? July 16, 2023. https://www.allthatsales.com/post/can-i-send-a-product-with-batteries-from-korea (accessed 10 August 2024).

81 Comply market, Product Compliance Requirements for Battery Electric Vehicles. https://www.complymarket.com/storage/25012024143717Battery_electrical_vehicle_South_Korea_60.pdf (accessed 4 August 2024).

82 Seoul Metropolitan Government (2017). Framework Act on Environmental Policy. https://elaw.klri.re.kr/eng_mobile/viewer.do?hseq=44666&type=part&key=39 (accessed 5 August 2024).

83 Korea Environment Corporation (2023). Extended Producer Responsibility (EPR). https://www.keco.or.kr/en/lay1/S295T386C400/contents.do (accessed 5 August 2024).

84 Chambers and Partners, Environmental Law, 2023, https://practiceguides.chambers.com/practice-guides/environmental-law-2023/south-korea (accessed 6 August 2024).

85 Brugato, T. (2023). EPA Clarifies Hazardous Waste Requirements Applicable to Lithium-ion Batteries. https://www.globalpolicywatch.com/2023/05/epa-clarifies-hazardous-waste-requirements-applicable-to-lithium-ion-batteries/ (accessed 4 August 2024).

86 Bird, R., Baum, Z.J., Yu, X., and Ma, J. (2022). The regulatory environment for lithium-ion battery recycling. *ACS Energy Letters* 7 (2): 736–740, https://pubs.acs.org/doi/10.1021/acsenergylett.1c02724 (accessed 4 August 2024).

87 Recycling Inside (2023). Lithium Battery Recycling Regulations And Opportunities. https://recyclinginside.com/e-waste-recycling/lithium-battery-recycling-regulations-and-opportunities/ (accessed 4 August 2024).

88 Nippon Recycle Center Corp., Business Overview. https://www.recycle21.co.jp/recycle-e/service/detail.html (accessed 4 August 2024).

89 PHMSA, Lithium Battery Packaging Evaluation. https://www.phmsa.dot.gov/sites/phmsa.dot.gov/files/2021-03/Lithium Battery Packaging Evaluation- Final Report.pdf (accessed 4 August 2024).

90 PHMSA (2023). LITHIUM BATTERY GUIDE FOR SHIPPERS: A compliance Tool for all modes of transportation. https://www.phmsa.dot.gov/sites/phmsa.dot.gov/files/2023-07/Lithium%20Battery%20Guide.pdf (accessed 5 August 2024).

91 Federal Aviation Administration (FAA) (2023). Interactive Guide to Shipping Lithium Batteries. https://www.faa.gov/sites/faa.gov/files/2023-Interactive-Guide-to-Shipping-Lithium-Batteries.pdf (accessed 5 August 2024).

13

Battery Safety and Reliability Standards

İlknur Baylakoglu and Yan Ning

Batteries have become the lifeblood of our modern world, powering everything from smartphones to electric vehicles. However, with this immense potential comes the responsibility to ensure their safe and reliable operation. This chapter delves into the intricate world of battery safety and reliability standards, exploring the organizations, regulations, and tests that underpin our trust in these ubiquitous energy sources.

The focus of this book is battery safety. In addition to safety standards, this chapter also covers battery reliability standards. Understanding and adhering to reliability standards are essential because a reliable battery reduces the likelihood of safety incidents. Reliability and safety are two sides of the same coin, both crucial for ensuring the overall performance and trustworthiness of batteries. Providing information on both safety and reliability standards gives a more comprehensive understanding of battery performance requirements. This helps designers, engineers, and users to appreciate the full spectrum of standards that contribute to the development of safe and reliable batteries.

Section 13.1 will navigate the landscape of international and regional organizations that set the foundation for battery safety and reliability benchmarks. It will also explore the role of regional and national regulatory bodies, which establish and enforce specific requirements tailored to local needs and considerations. In addition, this section will examine the importance of third-party battery certifications, highlighting how independent verification fosters trust and ensures adherence to established safety protocols. Together, these standards, regulatory, and certification bodies form a comprehensive framework that governs battery safety and reliability globally.

Sections 13.2 and 13.3 will dissect the various safety and reliability standards that govern different aspects of batteries. It will examine cell safety and reliability standards, addressing electrical and environmental hazards to ensure the safe operation of the fundamental building blocks of batteries. From there, it will transition to pack safety and reliability standards, covering transportation requirements, electrical, thermal, and mechanical abuse testing, and how these standards protect battery packs from environmental stresses. This section will also explore performance requirements at both the cell and pack levels, ensuring that batteries can maintain reliable functionality over their lifespans. Finally, it will delve into the functional requirements for battery management systems (BMSs) in battery packs, which play a crucial role in monitoring key parameters and safeguarding against dangerous failures like overcharging, overheating, and short

The Safety Challenges and Strategies of Using Lithium-Ion Batteries, First Edition. Edited by Michael G. Pecht.

circuit (note: in this chapter, BMS represents both the BMS and battery management unit, and the two concepts are not distinguished separately).

Section 13.4 will explore how safety and reliability standards adapt to the unique demands of various battery applications. It will cover portable devices like smartphones and laptops, automotive applications for electric and hybrid vehicles, and uninterruptible power supply (UPS) and power systems that ensure uninterrupted power. Additionally, this section will look at marine, navy, avionics, and space applications, each requiring specific safety considerations due to the unique and often extreme environments in which they operate.

Finally, this chapter will explore how safety and reliability standards are adapting to keep up with advancements in battery technology. As the focus shifts toward sustainability and smart battery systems, standards are evolving to support new materials, environmentally friendly practices, and improved BMSs with data analytics. It will also consider the growing trend of second-life applications, where used batteries are repurposed, and how international collaboration helps to shape future standards. Identifying gaps in current regulations will ensure that safety standards continue to evolve alongside technological innovation.

In summary, this chapter provides a comprehensive overview of the standards and guidelines that ensure battery safety and reliability. By examining both foundational and application-specific standards, it equips readers with the knowledge needed to navigate this complex landscape, fostering a deeper understanding of battery safety and reliability requirements. These insights not only highlight the regulations that protect users but also emphasize sustainable practices and support advancements in battery technology.

13.1 The Landscape of Battery Safety Standards

The safe and reliable operation of batteries is not a coincidence. It is the result of a meticulous global effort driven by a complex network of international and regional standards organizations, regulatory bodies, and certification bodies. This intricate landscape ensures consistent safety benchmarks across borders and fosters trust in the batteries that power our daily lives.

The first key players in this global effort are the international and regional standards organizations. These nongovernmental entities act as the architects, meticulously developing and publishing technical standards for various products, including batteries. Accordingly, these voluntary standards prioritize safety, quality, and performance, and they are widely adopted by industry as a foundation for responsible battery development.

Standards alone are not enough, however. Regional and national regulatory bodies, acting as the enforcers, translate these international standards or their own developed regulations into actionable frameworks within their jurisdictions. These regulations can encompass various aspects of battery safety, from the manufacturing process and transportation protocols to labeling requirements, proper disposal methods, and even consumer use guidelines.

Finally, we have the independent arbiters, certification bodies. These organizations conduct rigorous testing and verification processes to ensure that batteries comply with the established standards and regulations. While some certification bodies may publish their standards, their primary function is to provide certifications and marks that demonstrate a battery's adherence to safety and performance requirements. These certifications give consumers and manufacturers confidence in the safety and reliability of the battery products they encounter.

This section will delve deeper into each of these key players. It will explore the prominent international and regional standards organizations that set the foundation for battery safety. It will then examine the role of regulatory bodies who translate these industry standards or their developed regulations into actionable safety frameworks within their jurisdictions. Finally, this section will introduce the certification bodies that verify compliance. Throughout this exploration, a compendium of the battery safety and reliability standards associated with each of these key players will be provided.

13.1.1 International and Regional Standards Organizations

The intricate landscape of battery safety is built upon the strong foundation established by prominent international and regional standards organizations. These independent, nongovernmental organizations play a crucial role by establishing foundational technical specifications that serve as a blueprint for safe battery design, manufacturing, and testing. They act as the architects, setting the groundwork for consistent safety benchmarks across borders.

The foremost international standards organization in the battery safety domain is the International Electrotechnical Commission (IEC), established in 1906 [1]. The IEC boasts a long history of fostering international cooperation in areas of electrical, electronic, and related technologies. Their contributions to battery safety are substantial, with a comprehensive suite of standards that address various aspects of battery safety, encompassing chemical safety, electrical safety, environmental safety, and transportation safety. The IEC publishes a comprehensive set of battery safety and reliability standards, forming a cornerstone for global battery safety. Here is a compendium of some key IEC battery safety and reliability standards:

- **IEC 60086 series** standards specifically address primary battery safety and performance. They play a crucial role in ensuring the responsible development, manufacturing, and use of primary batteries across various applications [2–7].
- **IEC 62133 series** standards delve into the safety requirements for secondary cells and batteries containing alkaline or other nonacid electrolytes. They encompass various safety aspects, including chemical, electrical, thermal, and mechanical considerations. The series is further divided into IEC 62133-1 [8] for nickel-based batteries and IEC 62133-2 [9] for lithium-based batteries, reflecting the specific safety needs of different battery chemistries.
- **IEC 62281** specifically addresses the safe transportation of lithium batteries, outlining packaging, labeling, and testing requirements [10].
- **IEC 62619** tackles safety requirements for secondary lithium cells and batteries used in industrial applications [11].
- **IEC 62660 series**, consisting of two parts, provides a comprehensive framework for testing and evaluating the safety and performance of secondary lithium-ion cells used for electric vehicle propulsion. IEC 62660-1 [12] concentrates on evaluating the performance characteristics of secondary lithium-ion cells, while IEC 62660-2 [13] focuses on evaluating a battery's ability to withstand extreme conditions and potential abuse scenarios.
- **IEC 61960 series** currently consist of two parts, establishing performance tests, designations, markings, dimensions, and other requirements for secondary lithium cells and batteries used in portable applications. IEC 61960-3 [14] focuses on prismatic and cylindrical cells and batteries, while IEC 61960-4 [15] focuses on coin-shaped cells and batteries.

Complementing the foundational work of IEC, the Institute of Electrical and Electronics Engineers (IEEE) is another prominent international organization that shapes the landscape of battery safety and reliability standards. IEEE is a renowned international organization established in 1963 through the merger of the Institute of Radio Engineers (founded in 1907) and the American Institute of Electrical Engineers (founded in 1884) [16]. IEEE boasts a rich history of advancing electrical and electronic technologies, and its contributions to battery safety and reliability standards are significant [17]. Unlike IEC, which focuses on foundational standards, IEEE complements this groundwork by developing more specific standards tailored to distinct battery applications and emerging technologies. IEEE standards delve deeper into the unique safety challenges and requirements of various battery types and their end products. This targeted approach ensures a comprehensive safety net across the battery landscape. Here is a glimpse into some key IEEE standards that contribute to the battery safety and reliability landscape:

- **IEEE 1725** addresses the design, testing, and manufacturing requirements for safe and reliable lithium-ion batteries (LIBs) used in mobile phones. It emphasizes thermal management, electrical interface specifications, and end-user safety considerations [18].
- **IEEE 1625** focuses on safety practices for rechargeable batteries used in laptops and other portable electronic devices. It complements IEC standards by addressing specific safety concerns in the design, testing, and manufacturing processes of portable computing devices [19].

The International Organization for Standardization (ISO), founded in 1947, is the world's largest developer of voluntary international standards, encompassing various industries and technologies. ISO plays a crucial role in advancing battery safety, setting global benchmarks that complement the work of other organizations like IEC and IEEE. ISO's approach ensures that battery technologies are standardized with an emphasis on safety, quality, and interoperability, addressing challenges from design and manufacturing to end-user applications [20]. Below is an overview of some key ISO battery safety and reliability standards:

- **ISO 12405 series** is essential in ensuring the safety and reliability of LIBs used in electric vehicles (EVs). It defines specific test procedures for evaluating the performance, reliability, and abuse tolerance of LIB packs and systems designed for EV propulsion. The ISO 12405 series has undergone revisions: ISO 12405-1 [21] and ISO 12405-2 [22] have been withdrawn, and their requirements have been consolidated and incorporated into the current standard, ISO 12405-4 [23], which focuses on performance testing. Similarly, ISO 12405-3 [24] has been withdrawn, with its requirements revised and integrated into ISO 6469-1 [25].
- **ISO 6469 series** is a collection of standards that address safety specifications for electrically propelled road vehicles. It aims to ensure the safe operation of EVs by setting out requirements for various aspects of EV safety. ISO 6469-1 is directly related to the safety requirements and testing of onboard rechargeable energy storage systems (RESSs), which primarily consist of batteries. It includes specifications for the reliability and safety of these battery systems [25].

Adding to the international landscape of battery safety and reliability standards, the Society of Automotive Engineers International (SAE) is another prominent organization that plays a vital role. SAE International develops standards specifically focused on the automotive industry, including standards related to battery safety for EVs and hybrid electric vehicles (HEVs). These standards complement broader international standards by addressing the specific safety requirements and testing procedures unique to automotive battery applications [26]. Here is a glimpse into some key SAE standards that contribute to battery safety and reliability in automotive applications:

- **SAE J2929** defines a minimum set of acceptable safety criteria for LIB systems used in electric and hybrid vehicles. It focuses on the battery system, outlining requirements for design, testing, and manufacturing to ensure safety within the vehicle propulsion system [27].

- **SAE J2464** outlines specific test procedures for evaluating the performance and safety of electric and hybrid electric vehicle batteries under various abuse conditions, such as crush, nail penetration, and thermal runaway [28].
- **SAE J2380** provides a procedure to evaluate the effect of long-term, road-induced vibration on the performance and service life of EV batteries [29].
- **SAE J2288** focuses on battery performance testing for electric and hybrid vehicles. It specifies test procedures and requirements for evaluating the performance and life cycle of batteries used in these vehicles [30].
- **SAE J1798** is a recommended practice for the performance rating of EV battery modules. It provides common test and verification methods to determine the performance of EV battery modules [31].
- **SAE J2289** provides comprehensive guidelines for the design of battery systems used in electric-drive vehicles. It covers a wide array of aspects essential for ensuring the functionality and safety of battery packs [32].

The National Electrical Manufacturers Association (NEMA), established in 1926, is a key player in North America's electrical industry. NEMA collaborates with other organizations to develop and promote standards that ensure the safe design, manufacturing, and use of electrical products, including the rechargeable batteries that power many of our devices [33]. The notable battery standards under NEMA's purview are the American National Standards Institute (ANSI) C18.1M and ANSI C18.2M.

- **ANSI C18.1M** is a two-part standard that defines terminology and general requirements for portable primary (nonrechargeable) batteries. It covers various aspects, such as definitions of terms related to battery types, performance characteristics, labeling requirements, and testing methods [34].
- **ANSI C18.2M** is a two-part standard focusing on portable rechargeable cells and batteries. It specifies requirements for their design, construction, testing, and performance. Part 1 primarily covers nickel-metal hydride (Ni-MH) and nickel-cadmium (Ni-Cd) rechargeable cells and batteries. Part 2 expands the scope to include additional types of rechargeable batteries, including LIBs, beyond those addressed in Part 1 [35].

Telcordia Technologies, formerly known as Bell Communications Research (Bellcore), emerged in 1983 following the breakup of AT&T as a key player in the telecommunications industry. Over time, Telcordia evolved from its research roots to become a leading developer of standards and specifications for telecommunications equipment and services. In 2012, Telcordia was acquired by Ericsson and later rebranded as Iconectiv in 2013 [36, 37]. Despite these changes, the company has continued ensuring compatibility, performance, and safety across various network components. To address the specific needs of LIB technology, which is increasingly relied upon for telecom backup power, Telcordia developed the GR-3150-CORE standard. Today, Iconectiv continues to manage and develop Telcordia's legacy standards to ensure reliable network operations globally.

- **GR-3150-CORE** ensures the safety and performance of rechargeable LIBs for telecom backup power, fostering dependable network operation even in the face of power disruptions. By establishing these benchmarks, Telcordia promotes the reliable deployment of LIB technology within the telecommunications infrastructure [38].

The importance of harmonization in battery safety standards cannot be overstated. By collaborating extensively, standards organizations ensure consistency across international borders. This eliminates confusion and fosters seamless global trade while upholding rigorous safety benchmarks. Furthermore, regional and national regulatory bodies often leverage these standards as a foundation. They may reference, adopt, or even develop supplemental standards in alignment with this global approach, solidifying a unified front for battery safety. In the following section, we will explore how regional and national entities translate these international standards into actionable frameworks.

13.1.2 Regional and National Regulatory Bodies

Building upon the foundation established by international and regional battery safety standards, regional and national regulatory bodies play a crucial role in translating these standards into actionable frameworks. This section will explore how these entities adopt and implement international standards within their jurisdictions. For example, many regional and national regulatory bodies recognize the value of established international standards like IEC 62133 [8, 9] or IEC 62281 [10]. These comprehensive standards provide a robust framework for the safety testing and evaluation of LIBs. Accordingly, regulatory bodies can leverage these battery standards as a starting point for their regulations.

Some regulatory bodies might directly adopt international standards with minimal modifications, ensuring consistency with globally recognized safety benchmarks. Others might reference these international standards as a foundation and establish additional regional or national requirements specific to their jurisdictions. This allows them to address local considerations or emerging technologies while aligning with core safety principles. Some regulatory bodies may even develop their own standards and regulations.

The United Nations Economic Commission for Europe (UNECE) is one of the regional commissions of the United Nations, established in 1947 to promote economic integration and cooperation among its member states [39]. UNECE develops a wide range of norms, standards, and conventions to facilitate international trade, enhance environmental protection, and improve transportation safety, including safely transporting dangerous goods such as batteries.

- **UN 38.3**, formally known as Recommendations on the Transport of Dangerous Goods, Manual of Tests and Criteria, Part III, Section 38.3, is a crucial regulatory framework developed by UNECE for the safe transport of lithium batteries. This regulation outlines specific testing requirements to ensure that lithium batteries can withstand the rigors of transportation without posing safety risks [40].

The European Union (EU) has established a comprehensive regulatory framework for batteries, promoting safety, responsible use, and sustainability. This framework incorporates both international standards and specific EU regulations.

- **Battery Directive 2006/66/EC** focuses on the entire battery life cycle and labeling requirements. It does not directly mandate specific safety testing standards. However, it implicitly encourages manufacturers to follow established international standards like IEC 62133 to ensure the batteries they place on the market meet safety benchmarks. The directive establishes requirements for clear labeling of batteries, restricts the use of certain heavy metals, and promotes responsible life cycle management practices [41].
- **Battery Recycling Directive 2013/56/EU** specifically focuses on improving battery recycling efficiency and reducing the environmental impact of batteries. It sets ambitious targets for battery collection and recycling rates across EU member states, fostering a more sustainable battery ecosystem within the region. It does not directly reference safety standards. However, by promoting responsible handling and disposal of batteries, it indirectly contributes to overall battery safety [42].

- **Regulation (EU) 2019/1020, the Market Surveillance Regulation,** ensures that products, including batteries, entering the EU market comply with established EU safety, health, and environmental requirements. It strengthens market surveillance mechanisms to identify and address any noncompliant batteries that may pose a risk to consumers or the environment. While it does not specify battery safety standards, it allows for referencing established international standards during market surveillance activities. Here, batteries found noncompliant with internationally recognized safety standards, like IEC 62133 [8, 9] or UN 38.3 [40], would likely be flagged for corrective action [43].

The EU's approach to battery regulation offers a multifaceted benefit. By referencing international standards and establishing consistent requirements across member states, the framework promotes a unified approach to battery safety within the European market, simplifying compliance for manufacturers and fostering smoother trade. Furthermore, the focus on responsible recycling and collection through the Battery Recycling Directive strengthens the EU's commitment to a more sustainable battery life cycle. This encourages the development of efficient recycling infrastructure and promotes responsible battery management practices. Finally, strict safety requirements and market surveillance mechanisms work together to ensure the safety and quality of batteries available to European consumers.

In the United States, several regulatory bodies oversee various aspects of battery safety and environmental impact. The two primary entities are the Department of Transportation (DOT) and the Environmental Protection Agency (EPA). In addition, the Consumer Product Safety Commission (CPSC) plays a significant role in protecting consumers from hazards related to battery use in consumer products. The DOT is a federal agency responsible for ensuring the safe and efficient movement of people and goods across the United States. It oversees transportation regulations and standards for various modes of transportation, including air, rail, road, and sea. In the context of battery safety, the DOT regulates the transportation of lithium batteries to prevent accidents during transit and ensure compliance with safety standards.

- **49 CFR 173.185** specifically addresses the transportation of lithium batteries. It includes comprehensive requirements for the packaging, labeling, and documentation of lithium batteries during transit. These regulations ensure that lithium batteries are transported safely, minimizing the risk of incidents such as fires or leaks during shipping. It encompasses various aspects of battery transport, including testing protocols to simulate conditions during transportation, packaging standards to prevent short circuits, and labeling requirements to provide clear information about the battery's contents and handling instructions [44]. It explicitly requires the lithium cells and batteries to meet the UN 38.3 standard [40].

The EPA is a federal agency tasked with protecting human health and the environment. The EPA's responsibilities include developing and enforcing regulations that ensure the safe disposal, recycling, and management of hazardous materials, including batteries. By setting guidelines and standards for handling and disposing battery waste, the EPA aims to minimize environmental impact and prevent potential health risks associated with improper battery disposal.

- **The Resource Conservation and Recovery Act (RCRA)** governs the disposal and recycling of hazardous waste, including batteries, ensuring proper management and disposal to protect human health and the environment. This act sets guidelines for the treatment, storage, and disposal of hazardous waste, promoting safe and environmentally sound practices [45].

The CPSC, another key federal agency, is responsible for ensuring the safety of consumer products, including batteries used in consumer electronics. The CPSC monitors and enforces safety standards to protect consumers from potential hazards such as fires, explosions, and chemical burns associated with battery failures. In doing so, the CPSC adopts and refers to several well-established industry standards, including IEC 62133 [8, 9], ANSI/NEMA C18 [34, 35], and standards from Underwriters Laboratories (UL), which will be discussed in Section 13.1.3. These standards provide a comprehensive framework for battery safety in consumer products, addressing various risks associated with battery design, testing, and usage.

The US regulatory framework for batteries integrates transportation safety, environmental protection, and consumer safety. DOT's 49 CFR 173.185 ensures safe lithium battery transport, EPA's RCRA guidelines support environmentally safe disposal and recycling, and the CPSC enforces safety standards for consumer battery use, reducing risks like fires and chemical burns. Together, these regulations provide a comprehensive approach to battery safety throughout the life cycle, from transport to disposal.

In China, battery safety and environmental regulations are overseen by several key regulatory bodies. These agencies ensure that batteries manufactured, used, and disposed of within the country meet stringent safety and environmental standards. The key regulatory bodies in China include the China National Institute of Standardization (CNIS) and the Ministry of Industry and Information Technology (MIIT). The CNIS is responsible for developing national standards in various fields, including battery safety, and it works with other government bodies and industry stakeholders to establish and enforce standards that promote the safe use, transportation, and disposal of batteries [46]. The MIIT is a key regulatory body that oversees the industrial sector in China, including the production and use of batteries. It establishes regulations and policies to promote industrial safety, innovation, and environmental protection [47]. Below are some key battery standards and regulations developed by the CNIS under the jurisdiction of the MIIT:

- **GB/T 18287** specifies safety requirements for LIBs used in mobile phones. It includes guidelines for performance testing, safety evaluations, and labeling requirements to ensure that LIBs are safe for consumer use. The standard aligns with international best practices and incorporates specific requirements to address local conditions and technological developments in China. This is a voluntary regulation, not legally mandatory. However, adherence to such standards is strongly encouraged to ensure product quality and safety, and it is often required in practice by industry stakeholders and market expectations [48].
- **GB 31241** specifies safety requirements for lithium-ion cells and batteries used in portable electronic equipment. This standard is crucial for ensuring the safety and reliability of LIBs in consumer electronics, which are widely used in devices such as smartphones, tablets, laptops, and wearable technology. Compliance with GB 31241 is legally required to ensure the safe use of LIBs in consumer electronics [49].
- **GB/T 36972** outlines safety requirements for lithium-ion power battery packs and systems used in EVs. This standard is crucial for ensuring the safety, reliability, and performance of LIBs in the automotive sector, as EV adoption increases. This is a voluntary regulation, not legally mandatory. However, adherence to such standards is strongly encouraged to ensure product quality and safety, and it is often required in practice by industry stakeholders and market expectations [50].

In Japan, battery safety regulations are overseen by several key regulatory bodies to ensure that batteries meet rigorous safety standards for consumer protection and environmental sustainability. The Ministry of Economy, Trade and Industry (METI) administers these regulations under the Electrical Appliance and Material Safety Law (Denki Yōhin Anzen Hō, abbreviated as DENAN). This law mandates specific safety requirements for electrical products, including batteries, sold in Japan. Its primary goal is to protect consumers from hazards like overheating, fire, and explosion by enforcing strict safety criteria. Compliance with the DENAN Law is mandatory for manufacturers seeking to market batteries in Japan, fostering consumer confidence and safety [51]. Additionally, the Japanese Industrial Standards Committee (JISC) is responsible for developing and updating Japanese Industrial Standards (JIS), which govern various aspects of product safety and quality, including batteries [52]. Below, key battery standards and regulations developed by the JISC under the METI's oversight are highlighted:

- **JIS C 8712** sets safety requirements for LIBs used in various applications. It covers design specifications, testing procedures, and performance criteria to ensure the safe and reliable use of LIBs in consumer electronics, industrial equipment, and automotive applications. Compliance with JIS C 8712 is crucial for manufacturers to demonstrate that their LIBs meet rigorous safety standards in Japan. Although it is a voluntary regulation, JIS C 8712 is widely encouraged by industry stakeholders and manufacturers to ensure product safety and quality [53].
- **JIS C 8714** specifies safety tests for portable lithium-ion secondary batteries. It outlines testing procedures and conditions to evaluate the safety performance of portable LIBs under various environmental and operational stresses. Adherence to JIS C 8714 ensures that portable LIBs are safe for handheld devices, supporting both consumer safety and product reliability. Similarly, JIS C 8712 is a voluntary regulation, but it is widely encouraged by industry stakeholders and manufacturers to ensure product safety and quality [54].

In conclusion, regional and national regulatory bodies worldwide play a critical role in harmonizing and enforcing battery safety standards, translating international guidelines like IEC standards and UNECE recommendations into practical regulations tailored to local contexts. Organizations such as UNECE in Europe ensure safe battery transport through standards like UN 38.3. At the same time, the EU integrates comprehensive frameworks like the Battery Directive and Market Surveillance Regulation to uphold safety and sustainability. In the United States, agencies like DOT and EPA enforce stringent regulations on battery transportation and disposal, safeguarding public safety and environmental health. Similarly, China's CNIS and MIIT oversee standards such as GB/T 18287, GB 31241, and GB/T 36972, ensuring safety across mobile devices, electronics, and electric vehicles. In Japan, METI and JISC regulate under the DENAN Law and JIS standards, fostering confidence in battery safety and performance. Together, these regulatory efforts sustain global standards while accommodating local needs and technological advancements in battery safety.

13.1.3 Certification Bodies

The landscape of battery safety extends beyond the foundational work of international and regional standards organizations and the regulatory frameworks established by regional and national bodies. Independent certification bodies are crucial for ensuring that batteries comply with these standards and regulations. Operating independently from manufacturers and regulatory authorities, certification bodies provide an unbiased assessment of battery products through rigorous testing and verification processes. Their role encompasses testing, certification, and ongoing surveillance to ensure that batteries meet the highest safety and performance standards.

Certification bodies conduct comprehensive tests on batteries to evaluate their compliance with relevant standards and regulations. These tests cover various safety aspects, including chemical stability, electrical performance, mechanical durability, and resistance to extreme conditions. Through rigorous testing protocols, certification bodies identify any potential safety issues, ensuring that only batteries meeting stringent criteria reach the market.

Upon passing these tests, certification bodies issue marks or labels that manufacturers can display on their products. These marks signal to consumers and regulatory authorities that a battery has undergone thorough testing and meets the necessary safety and reliability criteria. This provides trust and assurance, helping consumers to make informed decisions and regulatory bodies to enforce compliance.

Certification is not a one-time event. Certification bodies conduct periodic audits and testing to ensure ongoing compliance with safety standards. This continuous surveillance helps maintain the integrity of certified products throughout their life cycle. By regularly verifying that batteries continue to meet established standards, certification bodies play a crucial role in sustaining long-term safety and reliability in the battery industry.

Several prominent certification bodies are recognized globally for their stringent testing protocols and reliable certification processes, including the key players in the field described in this section.

The IEC System for Conformity Testing and Certification of Electrical Equipment (IECEE) plays a pivotal role in ensuring the safety, performance, and reliability of batteries through its internationally recognized certification system. Operating under the auspices of the IEC, the IECEE provides a structured framework for testing and certifying batteries, helping manufacturers meet global standards and regulatory requirements [55]. The IECEE certification process is comprehensive, encompassing a range of rigorous tests designed to evaluate various aspects of battery safety and performance. These tests are conducted by following specific IEC standards, which are globally accepted benchmarks for quality and reliability. By adhering to these standards, the IECEE ensures that certified batteries meet the highest safety and performance criteria, assuring consumers, manufacturers, and regulatory bodies alike. The IECEE conducts various tests on batteries to evaluate their compliance with relevant IEC standards. Some of the key standards and tests include IEC 62133 [8, 9], IEC 62619 [11], IEC 61960 [14, 15], IEC 62281 [10], and IEC 62660 [12, 13].

Underwriters Laboratories (UL) is a globally recognized safety certification organization. Established in 1894, UL has a long history of developing safety standards and providing certification services for various products, including batteries [56]. UL's certification mark is one of the most trusted symbols of safety and quality worldwide. Some key battery safety and reliability standards that UL developed include:

- **UL 1642** specifically addresses the safety of lithium-ion cells used in consumer electronics. It encompasses tests for electrical, thermal, and mechanical abuse conditions [57].
- **UL 2054** covers the safety of lithium batteries used in various applications. It includes tests for overcharging, short circuiting, and thermal and physical abuse, ensuring that batteries can withstand real-world usage [58].
- **UL 2575** is a safety standard that covers LIB systems used in battery-operated power tools and motor-operated, heating, and lighting appliances. It primarily focuses on the risks of fire and explosion during battery usage and storage while requiring integration with end-product standards for comprehensive safety evaluations. This standard ensures that these battery systems meet basic safety requirements but does not cover toxicity, electric shock, or transportation hazards [59].
- **UL 2271** addresses the safety of LIBs for light electric vehicle (LEV) applications, such as e-bikes and scooters. It includes tests for overcharge, discharge, mechanical impact, and environmental exposure to ensure safe operation in these smaller vehicles [60].
- **UL 2580** focuses on the safety requirements for LIBs used in EVs. It includes tests for thermal management, electrical performance, and mechanical integrity to ensure that the batteries are safe and reliable for automotive applications [61].

- **UL 1973** covers batteries for stationary, vehicle auxiliary power, and light electric rail (LER) applications. It includes tests for electrical performance, thermal stability, and environmental durability to ensure the safety and reliability of batteries in these specific contexts [62].
- **UL 9540** focuses on the safety of energy storage systems (ESS) and equipment. It is designed to cover stationary ESS that store and deliver electric energy, such as large-scale LIB systems used in residential, commercial, and industrial applications. The standard helps ensure that these systems meet rigorous safety criteria related to performance, environmental durability, and resistance to hazardous conditions such as fire or explosion [63].

Beyond evaluating compliance with their own UL standards, UL offers certification services against a wide range of international standards. For instance, UL can certify products to meet IEC standards like 62133 [8, 9] and 62619 [11], IEEE standards like 1625 [18] and 1725 [19], ISO 12405 [23], and other regulations from various regions and countries. By offering certification against UL standards and international benchmarks, UL plays a critical role in fostering global safety and interoperability across diverse industries.

TÜV Rheinland [64] and TÜV SÜD [65] are globally recognized certification bodies that ensure batteries' safety, reliability, and compliance. They offer extensive experience in testing, inspecting, and certifying various battery types for applications like consumer electronics, electric vehicles, and ESS. Their comprehensive certification processes cover safety, performance, and quality aspects. Both entities certify batteries against well-known standards from IEC and UL, among others.

Intertek [66] is a leading total quality assurance provider to industries worldwide. With a history dating back to 1885, Intertek offers testing, inspection, and certification services that help to ensure the safety and reliability of battery products. Intertek's extensive experience covers a wide range of battery types, from the familiar AA batteries in remote controls to the high-powered LIBs powering electric vehicles. Intertek leverages its expertise to offer testing and certification services against a vast array of battery standards, such as UN 38.3 [40], well-known IEC and UL standards, and a wider range of national and regional battery standards, catering to specific market needs.

There are numerous globally recognized, regional, and national certification bodies, and this section cannot cover all of them. Certification bodies serve as essential guardians of battery safety and performance. Through rigorous testing and verification processes, they ensure that batteries meet established standards and regulations, cultivating trust among consumers, manufacturers, and regulatory bodies. As the battery industry continues to evolve, certification bodies will play an increasingly vital role in upholding the stringent standards of safety and performance for these essential power sources.

13.2 Battery Cell Safety and Reliability Standards

At the heart of every battery system lies the individual cell, the most fundamental unit that determines the overall performance, reliability, and safety of the battery. While it is easy to focus on the complete battery pack or module, understanding the critical role of cells within these larger systems is essential. Each cell serves as a self-contained energy source, and the aggregation of these cells forms the backbone of everything from the smallest portable electronics to the largest industrial ESS. The behavior of each cell directly impacts the integrity of the battery as a whole, making cell-level standards a cornerstone of battery safety and reliability.

The importance of cells becomes even more pronounced when considering the diverse range of applications they serve. For a consumer electronic device, a single faulty cell can render a smartphone unusable or even cause it to overheat and catch fire, as was seen in highly publicized incidents in recent years. In an electric vehicle, the stakes are even higher. A malfunctioning cell within a high-capacity battery pack could lead to a catastrophic failure that endangers not only the vehicle's occupants but also pedestrians and property. Therefore, standards that ensure the safety and reliability of individual cells are not merely guidelines—they are the protective barriers against potential disaster.

One of the key challenges of setting standards for cells is the sheer variety of battery chemistries and configurations in the market today. From traditional nickel-based chemistries to advanced lithium-ion cells, each type of cell has unique properties that influence its behavior under normal operation and during failure conditions. For instance, while nickel-based cells have long been considered safer due to their relatively stable chemistry, lithium-ion cells offer much higher energy density but come with increased risks, such as the potential for thermal runaway. This makes it crucial for standards to be tailored to the specific characteristics of each cell type, ensuring that safety measures are appropriate for the technology in question.

The evolution of cell standards over the past few decades reflects the rapid pace of technological innovation in the battery industry. Early standards focused on ensuring that cells could safely deliver energy without posing undue risks to users. However, as battery applications expanded, so too did the scope of these standards. Today, cell standards not only address safety concerns but also encompass aspects of reliability, performance, and environmental impact. These standards are continually updated to reflect new research and lessons learned from field experience, demonstrating the industry's commitment to advancing safety in parallel with technological progress.

This section delves into the key standards that shape rechargeable lithium-ion cell safety and reliability. While some standards focus specifically on cells, many also apply to cell blocks, modules, packs, and battery systems. The standards discussed in this section are valuable for companies and industry professionals involved in the development, manufacturing, and use of LIB cells.

13.2.1 Transportation Standards

Transportation of lithium-ion cells poses significant challenges due to the inherent risks associated with their chemistry, particularly during long-distance shipping by air, sea, or road. Over the years, international standards have been developed to address the unique safety requirements of transporting these energy-dense cells. This section explores the key transportation standards for batteries, while Chapter 12 provides a comprehensive discussion on battery transportation.

Among the most widely adopted transportation standards are UN 38.3 and IEC 62281, which provide comprehensive guidelines for mitigating the risks posed by electrical, thermal, mechanical, and other environmental stresses during transit. These standards apply to individual cells and battery packs, ensuring protection across the entire battery supply chain by subjecting batteries to rigorous testing that verifies their ability to withstand the often harsh and unpredictable conditions encountered during transportation.

UN 38.3 is a section within the UN Manual of Tests and Criteria, which is currently in its 8th edition [40]. UN 38.3 is an internationally recognized regulation that provides a comprehensive framework for testing lithium cells and batteries to ensure that they can be safely transported. The 8th edition includes new provisions for the testing of sodium-ion batteries. The UN 38.3 section specifies a series of tests designed to simulate conditions a battery cell or pack might experience during transportation. It covers a wide range of abuse conditions, such as short circuit, overcharging, and forced discharges (T5-T8), as well as environmental factors like altitude simulation (T1) and thermal cycling (T2). Additionally, mechanical stress tests such as vibration (T3), shock (T4),

Table 13.1 Cell and battery transportation standards.

Part 1 of 2

Standard	Title	Electrical abuse test	Thermal test	Mechanical abuse test
UN 38.3: 2019 [40]	UN Manual of Tests and Criteria, Rev.8 (2023)—Lithium metal, lithium-ion, and sodium-ion batteries	• T5: External short circuit • T7: Overcharge • T8: Forced discharge	• T2: Thermal cycling	• T3: Vibration • T4: Shock • T6: Impact/crush
IEC 62281: 2019 [10]	Safety of primary and secondary lithium cells and batteries during transportation	• T5: External short circuit • T7: Overcharge • T8: Forced discharge	• T2: Thermal cycling	• T3: Vibration • T4: Shock • T6: Impact/crush • P1: Package drop

Part 2 of 2

Standard	Other environmental test	Application area	Application battery level
UN 38.3: 2019 [40]	• T1: Altitude simulation	Various applications	Cell and pack
IEC 62281: 2019 [10]	• T1: Altitude simulation	Various applications	Cell and pack

and impact/crush (T6) are included to replicate physical stresses cells or batteries may encounter during shipment. UN 38.3 requires all cell types to undergo tests T1 to T6 and T8. These tests are critical for various applications, ensuring their safety from when they leave the manufacturing facility until they reach their destination.

Complementing UN 38.3, IEC 62281 specifies test methods and requirements for transporting primary and secondary lithium cells and battery packs, providing a harmonized set of safety requirements along with other IEC standards [10]. Originally developed in response to the growing commercial interest in lithium batteries, this standard emphasizes the same tests as UN 38.3 and additionally introduces P1, a package drop test, which ensures that batteries, when packed for transport, can endure the potential impact of being dropped during handling. Like UN 38.3, this standard applies to a broad range of applications, making it a comprehensive guideline for safe transportation.

The comparison of the two standards in Table 13.1 highlights their shared focus on electrical, thermal, and mechanical abuse tests, along with altitude simulation. While the two standards share many similarities, IEC 62281 has an additional focus on packaging safety, reflecting its approach to mitigating risks during the entire transportation process.

13.2.2 Abuse and Environmental Standards

Lithium-ion cells are subject to a range of potential hazards during their operation, making it critical to assess their behavior under abuse and environmental stress. Abuse tests simulate scenarios where the cell might be exposed to electrical, thermal, mechanical, or other environmental stresses that exceed normal operating conditions. These tests ensure that cells maintain safety even under extreme situations, protecting the user and the surrounding environment. A number of international standards have been developed to address these concerns, each focusing on different aspects of cell safety for various applications.

The earliest efforts to address cell safety were focused on primary batteries. The IEC 60086 series [2–7] sets the stage for the safe use of primary cells by establishing guidelines on labeling, marking, and safety operations. These standards play a foundational role in ensuring the safety of non-rechargeable batteries, which are generally considered less risky due to their simpler chemistry.

As rechargeable cells gained prominence, the need for more specific safety standards became apparent. The IEC 62133 standards, divided into two parts—IEC 62133-1 [8] for nickel systems and IEC 62133-2 [9] for lithium systems—were a critical development in ensuring the safety of portable sealed secondary cells and batteries made from them. These standards focus on preventing thermal runaway, which can occur in lithium-ion cells due to their high energy density. At the lithium cell level, the IEC 62133-2 standard includes electrical abuse tests, such as continuous charging for seven days, external short circuit, overcharge, and forced discharge. It addresses thermal abuse through high-temperature testing. It also requires mechanical abuse tests, including crush, freefall, vibration, and mechanical shock, all critical for portable devices that may be dropped or subjected to impact during use. These tests ensure that lithium cells in portable applications can safely operate under various real-world conditions.

The growing use of secondary lithium cells in industrial applications introduced new challenges that required additional safety standards. The IEC 62619 [11] standard was developed to ensure the safety of secondary lithium cells used in industrial applications, where they are often exposed to more demanding conditions than in consumer electronics. The standard emphasizes external short-circuit, overcharge, forced discharge, and internal short-circuit tests, crucial for preventing failures in large industrial cells. The thermal abuse test and mechanical tests, such as impact and drop, ensure that cells can withstand extreme environmental conditions typically encountered in industrial use. Note that IEC 62619 has additional propagation tests and battery system functional safety requirements for BMSs, which will be discussed in Section 5.3.

For EVs, the IEC 62660 series [12, 13] became instrumental in addressing the reliability and safety of lithium-ion cells used for propulsion. The IEC 62660-2 [13] standard applies to lithium-ion cells and cell blocks used in EV propulsion systems. Electric vehicle applications require cells to perform reliably under dynamic and often harsh conditions. This standard subjects' lithium-ion cells to electrical abuse tests, including external short-circuit, overcharge, and forced discharge. The standard also requires thermal tests, such as high-temperature endurance and temperature cycling, to ensure cell reliability under fluctuating temperature conditions. Mechanical abuse tests, including crush, impact, and penetration tests, ensure the cells' resilience to physical damage, which is crucial for vehicle safety.

In North America, UL standards are pivotal in ensuring the safety of lithium-ion cells and batteries across various industries. UL 1642 [14] is a key standard for assessing the safety of lithium cells across a variety of applications. It emphasizes electrical abuse tests—such as short-circuit, abnormal charging, and forced discharge—and thermal abuse tests, including heating, temperature cycling, and the projectile test. Mechanical and environmental abuse tests, such as crush, impact, shock, vibration, and altitude simulation, further ensure that cells can withstand harsh mechanical and environmental conditions. The broad application of UL 1642 makes it suitable for a wide range of lithium cells used in consumer electronics, medical devices, and other critical applications where cell failure could lead to safety hazards.

In parallel, SAE plays a crucial role in developing standards that govern the safety and performance of automotive batteries, particularly for electric and hybrid vehicles. These standards provide a framework for testing and validating battery systems, focusing on reliability and safety. SAE J2464 [28] focuses on abuse testing for RESSs used in electric and hybrid vehicles. The abuse tests in this standard are intended to cover individual RESS cells (of batteries or capacitors), modules, and packs. At the cell level or above, the standard introduces rigorous electrical abuse tests, including short-circuit, overcharge, forced discharge, and separator shutdown integrity, which test the cell's ability to prevent thermal runaway. The thermal abuse tests cover thermal stability and thermal shock cycling, ensuring that the cell remains stable even in extreme thermal conditions. Mechanical tests include shock, crush, and penetration, assessing the cell's structural integrity during vehicle accidents. Additionally, the standard requires a hazardous substance monitoring test for the cell level and above, ensuring safety even in case of hazardous material release. At the module level or above, the standard requires a cycling without thermal management test, high-temperature hazard test, passive propagation resistance test, roll-over simulation, and immersion test. Drop tests are required at the pack level only.

The comparison of the above standards in Table 13.2 highlights their focus on addressing electrical and thermal abuse and mechanical and other environmental stresses. While the standards share common goals, each focuses on specific application areas, from portable electronics to industrial systems and electric vehicles, ensuring that cell-level safety is maintained across a wide range of use cases.

In addition, IEEE has also contributed to the development of cell safety standards, particularly in the context of consumer electronics. IEEE 1725 [18] and IEEE 1625 [19] address the safety of rechargeable batteries in cellular telephones and multicell mobile computing devices, respectively. These standards provide guidelines for the design, manufacturing, and testing of battery cells and packs, ensuring that they can safely power the portable devices that have become ubiquitous in modern life. Outside of the application difference, the structures of the two standards are very similar. They discuss considerations of cells, packs, host devices, adapters, and total systems. This section focuses on the cell-level abuse and environmental requirements from these IEEE standards; pack-level requirements will be discussed in Section 13.3.

IEEE 1725 and IEEE 1625 emphasize the importance of pre-production testing, qualification testing, and ongoing reliability testing to ensure that the cells meet performance characteristics and are robust enough to withstand abuse conditions. The testing protocols rely on recognized standards such as UL 1642, IEC 62133, and others. The IEEE standards also cover regulatory requirements for transportation. Additionally, they specify tests like thermal tests and external short-circuit tests.

13.2.3 Performance and Durability Standards

In addition to transportation and abuse testing, ensuring the cell performance and durability are critical for maintaining safety, reliability, and efficiency in battery systems. Performance and durability standards assess a cell's ability to deliver consistent performance over its lifespan, enduring various conditions such as charge and discharge cycles, temperature fluctuations, and environmental stresses. These tests help to predict a cell's behavior during real-world use, allowing manufacturers and users to estimate service life, energy efficiency, and resistance to performance degradation.

Lithium-ion cells, especially those used in high-demand applications such as EVs, ESSs, and portable electronics, are subject to rigorous performance testing to ensure long-term viability. Standards focusing on performance and durability are essential to validate the robustness of the cells and their ability to withstand repeated charge-discharge cycles, high and low temperatures, and other operational stressors. This section outlines key standards for performance and durability testing in various applications. The tests required from the standards are listed in Table 13.3.

Table 13.2 Cell and battery abuse and environmental standards.

Part 1 of 2

Standards	Titles	Electrical abuse tests	Thermal abuse and environmental tests
IEC 62133-2: 2017 [9]	Secondary cells and batteries containing alkaline or other nonacid electrolytes—Safety requirements for portable sealed secondary cells and for batteries made from them, for use in portable applications; Part 2: Lithium systems	• Continuous charging at constant voltage • External short-circuit • Overcharging • Forced discharge • Forced short circuit	• Thermal abuse
IEC 62619: 2022 [11]	Secondary cells and batteries containing alkaline or other nonacid electrolytes—Safety requirements for secondary lithium cells and batteries, for use in industrial applications	• External short circuit • Overcharge • Forced discharge • Internal short circuit • Propagation test	• Thermal abuse
IEC 62660-2: 2018 [13]	Secondary lithium-ion cells for the propulsion of electric road vehicles—Part 2: Reliability and abuse testing	• External short circuit • Overcharge • Forced discharge	• High temperature endurance • Temperature cycling
UL 1642: 2020 [57]	Lithium batteries	• Short circuit test • Abnormal charging test • Forced discharge test	• Heating test • Temperature cycling • Projectile test
SAE J2464: 2021 [28]	Electric and hybrid electric vehicle rechargeable energy storage system (RESS) safety and abuse testing	• Short circuit • Overcharge • Forced discharge • Separator shutdown integrity test	• Thermal stability test • Thermal shock cycling • High temperature hazard test • Cycling without thermal management/thermal runaway test • Passive propagation resistance test

Part 2 of 2

Standards	Mechanical abuse and environmental tests	Application areas	Application battery levels
IEC 62133-2: 2017 [9]	• Crush • Freefall • Vibration • Mechanical shock • Case stress at high ambient temperature	Portable and various applications	Cell and pack
IEC 62619: 2022 [11]	• Impact • Drop	Stationary applications, motive applications with the exception of road vehicles	Cell, cell block, and battery system
IEC62660-2: 2018 [13]	• Vibration • Shock • Crush	EVs	Cell and cell block
UL 1642: 2020 [57]	• Crush • Impact • Shock • Vibration • Altitude simulation test	Consumer electronics, medical devices, and various applications	Cell
SAE J2464: 2021 [28]	• Shock • Penetration • Crush • Rollover • Immersion test • Drop • Hazardous substance monitoring test	Electric and hybrid electric vehicles	Cells (battery or capacitors), modules, and packs

Table 13.3 Cell and battery performance and durability standards.

Part 1 of 2

Standards	Titles	Performance and durability tests
IEC 61960-3: 2017 [14]	Secondary cells and batteries containing alkaline or other nonacid electrolytes—Secondary lithium cells and batteries for portable applications—Part 3: Prismatic and cylindrical lithium secondary cells and batteries made from them	• Discharge performance • Capacity retention and recovery after long-term storage • Endurance in cycles • Measure of internal resistance • Electrostatic discharge (ESD) test
IEC 61960-4: 2024 [15]	Secondary cells and batteries containing alkaline or other nonacid electrolytes—Secondary lithium cells and batteries for portable applications—Part 4: Coin secondary lithium cells, and batteries made from them	• Discharge performance • Capacity recovery after long-term storage • Endurance in cycles • Measure of internal resistance
IEC62660-1: 2018 [12]	Secondary lithium-ion cells for the propulsion of electric road vehicles—Part 1: Performance testing	• Capacity test • Power density test • Energy density test • Storage life test • Cycle life test with periodic performance evaluation • Energy efficiency test
SAE J2288: 2020 [30]	Life cycle testing of electric vehicle battery modules	• Capacity test • Dynamic capacity and power test • Cycle life test with periodic performance evaluation • Methodology to determine the expected service life (in number of cycles)
SAE J1798: 2019 [31]	Recommended practice for performance rating of electric vehicle battery modules	• Static capacity test • Dynamic capacity test • Peak power capability test • Charge retention test • Charge acceptance test • Storage test

Part 2 of 2

Standards	Application areas	Application battery levels
IEC 61960-3: 2017 [14]	Portable applications	Cell and pack
IEC 61960-4: 2024 [15]	Portable applications, watches, backup power supplies	Cell and pack
IEC62660-1: 2018 [12]	EVs	Cell and cell block
SAE J2288: 2020 [30]	EVs	Battery module
SAE J1798: 2019 [31]	EVs	Battery module

The IEC 61960 series, particularly Parts 3 [14] and 4 [15], focuses on performance testing of lithium cells and batteries used in portable applications. These standards are especially relevant for consumer electronics, wearables, and backup power applications. Part 3 emphasizes prismatic and cylindrical cells' discharge performance, capacity retention, and the recovery of capacity after long-term storage. It also includes endurance testing to evaluate the cells' ability to maintain their performance over repeated charge and discharge cycles. Additionally, the standard measures changes in internal resistance over time, which can provide insights into cells' efficiency and thermal management capabilities. Part 4, which applies to coin-type secondary lithium cells, addresses similar performance concerns but tailors the testing criteria to smaller-scale applications such as watches and backup power systems. Electrostatic discharge (ESD) testing is also included to ensure that cells can handle sudden voltage spikes, common in portable devices.

IEC 62660-1 [12] establishes performance testing protocols for lithium-ion cells and cell blocks used in EV propulsion systems. The standard reflects the need for cells to maintain high energy efficiency, cycle life, and capacity under demanding driving conditions. It also includes a comprehensive cycle life test, where cells are subjected to repeated charge and discharge cycles, with periodic performance evaluations to monitor any degradation. This allows manufacturers to predict the long-term behavior of cells, ensuring that they can withstand the demands of EV applications. Additionally, the standard includes energy efficiency testing, which assesses the cells' ability to convert stored energy into usable power—an essential factor for maximizing vehicle range and performance.

In North America, the SAE J2288 [30] and SAE J1798 [31] standards are key benchmarks for the performance testing of EV battery modules. These standards provide a structured methodology to estimate service life, helping manufacturers to ensure that battery modules can endure thousands of charge-discharge cycles while maintaining their performance. SAE J2288 outlines tests designed to evaluate the cycle life of battery modules, monitoring their ability to retain capacity and deliver power over repeated cycles. SAE J1798 adds dynamic testing protocols, simulating real-world driving conditions to assess battery modules' performance under variable loads, such as acceleration and deceleration. These standards also address the efficiency of charge retention and acceptance, which is critical for hybrid and electric vehicles that depend on both regenerative charging and conventional recharging. In addition, storage tests are included to evaluate how battery modules perform after periods of inactivity, ensuring that they can be stored without significant performance loss.

In addition to the standards discussed above, IEEE 1725 [18] and IEEE 1625 [19] are critical for ensuring the performance and durability of rechargeable batteries used in portable electronic devices. Both standards emphasize the importance of long-term reliability, requiring batteries to undergo extensive pre-production, qualification, and ongoing reliability tests. These include capacity and cycle life tests, which ensure that the battery maintains its rated capacity after repeated charge-discharge cycles, and charge retention and acceptance tests, which evaluate how well a battery retains its charge during periods of nonuse and how efficiently it accepts a charge during recharging. The requirements of these standards for ongoing reliability ensure that batteries used in mobile phones and computing devices deliver consistent performance over their operational life, supporting the high reliability needed for modern portable electronics.

13.3 Battery Pack and System Safety and Reliability Standards

While individual cells are the fundamental building blocks of batteries, the battery pack and battery system represent the integrated units that deliver power safely, reliably, and efficiently across various applications. The battery packs and systems involve assembling multiple cells into a cohesive unit, along with the associated control and protection systems. A failure at the pack or system levels can have significant consequences, potentially leading to thermal runaway events, system malfunctions, or even catastrophic failures that could compromise the safety of the entire application.

Battery pack and system standards are designed to address the unique challenges of integrating multiple cells, including thermal management, electrical safety, mechanical durability, and the functionality of the BMS. Unlike cell-level standards, which focus primarily on the individual cell's safety and reliability, battery pack and system standards encompass a broader scope, ensuring that the entire unit operates safely under various conditions, including transportation, everyday use, and extreme scenarios.

This section delves into the key standards governing the safety and reliability of battery packs and systems, exploring how these standards build upon cell-level requirements while addressing the additional complexities of pack and system integration. The key standards governing battery packs and systems will be explored, and they are organized into four main areas: transportation requirements, abuse and environmental requirements, performance requirements, and functional and BMS requirements.

13.3.1 Transportation Standards

Building on Section 13.2.1, this section focuses on the key transportation standards specific to battery packs, while Chapter 12 offers a comprehensive discussion on battery transportation. The transportation of battery packs presents significant safety challenges, particularly for LIBs, which are prone to thermal runaway under certain conditions. This section examines the transportation safety requirements specific to battery packs, building on the foundational knowledge of UN 38.3 [40] and IEC 62281 [10] discussed in the context of cell-level safety in Section 5.2. While the tests required under these standards for individual cells are critical, transporting battery packs introduces additional complexities that necessitate rigorous testing to ensure safety throughout the shipping process. The required safety tests from the two standards are listed in Table 13.1.

The UN 38.3 standard [40], widely recognized for governing the safe transportation of lithium batteries, extends its comprehensive testing protocol to battery packs. While the general principles of pack-level testing—addressing electrical, thermal, mechanical, and environmental stresses—remain consistent with cell-level requirements, the focus here is on how these packs perform under transportation conditions that could induce thermal runaway or other hazardous failures. Rechargeable battery packs, including those composed of previously tested cells, are required to undergo tests T1 to T5 and T7.

These tests are crucial because the interaction between multiple cells, the packaging, and the BMS can significantly influence overall safety. For instance, the T5 (external short circuit) test evaluates how well the pack as a whole can handle a direct short circuit without leading to dangerous failures such as fire or explosion. Similarly, the T7 (overcharge) test is critical for assessing whether the integrated BMS can effectively prevent overcharging.

While the IEC 62281 standard [10] builds upon the foundational testing protocols outlined in UN 38.3, it offers additional, targeted specifications that address the unique challenges of transporting lithium battery packs. At the pack level, this standard goes beyond merely replicating cell-level tests, incorporating provisions that are particularly relevant for ensuring the safe transit of larger, more complex battery packs.

A key aspect of IEC 62281 is its emphasis on the role of packaging for safeguarding batteries during transportation. While the standard includes the same rigorous T1 to T8 tests outlined in UN 38.3, it introduces P1, package drop test for battery packs. This test is designed to assess the integrity of the packaging by simulating the drops and impacts that are likely to occur during handling. The P1 test is crucial because it ensures that the protective packaging can effectively shield the battery from damage if it is accidentally dropped—a scenario that is too common in real-world transportation environments.

13.3.2 Abuse and Environmental Standards

Abuse and environmental testing of battery packs and systems are essential for assessing how these complex systems withstand extreme conditions during real-world operations. Abuse tests simulate scenarios where battery packs and systems might experience electrical, thermal, or mechanical stresses beyond their normal operating parameters, ensuring that failures under these conditions do not lead to hazardous events such as fires or explosions. Similarly, environmental tests ensure that battery packs maintain integrity when exposed to harsh conditions, including extreme temperatures, moisture, and vibration. These standards apply to various battery packs and systems, from those used in EVs to stationary ESSs.

First, some of the abuse and environmental standards discussed in Section 13.2.2 also apply to battery packs or systems. For instance, IEC 62133-2 [9] sets safety requirements for secondary lithium cells and packs, particularly those used in portable applications. It outlines critical tests such as external short-circuit and overcharge for LIB packs, ensuring that packs can operate safely under abusive conditions. These tests are crucial for preventing thermal runaway, which is a key concern at the pack level due to the higher energy and complexity of the packs. Mechanical tests such as vibration, shock, and freefall tests can be conducted at the cell or pack levels to ensure that the battery maintains its structural integrity when subjected to mechanical stresses. Additionally, the standard requires pack-level case stress tests at high ambient temperatures to assess the pack's ability to withstand deformation or rupture under extreme heat conditions, ensuring overall safety and durability in demanding environments.

IEC 62619 [11], tailored for industrial applications, extends safety principles to more demanding environments. While many tests focus on the cell or cell block levels, the drop test applies from the cell level to the full battery system, ensuring structural integrity after mechanical impacts. A key requirement at the system level is the propagation test, which evaluates the battery system's ability to prevent thermal runaway from spreading between cells. This test ensures that, if a single cell fails, the event is contained within that cell or module, preventing a chain reaction that could lead to catastrophic system failure, making it critical for large industrial battery systems.

SAE J2464 [28] focuses on abuse testing for RESSs. While this standard encompasses cells, it also applies at the system level. It mandates rigorous electrical abuse tests, including short-circuit and forced discharge, which ensure that cells, modules, and assembled packs, can resist catastrophic failure. The mechanical abuse and environmental tests above cell level, including mechanical shock, crush, penetration, and thermal shock cycling, simulate the stresses a battery system might experience during vehicle operation or accidents. Importantly, at the module or pack levels, SAE J2464 also requires additional tests such as cycling without thermal management, high-temperature hazard, passive propagation resistance, roll-over simulation, immersion test, and drop test, ensuring that the entire battery system, not just the cells, remains safe under various conditions.

IEEE 1625 [19] and IEEE 1725 [18] also define crucial abuse and environmental tests at the pack level for portable devices. IEEE 1625 emphasizes the safety of battery packs in mobile computing devices, such as a vibration test to simulate real-world mechanical stresses. The standard also includes the transportation test (UN 38.3) and altitude simulation (IEC 62281) to assess a battery pack's ability to withstand conditions encountered during transportation. A key feature of IEEE 1625 is the mitigation of thermal runaway, ensuring that thermal incidents at the cell level do not propagate throughout the entire pack. IEEE 1725, similarly, addresses battery pack safety for cellular devices. The standard requires tests such as external short-circuit, thermal protection, and ESD to ensure that battery packs maintain safety under various electrical abuse conditions. Additionally, the drop test evaluates a battery pack's structural integrity during handling and accidental falls. Together, IEEE 1625 and 1725 provide a robust framework for ensuring the safety and reliability of battery packs in consumer electronics, particularly under abusive or environmentally harsh conditions.

After reviewing the standards already addressed in Section 13.2.2, the focus is now on standards specifically targeting the battery pack and system levels. These additional standards, summarized in Table 13.4, address unique abuse and environmental requirements tailored to battery packs and systems. While UL 1642 [57] focuses specifically on the safety of secondary lithium cells used in consumer electronics, UL 2054 [58] expands the scope to cover battery packs, including a broader range of chemistries beyond lithium-ion. UL 2054 requires that individual cells comply with UL 1642, ensuring that they meet stringent safety criteria before integration into packs. UL 2054 addresses battery pack performance under various electrical abuse scenarios, such as short-circuit, overcharge, and forced discharge. It mandates thermal tests, including heating, temperature cycling, and fire exposure to evaluate how battery packs withstand extreme temperatures. Mechanical abuse tests, such as crush and impact assessments, simulate potential physical damage, ensuring battery safety under various conditions. UL 2054s comprehensive approach guarantees the safety of household and commercial battery packs, even in demanding operational environments.

Designed for LIB systems used in electric power tools and motor-operated appliances, UL 2575 [59] establishes stringent requirements for electrical and thermal abuse resistance. This standard subjects' battery systems to tests that simulate overcharge, forced discharge, and short circuit, ensuring their safety under conditions that might lead to electrical failure. Thermal tests, such as heating and temperature cycling, further ensure that the battery system can endure temperature fluctuations during high-demand operations. Additionally, mechanical abuse tests, including vibration and drop tests, ensure the structural integrity of the battery system during operation and transport. UL 2575 plays a key role in ensuring that battery systems in high-performance tools remain safe under operational and abusive stresses.

UL 2271 [60] focuses on battery packs used in LEV applications. These packs face unique challenges due to the dynamic and sometimes unpredictable nature of vehicle operation. UL 2271 establishes rigorous tests to assess how these battery systems withstand electrical abuse, such as overcharging and imbalanced charging, which can lead to dangerous conditions if not managed correctly. The thermal tests include temperature cycling, ensuring that battery packs remain reliable under fluctuating temperature conditions typical in LEV use. Mechanical abuse tests, including vibration, crush, and immersion tests, ensure that LEV battery packs maintain safety when subjected to mechanical shocks during everyday use.

UL 2580 [61] establishes safety standards for battery systems used in EVs, addressing electrical, thermal, and mechanical stresses. The standard requires battery packs to undergo rigorous electrical abuse testing, including short circuit, overcharge, and over-discharge protection. These tests assess the battery system's ability to handle extreme electrical conditions without catastrophic

Table 13.4 Battery pack and system abuse and environmental standards.

Part 1 of 2

Standards	Titles	Electrical abuse tests	Thermal abuse and environmental tests
UL 2054: 2021 [58]	Household and commercial batteries	• Short circuit test • Abnormal charging test • Abuse overcharge test • Forced discharge test • Limited power source test	• Heating test • Temperature cycling test • Fire exposure/projectile test
UL 2575: 2011 [59]	Lithium-ion battery systems for use in electric power tool and motor operated, heating, and lighting appliances	• Short circuit test • Overcharge test • Forced discharge test	• Heating test • Temperature cycling test
UL 2271: 2023 [60]	Batteries for use in light electric vehicle (LEV) applications	• Short circuit test • Overcharge test • Forced discharge test • Imbalanced charging test • Dielectric voltage withstand • Isolated resistance	• Temperature cycling test
UL 2580: 2020 [61]	Batteries for use in electric vehicles	• Short circuit test • Overcharge test • Over-discharge protection test • Isolated resistance • Imbalanced charge test • Dielectric voltage withstand • Continuity test	• Temperature cycling test • Fire exposure • Single-cell failure design tolerance • Failure of cooling/thermal stability system
UL 1973: 2022 [62]	Batteries for use in light electric rail (LER) applications and stationary applications	• Short circuit test • Overcharge test • Over-discharge protection test • Imbalanced charge test • Continuity test • Working voltage measurements • Dielectric voltage withstand test • Testing of electrical components	• Thermal cycling test • Temperature and operating limits check • Failure of cooling/thermal stability system • Fire exposure/projectile test • Single-cell failure design tolerance
ISO 6469-1: 2019 [25]	Electrically propelled road vehicles—safety specifications Part 1: Rechargeable energy storage system (RESS)	• Short circuit • Overcharge protection • Over-discharge protection • Protection against internal overheating	• Thermal shock cycling • Exposure to fire (simulated vehicle accident)
SAE J2929: 2013 [27]	Electric and hybrid vehicle propulsion battery system safety standard—lithium-based rechargeable cells	• External short circuit • Single point overcharge protection system failure • Single point over-discharge protection system failure • Protection against high voltage exposure	• Temperature cycling test • Exposure to simulated vehicle fire • Single point thermal control system failure
SAE J2380: 2021 [29]	Vibration testing of electric vehicle batteries		

Table 13.4 (Continued)

Part 2 of 2

Standards	Mechanical abuse and environmental tests	Application areas	Application battery levels
UL 2054: 2021 [58]	• Crush test • Impact test • Shock test • Vibration test • 250 N steady force test • Mold stress relief test • Drop impact test	Household and commercial batteries, including lithium batteries, portable applications	Battery pack
UL 2575: 2011 [59]	• Vibration test • Shock test • Crush test • Drop test • Altitude simulation test	Battery-operated electric power tool and motor operated, heating, and lighting appliance	Battery system
UL 2271: 2023 [60]	• Vibration test • Shock test • Crush test • Drop test • Rotation test • Immersion test	Light electric vehicle (LEV) applications	Battery pack and battery pack-electrochemical capacitor assembly
UL 2580: 2020 [61]	• Vibration test • Shock test • Crush test • Drop test • Rotation • Immersion test	EVs	Battery pack and battery pack-electrochemical capacitor assembly
UL 1973: 2022 [62]	• Static force test • Impact test • Drop impact test • Wall mount fixture/handle test • Vibration test • Shock test • Crush test • Mold stress test • Start-to-discharge test • Resistance to moisture test • Salt fog test	Stationary batteries for energy storage applications	Battery system
ISO 6469-1: 2019 [25]	• Vibration • Mechanical shock • Vehicle crash (simulated vehicle accident) • Immersion into water (simulated vehicle accident)	EVs and HEVs	Battery pack and system
SAE J2929: 2013 [27]	• Drop test • Mechanical shock • Immersion test • Battery enclosure integrity	EVs and HEVs	Battery system
SAE J2380: 2021 [29]	• Vibration test	EVs	Battery module and pack

failure. Thermal tests, such as temperature cycling and fire exposure, ensure that the battery system can withstand high temperatures without catching fire or exploding. Mechanical tests, including vibration and shock, ensure a battery's durability during everyday use and vehicle collisions. UL 2580 is critical for ensuring the safety and reliability of EV battery packs, where the stakes of failure are high due to the energy density and size of the battery systems.

UL 1973 [62] focuses on battery systems used in LER and stationary applications. These applications often require batteries to operate under long-term, harsh environmental conditions. UL 1973 subjects' battery systems to electrical abuse tests such as short circuit, overcharge, and imbalanced charging to assess a system's resilience to electrical failures. Additionally, thermal cycling tests evaluate a system's ability to endure repeated temperature fluctuations. Mechanical abuse tests, such as impact and drop tests, ensure that the battery system remains structurally sound when subjected to physical damage. The inclusion of environmental resilience tests, such as resistance to moisture and salt fog, ensures that a battery system performs reliably in outdoor and industrial environments. While UL 1973 generally applies to battery systems, in certain cases tests may be conducted on individual cells or subassemblies if they are deemed representative of the complete battery system. This allows for flexibility in testing scenarios where testing the entire system is not necessary or practical.

ISO 6469-1 [25] defines safety specifications for RESSs used in electric and hybrid vehicles. The standard addresses critical electrical abuse scenarios, such as short circuit, overcharge, and internal overheating, to prevent hazardous failures during vehicle operation. ISO 6469-1 also mandates thermal shock and fire exposure tests, simulating vehicle crash scenarios where battery packs might be exposed to extreme heat or fire. In addition to electrical and thermal tests, the standard includes mechanical shock, vibration, and immersion tests to ensure that a battery system remains safe and reliable under harsh operating conditions, such as road vibrations or accidents.

SAE J2929 [27] focuses on the safety of lithium-based rechargeable cells used in electric and hybrid vehicle propulsion systems. The standard mandates tests for electrical abuse, such as short circuit, overcharge, and over-discharge, ensuring that the battery system can withstand these events without failure. Thermal abuse tests, including temperature cycling and exposure to simulated vehicle fires, assess a battery's ability to endure extreme thermal conditions. Additionally, SAE J2929 requires mechanical tests such as shock, drop, and immersion tests, ensuring that a battery system remains structurally sound and safe during everyday vehicle use and in crash scenarios.

SAE J2380 [29] focuses on vibration testing for electric vehicle battery modules and packs, addressing the continuous mechanical stresses that batteries endure during vehicle operation. Vibration is a critical concern in automotive applications, as prolonged exposure can degrade battery performance and integrity. This standard simulates long-term exposure to vibration conditions by requiring tests that reflect real-world scenarios. A random vibration profile is typically used, though a specific vibration durability profile based on actual vehicle measurements is preferred to match the specific electric vehicle application better. By incorporating these detailed vibration tests, SAE J2380 ensures that battery systems can maintain their reliability and structural integrity throughout their life cycle, even under demanding operational conditions.

UL 9540 [63] is a system-level standard for energy storage systems and equipment. It addresses the overall safety of ESS by ensuring that these systems operate safely in residential, commercial, and utility settings. The standard covers electrical, mechanical, and environmental hazards, ensuring that all components of the ESS work together reliably and safely. Recent updates in the third edition, published in 2023, have introduced new requirements like explosion control and remote software update management, aligning the standard with modern ESS advancements and safety codes such as NFPA 855 [67, 68].

Together with UL 9540, UL 9540A [69] provides a specific test methodology for evaluating the potential for thermal runaway and fire propagation in battery energy storage systems. This test method is particularly important for large LIB installations, as it focuses on whether a failure in one cell can propagate to others, potentially causing a significant fire or explosion. Testing starts at the cell level, examining whether a cell can enter thermal runaway, and progresses to module, unit, and installation levels, assessing the ability of a system to contain or prevent fire propagation. The installation-level tests evaluate the effectiveness of fire suppression systems for controlling thermal events [70, 71]. The UL 9540 and UL 9540A standards are critical for ensuring the safety of modern ESSs, particularly as they are increasingly deployed in residential and industrial settings.

13.3.3 Performance and Durability Standards

The long-term performance and durability of battery packs and systems are essential for their safe and reliable operation, particularly in high-demand applications such as EVs and ESSs. Performance standards assess a battery's ability to maintain capacity, efficiency, and safety over its life cycle, when subjected to various stresses, such as charge-discharge cycles, temperature fluctuations, and other operational challenges.

Section 13.2.3 discussed IEC 61960-3 [14] and IEC 61960-4 [15], where the focus is on secondary lithium cells. These standards also apply to battery packs, and they provide critical guidance on performance testing, especially for prismatic, cylindrical, and coin-type secondary lithium batteries used in portable applications. The tests outlined in these standards evaluate key performance metrics, such as discharge capacity, cycle life endurance, internal resistance, capacity retention, and recovery after long-term storage. These factors are crucial for battery packs, ensuring their long-term durability.

In automotive applications, ISO 12405-4 [23] specifically focuses on the performance testing of lithium-ion traction battery packs and systems used in electric and hybrid vehicles. This standard evaluates a wide range of performance parameters, including the energy and capacity of the battery at various temperatures and discharge rates, power and internal resistance, and state-of-charge (SOC) loss during storage and operation. Additionally, ISO 12405-4 includes tests for cycle life, cranking power at low and high temperatures, and the energy efficiency of the battery during normal and fast charging. These tests ensure that the battery system can meet the demanding operational requirements of electric vehicles, where maintaining performance over extended periods and in diverse environmental conditions is critical. By addressing these performance parameters, ISO 12405-4 helps manufacturers predict and improve the longevity and reliability of battery systems in electric and hybrid vehicles, ensuring that they can meet the challenges of real-world usage. The performance and durability test requirements from ISO 12405-4 are summarized in Table 13.5.

These standards form the backbone of performance and durability testing for battery packs and systems, ensuring that they function efficiently under operational conditions and maintain their performance and safety throughout their operational lifespan. The rigorous testing outlined in these standards helps manufacturers and end users to assess how well the battery packs and systems will perform in the field, contributing to the overall safety and reliability of modern energy storage and propulsion systems.

13.3.4 BMS Functional Standards

BMUs and BMSs (BMS represents both BMUs and BMSs for simplicity in this chapter) play a critical role in ensuring the safety, reliability, and efficiency of battery packs and systems across various

Table 13.5 Battery pack and system performance and durability standards.

Standards	Title	Performance tests	Application areas	Application battery levels
ISO 12405-4: 2018 [23]	Electrically propelled road vehicles—test specification for lithium-ion traction battery packs and systems Part 4: Performance testing	• Energy and capacity at different temperatures and discharge rates • Power and internal resistance • No-load SOC loss • SOC loss at storage • Cranking power at low and high temperatures • Energy efficiency at normal and fast charging • Cycle life	EVs and HEVs	Battery pack and system

applications, from consumer electronics to EVs and industrial ESSs. The BMS is responsible for monitoring key parameters such as voltage, current, temperature, and SOC and implementing protective functions that prevent overcharging, over-discharging, and thermal runaway. Several key standards discussed throughout this chapter incorporate specific functional safety requirements for BMSs, ensuring that these systems operate effectively under various conditions.

IEC 62619 [11], which focuses on industrial LIBs, outlines comprehensive BMS functional requirements. At the system level, the BMS must monitor the electrical and thermal conditions of the battery pack and act when necessary to maintain safety. This includes managing overcharge, over-discharge, and temperature regulation, all of which are critical for preventing hazardous events such as thermal runaway. IEC 62619 also requires the BMS to detect an internal short circuit and respond promptly to avoid catastrophic system failures, especially in large-scale industrial applications.

For electric and hybrid vehicle applications, ISO 12405-4 [23] specifies detailed performance testing requirements for BMS functionality. The standard covers tests for SOC estimation, cycle life monitoring, and thermal management. These tests are essential to verify that the BMS can maintain safe operation across a wide range of temperatures and usage scenarios. This standard is particularly important for ensuring that battery systems in EVs can meet performance expectations while safeguarding against potential failures during operation.

ISO 6469-1 [25] defines safety requirements for high-voltage systems used in electric and hybrid vehicles, focusing on functional safety. The BMS must monitor SOC, temperature, and voltage levels, ensuring that the battery operates within safe limits during use. This standard also addresses electrical hazard protection—including overcharge protection, over-discharge protection, and protection against internal overheating—requiring the BMS to disconnect the battery in case of dangerous failures, which is critical for preventing electrical hazards in high-voltage applications.

SAE J2929 [27] outlines safety criteria for LIB systems in electric and hybrid vehicles, with specific functional safety requirements for BMSs. The standard emphasizes the need for the BMS to protect the battery system from overcharging, over-discharging, and overheating, ensuring safe operation in high-voltage vehicle propulsion systems. This includes monitoring and controlling electrical parameters to prevent dangerous failures during vehicle operation.

UL 1973 [62] is crucial for defining BMS functional requirements for stationary ESSs and LER applications. It mandates that the BMS actively monitors key metrics such as SOC, state-of-health (SOH), and temperature, ensuring that the battery system operates within safe parameters. The standard also requires the BMS to initiate protective measures in the event of abnormal conditions, such as disconnecting the battery pack if a hazardous situation arises.

UL 2580 [61] for EV battery systems also includes stringent BMS testing requirements. The BMS must protect against electrical and thermal abuse, ensuring that the system can safely handle conditions such as overcharging, short circuits, and extreme temperatures. Functional testing is a critical aspect of this standard, as it validates the BMS's ability to maintain system stability under demanding conditions, ensuring long-term reliability in EV applications.

IEEE 1625 [19] and IEEE 1725 [18] also address functional safety aspects of the BMS, particularly for battery packs used in mobile computing and cellular devices. IEEE 1625, which focuses on multicell batteries for mobile computing devices, outlines key BMS responsibilities such as charge and discharge protection, cell monitoring, and over-temperature protection. The standard also emphasizes traceability for system components, ensuring that battery management can be tracked for diagnostic purposes. These functional safety measures ensure that the BMS can effectively manage operational conditions, safeguarding batteries from overheating, overcharging, and electrical failure.

Similarly, IEEE 1725, specific to rechargeable LIBs used in mobile phones, includes stringent functional safety requirements. The BMS must ensure cell block overvoltage and undervoltage protection, preventing the battery from operating outside its safe voltage range. The standard also requires the validation of maximum voltage protection, ensuring that the BMS can control and limit voltage levels within the battery pack. This ensures that mobile phone batteries are protected from voltage-related risks, maintaining the overall integrity and safety of the batteries.

Collectively, these standards ensure that BMSs are equipped to handle the various challenges posed by modern battery applications. By monitoring, protecting, and controlling the battery pack or system, the BMS plays a key role in preventing hazardous conditions, extending the life of the battery, and ensuring that it operates safely and efficiently.

13.4 Safety Standards and Regulations Incorporating Batteries for Different Applications

Since electronic devices, appliances, and systems rely on batteries for various applications, specific safety considerations must be addressed to ensure the secure integration of batteries into these products. Each battery-powered system presents unique operational environments and challenges, which in turn require tailored safety protocols and performance standards. Batteries, as crucial components in everything from handheld electronics to large-scale energy storage and propulsion systems, must be designed and tested to meet stringent safety and reliability requirements specific to their use case.

Batteries used in portable consumer devices, such as smartphones or laptops, face different safety hazards than those used in electric vehicles or industrial applications. While portable devices prioritize size, weight, and energy density, the focus on automotive batteries shifts to long-term durability, thermal management, and high-energy performance. Meanwhile, in extreme environments like marine, aviation, and space applications, batteries must endure harsh conditions such as intense vibrations, temperature fluctuations, and potential exposure to corrosive materials.

To address these varied safety and operational demands, international and regional standards organizations, such as the IEC, UL, SAE, and NASA, have developed application-specific guidelines. These standards encompass everything from design considerations, such as cell chemistry and system architecture, to rigorous testing protocols that simulate real-world stresses like electrical abuse, mechanical impact, and thermal extremes. As technology evolves, these safety standards continue to adapt, ensuring that batteries can be safely integrated into emerging products and systems while minimizing the risk of failures.

This section explores the specific safety standards and regulations that govern battery applications across diverse industries. Whether powering portable electronics, propelling electric vehicles, ensuring uninterrupted power in critical systems, or enabling space exploration, batteries play a vital role in modern technology. The safety standards discussed here are tailored to each application's unique needs, ensuring that batteries are not only efficient but also secure and reliable under a wide range of operating conditions.

13.4.1 Portable Devices (e.g., Smartphones, Laptops)

The standards for portable devices encompass LIB safety, performance, and testing to ensure their safe use in portable devices. A rechargeable battery is an electronic device's heart, and, while they do not pump anything, batteries supply laptops, smartphones, and other electronics with their electric life force. Most batteries utilized for portable devices are lithium-ion, a technology so important to modern life that the 2019 Nobel Prize in Chemistry was awarded to three of its pioneers.

ANSI standards, which are referenced by the FDA for the safe design of medical products using Li-ion batteries (LIBs), are explained below [72]:

Published by NEMA, ANSI C18.2M Part 1-2019—Portable Rechargeable Cells and Batteries—General and Specifications [35] aids in assuring the electrical and physical interchangeability of rechargeable battery products from different manufacturers. It minimizes the proliferation of cell and battery types by defining standard performance tests and providing guidance to consumers, manufacturers, designers, and other end-users. It addresses battery dimensions, terminals, a standard charge method, polarity, and test conditions and procedures. The NEMA BS 30000-2021 guidance document provides information on the issues related to the use of LIBs, how fires start in batteries, and how they may be detected, controlled, suppressed, and extinguished. It also provides guidance on post-fire management.

In addition, IEC 62133-2:2017 [9] specifies requirements and tests for the safe operation of portable sealed secondary lithium cells and batteries containing nonacid electrolyte, under intended use and reasonably foreseeable misuse.

IEC 60086 Part 4 [5] provides safety requirements for power supply devices producing DC, such as cell phone chargers, laptop adapters, and PC power supplies. Similarly, UL 1642 [57] covers safety requirements for primary and secondary lithium battery cells used as a power source in electronic products.

13.4.2 Automotive (Electric Vehicles, Hybrid Electric Vehicles)

Automotive batteries for electric vehicles (EVs) are subject to various safety standards and regulations to ensure their safe operation and minimize risks associated with their use. Safety standards for EV batteries are extensive and complex. The large number of standards arises from regional and national differences and the inherent complexity of the Li-ion battery technology used in the pack.

The US National Highway Traffic Safety Administration (NHTSA) [73]. has established its ongoing Battery Safety Initiative for Electric Vehicles to coordinate research and other activities to address safety risks relating to EV batteries. The goal of the activity is to converge and simplify the regulatory requirements related to EV battery safety [73].

The NHTSA chaired the development of the initial United Nations (UN) Global Technical Regulation (GTR) for electric vehicle safety, which was established under the UN World Forum for the Harmonization of Vehicle Regulations in 2018. Phase 2 is currently underway to develop GTR No. 20 for EV safety. Some of the considerations to be included in GTR No. 20 are issues related to battery thermal runaway, water immersion, and vibration resistance [73].

SAE J2464 [28], the standard for Electric and Hybrid Electric Vehicle RESS Safety and Abuse Testing that evaluates the abuse tolerance of cells and battery packs for electric vehicles. It assesses the behavior of batteries during abuse scenarios and establishes safety guidelines for their selection, design, testing, evaluation, use, packaging, storage, transportation, and disposal. The UL 2580 standard [61] focuses on the safety of batteries used in electric vehicles, covering both lithium-ion and nickel-based batteries. It sets requirements for battery construction, materials, and electrical characteristics to ensure safe operation. ISO/CD 12405-2 aims to establish a test specification for lithium-ion traction battery packs and systems used in electric vehicles.

The European Council for Automotive (EUCAR) R&D has established performance standards, including the EUCAR Severity Hazard Levels and Descriptions, which are widely used to characterize the response of test articles to abuse tests [85]. Additionally, the Verband der Automobilindustrie e.V. (VDA) is a German interest group supporting the German automobile industry. It publishes a series of standards related to battery safety and abuse testing, which cover various aspects of battery safety, including physical, thermal, mechanical, and electrical stresses, as well as safety in transit and use in vehicles. The VDA is a member of the European Automobile Manufacturers Association (ACEA).

IEC 62660 series standards [12, 13] deal with secondary lithium-ion cells for the propulsion of electric road vehicles, providing details for performance testing, reliability abuse testing, and safety. This standard defines a set of requirements for an electric vehicle battery cell for performance, safety, remaining useful life estimation, and a set of tests to validate these functionalities.

Besides them, SAE J1772 [82] provides the general physical, electrical, functional, and performance requirements to facilitate conductive charging of plug-in hybrid electric vehicles (EV/PHEVs) in North America. It defines a common EV/PHEV and supply equipment vehicle conductive charging method, including operational requirements and the functional and dimensional requirements for the vehicle inlet and mating connector.

There are several standards for electric vehicle battery testing, such as the Nordic Ecolabel Testing (White Swan), Sandia National Laboratories Electrochemical Storage System Abuse Test Procedure Manual Requirements (SAND 99-0497), and UN Transportation Testing for Lithium Batteries (UN 38.3). Likewise, there are several for manufacturing quality control standards for automotive electric vehicle (EV) batteries, including IEC 62228, USABC (United States Advanced Battery Consortium), Electric Vehicle Battery Test Procedure Manual Requirements. There also exist UL standards for functional safety: UL 1998, ISO 26262 series, UL 1642/2054, UL 2580 (The Standard for Batteries for Use in Electric Vehicles), UL 1973 (The Standard for Batteries for Use in Stationary and Motive Auxiliary Power Applications), and UL 2271 (The Standard for Batteries for Use in LEV Applications). The USABC test procedure manual defines a set of safety, abuse, and performance tests required for validating the battery systems for electric vehicle batteries.

In addition, GB 38031 [83] (National Standard of the People's Republic of China, Electric Vehicles Traction Battery Safety Requirements) and IEC 63057 [84] (Secondary Cells and Batteries

Containing Alkaline or Other Non-acid Electrolytes—Safety Requirements for Secondary Lithium-ion Batteries) provide necessary or obligatory rules. ISO also provides specifications for electrically propelled road vehicles, such as ISO 6469-1 (Safety Specifications for RESS) [25].

IEC JWG 69 Li TC69/21/5C21A defines a set of general requirements for the LIB systems used in automotive applications. A set of tests are specified for testing safety, performance functionalities, and the battery pack's eligibility for serving a second-life application [77].

SAE J1766 defines a set of crash integrity requirements for electric vehicles, fuel cells, and HEVs. In the scope of passenger vehicles, the tests for assuring safety aspects, like prevention of electric shock, exposure to high voltage (HV), exposure to electrolyte spillage, etc., are treated. SAE J2464 defines a set of safety and abuse tests for the validation of RESSs used in electric and HEVs. SAE J2380 defines a set of requirements, test profiles, and test procedures for performing vibration tests on electric vehicle batteries. SAE J2929 defines a set of safety standard functionalities that must be satisfied by LIB systems used in electric and hybrid vehicle propulsion. EN 50604-1 defines a set of test procedures for validating the safety and performance functionalities of battery systems for implementation in LEV batteries. IEC 61982 [86] defines a set of performance and endurance tests for validation of the implementation of secondary batteries (except lithium) for the propulsion of electric road vehicles. IEC 62619 [11] defines a set of tests for validating the safe operation of industrial battery systems implemented in both stationary and motive applications (forklifts, line movers, etc.). It contains the requirement that BMSs should be designed according to the functional safety approach.

For electric vehicle safety, the SAE J2344_202010 Information Report identifies and defines the preferred technical guidelines relating to safety for vehicles that contain HV, such as EVs, HEVs, PHEVs, fuel cell vehicles (FCVs), and plug-in fuel cell vehicles (PFCVs) during normal operation and charging, as applicable [88].

UL 2580 defines a set of abuse tests to ensure the safe operation of the energy storage assembly (battery pack/battery pack, ultra-capacitor, etc.) in an electric vehicle. However, performance and reliability tests of energy storage systems are not in the scope of this standard. It applies to medium and heavy electric vehicles like passenger cars and trucks, and it does not apply to LEVs such as e-bikes, e-wheelchairs, and e-scooters.

UL 2271 is similar to UL 2580, defining a set of abuse tests to ensure the safe operation of energy storage assemblies. It applies to only LEV applications, like e-bikes, e-wheelchairs, and e-scooters, which are not intended to be driven on highways.

SAE J2288 defines life cycle test methods to predict EV battery cycle life. This is done by characterizing the expected degradation in its electrical performance as a function of life and by identifying the relevant failure mechanisms under nominal operating conditions. However, aging due to operating at lower or higher temperatures in comparison to the target application is not in the scope of the tests. Modules are continuously cycled (20% DOD + complete charging) until their capacity and power capabilities fall below 80% of their start-of-life values.

SAE J1798 defines a set of performance tests to estimate the parameters of an electric vehicle battery pack, like deliverable capacity at various discharge currents, discharge power, operating temperatures, and dynamic load profiles. These tests are for estimating factors like dependence of self-discharge on battery pack rest time and its ambient temperature and the dependence of charge acceptance and peak power performance capabilities on ambient temperature.

13.4.3 Uninterruptible Power Supplies and Power Systems

Batteries for uninterruptible power supplies (UPSs) and power systems are subject to various safety standards and regulations to ensure their safe operation and minimize risks associated with their use. Some of the key safety standards and organizations that contribute to the development of these standards include NFPA 110, which guides the installation, inspection, testing, and maintenance of emergency and standby power systems, including UPS equipment. It also outlines requirements for battery replacement and interchangeability. UL 1778 covers UPS equipment, including requirements for battery systems and their compatibility with UPS equipment. Similarly, UL 2580 applies to LIBs used in UPS applications, setting requirements for battery construction, materials, and electrical characteristics to ensure safe operation.

UL 1973, Batteries for Use in LER Applications and Stationary Applications, covers the use of battery systems for "special mobile service" applications, which may include UPS systems. These batteries are intended for installation within either a rail car or within a sheltered stationary location such as a rail substation. They may utilize regenerative braking from the trains as a source of energy for recharging, and they are intended for direct connection to the rail power lines [62].

In addition to these standards, it is essential to follow local codes and regulations, such as the National Electric Code (NEC) and the International Building Code (IBC), when installing and maintaining UPS systems.

13.4.4 Marine and Navy Applications

Standards for batteries used in marine and naval applications are primarily focused on ensuring the safety of personnel and equipment, as well as minimizing environmental impact. Key safety standards and organizations involved in developing these standards are described below.

The American Bureau of Shipping (ABS) document titled Requirements for Use of LIBs in the Marine and Offshore Industries provides requirements and reference standards to facilitate effective installation and operation of LIB systems in marine and offshore environments. It covers various LIB types and addresses safety concerns relevant to these applications.

Manufacturing quality control testing of navy batteries involves several important steps and considerations. The Navy Lithium Battery Safety Program assigns authority to Naval Sea Systems Command (NAVSEA) for all lithium battery safety in the Department of the Navy (DON), and the Naval Sea Systems Command Instruction (NAVSEAINST) 9310.1c Naval Lithium Battery Safety Program assigns certification authority to SYSCOMs with appropriate subject matter expertise. The Technical Publication S9310-AQ-SAF-010, Navy Lithium Battery Safety Program Responsibilities and Procedures, provides guidelines for design and includes procedures for safety testing.

During individual cell inspection, counterfeit or damaged cells, dimensionally out-of-tolerance cells, gross electrical problem cells, and other issues should be checked for quality assurance purposes. Functional, environmental, and safety test data; safety testing program plans; and system-level testing are also important parts of the safety testing process. Navy battery safety testing informs risk acceptance and helps to minimize risk to personnel and platforms while allowing the use of lithium batteries on ships, aircraft, and submarines to advance military capabilities. Full-scale submarine battery and energy storage testing and modeling capabilities unique to the Navy are provided by the Submarine Battery Explosion and Criticality Facility (SUBBEC) at the Naval Surface Warfare Center (NSWC), Crane Division.

Marine applications require high-quality and safe batteries to operate effectively and protect against potential risks. Manufacturers should follow strict guidelines and testing procedures to ensure the safety and reliability of lithium batteries used in marine environments.

The Naval Postgraduate School (NPS) Lithium Battery Safety Program ensures a robust and transparent approach to acquiring, managing, and operating lithium battery systems. It includes acquisition review, data recording, and testing to ensure compliance with safety standards. Besides this, the DON Lithium Battery Safety Program (LBSP) is responsible for assessing, evaluating, and minimizing risk to personnel and platforms while allowing the use of lithium batteries on ships, aircraft, and submarines. It includes testing, certification, and safety management procedures. The program follows strict guidelines outlined in various manuals and publications, such as the OPNAV INST 5100.23g Navy Safety and Occupational Health Program Manual, which assigns authority to NAVSEA for all DoN Lithium Battery Safety, and the NAVSEAINST 9310.1c Naval Lithium Battery Safety Program, which assigns certification authority to the systems commands (SYSCOM). These are the materiel agencies of the United States Department of the Navy, responsible for the design, construction, and maintenance of military systems such as ships, aircraft, facilities, and weapons with appropriate subject matter expertise.

In addition, technical publication S9310-AQ-SAF-010, Navy Lithium Battery Safety Program Responsibilities and Procedures, describes the safety review process and includes procedures for safety testing. NAVSEAINST 5100.12M, the System Safety Engineering Manual, invokes MIL-STD-882E for risk assessment, supporting processes outlined in LBSP documents. NAVSEAINST 5000.8, the Naval SYSCOM Risk Management Policy, gives overarching risk acceptance requirements used to support the LBSP process. The LBSP consists of several steps, including submission of a safety data package for the lithium battery to NAVSEA and safety testing. The safety testing includes tests by LBSP-approved personnel using specified test procedures and equipment. Later, the data package and test results are reviewed by the LBSP's designated technical agent, and upon successful completion of the previous steps, a formal recommendation for the proposed battery's use and environment is provided by NAVSEA.

The Navy's Lithium Battery Safety Program aims to minimize risk to personnel and platforms while enabling the use of lithium batteries on ships, aircraft, and submarines to advance military capabilities. It considers various hazards associated with lithium batteries, including venting of noxious and/or hazardous gases, fire, and high-pressure events.

Naval Surface Warfare Center, Carderock Division (NSWCCD), is involved in testing and evaluating LIBs and container solutions for safe transportation, storage, and charging. They work on developing common standards for lithium battery containment systems. These standards cover various aspects of battery safety, including physical, thermal, mechanical, and electrical stresses, as well as safety in transit and use in marine and naval applications. They aim to ensure that LIBs can resist these stresses and minimize the risk of fires, explosions, or other hazardous events. Additionally, local codes and regulations, such as the International Marine Organization (IMO) and the US Coast Guard, should be followed when installing and maintaining LIB systems in marine and naval applications. The American Boat & Yacht Council (ABYC) standard is a voluntary global safety standard for the design, construction, maintenance, and repair of recreational boats.

The ISO 14001:2004 Environmental Management System is a certification that ensures that the manufacturing process is compliant and licensed on an international level. It covers functional safety, impact, ignition protection, water ingress, and salt spray corrosion, which are common on pleasure craft and boats. The UL 2054 certification ensures that select battery models meet safety and quality control standards, which is essential for lithium batteries used in marine applications.

The UN/DOT 38.3 certification ensures that all Dakota Lithium batteries are tested and certified per section 38.3 of the UN Manual of Tests and Criteria (UN Transportation Testing). In summary, manufacturers of lithium batteries for marine applications should follow strict guidelines and testing procedures to ensure the safety and reliability of their products. Compliance with relevant safety standards and certifications, such as UN 38.3, the ABYC standard, UL 2054 certification, and UN/DOT 38.3 certification, is essential to ensure the safe and efficient use of lithium batteries in marine environments.

13.4.5 Avionics

Batteries used in avionics applications are subject to stringent safety standards to ensure their reliable operation and minimize risks associated with their use. The Federal Aviation Administration (FAA) has established several Technical Standard Orders (TSOs) and minimum performance standards (MPSs) for lithium battery systems in aviation applications. Some key safety standards and organizations involved in developing these standards include TSOs, the Radio Technical Commission for Aeronautics (RTCA), and the FAA. RTCA Special Committee 235 (SC235) is responsible for developing standards for nonrechargeable lithium batteries used in aviation applications. Additionally, the RTCA DO-227A document outlines the minimum operational performance standards (MOPSs) for nonrechargeable lithium batteries used in aviation applications. The FAA TSO-C179b document covers rechargeable lithium batteries and battery systems, using RTCA DO-311A as the MPS. FAA TSO-C142b covers nonrechargeable lithium cells and batteries, using RTCA DO-227A as the MPS.

These standards cover various aspects of battery safety, including physical, thermal, mechanical, and electrical stresses, in addition to safety in transit and use in avionics applications. They aim to ensure that LIBs can resist these stresses and minimize the risk of fires, explosions, or other hazardous events. Additionally, local codes and regulations, such as those from the International Civil Aviation Organization (ICAO) and the European Aviation Safety Agency (EASA), should be followed when installing and maintaining LIB systems in avionics applications.

SAE AIR6343, Design and Development of Rechargeable Lithium Battery Systems for Aerospace Applications, provides a basic overview of methodologies that should be utilized when undertaking the design and development of rechargeable lithium battery systems for aerospace applications.

A minimum requirement list for the battery-level design and development process includes:

- Functional hazard assessment (FHA)
- System safety assessment (SSA)
- Fault tree analysis (FTA)
- Failure mode and effects analysis (FMEA)
- Common cause analysis (CCA)
- Common mode analysis (CMA)
- Particular risk analysis (PRA)

In addition, some types of plans and reports may also include:

- Safety and reliability assessment
- Equipment reliability prediction (ERP)
- Safety function test report
- Cell performance and abuse evaluation test report
- Cell calendar/cycle life report
- System abuse test report
- Thermal analyses
- Exhaust gas analyses (composition, rate of release, total amount)
- Exhaust gas pressure analyses
- External short-circuit analyses
- Overcharge analyses
- Over-discharge analyses
- Enclosure validation test report
- Health monitoring strategy

- Component maintenance manual
- Installation guidelines
- Interface control drawing

13.4.6 Space Applications

Standards for battery safety in space applications are crucial for ensuring the reliability and safety of space missions. Several organizations and standards exist to address the unique challenges posed by space environments. Some of the key safety standards and organizations involved in developing these standards are described below.

The NASA Guidelines on Lithium-ion Battery Use in Space Applications provides a standard approach for defining, determining, and addressing safety, handling, and qualification standards for LIBs used in space applications. This guideline provides valuable information on the hazards and safety considerations associated with LIBs in space applications, and it collects information from various sources related to LIBs and their aerospace uses. NASA's guidelines and standards for LIBs in space applications are designed to minimize risks associated with the unique environment and demands of space travel, ensuring the safety and reliability of these batteries for extended periods.

JSC 20793 serves as the official battery safety standard within NASA, covering crewed space vehicle battery requirements. It began as a handbook in 1985 and was later converted to a requirements document in 2005. Revisions to JSC 20793 were made to incorporate lessons learned from the Boeing Dreamliner 787 and the extensive use of high-voltage and high-capacity LIBs. JWI 8705.3 is another battery processing document within NASA, focusing on energy and toxicity aspects of battery materials. Besides this, the ISO 17546:2016 standard addresses the safety of LIBs for space applications, focusing on obtaining sustainable development and preventing incidents related to lithium batteries.

American Institute of Aeronautics and Astronautics (AIAA) G-136-2022, Guide to Lithium Battery Safety for Space Applications, serves as a comprehensive guide to lithium battery safety for space applications. It includes human-rated space vehicles, satellites, launch vehicles, interplanetary probes, rovers, and landers.

The standards for product-level manufacturing quality control testing for batteries used in space applications involve a combination of regulatory requirements and best practices. Some key elements of these standards include regulatory compliance, such as IEC 62133, which specifies requirements and tests for the safety and performance of LIBs used in various applications:

- Space exploration
- Design and manufacturing quality control
- Functional baseline testing
- Environmental and vibration testing
- Charging and discharging cycles
- Leakage and vacuum testing
- Cell selection criteria
- In-process inspections and tests

These standards cover various aspects of battery safety, including physical, thermal, mechanical, and electrical stresses; safety in transit; and use in space applications. They aim to ensure that LIBs can resist these stresses and minimize the risk of fires, explosions, or other hazardous events in the unique space environment. Additionally, local codes and regulations, such as those established by NASA and the European Space Agency (ESA), should be followed when installing and maintaining LIB systems in space applications.

The certification process for batteries used in space applications is critical for ensuring the safety and reliability of these batteries in the challenging space environment. NASA has a comprehensive certification process for batteries used in space applications, which includes several phases. The engineering design/test phase involves cell- and battery-level testing to determine performance and safety characteristics. Initial testing is done on cells from different manufacturers to choose the most suitable candidate. Cell-level tests include rate capability, pulse capability, performance at different temperatures, and determination of vent-to-burst ratio. It also includes a test for safety tolerance under conditions of overcharge, over-discharge into reversal, external short, simulated internal short, and extreme temperatures. NASA-JSC [87] uses a vibration screening process and a stringent flight battery acceptance test program to screen cells and batteries for internal short defects.

The battery qualification phase includes extensive testing performed at the cell and battery levels in the relevant environment, considering unique requirements such as vacuum and extreme thermal environments. This phase confirms the safety of the battery under different abusive conditions. The flight acceptance phase involves testing the battery in the actual flight configuration to verify its performance and safety characteristics. For every new lot manufactured for a flight program, lot sample testing is required to confirm that each new lot has the same performance and safety characteristics as previously determined. Lot sample testing will not be necessary if only one lot is being used for flight and all of the testing has been carried out with adequate samples from the same single lot.

NASA employs a comprehensive certification process for batteries used in space applications. The process includes various stages and tests to ensure the safety, reliability, and performance of the batteries. A key component of NASA's LIB certification process includes safety testing and the development of requirements documents such as JSC 20793 and EA-CWI-033. NASA conducts validation of battery safety for space missions, involving safety certification processes, safety testing for LIBs, and the identification of hazards associated with using commercial lithium-ion cells in high-voltage and high-capacity configurations.

NASA adheres to various testing standards, such as UL 1642 for testing lithium cells and UL 2054 for battery pack-level tests. NASA follows a process for introducing a new lot or batch of certified batteries, which includes verifying qualification tests and performing abuse tests on batteries and cells. Additionally, NASA has a process for certifying noncommercially available off-the-shelf (non-COTS) batteries, which includes cell selection tests, battery design and assembly, battery engineering/certification tests, battery qualification tests, and battery flight tests. NASA has conducted numerous tests on the safety of lithium-ion cells, modules, and battery packs, considering the differences between high-voltage and high-capacity batteries and lower-voltage and lower-capacity packs.

ESA standard ECSS-E-HB-20-02A Li-ion battery testing handbook establishes the testing of Li-ion battery in order to assess their performance and their suitability to meet mission requirements and it includes safety tests section also for the potential hazards of venting, fire, burst, explosion and electrolyte leakage.

13.5 Trends in New Battery Safety Standards

International collaboration is crucial for developing and harmonizing global battery safety standards. As the demand for batteries grows, driven by the rise of electric vehicles, renewable energy storage, and portable electronics, ensuring safe and sustainable battery production and use has become a global priority. The development of new safety standards involves various stakeholders, including policymakers, industry leaders, and international organizations, all working together to address challenges related to supply chains, safety, and sustainability gaps in critical battery minerals.

One prominent initiative, the Global Battery Alliance (GBA), brings together over 150 businesses, governments, academics, and nongovernmental organizations to foster a sustainable and responsible battery value chain. The GBA's flagship program, the Battery Passport, aims to create a "quality seal" for batteries by providing trusted data on sustainable practices, responsible sourcing, and other indicators. Similarly, the EU Battery Regulation 2023/1542 outlines plans to create a sustainable battery industry, focusing on areas such as recycling, supply chain transparency, and reducing greenhouse gas (GHG) emissions.

These collaborative efforts are essential for keeping pace with the rapidly evolving battery industry. New advancements in battery technology, including higher energy density cells, solid-state batteries, and improved cell and packaging designs, have necessitated updated safety standards. As innovations such as lithium–sulfur batteries emerge, new safety and performance benchmarks must be established to ensure these next-generation batteries can be safely integrated into a wide range of applications.

Moreover, sustainability is becoming a key driver in shaping battery standards, with governments and organizations focusing on recycling technologies that recover materials like lithium, cobalt, and nickel from used batteries. Companies are exploring new business models to localize battery supply chains and reduce environmental impacts. The EU and the United States are expected to lead global growth in battery recycling, driven by recent regulatory changes and a trend toward localization of supply chains. China is already a mature market and will continue to play a significant role, projected to account for around 40% of total LIB demand by 2030. Despite progress, the global battery value chain still faces significant environmental, social, and governance (ESG) challenges [74], requiring continued international collaboration.

Advances in BMSs are another critical trend influencing safety standards. Leveraging artificial intelligence (AI) and data analytics, BMS technology is becoming more sophisticated, enabling real-time monitoring of battery health, optimizing performance, and improving safety measures like SOC estimation, voltage balance, and thermal management. This growing reliance on BMSs calls for updated standards that reflect the need for advanced control systems to manage battery safety across diverse applications.

As the industry continues to grow, another trend is the increasing focus on second-life applications for batteries. The reusing of batteries for energy storage and other applications requires a fresh look at safety standards to ensure that reused batteries meet the same high standards of safety and reliability as new ones.

Regarding the fire protection system design, the gaps in the available information for lithium-ion fire hazards can be generally categorized as such [89]:

- Limited understanding of the composition and flammability of leaked cell electrolyte and cell vent gases
- Lack of a fire protection commodity specification for bulk packaged lithium-ion cells or large format LIB packs
- Limited data regarding the effectiveness of potential suppressants, specifically water and incident clean-up

This section will explore key trends in new battery safety standards, examining how international collaboration, technological innovation, and sustainability efforts are shaping the future of the battery industry. Additionally, it will identify gaps in current regulations and discuss the importance of standardization in fostering safe and sustainable growth across the global battery value chain.

13.5.1 Evolving Battery Technologies

Battery safety is a multidisciplinary field that involves addressing challenges at the individual component, cell, and system levels. The considerable heat released during their operation and potential malfunctions pose fire risks, raising safety concerns for energy storage facilities. Early detection of potential risks allows for timely intervention and elimination, mitigating safety accidents, which is a crucial consideration in the large-scale construction and application of batteries.

Standards like ISO 61508 and ISO 26262, for battery packs in different applications of all industries, define functional safety as part of overall safety. ISO 61508 defines functional safety as the part of overall safety that depends on the correct functioning of electrical/electronic/programmable electronic (E/E/PE) safety-related systems and other technology safety-related systems. ISO 26262 is specifically designed for the automotive industry, focusing on safety in automotive systems, including passenger cars, trucks, buses, and motorcycles.

The CENELEC CLC/TC 21X standard defines a set of tests for validating the performance and safety requirements of batteries used across applications. The EN IEC 62485 standard defines a set of tests to validate the safe operation of battery systems, including safety precautions required to be present in the installations of the battery system application.

Sodium-ion batteries are now being specifically addressed in safety protocols, recognizing their unique safety considerations. Regulatory frameworks governing hazardous materials transportation have been updated to incorporate sodium-ion testing alongside LIBs. Additionally, new battery technologies, such as zinc-manganese oxide and solid-state batteries, are being developed. These technologies pose specific challenges, particularly in off-highway/off-road markets, where higher energy batteries must be kept safe to protect both property and human lives. Material development is also a focal point, with efforts to enhance heat absorption capabilities and introduce lightweight barriers.

As new technologies emerge, it is essential to understand their aging mechanisms and failure modes to assess possible hazards and develop appropriate safety management strategies. Frameworks like IEEE 1679, Recommended Practice for the Characterization and Evaluation of Energy Storage Technologies in Stationary Applications, guide the evaluation of emerging technologies. The document provides a common basis for the expression of performance characteristics and the treatment of life-testing data. A standard approach for analysis of failure modes is also provided, including assessment of safety attributes. UL qualification standards are designed to be generic, requiring a failure mode and effects analysis (FMEA) and hazardous materials assessment (HMA) to be performed. However, new test methods may be needed to assess the tolerance of system designs to unique failure modes in emerging technologies.

13.5.2 Sustainability

A new law to ensure that batteries are collected, reused, and recycled in Europe entered into force in 2023. The new EU Battery Regulation 2023/1542 will ensure that, in the future, batteries have a low carbon footprint, use minimal harmful substances, need fewer raw materials from non-EU countries, and are collected, reused, and recycled to a high degree. This will support the shift to a circular economy, increase the security of raw materials supplies and energy, and enhance the EU's strategic autonomy [75, 76]. Table 13.6 shows the EU Battery Regulation timeline.

Supply chain regulations govern the sourcing of battery materials, including the mining, processing, and transportation of these materials. They aim to ensure that battery materials are sourced sustainably and responsibly and that the workers involved in the supply chain are treated fairly.

Table 13.6 Timeline for EU Battery Regulation 2023/1542 compliance aspects.

Requirement	Timeline	Compliance
Carbon footprint declaration	July 2024	Mandatory for battery manufacturers
Performance classes	January 2026	Labeling requirement for batteries
Carbon footprint thresholds	July 2027	Minimum carbon footprint thresholds; by Jan 2028, prove that the life cycle carbon footprint is below maximum thresholds
Responsible sourcing due diligence	July 2025	Mandatory for economic operators; comply with supply chain due diligence on social and environmental risks
Digital battery passport	2026–2027	Mandatory for battery manufacturers; full labeling connected to each battery with QR code for product info and battery passport with carbon footprint; responsible sourcing for battery composition; recycled content; state-of-health performance; durability; and others
Material recovery targets	2027	90% for cobalt, nickel, copper, and lead; 50% for lithium
Collection rate for portable batteries	2027	63%
Collection rate for portable batteries	2030	73%
Recycling requirements	2025	Recycling efficiency: 75% for lead-acid batteries, 65% for lithium-based batteries, 80% for nickel-cadmium batteries, and 50% for other waste batteries
	July 2028	Declare recycled content of lithium, nickel, cobalt, and lead
	Dec 2030	Increase recycling efficiency to 70% for lithium-based batteries
	July 2031	Minimum levels of recycled content: lithium 6%, nickel 6%, cobalt 16%, and lead 85%
	Dec 2031	Increase material recovery: lithium 80%, nickel 95%, cobalt 95%, and copper 95%
	July 2038	Minimum levels of recycled content: lithium 12%, nickel 15%, cobalt 26%, and lead 85%

Source: Adapted from [76].

One of the most important aspects of battery supply chain regulation is the use of conflict minerals, which are minerals extracted in conflict-affected areas, helping to finance armed conflict. Another important aspect of battery supply chain regulation is the use of child labor, the employment of children under the age of 18 in hazardous or exploitative conditions. The International Labor Organization (ILO) has adopted several conventions that prohibit child labor, and many countries have laws that ban or restrict the employment of children. Battery companies are subject to these laws, and they must take steps to ensure that their products are not made with child labor.

Recycling and product stewardship regulations targeted at used batteries are becoming more common in the United States. At present, in California, all battery types must be recycled, and they may not be disposed of as solid waste. A new mandate requires that battery manufacturers develop collection programs. Similarly, New York has mandated that retailers collect used batteries starting in June 2011 [89].

The manufacturing of batteries can also have a significant impact on the environment. For example, the production of LIBs requires the use of hazardous chemicals, and the disposal of the batteries can release these chemicals into the environment. Battery companies are subject to environmental regulations that aim to minimize the environmental impact of battery manufacturing. These regulations can include requirements for companies to use environmentally friendly manufacturing processes, recycle hazardous materials, and dispose of batteries in an environmentally friendly manner [76].

13.5.3 Battery Management Systems and Data Analytics

BMSs can be defined as a safety control system required for managing individual cells of a battery pack and an entire battery pack. In the case of an EV battery pack status, information like pack SoC, SoH, voltage, current, available energy, distance to empty, voltage, and temperature differences between cells in the pack, among others, can be collected via OBD-II (On-Board Diagnostics) server-based mobile applications. With that said, possible standards for BMS feature extension are described below.

ISO 17826 defines considerations for implementing an interface for data transfer from a system to a cloud environment. The data collected through the OBD-II interface must be transferred to a cloud platform for postprocessing and storage.

IEC 27018 establishes control objectives, controls, and guidelines for implementing measures to protect information processed and stored in a public cloud computing environment. Storing and processing such large data on board may incur large computational and memory requirements. This data is highly significant while deciding its eligibility for serving a second-life application. When considering the importance and confidentiality of this data, cyber security standards should also be adopted for better application-related cloud storage.

It was developed jointly by ISO/IEC and ITU-T and is dual-numbered as both ISO/IEC 27017 and ITU-T X.1631, with identical content. The standard provides guidelines for information security controls applicable to the provision and use of cloud services by providing additional implementation guidance for relevant controls specified in ISO/IEC 27002 and additional controls with implementation guidance that specifically relate to cloud services.

For data collection, transmission, cloud-based post-processing, and storage, an architecture needs to be developed for connecting the battery system, transmission medium, and cloud platform in a network. The IEEE P2413, IoT standard should be adopted for battery system-related cloud services.

The AEC-Q100 generic standard defines the possible failure mechanisms for common electrical component qualification requirements. This standard should be adopted for failure analysis of integrated circuits (ICs) used in BMSs.

13.5.4 Second-Life Applications

The regulatory framework for second-life batteries is still in development, with a need for defined norms, technical standards, and guidelines. Currently, there are no guarantees regarding the quality or performance of second-life batteries, and few industry standards focus on BMSs or SOH disclosures. This lack of standardization creates challenges in refurbishing and repurposing EV batteries for stationary applications, as each battery design varies in size, electrode chemistry, and format.

The EU-funded project Electric Vehicle Enhanced Range, Lifetime and Safety Through Ingenious Battery Management (EVERLASTING) will develop technologies for estimating the remaining useful life of a battery and the collection of usage history data through the OBD-II interface for further post-processing. Later, a scenario of integrating an EV battery system (battery pack + BMS) into a second-life stationary application is described, along with the challenges faced and possible solutions [77].

When implementing an EV battery system in a second-life application, the communication protocol between subsystems in the application may vary from that of an EV application. A bridging communication protocol is required, which can communicate only the needed nonconfidential information from the EV BMS. Notably, a controller area network (CAN) open protocol as defined in the ISO 11898 standard could be a potential solution. This standard comprises several sections, defining both the physical and data link layers for transmission communication. The protocol was developed for and is used primarily in automotive applications to link sensors, actuators, and other control devices within a contained environment. Such an environment is necessary because transmission speeds are limited by cable length. The standard defines the arbitration protocols and electrical characteristics of the physical layer, including voltage, current, and several conductors necessary for transmission.

There are limited specific United Kingdom or European standards for reusing LIBs in the current context. This represents a significant regulatory void and may have far-reaching consequences. On the one hand, the absence of established standards may hinder the growth and development of the second-life battery industry, as safety, performance, and accountability may be dubious for businesses. Alternatively, the absence of regulations could inadvertently encourage unsustainable or unsafe practices, posing potential environmental and safety hazards. In addition, this regulatory void could restrict opportunities for maximizing the total life cycle value of LIBs, resulting in the loss of economic and environmental efficiencies. To encourage the safe, sustainable, and economically viable second-life use of LIBs in the United Kingdom and Europe, it is crucial to develop specific and exhaustive standards. The existence of several different EV battery chemicals imposes uncertainty on the market and discourages producers from investing to make their recycling processes more efficient. New standards aimed at reusing and recycling batteries should consider the different chemicals.

UL 1974 defines a set of potential 2nd life applications for EV batteries and lists the requirements that should be satisfied by an EV battery to efficiently serve the intended second-life application. In addition, the market for fire protection materials in electric vehicle batteries is expected to grow significantly, driven by the need for improved safety measures to prevent thermal runaway and fires in EV batteries.

13.5.5 International Collaboration

International collaboration is crucial for developing and harmonizing global battery safety standards. This collaboration involves policymakers, industry actors, and international organizations working together to address the supply, financing, and sustainability gaps in critical battery minerals. Key initiatives include the GBA, which is a partnership of over 150 businesses, governments, academics, and nongovernmental organizations that collaborate on ensuring a sustainable and responsible battery value chain. The GBA's flagship program, the Battery Passport, aims to provide trusted information on indicators related to responsible and sustainable practices, resulting in a quality seal for batteries. In addition, the EU Battery Regulation outlines plans to create a sustainable battery industry, including structures for more effective battery recycling, supply chain due diligence, and GHG emissions reduction. The Battery Passport will initially require only basic battery information and statistics but will expand to include more detailed data over time.

The United States Department of Defense (DOD) is working to align industry and military battery standards wherever practical, focusing on standardization and collaboration to ensure the effective use of batteries in various applications.hj,

The Faraday Institution is actively involved in various initiatives to improve battery safety standards across different applications. It aims to address standardization gaps in areas such as fire risk management, recyclability, and performance and abuse testing. This collaboration involves workshops with industry stakeholders to gather insights and inform the development of standards. The Faraday Institution emphasizes the need for improved safety standards in micromobility vehicles (MMVs), including the adoption of robust BMSs and dedicated charging infrastructure. They also recommend consumer education campaigns to raise awareness of battery safety and the hazards of battery failure [78].

IEC has published the IEC 62619 standard, which specifies requirements and tests for the safe production of secondary lithium cells and batteries used in industrial applications. This standard addresses safety concerns such as short circuits, overcharging, thermal abuse, and functional safety for BMSs.

The International Battery Seminar & Exhibit bills itself as "the premier event showcasing the state of the art of worldwide energy storage technology developments for consumer, automotive, military, grid, and industrial applications." This event includes a battery safety track that covers important battery regulations and safety testing methodologies, and it brings together experts from regulatory agencies, associations, and industry professionals to share knowledge and insights on ensuring battery safety.

Various international safety standards have been developed for batteries, including IEC 60086 for primary batteries and IEC 62133 for secondary batteries. These standards are adopted and harmonized by different countries, ensuring a consistent approach for global battery safety.

Multiple EU-funded battery projects are ongoing and applicable to a wide array of battery aspects [79]:

- BAT4EVER and HIDDEN: for the self-healing mechanism of LIBs
- BIG-MAP: for R&D infrastructure for the battery value chain
- Spartacus: to develop sensor solutions to detect degradation and failure mechanisms
- HEALING BAT: to create a toolbox consisting of self-healing materials, battery sensors, and bespoke BMSs
- OPERA: to use advanced experimental and computational methods for a new generation of environmentally friendly batteries with high energy density, improved safety, and higher cost-efficiency
- OPINCHARGE: to analyze and understand interactions between the interfacial structure, its chemical evolution and charge dynamics in an unprecedented level of detail
- PHOENIX: to build more reliable and performance batteries by embedding sensors and self-healing functionalities to detect degradation and repair damage via an advanced BMS
- SALAMANDER: uses smart sensors and self-healing functionalities embedded for battery longevity with manufacturability and economical recyclability with BMS

13.5.6 Standardization Gap Analysis

Lithium-ion batteries release a considerable amount of heat in their operation, and malfunctions cause fire risks, raising safety concerns for energy storage facilities. Improper use of batteries, such as overcharging, over-discharging, and overcurrent, can lead to thermal runaway of the batteries, occasionally evolving into combustions in energy storage systems.

As mentioned in previous sections, most of the standards development is shaped by international technical organizations, including ISO, IEC, and IEEE. In addition, influential entities such as the American Fire Protection Association (NFPA), Underwriters Laboratories (UL), and the German Technical Supervision Association (TÜV) significantly contribute to the formulation of technical regulations. In the United States, the Department of Energy (DOE), spearheaded by the DOE Sandia National Laboratories, collaborates with entities like NFPA, IEEE, and UL to develop a roadmap for energy storage safety standards.

However, these standards rarely discuss the use of BMSs, fire early warning technology, gas monitoring, particle analysis, and other parameters for energy storage safety online monitoring. Yuan Y et al. discussed the processes of early warning detection techniques, and as future work, the establishment of an IEEE standard working group is proposed to address this standard gap [80].

Early warning technology for batteries is a rapidly evolving field, with various techniques being developed to detect thermal runaway and prevent battery fires. The key to effective early warning systems lies in the strategic placement of sensors, standardization of testing, and the development of multilevel warning systems.

Current energy storage technology grapples with three pivotal issues: the limited scope of BMSs confined to battery units, the disjointed and nontargeted monitoring of conventional electrical equipment, and the passive and offline nature of existing fire protection measures. The novel approach establishes new standards and indicators for online monitoring and early warning technology, as shown in Table 13.7.

Table 13.7 Process for early warning technique for lithium-ion battery hazard.

Gap in standardization	Latent issue	Battery thermal runaway detection	Thermal runaway is a critical safety issue in lithium-ion batteries.
	Early warning	Early detection of latent issues allow time to address issue	Early warning systems can detect thermal runaway by monitoring parameters such as temperature, voltage, and current. Techniques include acoustic signal monitoring, thermal imaging, and gas detection.
	Danger	Thermal runaway, smoke, and fire development	Gas detection: Gas venting is a key indicator of thermal runaway. Early warning systems can detect gas venting and provide an early warning of battery failure. Techniques include gas sensors to detect flammable gases such as hydrogen and carbon dioxide.
Emergency response	Accident	Equipment damage, fire, causalities	Sensor placement: The placement of sensors is critical for accurate early warning. Sensors should be strategically placed to detect gas venting and thermal runaway effectively. Multilevel warning: A multilevel warning system can provide early warnings for different stages of thermal runaway, allowing for more effective mitigation strategies.

Source: Adapted from Yuan et al. [80].

Early warning technology is designed to exert control over scenarios preceding accidents, encompassing aspects like battery thermal runaway, equipment overheating, and emissions.

Standardization of thermal runaway testing is essential for developing reliable early warning systems. Companies like Amphenol Advanced Sensors are developing early warning systems for commercial electric vehicles, providing critical safety features for the industry.

A novel online monitoring system employs active inhalation and 24-hour cyclic sampling, utilizing various test technologies, including particle analysis, characteristic gases measurement, volatiles tracking, and assessing temperature and humidity levels. To enhance robustness, the system incorporates anti-interference measures to combat external factors like dust, oil mist, water mist, and fog. Although there are a considerable number of standards for battery-served applications in existence, adopting and interpolating them with a focus on BMS seems challenging [80]. The gaps in available standards for BMS are summarized in Table 13.8.

The functional safety of LIB systems on vehicles can be achieved by applying ISO 26262 guidelines at different development stages. The ISO 26262 standard provides regulations and recommendations to achieve automotive safety lifecycle for function or system (design, development, validation, production, function, and service) and supports the essential activities during component-, system-, and vehicle-level development phases. It also addresses the possible hazards

Table 13.8 Identified gaps in available standards for battery management system.

Focus	Identified gap
Cell monitoring	Voltage, current, and temperature monitoring of cells in the battery pack
Communication protocol	Internal and external communication protocols (type, transmission rate, etc.) must be defined
Data format	Data transmission and storage formats must be defined for both on-board and server-based processing and storage purposes
State estimation	Unified definition of state estimation parameters like SOC, SOH, SOP, SOE, and SOF
Test procedure	Real-world load cycle-based aging cycle definition for accurate life testing; standard full-charge and standard cycles should be clearly defined; remaining useful life (RUL) definition and tests
Vehicle-cloud integration	Set of data to be transferred by the BMS to cloud server for post processing or usage monitoring must be defined; storage database format and duration of storage must be defined; standardization of cybersecurity privacy policy, type, and duration of data storage in the cloud
Second-life applications	In addition to stationary applications, more applications identified as potential second-life applications for EV batteries; application-specific requirements and standards defined for the identified second-life applications; standardization of communication protocols for various subcomponents of second-life applications; switching of first-life BMS to second-life application, without need of additional requirements or major modifications
Data sheet specifications	Nominal battery test conditions and ratings should be standardized across manufacturers; test procedures for obtaining data sheet parameters and the tolerances in the estimated parameter values should be standardized across manufacturers
Open BMS platform	Requirements and standards should be defined for the development of open, flexible BMS platform

caused by the malfunctioning behavior of all electrical and electronic systems of the vehicle, including their communication. It discusses the application of hazard analysis and risk analysis (HARA) in terms of severity rating, controllability rating, and exposure rating during the design conceptual stage of the battery system. Additionally, it explains the procedure for defining the automotive safety integrity level (ASIL) rating for battery systems. Future work should be focused on the integration of failsafe strategies for software and hardware parts of LIB systems to achieve fail safety goals. Also, part- and system-level validation of the failsafe strategies needs to be determined [81].

13.5.7 Fire Hazard Gap Analysis

Sandia provides data regarding gas composition that can be used to detect cell leakage or vent gas products, and they are working on a sensor or sensor package for detecting leaking of venting cells [90]. At present, there is no publicly available large-scale fire test data for lithium-ion cells that can be used to fully assess the storage hazards of lithium-ion cells/batteries or to determine an appropriate commodity classification that could be used to provide an overall fire protection suppression strategy. Such commodity classifications for water-based suppression strategies are described in NFPA 13, the Standard for the Installation of Sprinkler Systems. However, a number of sources, including the FAA and the US Navy, recommend the use of water as a cooling and extinguishing agent. Water-based automatic sprinklers are the most widely used fire suppression equipment, and they have proved their efficiency and reliability over the years [89].

13.6 Summary

Battery safety and reliability are upheld by an intricate network of safety standards, regulations, and certification bodies, all working in tandem to ensure that batteries perform safely across a wide range of applications. Key players in developing and harmonizing these safety protocols include international and regional standard organizations, such as the IEC, IEEE, ISO, SAE, NEMA, and Telcordia. Meanwhile, regulatory bodies like the UNECE, EU, DOT, EPA, and CPSC in the Unites States; CNIS and MIIT in China; and METI and JISC in Japan play crucial roles in enforcing these standards. Certification bodies, including IECEE, UL, TÜV SÜD, and INTERTEK, ensure ongoing compliance through periodic audits and testing. This interconnected framework addresses critical safety and reliability requirements for battery cells, packs, and systems, ensuring they meet stringent criteria under varied operational conditions.

This chapter emphasizes the critical need for batteries to meet not only performance specifications but also stringent storage, transportation, and safety standards, especially in scenarios where misuse is possible. Safety mechanisms are embedded at every level—from individual cells and cell blocks to complete battery packs and systems—while quality control throughout the manufacturing process plays a crucial role in mitigating risks and preventing failures. However, it is essential to recognize that test standards are applied to representative samples and cannot guarantee universal compliance, making rigorous quality control vital across production.

Rigorous testing evaluates key factors such as thermal management, electrical stability, and mechanical durability, with standards tailored to specific environments like portable electronics, electric vehicles, marine systems, and aerospace applications. These safety protocols are instrumental in minimizing risks such as thermal runaway, overcharging, or mechanical stress, all of which could lead to serious failures.

As batteries are increasingly integrated into sectors like automotive, aerospace, and renewable energy, managing the complex thermal, electrical, and mechanical risks at the system level has become paramount. BMSs, which are responsible for monitoring and controlling battery safety, are evolving with the integration of artificial intelligence (AI). AI enhances BMS functionality by optimizing performance, predicting potential failures, and enabling real-time adjustments. However, as new technologies emerge and second-life battery applications grow, there is an increasing need for more comprehensive BMS standards.

Sustainability has become a central focus in battery development, with regulations addressing the entire life cycle—from the sourcing of materials to the recycling of used batteries. The EU's regulatory framework for battery recycling, for example, emphasizes reducing environmental impact and increasing the recovery of valuable materials like lithium, cobalt, and nickel. These efforts contribute to a circular economy, reducing the environmental footprint while also meeting the growing demand for raw materials. However, challenges remain, particularly in regulating conflict minerals and ensuring sustainable sourcing practices, both of which require ongoing global collaboration.

Emerging battery technologies, such as solid-state and lithium-sulfur batteries, offer promising new performance benefits, but they also present unique risks that demand updated safety standards. International collaboration is key for developing these standards, as evidenced by initiatives like the GBA, which promotes responsible production and use of batteries. As the battery industry continues to evolve, the global framework governing safety, performance, and sustainability must keep pace, ensuring that batteries remain secure and reliable in an ever-expanding array of applications.

Despite the strong network of safety standards, gaps remain, particularly with evolving technologies. Identified gaps include limitations in BMSs, which are often confined to individual units without comprehensive system integration or real-time monitoring. Fire protection measures remain largely passive, and key areas such as cell monitoring, communication protocols, data formats, and second-life applications need more robust standards. As batteries become more complex, these gaps highlight the need for updated standards that can adapt to new technologies and ensure better safety and integration across applications.

References

1 Thompson, S.P. (2011). *The Life of William Thomson, Baron Kelvin of Largs*. Cambridge University Press. ISBN: 9781108027182.

2 IEC Standard 60086-1 (2021). *Primary batteries – Part I: General*. International Electrotechnical Commission.

3 IEC Standard 60086-2 (2021). *Primary batteries – Part 2: Physical and electrical specifications*. International Electrotechnical Commission.

4 IEC Standard 60086-3 (2021). *Primary batteries – Part 3: Watch batteries*. International Electrotechnical Commission.

5 IEC Standard 60086-4 (2019). *Primary batteries – Part 4: Safety of lithium batteries*. International Electrotechnical Commission.

6 IEC Standard 60086-5 (2021). *Primary batteries – Part 5: Safety of batteries with aqueous electrolyte*. International Electrotechnical Commission.

7 IEC Standard 60086-6 (2020). *Primary batteries – Part 6: Guidance on environmental aspects*. International Electrotechnical Commission.

8 IEC Standard 62133-1 (2017). *Secondary cells and batteries containing alkaline or other non-acid electrolytes – Safety requirements for portable sealed secondary cells – Part 1: Nickel systems.* International Electrotechnical Commission.

9 IEC Standard 62133-2 (2017). *Secondary cells and batteries containing alkaline or other non-acid electrolytes – Safety requirements for portable sealed secondary cells – Part 2: Lithium systems.* International Electrotechnical Commission.

10 IEC Standard 62281 (2019). *Safety of primary and secondary lithium cells and batteries during transport.* International Electrotechnical Commission.

11 IEC Standard 62619 (2022). *Secondary cells and batteries containing alkaline or other non-acid electrolytes – Safety requirements for secondary lithium cells and batteries, for use in industrial applications.* International Electrotechnical Commission.

12 IEC Standard 62660-1 (2018). *Secondary lithium-ion cells for the propulsion of electric road vehicles – Part 1: Performance testing.* International Electrotechnical Commission.

13 IEC Standard 62660-2 (2018). *Secondary lithium-ion cells for the propulsion of electric road vehicles – Part 2: Reliability and abuse testing.* International Electrotechnical Commission.

14 IEC Standard 61960-3 (2017). *Secondary lithium cells and batteries for portable applications – Prismatic and cylindrical lithium cells.* International Electrotechnical Commission.

15 IEC Standard 61960-4 (2024). *Secondary lithium cells and batteries for portable applications – Coin cells.* International Electrotechnical Commission.

16 IEEE. History of IEEE: A Brief History of IEEE. https://www.ieee.org/about/ieee-history.html (accessed 22 September 2024).

17 IEEE. About IEEE. https://www.ieee.org/about/index.html (accessed 22 September 2024).

18 IEEE Std 1725-2021 (2021). *IEEE standard for rechargeable batteries for mobile telephones.* Institute of Electrical and Electronics Engineers.

19 IEEE Std 1625-2021 (2021). *IEEE standard for rechargeable batteries for multi-cell mobile computing devices.* Institute of Electrical and Electronics Engineers.

20 *The ISO Story.* International Organization for Standardization. https://www.iso.org/about-us.html (accessed 22 September 2024).

21 ISO Standard 12405-1 (2011). *Electrically propelled road vehicles – Test specification for lithium-ion traction battery packs and systems – Part 1: High power applications.* International Organization for Standardization (withdrawn).

22 ISO Standard 12405-2 (2012). *Electrically propelled road vehicles – Test specification for lithium-ion traction battery packs and systems – Part 2: High energy applications.* International Organization for Standardization (withdrawn).

23 ISO Standard 12405-4 (2018). *Electrically propelled road vehicles – Test specification for lithium-ion traction battery packs and systems.* International Organzation for Standardization.

24 ISO Standard 12405-3 (2014). *Electrically propelled road vehicles – Test specification for lithium-ion traction battery packs and systems – Part 3: Safety performance requirements.* International Organization for Standardization (withdrawn).

25 ISO Standard 6469-1 (2019). *Electrically propelled road vehicles – Safety specifications – Part 1: Rechargeable energy storage system (RESS) safety requirements.* International Organization for Standardization.

26 SAE International. About SAE. https://www.sae.org/about (accessed 22 September 2024).

27 SAE J2929 (2013). *SAE Standard for Electric and hybrid vehicle propulsion battery system safety.* SAE International.

28 SAE J2464 (2021). *SAE Standard for Electric and hybrid vehicle battery abuse testing.* SAE International.

29 SAE J2380 (2021). *SAE Standard for Electric vehicle battery vibration testing*. SAE International.

30 SAE J2288 (2020). *SAE Standard for Electric vehicle battery performance and life testing*. SAE International.

31 SAE J1798 (2019). *SAE recommended practice for performance rating of electric vehicle battery modules*. SAE International.

32 SAE J2289 (2008). *SAE recommended practice for electric vehicle battery system design guidelines*. SAE International.

33 National Electrical Manufacturers Association (NEMA). About NEMA. https://www.nema.org (accessed 22 September 2024).

34 ANSI C18.1M (2019). *American National Standard for portable primary cells and batteries – General and Specifications*. National Electrical Manufacturers Association.

35 ANSI C18.2M (2019). *American National Standard for portable rechargeable cells and batteries – General and Specifications*. National Electrical Manufacturers Association.

36 Ericsson. Ericsson Acquires Telcordia. https://www.ericsson.com/en/about-us/history/company/the-consequences-of-expansion/ericsson-acquired-telcordia (accessed 22 September 22).

37 Iconectiv. Corporate Fact Sheet. https://iconectiv.com/sites/default/files/documents/iconectiv%20corporate%20fact%20sheet.pdf (accessed 22 September 2024).

38 GR-3150-CORE (2012). *Generic Requirements for Rechargeable Lithium-Ion Batteries Used in Telecommunications Backup Power Applications*. Telcordia Technologies.

39 UNECE. About the United Nations Economic Commission for Europe. https://unece.org/mission (accessed 22 September 2024).

40 United Nations Economic Commission for Europe (2019). *Recommendations on the Transport of Dangerous Goods, Manual of Tests and Criteria, Part III, Section 38.3*. United Nations Economic Commission for Europe.

41 European Commission (2006). *Directive 2006/66/EC on Batteries and Accumulators and Waste Batteries and Accumulators*. European Commission.

42 European Commission (2013). *Directive 2013/56/EU of the European Parliament and of the Council*. European Commission.

43 European Union (2019). *Regulation (EU) 2019/1020 of the European Parliament and of the Council on Market Surveillance and Compliance of Products*. European Union.

44 U.S. Department of Transportation (2021). *Title 49 CFR 173.185 – Shippers–General Requirements for Shipments and Packagings*. U.S. Department of Transportation.

45 Environmental Protection Agency (1976). *Resource Conservation and Recovery Act (RCRA)*. Environmental Protection Agency.

46 CNIS. China National Institute of Standardization. https://en.cnis.ac.cn/bzygk_329/kyly/ (accessed 22 September 2024).

47 Ministry of Industry and Information Technology (MIIT). Regulations and Standards. https://english.www.gov.cn/state_council/2014/08/23/content_281474983035940.htm (accessed 22 September 2024).

48 GB/T 18287-2013 (2013). *General specification for lithium-ion batteries for mobile phones*. China National Institute of Standardization.

49 GB 31241-2014 (2014). *Safety requirements for lithium-ion cells and batteries for portable devices*. China National Institute of Standardization.

50 GB/T 36972-2018 (2018). *Safety requirements for lithium-ion power battery packs and systems for electric vehicles*. China National Institute of Standardization.

51 METI. DENAN Law (Electrical Appliance and Material Safety Law). https://www.mofa.go.jp/region/europe/eu/electrical.pdf (accessed 22 September 2024).

52 Japanese Industrial Standards Committee (JISC). About JISC. https://www.jisc.go.jp/eng/jisc/index_e.html (accessed 22 September 2024).
53 JIS C 8712 (2015). *Lithium-ion batteries for portable applications – Safety requirements*. Japanese Industrial Standards Committee.
54 JIS C 8714 (2015). *Safety tests for portable lithium-ion secondary batteries*. Japanese Industrial Standards Committee.
55 IECEE. About the IECEE. https://www.iecee.org/about/ (accessed 22 September 2024).
56 UL. About UL. https://www.ul.com/about (accessed 22 September 2024).
57 UL 1642 (2020). *Standard for lithium batteries*. UL.
58 UL 2054 (2021). *Standard for household and commercial batteries*. UL.
59 UL 2575 (2011). *Safety for battery systems in tools and appliances*. UL.
60 UL 2271 (2023). *Safety standard for light electric vehicle batteries*. UL.
61 UL 2580 (2020). *Safety standard for batteries in electric vehicles*. UL.
62 UL 1973 (2022). *Safety standard for stationary batteries*. UL.
63 UL 9540 (2023). *Energy storage systems and equipment*. UL.
64 TÜV Rheinland. Battery Testing Center. https://batterylab.tuv.com/ (accessed 22 September 2024).
65 TÜV SÜD. Battery Testing and Certification. https://www.tuvsud.com/en-us/services/testing/battery (accessed 22 September 2024).
66 Intertek. Battery Testing and Energy Storage Solutions. https://www.intertek.com/battery/ (accessed 22 September 2024).
67 UL 9540. *Energy Storage System (ESS) requirements evolving to meet industry and regulatory needs*. UL. https://www.ul.com/news/ul-9540-energy-storage-system-ess-requirements-evolving-meet-industry-and-regulatory-needs (accessed 22 September 2024).
68 ULSE. ULSE Publishes Third Edition of UL 9540, Energy Storage Systems and Equipment. https://ulse.org/news/ulse-publishes-third-edition-ul-9540-energy-storage-systems-and-equipment (accessed 22 September 2024).
69 UL 9540A (2019). *Test method for evaluating thermal runaway fire propagation in battery energy storage systems*. UL.
70 UL 9540A. *Test method*. UL. https://www.ul.com/services/ul-9540a-test-method (accessed 22 September 2024).
71 UL 9540A. *Battery Energy Storage System (ESS) test method*. UL. https://www.ul.com/news/ul-9540a-battery-energy-storage-system-ess-test-method (accessed 22 September 2024].
72 ANSI C18.2M. *Portable rechargeable cells and batteries*. ANSI. https://blog.ansi.org/2020/06/ansi-c182m-portable-rechargeable-cell-batteries/ (accessed 6 October 2024).
73 National Highway Traffic Safety Administration. Battery Safety Initiative. NHTSA. https://www.nhtsa.gov/battery-safety-initiative (accessed 20 October 2024).
74 McKinsey & Company. Battery 2030: Resilient, sustainable, and circular. https://www.mckinsey.com/industries/automotive-and-assembly/our-insights/battery-2030-resilient-sustainable-and-circular (accessed 1 May 2024).
75 European Commission. New law: more sustainable, circular, and safe batteries enters force. https://environment.ec.europa.eu/news/new-law-more-sustainable-circular-and-safe-batteries-enters-force-2023-08-17_en (accessed 1 May 2024).
76 Battery Regulation. https://www.batteryregulation.eu/ (accessed 1 May 2024).
77 Everlasting Project. https://everlasting-project.eu/ (accessed 1 May 2024).
78 The Faraday Institution. Faraday Institution webpage. https://www.faraday.ac.uk/ (accessed 1 May 2024).

79 Battery2030. Battery2030 projects. https://battery2030.eu/battery2030/projects/ (accessed 1 May 2024).

80 Yuan, Y., Ji, X., Dong, Z., and Lai, C.S. (2024). A gap analysis of technical standards for active safety online monitoring and fire hazards for lithium-ion batteries. *2024 IEEE Conference on Technologies for Sustainability (SusTech)*, Portland, OR, USA, 371–377.

81 Tikar, S.S. (2017). Compliance of ISO 26262 safety standard for lithium ion battery and its battery management system in hybrid electric vehicle. *2017 IEEE Transportation Electrification Conference (ITEC-India)*, Pune, India, 1–5.

82 SAE Standard J1772 (2017). *SAE electric vehicle and plug in hybrid electric vehicle conductive charge coupler*. SAE International.

83 GB 38031-2020 (2020). *Lithium-ion traction battery for electric vehicles–Safety requirements*. Standardization Administration of China.

84 IEC Standard 63057 (2020). *Secondary cells and batteries containing alkaline or other non-acid electrolytes – Safety requirements for lithium-ion batteries for use in road vehicles (IEC 63057:2020)*. International Electrotechnical Commission.

85 EUCAR. Battery requirements for future automotive applications. https://eucar.be/wp-content/uploads/2019/08/20190710-EG-BEV-FCEV-Battery-requirements-FINAL.pdf (accessed 20 October 2024).

86 IEC Standard 61982 (2012). *Secondary batteries (except lithium) for the propulsion of electric road vehicles – Performance and endurance tests*. International Electrotechnical Commission.

87 NASA. Vibration and Shock. https://www.nasa.gov/reference/jsc-vibration-shock/ (accessed 20 October 2024).

88 SAE International. Guidelines for Electric Vehicle Safety J2344_202010. https://www.sae.org/standards/content/j2344_202010/ (accessed 10 November 2024).

89 Mikolajczak, C., Kahn, M., White, K., and Thomas, R. (2011). *Long Lithium-Ion Batteries Hazard and Use Assessment*. Springer. ISBN: 978-1-4614-3485-6.

90 Vigil, K. (2024). Detecting battery failures quicker. https://www.sandia.gov/labnews/2024/10/31/detecting-battery-failures-quicker/ (accessed 10 November 2024).

14

Battery Rewrapping and Counterfeits

Lingxi Kong and Michael G. Pecht

The estimated global lithium-ion battery market size is $54.4 billion in 2023 [1]. However, a portion of this market is associated with counterfeited batteries. Lithium-ion batteries were first commercialized in 1991 [2]. Over a decade later, in 2003, about five million counterfeit cellular phone batteries were seized worldwide [3]. Counterfeit and rewrapped batteries are mingled with genuine batteries in the market, offering attractive prices at the expense of reliability or safety, thus exposing users to the risk of bodily injury and property damage. This risk is taken by individual customers in the retail market and any original equipment manufacturers who seek battery suppliers from the open market.

Reputable and well-known battery suppliers exist, but they generally provide batteries directly to companies that need significant volumes for installation and replacement, such as manufacturers of electric vehicles and energy storage systems. For low-volume and less-expensive products, such as e-bikes, e-scooters, and vaping devices, companies often buy from distributors or less reputable battery manufacturers to get better prices, but this can introduce risks. For example, in 2016, US Customs and Border Protection seized more than 16,000 hoverboards containing counterfeit batteries, with the value of the seized hoverboards totaling over $6 million [4].

Counterfeiting activities vary from re-marking an original product to illegally manufacturing the product based on the original design, reverse engineering, or a new design [5]. Typical counterfeiting behavior involves marketing batteries that were rewrapped batteries under false branding, recycled batteries without informing the customer, or defective cells.

A battery's internal structure is easy to replicate; a trained worker can make a lithium-ion battery in minutes with readily available equipment and apply a label to the cell cover or battery pack. However, the exact electrolyte and electrode formula cannot be easily replicated. Although there is almost no barrier to prevent fraudulent manufacturers from making functional counterfeit lithium-ion batteries, the safety and reliability of these batteries cannot be guaranteed. Figure 14.1 shows an example of a high-risk lithium-ion battery supply chain.

14.1 Counterfeiting

Battery counterfeiting activities exploit people's trust in well-known brands. A typical market affected by the problem of counterfeit lithium-ion batteries is e-cigarettes [7]. The e-cigarette

The Safety Challenges and Strategies of Using Lithium-Ion Batteries, First Edition. Edited by Michael G. Pecht.

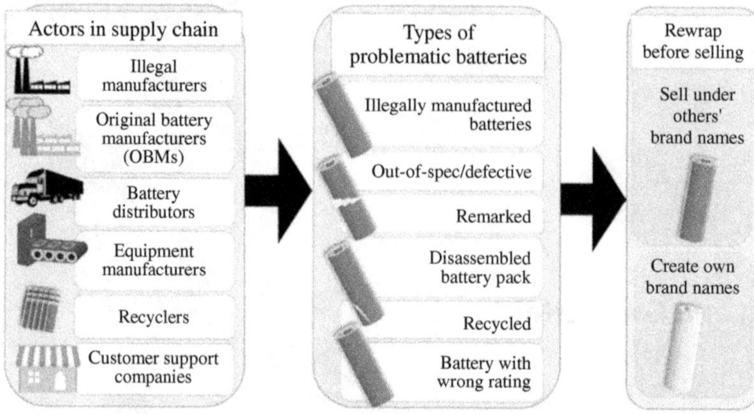

Figure 14.1 The supply chain of high-risk lithium-ion batteries. Source: Kong et al. [6]/MDPI/CC BY 4.0.

market is newer and smaller than the cell phone and laptop markets. Compared to the embedded battery design in cell phones and laptops, one significant difference in e-cigarettes is that many have replaceable cell designs, allowing end users to install spares or replacement batteries. However, the replaceable design enables end users to select counterfeit or rewrapped batteries in the retail market because they cannot distinguish them from genuine batteries. Batteries capable of high-current discharge are popular among e-cigarette users because higher currents can generate more vapor and stronger stimulation. However, the dilemma is that many original battery manufacturers (OBMs) state that their products are not designed for such an application.

As a major OBM, Sony asserted that its 18650 battery is not intended for e-cigarette usage but for power products [8], even though Sony's VTC 18650 battery is highly recommended on almost every e-cigarette consumer online forum. In 2016, Sony transferred its battery business to Murata Manufacturing [9], but 18650 batteries under the Sony brand are still entering the e-cigarette market. Some of these batteries may be genuine, but they may have been classified as lower-grade batteries by the OBM. In 2021, the US Consumer Product Safety Commission issued this warning regarding 18650 batteries:

The US Consumer Product Safety Commission warns consumers not to buy or use loose 18650 lithium-ion battery cells. These cells are manufactured as industrial components of battery packs and are not intended for individual sale to consumers. However, they are being separated, rewrapped, and sold as new consumer batteries, typically on the Internet [10].

Brand manufacturers are aware of the counterfeit issues and have brought them to the public's attention. Sony warns that "misuse of Sony lithium-ion cells in e-cigarettes and vape pens may cause a serious risk to personal property and safety" [8]. As a countermeasure, Sony has established retail stores [11], which source products directly from the manufacturer and significantly reduce the possibility of counterfeit batteries entering the market. Another popular brand among e-cigarette users is LG Chem, which has also warned against the use of its batteries in e-cigarette applications. LG Energy Solution stated that their "Lithium ion Battery Cell must not be indirectly or directly sold to, distributed, or used as a stand-alone, removable, or consumer-replaceable power source for any electronic devices (e.g., e-cigarettes, lanterns, etc.) LG Energy Solution's policy does not permit, approve or recommend such act" [12]. Counterfeit batteries are not limited to the e-cigarette market; Nikon and Canon also face problems with counterfeit batteries in camera applications.

Without establishing a universal battery tracking system along the supply chain, the impact of the original brand holders' isolated activities may be limited. Counterfeit batteries can still reach consumers from other nodes in the supply chain since most OBMs do not directly interact with their final customers. OBMs most often sell batteries to equipment manufacturers who incorporate them into devices to serve as power sources. A typical lithium-ion battery-powered electronic device supply chain has several nodes before the battery reaches the end user. Due to the nature of battery logistics, various entities may be held responsible for a battery incident. For example, in a 2015 lawsuit, a woman was injured by an exploding battery in an e-cigarette. Although it was not mentioned whether the battery was identified as counterfeit, the victim was awarded nearly $1.9 million from the local e-cigarette store, the distributor, and the wholesaler, all of whom were found responsible for the injury [13]. Although, in this specific case, the OBM was not included in the liability list, battery manufacturers are usually sued when they can be identified. The seeming manufacturer will become the scapegoat if the problem-causing batteries are counterfeit. Moreover, the actual manufacturer and the seeming one may not be identified for rewrapped batteries, which will be discussed in the following sections. The battery business is internationalized. The difference in the local laws and regulations can be used to gain illegal profits. Battery quality issues matter to the entire supply chain.

14.2 Rewrapping

Rewrapped and illegally manufactured batteries fall into different categories of possible fraudulent or unlawful actions. Battery manufacturers, such as Sony, LG Chem, and Samsung, are facing the issue of rewrapped batteries, where battery cells are pulled from larger battery packs and rewrapped [14]. Compared to fraudulently made counterfeit batteries, rewrapped batteries may have been initially sold to a company by the OBM for battery pack usage. Although these batteries are not intended to be sold directly to individual customers, battery packs can be separated and rewrapped as individual cells to sell to customers. Some of these used batteries are wrapped with a new covering to make them appear new. Rewrapped batteries may also be defective or substandard, meant to be discarded but instead sold with original or off-brand labels. Because of the mislabeling, a rewrapped battery may be used in unexpected conditions, increasing risks to the consumer.

The outer wrapper of a cylindrical battery covers the outer metal case and prevents potential short circuits [15]. Typically, battery wrappers contain information about the brand name, nominal voltage, capacity, and notices to consumers. However, some distributors use wrappers to build up their brands. These distributors do not manufacture batteries but instead rewrap batteries from other manufacturers [16]. Rewrapped batteries come from various sources and include recycled and defective batteries. There may be scratches on the battery's metal outer case, but they can appear new once rewrapped with a fresh cover. Counterfeit batteries may not meet the designed capacity or electrical specifications [17], and even if they do, they are unlikely to meet other design and material specifications of the genuine design.

For example, counterfeit batteries made through rewrapping can make an 18350 cell look like an 18650 cell. Compared with 18650 batteries (65 mm length), 18350 batteries are much shorter (35 mm length). During rewrapping, an 18350 cell can be connected to a step-up circuit, and the gap between the 18350 and 18650 is filled with materials such as sand and paper to make it appear the same size as an 18650 cell [18]. Figure 14.2 schematically illustrates how this kind of counterfeit battery is constructed.

Manufacturers' guidance for identifying counterfeit batteries primarily focuses on the printing quality of the wrapper or the label on the batteries. Canon's website provides customers with information about how to avoid counterfeit Canon lithium-ion batteries [19]. Sony also warns customers

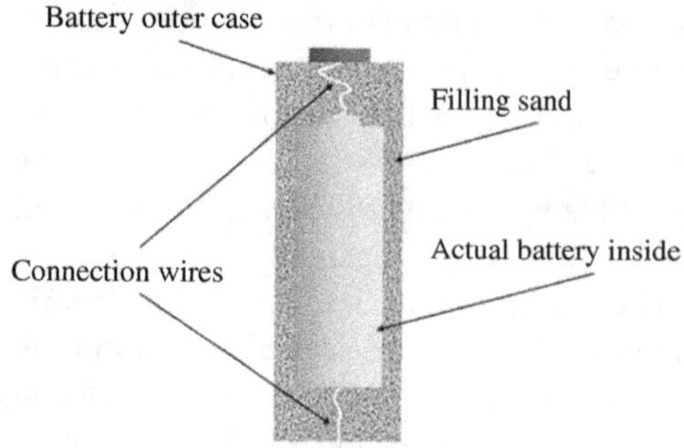

Figure 14.2 Illustration of a counterfeit 18650 battery made from a smaller 18350 battery. Source: Kong et al. [6]/MDPI/CC BY 4.0.

that a genuine battery always has a clearly printed logo, whereas the logo on a counterfeit battery may be "blurry, sloppy, off-color, or misplaced" [20]. However, these methods are unreliable since the wrapper or the label can also be counterfeited.

14.3 Counterfeit Batteries in the Market

The supply chain can be divided into the battery supply chain and the device supply chain. After manufacturing lithium-ion batteries, the first step in the battery supply chain is the transfer from the OBMs to the device manufacturers. LG, Samsung, Sony, and Maxell are examples of OBMs, and the batteries they produce are supplied to device manufacturers. For device manufacturers, batteries are either embedded in their products or replaceable. The replaceable design allows consumers to switch to third-party batteries later. In one scenario, the OBM sells batteries to a battery distributor (for example, Sekorm is the distributor for Maxell), and then the battery distributor sells the batteries to device manufacturers. In another scenario, the OBM can act as a battery distributor and directly sell its batteries to device manufacturers. In both situations, the batteries are sold in large volumes, often hundreds of thousands of cells. The second part of the supply chain involves the device supply chain, which covers the steps from the device manufacturers to the consumers.

Counterfeit batteries can enter the supply chain at various points, including OBMs, battery distributors, device manufacturers, and retailers. Some devices with embedded batteries prevent the consumer from replacing the battery. However, devices with replaceable batteries can end up with counterfeits when replaced by a consumer who wants a more affordable battery.

Counterfeit battery sellers use online trading platforms to further spread counterfeit batteries, with documented involvement in Amazon and eBay [21, 22]. Worldwide, it was reported that 3.3% of the global trade of all goods in 2019 consisted of counterfeit and pirated goods, with the United States being the most affected country, where seized fake products accounted for 24% of the global total of seized counterfeit goods in 2016 [23]. In the online selling business model, even a single person can sell a large volume of counterfeit batteries. In 2019, one counterfeit scheme involved around $24 million worth of counterfeit batteries for laptop computers [24].

Distributors who do not manufacture lithium-ion batteries can build their brands by rewrapping batteries from various sources. For electronic components, standards like IPC-1752 require companies in the supply chain to share information on constituent materials. However, similar standards do not exist for dealing with counterfeit lithium-ion batteries. Additionally, existing regulations may be difficult to enforce with lithium-ion batteries because the battery components are sealed within cases. As a result, the materials analyses necessary for authentication are costly and time-consuming.

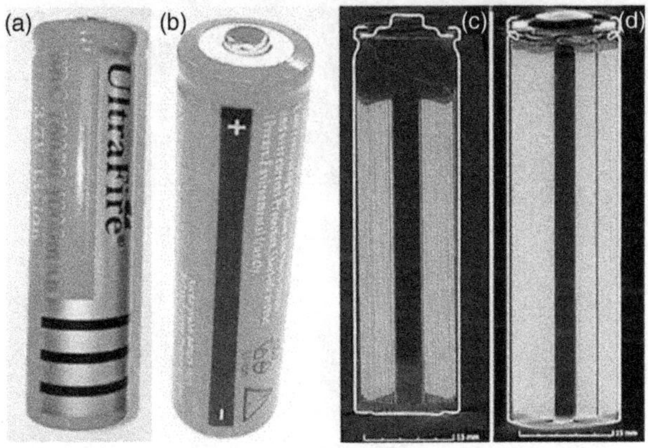

Figure 14.3 Counterfeit 18650 battery with a label showing (a) an exaggerated capacity of 4000 mAh and (b) misspelled words. Under X-ray computed tomography, (c) the counterfeit battery shows a shorter internal "jelly roll" than the (d) genuine product (Source: Photos courtesy of DfR Solutions) [6].

Specifications of counterfeit batteries can be inaccurate. For example, many batteries are available online under the brand name UltraFire. The labeled battery capacity varies significantly, with some batteries even claiming a capacity as high as 4000 mAh. However, according to UltraFire [25], they have never manufactured any battery that can reach a capacity as high as 4000 mAh, so the advertised batteries are counterfeit. Figure 14.3 shows the internal structural differences between counterfeit and genuine batteries [6]. These rewrapped batteries are unlikely to meet the specifications of the authentic product and could be prone to thermal runaway, where the thermal energy released from a battery [26] can result in a battery fire or explosion.

Numerous fire and explosion incidents related to MXJO 18650 cylindrical lithium-ion batteries have raised concerns regarding the company's reliability and safety capability maturity. An assessment of existing MXJO batteries showed there were burrs on the battery's tabs that could lead to an internal short circuit [27]. There was also no positive temperature coefficient (PTC) thermistor, a safety component required by battery standards IEEE 1625, IEEE 1725, and UL 2054. Furthermore, while the MXJO batteries studied were equipped with a current interrupt device (CID) and vents, numerous incidents were identified where these safety mechanisms failed to prevent fires and explosions. The missing PTC in the MXJO battery indicated that the company did not use all the available methods required by the standards to provide proper product reliability and safety.

The company, which is related to the brand name "MXJO," was liquidated in China in 2019, and no valid address could be found to indicate that the company was conducting any design, manufacturing, or testing activities. In addition, even during the company's operational period, it did not appear to have the capital capability to support mass production or the appropriate quality process controls [28]. As a result, although the MXJO batteries are still on the market, there is no way to audit the company to assess its design, manufacturing, testing, quality, reliability, and safety assurance processes.

Because of the various sources of batteries and changes made by OBMs among different batches, even batteries under the same brand name can have notable differences. A case study involving three LG 18650 batteries presents an example of inconsistent lithium-ion battery structures sharing the same brand name, illustrating the efforts required to identify counterfeit lithium-ion batteries. Three LG 18650 batteries are shown in Figure 14.4; they were from different distributors, with brown outer casings on all three batteries. According to the owner, who is also a victim of e-cigarette explosion incidents, these batteries were sold under the same listing and were purchased from a general market. However, since LG claims that it does not sell 18650 batteries to individual customers, it cannot be assumed that any of the batteries are authentic.

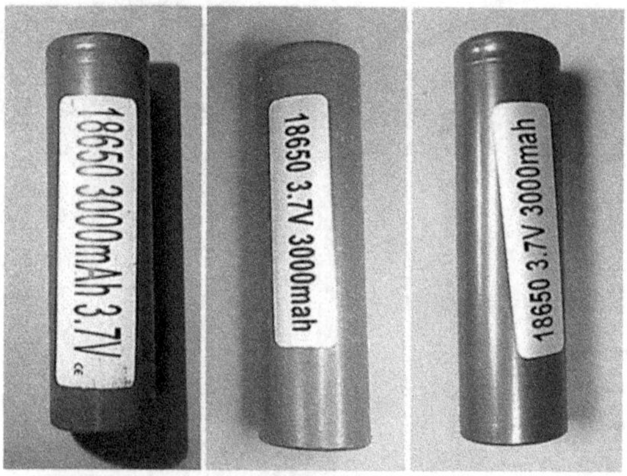

Cell L-1 Cell L-2 Cell L-3

Figure 14.4 LG Chem 18650 batteries with different labels. Source: Kong et al. [6]/MDPI/CC BY 4.0.

When specifications—including dimensions, weight, impedance, and capacity—are available for authentic batteries, battery parameters from questionable batteries can be compared to the specifications as indicators of counterfeiting. The battery materials and marking codes can also be compared if an authentic battery is available. Although the text font and size on the labels are different in this LG example, all of the batteries claim a 3000 mAh capacity and a 3.7 V output voltage.

As shown in Figure 14.5, the battery serial number is printed directly on the outer casing under the label. The code is assigned and printed by the battery manufacturer and explained in the battery specification, which helps users understand manufacturing information about these samples. Figure 14.6 is an example of the LG battery specification, which shows the meaning of the printed numbers and letters on the battery wrapper. Numbers and letters in different sequences provide information about the battery.

The first line of printed text on cell L-1 reads "LGDBHG21865." Compared to the given example, "LG" indicates the company name. "D" indicates the factory/complex site. "B" is an indicator of battery quality, assigned based on comparing the battery's actual capacities and the designed value. "B" means the battery capacity is higher than 2400 mAh. "HG2" indicates the type of this specific battery. The last four numbers, "1865," refer to the battery size/model size: "18" refers to the diameter of the cylindrical cell, which is 18 mm, and "65" refers to the length of the entire cell, which is 65 mm.

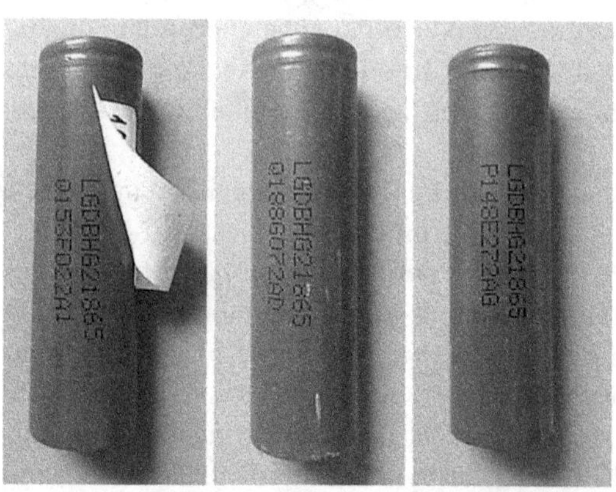

Cell L-1 Cell L-2 Cell L-3

Figure 14.5 Serial numbers printed on the casing (under the wrapper). Source: [6]/MDPI/CC BY 4.0.

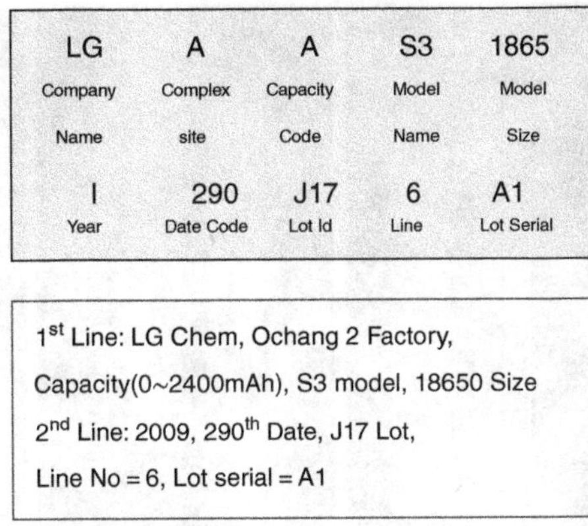

Figure 14.6 Interpretation of text printed on the outer casing, an example given in the battery specification. Source: Chem [29]/LG Chem.

Regarding the second line: "Q" refers to the manufacturing year. The letter is chosen in alphabetic order. As mentioned in the product specifications [29], the letter "I" indicates the year 2009 (shown in Figure 14.6). Thus, the "Q" on cell L-1 in Figure 14.5 indicates 2017. The number "153" after the letter "Q" refers to the date code for when this battery was manufactured. Thus, "Q153" means this battery was manufactured on the 153rd day of 2017. "F02" is the lot ID. The second number, "2," indicates the line number in the factory. "A1" at the end of the second line refers to the lot serial number.

For the three batteries, the printed text on each battery's outer casing is listed in Table 14.1. The first lines of text are the same for all three cells, indicating they are all LG HG2 18650 batteries manufactured in the same factory. The second lines of text are different, meaning the batteries were manufactured in different years and dates. The lot ID and lot serial number are also different. However, because the serial numbers on all three batteries follow the standard format for identifying cells, as illustrated in Figure 14.6, it is difficult to determine if these batteries are counterfeit or genuine.

Based on a visual inspection of the three batteries, the LG batteries show little difference in the outer casing. Even the serial numbers follow the same format, making it difficult to distinguish them. However, the computed tomography (CT) scan images reveal their differences. Unlike destructive methods, CT allows a battery's internal structure to be examined without disassembly. Although the three batteries show little difference in outward appearance, the CT images (Figure 14.7) clearly show different inner structures. Cell L-1 has two negative electrode tabs, while cells L-2 and L-3 have only one negative electrode tab. An increased number of tabs is beneficial for reducing electron transport loss and achieving a uniform current density and temperature distribution, especially under a high C-rate. However, an increased number of tabs can also increase opportunities for manufacturing defects and raise the battery manufacturing cost. Another difference is that only cell L-3 has a mandrel in the center, while the other two have nothing in the center.

Table 14.1 Comparison of the text printed on the outer casing of the three batteries.

Cell no.	Company	Complex code	Capacity code	Model name	Model size	Year	Date code	Lot id	Line	Lot serial
L-1	LG	D	B	HG2	1865	Q	153	F02	2	A1
L-2	LG	D	B	HG2	1865	Q	188	G07	2	AD
L-3	LG	D	B	HG2	1865	P	148	E27	2	AG

Figure 14.7 CT images showing the inner structures of cells L-1, L-2, and L-3. Source: Kong et al. [6]/MDPI/CC BY 4.0.

OBMs should apply better quality control to prevent defective batteries from entering the market. They should also use permanent markings on their batteries, such as engraving on the metal case. The UL standard for e-cigarettes, UL 8139, has incorporated such a request.

14.4 Hazards of Counterfeit Batteries

Counterfeit batteries in the market are a threat to consumer confidence. With the popularity of lithium-ion batteries, there are increasing concerns about battery quality, safety, and durability due to the rising number of incidents reported. Counterfeit lithium-ion batteries are usually less expensive than genuine ones [30] because it is unlikely that a counterfeiter would spend the resources for proper design, qualification, quality control, or supply chain management. On the one hand, counterfeit batteries often have exaggerated capacity specifications. This can cause the battery to operate under overstressed conditions, increasing the risk of fire or explosion. On the other hand, counterfeit batteries may lack protective devices [31] against overcharging or thermal runaway. Such batteries are typically made from defective, used, or illegally manufactured cells with little or no quality control. As a result, compared to legitimate and quality products, counterfeit batteries pose a higher threat to users' safety during potential overheating or fire incidents [20].

Counterfeit batteries lack active energy storage materials, which means they cannot meet the specified capacity they claim. Using counterfeit batteries with exaggerated capacity claims can cause the battery to work beyond tolerable stress conditions. In the worst case, if a load is applied to a counterfeit battery based on its exaggerated rating, the battery may be abused and create unsafe conditions. A battery's maximum charge/discharge current is usually determined by its available capacity, but with an exaggerated capacity, the battery may be used under overcurrent conditions, leading to overheating. Due to the flammable nature of organic electrolytes, overstressed lithium-ion batteries can potentially result in a fire or an explosion. During a thermal runaway event, exothermic reactions rapidly increase the internal battery temperature, reaching as high as 943 °C [32].

Counterfeit batteries may be offered at an attractive price, but this comes at the cost of lower quality control and fewer safety features (against fire or explosion). For example, the counterfeit UltraFire battery had no vent holes or other safety mechanisms, making it more vulnerable to potential thermal runaway hazards than the genuine product. Nikon [33] also warns that "counterfeit products may not be equipped with safety mechanisms such as the protection device." Genuine lithium-ion batteries have protective devices, including a PTC, a CID, and a safety vent [34]. However, counterfeit batteries often omit these protective devices to reduce costs. In practice, counterfeit batteries may be overstressed even under normal use conditions. Counterfeit batteries are not a new problem; there have been bodily injury incidents caused by counterfeit batteries, which were reported before 2009. In one case, a man was injured when an inexpensive counterfeit phone battery exploded during a conversation. The battery explosion resulted in deep burns on the man's cheek and neck, and it fractured his skull [35].

Without thorough characterization and a correct battery rating label, the degradation level and the remaining useful life of a rewrapped battery are unknown. There may be no life tests conducted on these problematic batteries. It is also not guaranteed that all the batteries operate within their safe range. Thus, using these batteries can be dangerous. The actual manufacturer of rewrapped batteries sold under others' brand names is unknown. If such a battery causes any damage, the marketplace typically refers to the company whose name appears on the label. As a result, OBMs involved in legal cases related to injury or property damage from counterfeit batteries have issued warnings and guidance for consumers to detect counterfeit batteries, attempting to shift the responsibility to consumers. However, these activities by OBMs have limited effectiveness in preventing or eliminating counterfeit batteries.

Identifying counterfeit lithium-ion batteries requires expertise and knowledge that are unlikely to be accessible to individual customers. The current method for distinguishing counterfeit batteries involves inspecting the wrapper. Readily available guidance on identifying counterfeit batteries involves simple visual inspections, such as comparing the outer case color and labels to those of genuine products [36]. These methods focus on inspecting the outer battery wrapper, which can also be counterfeited.

Third-party organizations, independent of OBMs, can provide authentication markings for products. The South Korean Electrical Appliances and Consumer Products Safety Control Act 19 [37] even banned products without the "safety assurance mark" from distribution. However, even the authentication of qualified battery markings can be counterfeited, so determining the authenticity of batteries by inspecting the accompanying markings is unreliable. In 2013, Underwriters Laboratories (UL) released a list and photos of counterfeit lithium-ion batteries labeled with a forged UL marking [38]. The UL standard 8139 (2018) requires that markings for electrical systems of electronic cigarettes "shall be permanent" [39]. Although UL-8139 focuses on marking batteries for e-cigarettes, the same concepts are beneficial if applied to the entire battery market.

14.5 Summary

There is always a risk of getting a counterfeit, rewrapped, or scrap battery in the procurement process, especially when the battery is selected from a distributor, retailer, warehouse, or a third-party entity other than a trustworthy battery manufacturer. Unfortunately, because of the high-energy nature of lithium-ion batteries, a counterfeit, rewrapped, or scrap battery is more likely to result in performance issues and thermal runaway.

A purchaser will generally find it difficult to visually determine whether a lithium-ion battery is counterfeit if they do not buy it directly from the manufacturer [40]. The methods used to identify counterfeit batteries require examination of the internal battery structure and comparison to known good samples. Internal nondestructive examination typically involves some form of X-ray, with the best results achieved using CT scan methods. Some additional methods include: measuring and weighing the batteries and comparing the measurements to the datasheet; looking for faded codes and unusual wear and tear like burns, rust, or excessive scratches; checking for quality control markings; and conducting voltage tests while discharging the battery to look for spikes or jumpy voltage readings.

IEEE Standard for Organizational Reliability Capability (IEEE-1624) defines the reliability capability of organizations and sets the criteria for organizational reliability capability assessment. Device manufacturers can use this standard to assess key practices a battery supplier should have, including reliability requirements and planning, training and development, reliability analysis, reliability testing, supply chain management, failure data tracking, analysis, verification, validation, and reliability improvements. The standard should be implemented by conducting battery manufacturers' audits.

References

1 Grand View Research (2023). Lithium-ion Battery Market Size, Share & Trends Analysis Report by Product (LCO, LFP, NCA, LMO, LTO, NMC), by Application (Consumer Electronics, Energy Storage Systems, Industrial), by Region, and Segment Forecasts, 2022–2030. https://www.grandviewresearch.com/industry-analysis/lithium-ion-battery-market (accessed 06 August 2024).

2 Ziegler, M.S. and Trancik, J.E. (2021). Re-examining rates of lithium-ion battery technology improvement and cost decline. *Energy & Environmental Science* 14 (4): 1635–1651.

3 Lemon, S. (2003). Nokia takes steps to fight mobile phone explosions. https://www.infoworld.com/article/2232172/nokia-takes-steps-to-fight-mobile-phone-explosions-2.html (accessed 06 August 2024).

4 U.S. Customs and Border Protection (2016). CBP Seizes Record Amount of Counterfeit Hoverboards. https://www.cbp.gov/newsroom/local-media-release/cbp-seizes-record-amount-counterfeit-hoverboards (accessed 20 January 2022).

5 Pecht, M. and Tiku, S. (2006). Bogus: electronic manufacturing and consumers confront a rising tide of counterfeit electronics. *IEEE Spectrum* 43 (5): 37–46.

6 Kong, L., Das, D., and Pecht, M.G. (2022). The distribution and detection issues of counterfeit lithium-ion batteries. *Energies* 15 (10): 3798.

7 Rosales, I. (2018). Battery Expert Warns Counterfeit Batteries have Invaded the U.S Vape Pen Industry. ABC Action News. https://www.abcactionnews.com/news/region-south-pinellas/st-petersburg/battery-expert-warns-counterfeit-batteries-have-invaded-the-us-vape-pen-industry (accessed on 16 October 2018).

8 Sony (2015). WARNING: ECigs, Vape Pens, and Lithium-Ion Batteries. https://www.sony.com/electronics/support/articles/S1F1621 (accessed 9 March 2021).

9 Murata (2016). Murata Manufacturing and Sony Sign Definitive Agreement for the Transfer of Battery Business. https://www.murata.com/about/newsroom/news/company/general/2016/1031 (accessed 18 February 2018).

10 The U.S. Consumer Product Safety Commission (CPSC) (2021). CPSC Issues Consumer Safety Warning: Serious Injury or Death Can Occur if Lithium-Ion Battery Cells Are Separated from Battery Packs and Used to Power Devices. https://www.cpsc.gov/zhT-CN/node/48082 (accessed 10 March 2021).

11 Sony. Where to buy. https://www.sony-asia.com/microsite/retailshops/ (accessed 16 October 2018).

12 LG EEEnergy Solution. Li-ion Battery Safety Guide. https://www.lgensol.com/en/cs-safetyguide-caution-tab1 (accessed 06 August 2024).

13 Branson-Potts, H. (2015). Woman Burned by Exploding E-Cigarette Battery Awarded $1.9 Million. Los Angeles Times. http://www.latimes.com/local/crime/la-me-ecigarette-burns-verdict-20151001-story.html (accessed 18 February 2018).

14 Olinger, M. (2017). SC Man Sued Sony after Exploding E-cig Battery Disfigured Leg. The State. http://www.thestate.com/news/local/article158845194.html (accessed 18 February 2018).

15 Zamplebox (2015). Mechanical Mods | Single Battery Devices | Battery Safety. https://www.zamplebox.com/community/mechanical-mods-single-battery-devices-battery-safety (accessed 18 February 2018).

16 Lutz, L. (2017). Are You Using a "Rewrap" Battery? Exposing The WORST Brands In Vaping. Vapebeat. https://vapebeat.com/features/are-you-using-a-re-wrap-battery (accessed 18 February 2018).

17 Wu, C.-M.J., Winger, L.K., and Book, C.S. (2015). System and method for detecting counterfeit and defective batteries using battery characteristic profiles. US Patent 9,146,282, issued 29 September 2015.

18 Spinfuel Vape (2016). Counterfeit 18,650 Batteries Are Everywhere. https://spinfuel.com/counterfeit-18,650-batteries-everywhere/ (accessed 18 February 2018).

19 Canon. About Counterfeits: Understanding the Problem. https://www.usa.canon.com/internet/portal/us/home/explore/product-showcases/about-counterfeits (accessed 18 February 2018).

20 Sony (2018). Watch Out for Counterfeit Products. https://www.sony.com.sg/electronics/support/articles/00200902 (accessed 16 October 2018).

21 U.S. Attorney's Office, Southern District of Texas (2015). Woman Sentenced For Selling Counterfeit Samsung Batteries. https://www.justice.gov/usao-sdtx/pr/woman-sentenced-selling-counterfeit-samsung-batteries (accessed 22 April 2022).

22 Semuels (2019). When Your Amazon Purchase Explodes. The Atlantic. https://www.theatlantic.com/technology/archive/2019/04/lithium-ion-batteries-amazon-are-exploding/587005/ (accessed 22 April 2022).

23 Florence, M. (2019). Trends in Trade in Counterfeit and Pirated Goods, The European Agency for Safety and Health at Work. Spain. https://policycommons.net/artifacts/4488673/trends-in-trade-in-counterfeit-and-pirated-goods/5291292/ (accessed 07 August 2024). CID: 20.500.12592/cx6072.

24 U.S. News (2019). Man Arrested in $24M Scheme to Sell Bogus Laptop Batteries. https://www.usnews.com/news/best-states/california/articles/2019-12-19/man-arrested-in-24m-scheme-to-sell-bogus-laptop-batteries (accessed 22 April 2022).

25 The UltraFire SCAM. (Fake 18650 Li-Ion Batteries Rated 4200mAh) BEWARE, ChrisViral Vlogs. https://www.ultrafire.com/collections/batteries (accessed 06 August 2024).

26 Said, A.O., Lee, C., and Stoliarov, S.I. (2019). Experimental investigation of cascading failure in 18,650 lithium ion cell arrays: impact of cathode chemistry. *Journal of Power Sources* 446: 227347.

27 Wu, Y., Saxena, S., Xing, Y. et al. (2018). Analysis of manufacturing-induced defects and structural deformations in lithium-ion batteries using computed tomography. *Energies* 11 (4): 925.

28 Kong, L. and Pecht, M.G. (2023). A case study into a battery company and their 18650 batteries. *e-Prime-Advances in Electrical Engineering, Electronics and Energy* 6: 100294.

29 Chem, L. (2011). Product Specification Rechargeable Lithium Ion Battery Model: ICR 18,650 S3 2200mAh. Seoul. http://www.batteryspace.com/prod-specs/4869.pdf (accessed 18 February 2018).

30 Underwriters Laboratories. Counterfeit Lithium-ion Cells and Batteries. https://besafebuyreal.ul.org/sites/default/files/2020-11/UL%20Counterfeit%20Batteries%20eGuide_Interactive.pdf#:~:text=Counterfeit%20Lithi-um-ion%20Cells%20and%20Batteries%20Widespread%20demand%20of,by%20copying%20high-quality%20products%20and%20deceive%2 (accessed 22 April 2022).

31 Camera & Imaging Products Association, CIPA. Notice Concerning Counterfeit Battery Packs. Camera & Imaging Products Association, CIPA. http://www.cipa.jp/battery/index_e.html (accessed 16 October 2018).

32 Fu, Y., Lu, S., Li, K. et al. (2015). An experimental study on burning behaviors of 18,650 lithium ion batteries using a cone calorimeter. *Journal of Power Sources* 273: 216–222.

33 Nikon (2019). Counterfeit batteries and accessories. https://www.nikonimgsupport.com/na/NSG_article?articleNo=000044699&configured=1&lang=en_SG#:~:text=Counterfeit%20Nikon%20Product,-The%20Counterfeit%20packaging&text=%22Nikon%22%20and%20%221100mAh%22,as%20the%20country%20of%20manufacture (accessed 06 August 2024).

34 Kong, L., Li, C., Jiang, J., and Pecht, M.G. (2018). Li-ion battery fire hazards and safety strategies. *Energies* 11: 2191.

35 Ben, D., Ma, B., Liu, L. et al. (2009). Unusual burns with combined injuries caused by mobile phone explosion: watch out for the "mini-bomb"! *Journal of Burn Care & Research* 30: 1048.

36 Battery Bro (2015). Is It a Fake 18,650 Battery? (Featuring the Samsung 25R). https://batterybro.com/blogs/18650-wholesale-battery-reviews/30828867-is-it-a-fake-18650-battery-featuring-the-samsung-25r (accessed on 06 August 2024).

37 KLRI (2017). Electrical Appliances and Consumer Products Safety Control Act. Law Viewer. https://elaw.klri.re.kr/eng_mobile/viewer.do?hseq=45624&type=sogan&key=13 (accessed 25 February 2020).

38 UL (2018). UL Warns of Counterfeit UL Mark on DULEX Lithium-ion Rechargeable Batteries (Release 18PN12). https://www.ul.com/news/ul-warns-counterfeit-ul-mark-dulex-lithium-ion-rechargeable-batteries-release-18pn12 (accessed 06 August 2024).

39 UL 8139 (2018). *Standard for electrical systems of electronic cigarettes and vaping devices.* Northbrook, IL, USA: Underwriters Laboratories.

40 The Load Star (2015). Lennane. Banning Lithium Batteries from Aircraft Bellies Won't Stop Fake or Smuggled Shipments—IATA. https://theloadstar.co.uk/banning-lithium-batteries-aircraft-bellies-wont-stop-fake-smuggled-shipments-says-iata/ (accessed 18 February 2018).

15

Supply Chain Battery Regulations

Shalini Dwivedi and Aparna Akula

Supply chain regulations ensure the ethical sourcing and fair treatment of workers involved in battery material extraction. Such regulations address concerns including conflict mineral usage and child labor, with precedents like the US Dodd-Frank Act and the EU Conflict Minerals Regulation [1]. Lithium-ion battery (LIB) regulations aim to minimize environmental and safety costs, notably transportation, while regulating access to resources essential for LIB production. Simultaneously, they seek to mitigate the economic consequences of LIB usage and production while also fostering technological progress. They help protect both consumers and the environment.

15.1 EU Battery Regulation 2023

To create a global system where products are reused or recycled, and resources are used responsibly, the European Union (EU), a big market (USD 28.126 billion) [2] for batteries, has established the EU Battery Regulation 2023/1542, which supersedes EU Battery Directive 2006/66/EC. The EU Battery Regulation establishes a framework that spans the entire battery life cycle, encompassing raw material extraction, product design (including electric vehicles and mobile phones), and end-of-life management [3].

The EU Directive 1991/157/EEC, enacted on March 18, 1991, established the first regulatory framework for the management and disposal of batteries and accumulators containing hazardous substances [4]. Aimed at reducing environmental impact, it imposed limits on toxic metals like lead, mercury, and cadmium, mandated efficient collection and recycling systems, and placed producer responsibility for clear labeling and disposal information. The directive also required member states to conduct consumer education campaigns and harmonize national regulations, fostering public participation and awareness. Its implementation led to significant advancements in battery recycling and the reduction of hazardous substances, laying the groundwork for subsequent legislation, such as Directive 2006/66/EC, to further enhance environmental protection.

The EU Battery Directive 2006/66/EC regulates the life cycle of batteries and accumulators, and mandates collection, recycling, treatment, and disposal protocols [5]. It restricted the marketing of most mercury- and cadmium-containing batteries. Furthermore, it established minimum collection targets of 25% of all waste portable batteries by 2012 and 45% by 2016. Recycling targets are defined based on average weight, with 65% for lead-acid, 75% for nickel-cadmium, and 50% for other battery chemistries [6]. Additionally, the directive promotes producer responsibility,

The Safety Challenges and Strategies of Using Lithium-Ion Batteries, First Edition. Edited by Michael G. Pecht.

requiring manufacturers to play a role in battery end-of-life management. Finally, it mandated battery labeling and facilitated their removal from equipment.

The EU leads the global shift toward a circular economy, as evidenced by the enactment of the EU Battery Regulation 2023/1542. This regulation, a crucial component of the European Green Deal Circular Economy Action Plan (CEAP), directly confronts the mounting environmental and ethical concerns arising from the exponential growth in battery utilization for electric vehicles and energy storage [7]. The European Green Deal, inaugurated in 2019, encompasses a multifaceted array of policy initiatives aimed at several key objectives. These objectives include attaining climate neutrality within the European Union by the year 2050, preserving and enhancing biodiversity, fostering the establishment of a circular economy, mitigating pollution, enhancing competitiveness, and facilitating a transition that is equitable and fair for all stakeholders involved [8].

EU Battery Regulation 2023/1542, initially proposed in 2020, complements the Strategic Action Plan on Batteries. It entered into force in August 2023, replacing EU Battery Directive 2006/66/EC. The regulation establishes a detailed framework that spans the entire battery life cycle, encompassing critical phases such as raw material extraction, product design (including electric vehicles and mobile phones), and end-of-life management. It represents a significant advancement in the EU's approach to battery management, and it prioritizes safety and environmental sustainability by promoting a holistic life-cycle approach for batteries. Given its far-reaching impact across the battery supply chain, a rigorous understanding of the regulation's provisions is essential for all stakeholders involved in battery production, consumption, and disposal. This framework applies to all batteries placed on the EU market, except batteries specifically designed for military and space applications and safety-critical equipment in nuclear installations [3].

15.2 Unveiling the Regulatory Framework: Key Features and Insights

EU Battery Regulation 2023/1542 addresses shortcomings of the previous EU Battery Directive 2006/66/EC. Those shortcomings include limited applicability to technological advancements and novel battery applications, inefficiencies in waste battery collection, and inadequate material recovery from spent batteries. As a CE marking regulation, Regulation 2023/1542 mandates adherence to a series of producer obligations [3]. It prioritizes three objectives. First, it aims to enhance the internal market's operations concerning battery products, processes, waste batteries, and recyclates through standardized regulations. Second, it fosters a circular economy. Third, it seeks to diminish the environmental and societal consequences of batteries throughout their life cycle, particularly within supply chains [9]. Key features of the framework are discussed in Figure 15.1. In addition to the features listed in Figure 15.1, minimum requirements for differentiating used batteries from waste batteries in export are also highlighted in the regulation.

EU Battery Regulation 2023/1542 supersedes previous legislation in several key areas. It streamlines regulations by repealing the earlier Battery Directive 2006/66/EC. It strengthens Waste Directive 2008/98/EC by revising extended producer responsibility (EPR) battery requirements. The regulation integrates with the EU Market Surveillance Regulation 2019/1020, ensuring compliance with market access requirements for batteries.

EU Battery Regulation 2023/1542 signifies the EU's commitment to a more sustainable and responsible battery ecosystem by establishing a detailed framework encompassing producer obligations, life-cycle management, and revised directives. The battery passport initiative [8], a primary inclusion in the regulation, addresses the need for a digital product passport ecosystem, as

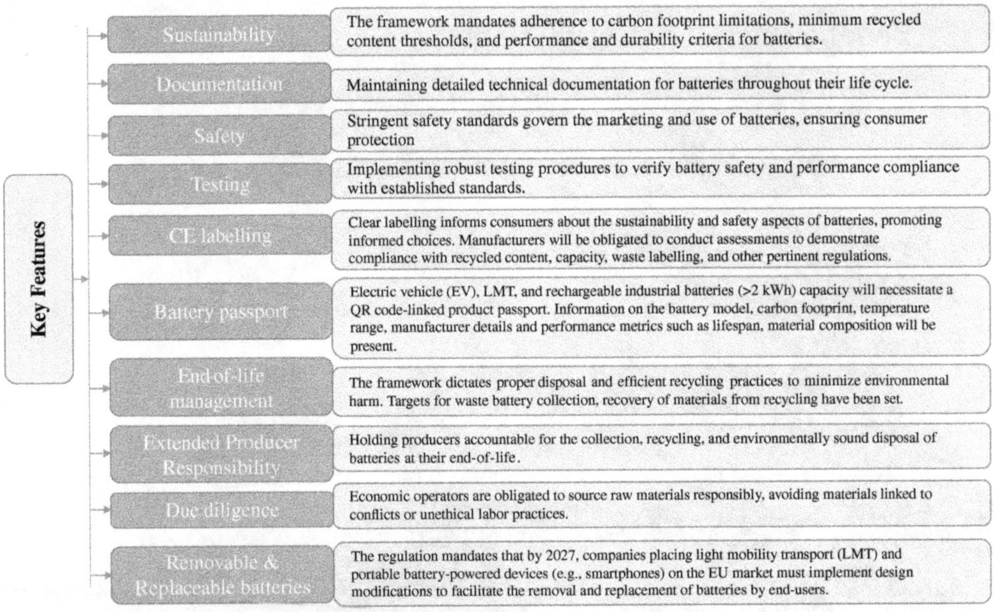

Figure 15.1 Key features of EU Battery Regulation 2023.

emphasized by King et al. [10]. The battery passport encompasses furnishing essential situational insights into batteries, shifting the paradigm from perceiving products solely as consumables to adopting a "product-as-a-service" approach. This transition aims to enhance transparency in decision-making processes and facilitate formulating the business models predicated on value creation, capture, and conservation [11]. Figure 15.2 illustrates the concept of the battery passport across the entire supply chain.

15.2.1 Evolutionary Shift: Battery Regulation 2023 Versus Battery Directive 2006

EU Battery Regulation 2023/1542 adopts a CE marking approach, significantly departing from the framework established by the previous Battery Directive 2006/66/EC [3]. The differences are shown in Table 15.1.

15.2.2 A Forward Look at EU Battery Regulation 2023/1542

Achieving compliance with EU Battery Regulation 2023/1542 may present significant challenges for some companies, potentially necessitating the adoption of unfamiliar processes or, in more complex scenarios, a complete product redesign [7, 12, 13]. Critical areas of challenges are tabulated in Table 15.2.

The regulation's new standards and requirements could significantly impact the sale of batteries and battery-operated products. Products failing to meet these standards may be ineligible for sale in the EU, potentially reducing product variety. Moreover, complying with the new standards could escalate battery manufacturing and recycling costs. Consequently, affected products may see price increases, influencing customer purchasing behaviors. A price increase may stem from various factors. First, stricter manufacturing and environmental standards may elevate the procurement costs of batteries and battery-operated products. Consequently, retailers must reevaluate their

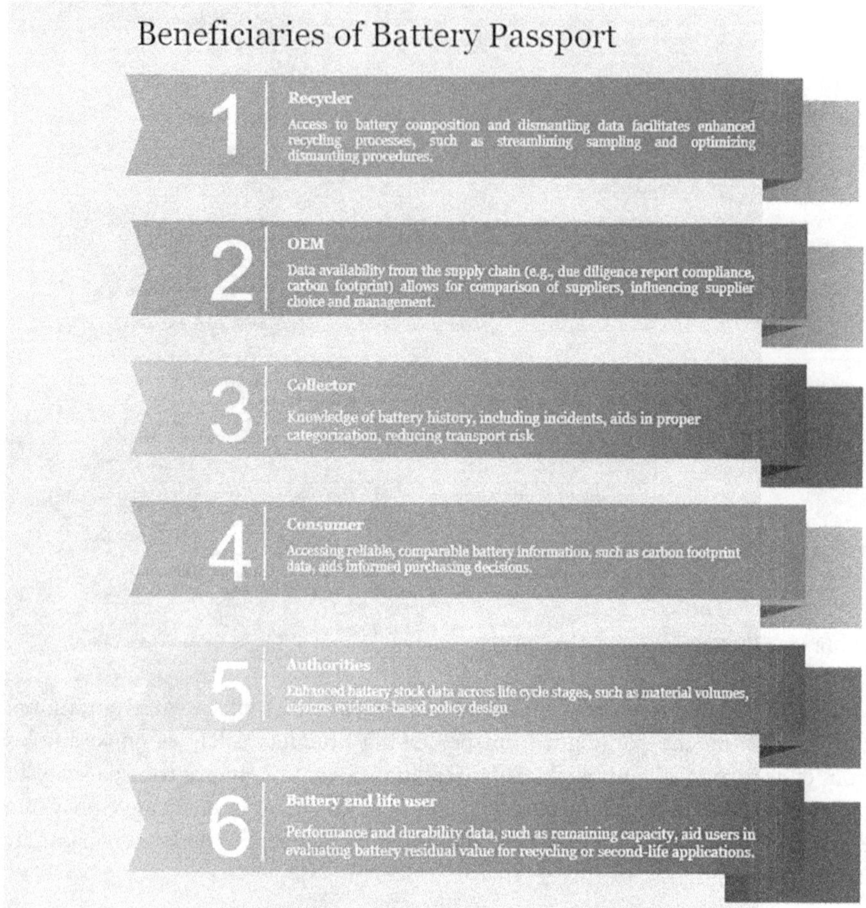

Figure 15.2 Beneficiaries of Battery Passport along the supply chain.

Table 15.1 Differences between Battery Regulation 2023/1542 and Battery Directive 2006/66/EC.

Requirement	Battery Regulation 2023/1542	Battery Directive 2006/66/EC
Battery categorization	(a) Portable (b) Automotive (c) Industrial (d) EV (e) Light means of transport	(a) Portable (b) Automotive (c) Industrial
Maximum allowed regulated substances content in batteries (% by weight)	(a) Mercury (0.0005%) (b) Cadmium (0.002%) (c) Lead (0.01%)	(a) Mercury (0.0005%) (b) Cadmium (0.002%)
Labeling and packaging information	(a) Product information (b) Manufacturer information (c) Separate collection symbol (d) QR code (e) Carbon footprint label (f) Battery capacity (g) Presence of any hazardous substance	(a) Separate collection symbol (b) Chemical symbol (c) Battery capacity
Declaration of Conformity	Required	Not required

Table 15.2 Challenges existing in the current EU Battery Regulation 2023.

Challenge	Description
Carbon footprint transparency	• The regulation mandates enhanced transparency regarding a battery's carbon footprint across its entire life cycle. • This necessitates labeling that reflects the environmental impact across all phases, excluding the use phase. • Achieving this requires implementing the complex data collection processes and meticulous emissions calculations to determine the final footprint.
Ethical sourcing	• Stringent due diligence obligations require companies to proactively address social and environmental risks associated with the sourcing, processing, and trading of primary and secondary raw materials. • Adhering to the regulations will be subject to rigorous third-party audits.
Design and information requirements	• The regulation dictates stricter design and information requirements for batteries and battery-powered products. • For some companies, achieving compliance may necessitate a complete product redesign to remain competitive within the EU market.
Compliance complexity	• Striking a balance between sustainability and performance can be difficult. • Strict carbon footprint limitations and recycled content thresholds may limit innovation and efficiency.
Supply chain concerns	• The due diligence policy might not fully mitigate environmental and social risks associated with battery production.
Technological advancements	• Rapid advancements in battery technology could outpace regulations, creating safety coverage gaps. • Regulations specific to certain battery types may discourage research into more sustainable battery technologies.
Competition	• Higher production costs due to compliance could disadvantage manufacturers compared to those operating in less regulated regions.
Removable batteries	• Designing products with readily removable and replaceable batteries for consumers might not be feasible for all devices due to size or functionality constraints. • Manufacturers might prioritize aesthetics over user-friendliness.
Second-life batteries	• The current framework outlines regulations for repurposing industrial and electric vehicle (EV) batteries; portable devices are not included. • Exempting second-life batteries from carbon footprint and recycled content obligations might lead to suboptimal reuse practices.
Recycling challenges	• While setting minimum recycled content thresholds is essential, there is a risk of manufacturers using low-quality recycled materials to meet these thresholds.
The waste battery collection target	• Collection targets should be proportional to the availability of waste batteries, not to the volume of batteries placed on the market. • This will provide flexibility for suppliers and manufacturers, allowing them to adapt to actual market conditions and consumer behavior rather than being tied to rigid targets that may not reflect the current state of battery waste.
Financial obligation of manufacturers	• The regulation lacks clear directives on the financial contribution manufacturers are obligated to make towards the collection and recycling of spent batteries. • This ambiguity could lead to insufficient compensation for recyclers, potentially hindering the efficiency of waste collection.

supply chains and potentially make adjustments. Second, significant costs arise from establishing and operating return and recycling systems, which include setting up collection points, collaborating with recycling firms, providing return labels, and organizing pickups. Additionally, increased costs are associated with delivering and updating product information and launching customer awareness campaigns. These additional expenses must be meticulously factored into the pricing and business planning of online retailers [14].

The costs resulting from compliance with the new regulation could impact prices for end consumers. Retailers face the challenge of covering these costs without substantially raising prices and jeopardizing their competitiveness. To mitigate price increases, retailers may consider distributing costs across a range of products or framing the additional expenses as an investment in more sustainable business practices. Effective communication with customers regarding the benefits of the new battery standards and the company's commitment to sustainability can help increase acceptance of potential price adjustments.

15.2.3 Navigating Challenges and Solutions

EU Battery Regulation 2023/1542 focuses on addressing safety risks associated with batteries in a general sense [12]. Safety considerations must encompass the entire battery supply chain, covering various stages such as material handling and cell manufacturing. Improving safety at any stage would positively impact the supply chain, including product utilization, maintenance, repair, and potential secondary uses. Furthermore, safety protocols must extend to activities like dismantling, transportation, storage, and recycling of batteries, encompassing waste, damaged, or defective units [15].

The regulation lacks specifications regarding developing a safety assessment methodology at the material level. Also, it does not address methods that could enhance safety management during manufacturing. These methods involve handling nanoparticles, scaling processes, and devising secure designs.

Safety concerns exist surrounding mechanical, thermal, and electrical damage due to car crashes, battery leakage, overheating, overcharging, and electrical system failure. Mandatory test standards such as SAE J2464-2009 and SAE J2929-2011 are in place to simulate these conditions and assess lithium-ion batteries before their use in EVs [16]. However, to foster wider acceptance of batteries meeting established safety standards, manufacturers could offer insurance coverage to users for internal battery damage, even if it arises within safe operating parameters—a facet not addressed in the new EU guidelines. Current safety guidelines do not specifically address battery risks like thermal runaway [17]. It is similarly important to implement safety guidelines for battery chargers to warn users about faulty chargers, an aspect not explicitly covered in the EU guidelines [18].

15.3 Battery Sustainability Practices Worldwide

As society increasingly relies on batteries for energy storage, electric vehicles, and portable devices, it is imperative to prioritize safety, environmental protection, and efficient battery usage. Consequently, countries outside the EU have proactively established regulations on battery safety issues and the disposal or recycling of spent batteries.

15.3.1 United States of America (USA)

In the USA, Underwriters Laboratories (UL) standards address the safety and performance aspects of lithium-ion batteries. These standards include UL 1642, which evaluates safety features, including electrical, mechanical, and thermal aspects of lithium batteries. UL 2054 ensures battery safety in household and commercial applications, such as consumer electronics and small appliances. UL 2580 addresses electric vehicle battery systems, covering safety, performance, and environmental considerations, among others [19].

Hazardous Materials Regulations (HMR) stipulate specific transport requirements for lithium-ion batteries due to their potential hazards. These regulations encompass various aspects, including the obligation for importers to affix the lithium-ion battery mark on product packaging according to Title 49 of the Code of Federal Regulations (CFR) [20].

The Environmental Protection Agency (EPA) has developed a proposal for universal waste standards specifically designed for lithium-ion batteries and distinct from the current general battery universal waste category. In May 2023, the EPA issued a memorandum along with FAQs regarding the classification of lithium-ion batteries under the Resource Conservation and Recovery Act (RCRA). While not introducing new regulations, the EPA clarified its interpretation of existing hazardous waste guidelines concerning lithium-ion batteries. The memorandum emphasizes that, in most cases, lithium-ion batteries are considered hazardous waste at the end of their lifespan and can be handled according to universal waste standards until they reach a designated facility [21].

The establishment of a universal waste category specifically for lithium-ion batteries was meant to enhance safety standards and decrease fires resulting from mismanaged end-of-life batteries, all while encouraging battery recycling [22]. The EPA has published a list of recycling centers and provides details about certified recyclers on their website. This initiative encourages consumers to responsibly dispose of used batteries, which can then be efficiently recycled.

The current status of US guidelines on batteries reflects a multifaceted landscape shaped by various regulatory bodies and initiatives. At the federal level, agencies such as the EPA and the Department of Transportation (DOT) have established frameworks addressing battery production, use, and disposal, primarily focusing on environmental and safety considerations. However, battery recycling laws are different in each state [23]. The absence of comprehensive federal legislation dedicated to batteries has resulted in a patchwork of regulations across states, leading to inconsistencies and challenges in ensuring uniform standards nationwide.

15.3.2 China

In China, several battery regulations have been implemented by the Ministry of Industry and Information Technology (MIIT) recently. The Interim Measures for the Management of Recycling and Utilization of Power Batteries for New Energy Vehicles [24] was implemented in February 2018, and the Pilot Work on Recycling and Utilization of Power Batteries for Electric Vehicles [25] was put into effect in August 2018.

These two MIIT policies require manufacturers of electric vehicles (EVs) to establish facilities for the collection and recycling of spent batteries as part of its strategy to address the increasing waste generated by the sector. Under these regulations, EV manufacturers are mandated to establish recycling channels and service outlets for the collection, storage, and transfer of old batteries to specialized recyclers. Additionally, carmakers, in collaboration with battery manufacturers and their sales units, are tasked with developing a traceability system to identify owners of discarded batteries [26]. The regulations establish parameters for battery recycling, delineating pilot regions and specifying the obligations of battery producers, consumers, and recyclers.

The Ministry of Environmental Protection (MEP) issued the New Policy on Pollution Prevention Techniques from Waste Batteries in December 2016. This policy governs the collection, transportation, storage, utilization, and treatment of waste lithium-ion batteries, aiming to mitigate pollution resulting from their improper handling and disposal [27].

The new MEP policy addresses the management of various types of waste batteries throughout their life cycle, including production, transportation, sale, storage, use, maintenance, and remanufacturing processes. It specifically targets waste lead-acid batteries, lithium-ion batteries, and mercury-containing cells. Recycling enterprises are encouraged to adopt information technology, such as the "Internet of Things Plus," to establish comprehensive collection systems. At the same time, battery manufacturers are prompted to fulfill extended producer responsibility obligations. Consumers are advised to deposit waste batteries at designated collection points. Stringent measures are outlined to ensure safe handling and transportation, including effective packing and prohibitions against open burning and crude metal extraction. The policy promotes intelligent dismantling, sorting, and utilization of battery components alongside developing advanced technologies for material utilization and electrolyte extraction. These measures aim to improve the efficiency and sustainability of battery recycling, reducing environmental impact and maximizing resource recovery.

On August 1, 2024, China decided to implement the CCC (China Compulsory Certification) for lithium-ion batteries [28]. This certification mandates that all lithium-ion batteries manufactured, sold, or imported into China must comply with stringent safety, performance, and environmental standards. The CCC certification process involves rigorous testing and evaluation to ensure that batteries meet these criteria, thereby enhancing product safety and reliability. The implementation of CCC certification aims to mitigate risks associated with lithium-ion batteries, such as overheating and fire hazards, while promoting higher quality standards within the industry. This move is expected to significantly impact manufacturers and suppliers, who must now adhere to these comprehensive regulatory requirements to access the Chinese market.

15.3.3 Japan

Japanese organizations also set safety standards for lithium-ion batteries. The Japanese Industrial Standards Committee (JISC) developed JIS C 8712 in 2006, closely resembling International Electrotechnical Commission standard IEC 62133, which defines the requirements and tests for batteries containing alkaline or other nonacid electrolytes [24]. In response to safety concerns such as overheating and fires, JIS C 8714 was introduced in 2007, focusing on safety tests for portable lithium-ion batteries, especially addressing the risk of internal short circuits. In 2001, the Japanese Electrical Appliance and Material Safety Law was implemented. Known as the DENAN law, it mandates safety testing for batteries and cells, particularly lithium-ion cells, with a capacity exceeding 400 Wh/l [29].

Japan's Ministry of Economy, Trade, and Industry (METI) announced that effective December 27, 2024, it will replace the DENAN Standard J62133-2 (2021) Appendix 9 with Appendix 12 of the International Electrotechnical Commission (IEC) standard 62,133-2 (2017). This change, supported by UL Solutions Japan, aims to enhance the safety of lithium-ion batteries by incorporating rigorous monitoring methods and safety requirements, including comprehensive testing measures for constant charging and temperature cycling, under the Electrical Appliances and Materials Safety Act. The adoption of Appendix 12 introduces a systematic method for monitoring each cell or cell block within a battery pack, which was not covered in the previous Appendix 9. Appendix 12 is specifically developed to address the risks of overcharging lithium-ion batteries, targeting a crucial safety issue [30].

In Japan, there are no dedicated guidelines in place for the disassembly, disposal, or sorting of lithium-ion batteries for recycling purposes. Instead, they adhere to existing international regulations tailored specifically to the recycling of lithium-ion batteries [30]. Japan and Europe have planned to collaborate on a battery recycling system aimed to facilitate the exchange of information concerning mining sites and suppliers of minerals utilized in battery production and recycling [31].

15.3.4 India

In India, The Batteries (Management and Handling) Rules, 2001, issued by the Ministry of Environment and Forests (MoEF) on May 16, 2001, aim to regulate the collection and recycling of used lead-acid batteries nationwide. These rules pertain to various stakeholders involved in battery life-cycle activities, including manufacturers, importers, recyclers, auctioneers, and consumers [32].

The Ministry of Environment, Forest, and Climate Change (MoEFCC) introduced the Battery Waste Management Rules on August 24, 2022, superseding the previous Batteries (Management and Handling) Rules which exclusively addressed lead-acid batteries. Unlike its predecessor, the 2022 rules encompass a broader spectrum of battery types, incorporating electric vehicle batteries, portable batteries, automotive batteries, and industrial batteries [33]. Additionally, it extends coverage to entities engaged in activities related to the collection, segregation, transportation, refurbishment, and recycling of waste batteries. The key provision is the minimum collection targets, mandating that at least 50% of batteries placed in the market in the fiscal year 2022–2023 must be collected, with a requirement of 100% collection and refurbishment or recycling by the end of a ten-year compliance cycle [33].

The 2001 rules were primarily focused on the roles and obligations of manufacturers, dealers, and importers within the battery industry. In the 2022 rules, the responsibilities and functions of the stakeholders were redefined. It introduces Extended Producer Responsibility (EPR), which mandates that Indian battery manufacturers and importers take additional responsibility for managing batteries throughout their life cycle. This regulation ensures that batteries are collected at the end of their lifespan and recycled by registered recyclers, encouraging stakeholders to adopt eco-friendly practices and invest in recycling infrastructure. This emphasizes establishing authorized recycling facilities, drop-off points, and collection centers across the country.

The 2022 rules stress the importance of public awareness campaigns and educational programs to promote responsible battery disposal and recycling practices. These initiatives aim to raise environmental awareness and encourage active participation in battery collection efforts [34]. Consumers, as the end users of batteries, are tasked with ensuring proper disposal by separating waste batteries from other streams and delivering them to organizations involved in collection, refurbishment, or recycling. Refurbishers and recyclers are required to meet mandatory recovery targets for battery materials, with percentages increasing annually until reaching 90% recovery for all battery types by 2026–2027. Additionally, they need to provide certificates for waste battery processing [35].

The introduction of EPR within the new regulations is anticipated to bolster the circular economy by mandating the incorporation of recycled materials in producing new batteries, thereby reducing reliance on raw materials and conserving natural resources. Furthermore, environmental conservation is emphasized through prohibitions against battery disposal in landfills and incineration. These regulations signify a significant stride toward sustainable battery management, addressing pertinent environmental concerns linked to battery waste [11, 13].

Despite India's strides toward implementing policies and regulations for battery recycling, several challenges need to be addressed [34]. First, the absence of regulatory standards poses a significant hurdle. The lack of guidelines for testing and classifying used batteries that could still be viable for secondary use complicates the process of identifying suitable disposal methods. Second, the rules overlook the importance of battery labeling. Without clear labeling on battery packaging, distinguishing between different types of batteries and identifying appropriate disposal methods becomes problematic, contributing to confusion in the recycling process. Third, the absence of incentives for compliance presents another obstacle. The rules do not offer explicit incentives for stakeholders to adhere to recycling regulations, potentially dampening their motivation to participate in recycling efforts actively. Addressing these challenges is crucial for effectively implementing battery recycling policies in India. However, on March 14, 2024, MoEFCC issued the Battery Waste Management (Amendment) Rules, 2024, updating the Battery Waste Management Rules, 2022. Key changes include Rule 10, which sets the price for Extended Producer Responsibility certificates between 30% and 100% of the Environmental Compensation for noncompliance, and Rule 13, which revises guidelines for imposing and collecting Environmental Compensation for waste battery management [36].

15.4 Summary

EU Battery Regulation 2023/1542 seeks to govern and enhance the utilization of batteries within the European Union. The regulation establishes benchmarks for the production, labeling, utilization, and disposal of batteries, intending to mitigate environmental consequences and foster the circular economy. It focuses on alternative ways to give used batteries a second life and extend their lifespan. However, people need to be educated to ensure safety with respect to each alternative way of sustainable use of retired batteries. Standardized data collection practices will also be necessary to reduce fire incidents and risky suppliers, as well as address emerging risks associated with new battery chemistries.

In the USA, lithium-ion battery safety is governed by Underwriters Laboratories (UL) standards, including UL 1642, UL 2054, and UL 2580, covering various applications from consumer electronics to electric vehicles. Hazardous Materials Regulations mandate specific transport requirements, while the EPA's May 2023 memorandum clarifies lithium-ion batteries as hazardous waste under universal waste standards, promoting safe disposal and recycling. In China, new regulations from the Ministry of Industry and Information Technology require EV manufacturers to set up recycling facilities and tracking systems for spent batteries. The Ministry of Environmental Protection's 2016 policy and upcoming CCC certification, effective August 2024, aim to enhance safety and quality standards. Japan is updating its standards with Appendix 12 of IEC 62133-2 to address overcharging risks, effective December 2024. In India, the 2022 Battery Waste Management Rules expand regulations beyond lead-acid batteries to include all types, focusing on Extended Producer Responsibility (EPR), recycling targets, and public awareness. Recent amendments to these rules set prices for EPR certificates and refine environmental compensation guidelines.

References

1 Battery Regulations: A Comprehensive Guide, *Pangea Strategic Intelligence*, 2023. https://www.pangea-si.com/intelligence/battery-regulations-a-comprehensive-guide/ (accessed 23 April 2024)

2 E. & Power (2023). Europe Battery Market, 2023. https://www.nextmsc.com/report/europe-battery-market (accessed 7 August 2024).

3 Yvette, S. EU Batteries and Waste Batteries Regulation: An Overview. https://www.compliancegate.com/batteries-regulation-european-union/ (accessed 21 March 2024).

4 EU Monitor, Directive 1991/157 - Batteries and accumulators containing certain dangerous substances. https://www.eumonitor.eu/9353000/1/j9vvik7m1c3gyxp/vitgbghlytze (accessed 7 August 2024).

5 Intertek, European Union Battery Directives. https://www.intertek.com/batteries/eu-battery-directive-services/ (accessed 21 March 2024).

6 Waste statistics - recycling of batteries and accumulators, *Eurostat*, 2023. https://ec.europa.eu/eurostat/statistics-explained/index.php?title=Waste_statistics_-_recycling_of_batteries_and_accumulators&stable=0#Recycling_of_batteries_and_accumulators (accessed 22 April 2024).

7 Ted, G. (2024). Complying with the new EU Batteries Regulation, February 2024.

8 Battery Pass Consortium (2023). Battery Passport Content Guidance V1.1, no. December, 2023 [Online]. https://thebatterypass.eu/resources/.

9 Watkins, K. (2023). New EU sustainable batteries regulation becomes law – what you need to know, *Freshfields Bruckhaus Deringer*. https://sustainability.freshfields.com/post/102ilw9/new-eu-sustainable-batteries-regulation-becomes-law-what-you-need-to-know (accessed 21 April 2024).

10 King, M.R.N., Timms, P.D., and Mountney, S. (2023). A proposed universal definition of a Digital Product Passport Ecosystem (DPPE): worldviews, discrete capabilities, stakeholder requirements and concerns. *Journal of Cleaner Production* 384 (November 2022): 135538. https://doi.org/10.1016/j.jclepro.2022.135538.

11 Stretton, C. (2023). EU battery passport regulation requirements, *CIRCULARISE*. https://www.circularise.com/blogs/eu-battery-passport-regulation-requirements (accessed 22 April 2024).

12 New EU regulatory framework for batteries, 2023.

13 China hopes new EU Batteries Regulation not become trade barriers: MOFCOM, *Global Times*, 2023. https://www.globaltimes.cn/page/202308/1296923.shtml (accessed 10 April 2024).

14 Barsony, T. (2023). EU Battery Regulation 2023 BATT2 – Important for Online Commerce, *ecosistant*. https://www.ecosistant.eu/en/eu-battery-regulation-2023-batt2-important-for-online-commerce/ (accessed 15 April 2024).

15 Summary - Position paper on safety [Online]. https://energy.ec.europa.eu/system/files/2021-07/summary_-_position_paper_on_safety_0.pdf.

16 Duan, J., Tang, X., Dai, H. et al. (2020). *Building Safe Lithium-Ion Batteries for Electric Vehicles: A Review*, vol. 3, no. 1. Springer Singapore.

17 Feng, X., Ouyang, M., Liu, X. et al. (2018). Thermal runaway mechanism of lithium ion battery for electric vehicles: a review. *Energy Storage Materials* 10 (December 2016): 246–267. https://doi.org/10.1016/j.ensm.2017.05.013.

18 ACCC (2023). Lithium-ion batteries and consumer product safety [Online]. https://www.accc.gov.au/system/files/Lithium-ion Batteries report_3_0.pdf.

19 Liu, J. (2023). Types of International Battery Safety Standards and Regulations. https://www.mokoenergy.com/battery-safety-standards/ (accessed 22 March 2024).

20 Shen, Y. (2022). Lithium Battery Regulations and Standards in the US: An Overview, *CompliianceGate*. https://www.compliancegate.com/lithium-battery-safety-standards-united-states/ (accessed 24 April 2024).

21 Boxerman, S. and Langan, D. (2023). U.S. EPA Memo Offers Direction on Lithium Battery Recycling—Essential to the Future Supply Chain for Battery Electric Vehicles, *SIDLEY*. https://environmentalenergybrief.sidley.com/2023/08/16/u-s-epa-memo-offers-direction-on-lithium-battery-recycling-essential-to-the-future-supply-chain-for-electric-vehicle-batteries/ (accessed 6 May 2024).

22 Improving Recycling and Management of Renewable Energy Wastes: Universal Waste Regulations for Solar Panels and Lithium Batteries, *EPA*. https://www.epa.gov/hw/improving-recycling-and-management-renewable-energy-wastes-universal-waste-regulations-solar (accessed 6 May 2024).

23 State Recycling Laws, *Battery Council International*. https://batterycouncil.org/recycling-sustainability/state-recycling-laws/ (accessed 6 May 2024).

24 Arima, K. (2016). Ensuring Lithium Ion Battery Safety in Japan, *Japan Industry News*. https://www.japanindustrynews.com/2016/11/ensuring-lithium-ion-battery-safety-japan/ (accessed 15 April 2024).

25 Bird, R., Baum, Z.J., Yu, X., and Ma, J. (2022). The Regulatory Environment for Lithium-Ion Battery Recycling. *ACS Energy Letters* 7 (2): 736–740. https://doi.org/10.1021/acsenergylett.1c02724.

26 Kalson, P. (2019). Study of large format EV Lithium-ion battery recycling in China, *NAATBatt (Avicenne Energy)*. https://naatbatt.org/wp-content/uploads/2019/03/Executive-Summary-of-China-Recycling-Study-v2-1.pdf (accessed 26 April 2024).

27 Zhang, W., Xu, C., He, W. et al. (2018). A review on management of spent lithium ion batteries and strategy for resource recycling of all components from them. *Waste Management & Research* 36 (2): 99–112. https://doi.org/10.1177/0734242X17744655.

28 TUV (2023). China: Announcement of Compulsory Product Certification Management for Lithium-ion Batteries and Other Related Products. https://www.tuvsud.com/en/e-ssentials-newsletter/consumer-products-and-retail-essentials/e-ssentials-3-2023/china-announcement-of-compulsory-product-certification-management-for-lithium-ion-batteries (accessed 7 August 2024).

29 International Compliance Series: JAPAN, *Energy Assurance*, 2021. https://blog.energy-assurance.com/blog-and-news/international-compliance-series-japan (accessed 29 April 2024).

30 Jiang, M. (2022). Japan's Battery Industry Policy and Battery Recycling Development, *SMM*. https://news.metal.com/newscontent/101900868/Japan\stquotes-Battery-Industry-Policy-and-Battery-Recycling-Development/ (accessed 5 May 2024).

31 Bockey, G. (2024). Japan and Europe to Collaborate on Battery Recycling, *BATTERY-NEWS.COM*. https://battery-news.de/en/2024/04/22/japan-and-europe-to-collaborate-on-battery-recycling/ (accessed 5 May 2024).

32 PIB (2001). BATTERIES (MANAGEMENT AND HANDLING) RULES, 2001 NOTIFIED. https://archive.pib.gov.in/archive/releases98/lyr2001/rjun2001/07062001/r0706200110.html (accessed 24 April 2024).

33 AZB & Partners (2023). India: Battery Waste Management Rules, 2022: A New Era For Battery Recycling, *mondaq*. https://www.mondaq.com/india/waste-management/1269730/battery-waste-management-rules-2022-a-new-era-for-battery-recycling (accessed 25 April 2024).

34 The Policy Landscape for Lithium-ion Battery Recycling in India: Advancement in Policy Implementation, *Batx Energies*, 2023. https://batxenergies.com/advancing-lithium-ion-battery-recycling-policy-in-india/ (accessed 4 May 2024).

35 Goenka, V. (2023). Change In Dynamics Of Battery Management In India - Battery Waste Management Rules, 2022, *mondaq*. https://www.mondaq.com/india/waste-management/1326772/change-in-dynamics-of--battery-management--in-india---battery--waste-management--rules-2022 (accessed 29 April 2024).

36 M. of Environment (2024). Battery Waste Management (Amendment) Rules, 2024. http://www.indiaenvironmentportal.org.in/content/476868/battery-waste-management-amendment-rules-2024/ (accessed 7 August 2024).

16

Right to Repair Legislation and the Implications on Battery Safety in the EU

Simin Peng, Quanqing Yu, and Yuwei Nie

A circular economy, also known as a resource-cycling economy, is a sustainable economic development approach, characterized by resource conservation, recycling, and environmental harmony. It emphasizes organizing economic activities into a feedback loop of "resource-product-recycled-resource," involving production and consumption patterns that maximize the reuse, repair, and recycling of existing material and products, thus extending the life cycle of products [1]. A circular economy serves as the foundation for building a sustainable society. The EU's right-to-repair rules and measures to enhance battery replaceability in portable devices are important initiatives to promote the circular economy. However, these regulations have triggered controversies among multiple stakeholders.

A circular economy is guided by three principles [2], which were developed in response to the need for sustainable development:

- Elimination of waste and pollution: a circular economy identifies and eliminates the negative impacts of economic activities on human health and natural systems. This includes emissions of greenhouse gases and harmful substances, structural losses like air, soil, and water pollution, as well as traffic congestion.
- Maintenance of product and material utility: a circular economy promotes activities that preserve value in the form of energy, labor, and materials. This involves focusing on sustainability, reuse, remanufacturing, and recycling to maintain the circulation of products, components, and materials in the economy.
- Regeneration and restoration of natural systems: a circular economy avoids using nonrenewable resources and protects or enhances renewable resources. This is accomplished by returning valuable nutrients to the soil to support regeneration or by using renewable energy sources instead of fossil fuel energy sources.

The European Union (EU) has shown a long-term commitment to a circular economy, with sustainable development and environmental protection. In 2015, the EU launched its first Circular Economy Action Plan (CEAP), which outlined 54 measures to be implemented and proposed four waste-related legislations, thus laying the foundation for the EU's circular economy agenda [3]. In December 2019, the EU introduced the European Green Deal, aimed at transforming the EU economy into a more sustainable one. Consequently, the EU aims to leverage energy-saving practices, improve resource efficiency, and transition to renewable energy sources. The European Green

The Safety Challenges and Strategies of Using Lithium-Ion Batteries, First Edition. Edited by Michael G. Pecht.

Deal also seeks to address climate change, aiming for European carbon neutrality by 2050 [4]. Recognizing the importance and necessity of implementing circular economy models, the EU has adopted plans and action programs concerning the circular economy and waste reduction.

In March 2020, the European Commission introduced a new CEAP, which includes goals such as sustainable product design and reducing electronic and electrical waste. The plan sets specific targets, such as right to repair, enhancing reuse capabilities, introducing universal chargers, and incentivizing electronic product recycling schemes [5].

In February 2021, the European Parliament passed a resolution on the CEAP, calling for further action to achieve a fully circular, carbon-neutral, and sustainable economy by 2050. With this action plan, the EU established stricter recycling rules and encouraged extending the lifespan of electronic products through reuse and repair. Starting in March 2021, new EU energy labels went into effect, aiming to improve the lifespan and repair time of electronic products. Discarded electronic and electrical equipment contains potentially harmful substances that pollute the environment and pose health risks during recycling. To address this, the EU has legislated to prevent the use of certain chemicals [6]. In March 2022, as part of CEAP, the European Commission introduced the first set of measures to accelerate progress toward a circular economy. The proposals include promoting sustainable products, encouraging consumers to embrace green consumption, reviewing building materials regulations, and implementing a sustainable textiles strategy [7]. This holds significance for promoting sustainable development and an overall circular economy.

However, with the acceleration of the renewal and iteration cycle of electronic products, the quantity of electronic and electrical waste has surged. The production of electronic and electrical equipment relies heavily on the use of precious metals and other scarce materials. The proper handling of these waste materials directly affects both environmental and economic issues. Therefore, the generation and treatment of electronic and electrical waste pose significant challenges to the steady advancement of a circular economy and the goals set out in the Green Deal.

16.1 Generation and Treatment of Electronic Waste in Europe

According to data from the United Nations Environment Program (UNEP), the world generates 20–50 million tons of electronic waste annually, a figure expected to grow at 3–5% per year [8]. While developed countries produce the most electronic waste per capita, approximately 70–80% of global electronic waste is shipped to landfill sites in developing countries, where it is sorted, sold, or incinerated to extract materials. In 2019, 2.4–9.7 million tons of electronic wastes were exported to developing countries [9]. According to Figure 16.1, in 2022, the global electronic waste management system market was valued at $52 billion, which is projected to reach approximately $160 billion by 2032. The electronic waste market is growing at a rapid pace [9].

Europe is one of the largest producers of electronic waste globally, with quantities difficult to accurately estimate. However, reports from various sources [11], such as the World Customs Organization and Interpol, suggest that Europe exports at least 1.1 million tons of electronic waste to Africa and Asia annually. Exporting electronic waste can lead to significant economic losses. In 2019 alone, the loss of precious metals and critical raw materials from electronic waste amounted to $13 billion [12]. Despite the high value of electronic waste, the recovery rates of precious metals are low during recycling processes, leading to losses of these metals [13]. In the EU market, electrical and electronic equipment increased from 7.6 million tons in 2012 to 13.5 million tons in 2021. At the same time, the total amount of electrical and electronic equipment collected rose from 3 million tons in 2012 to 4.9 million tons in 2021.

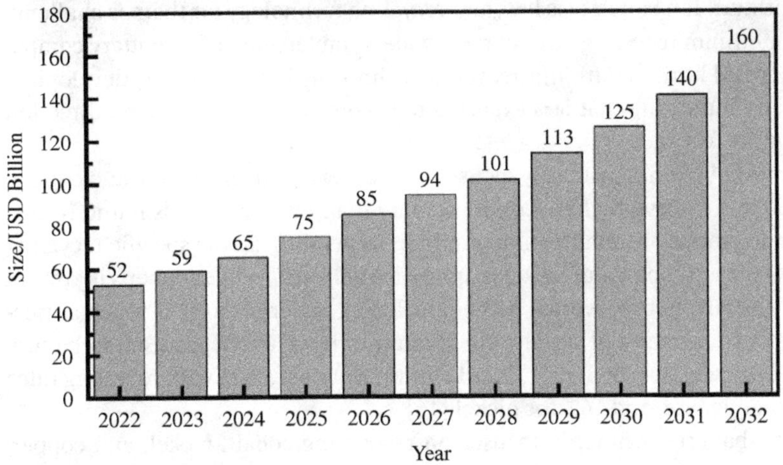

Figure 16.1 Global e-waste management system market size. Source: Adapted from [10].

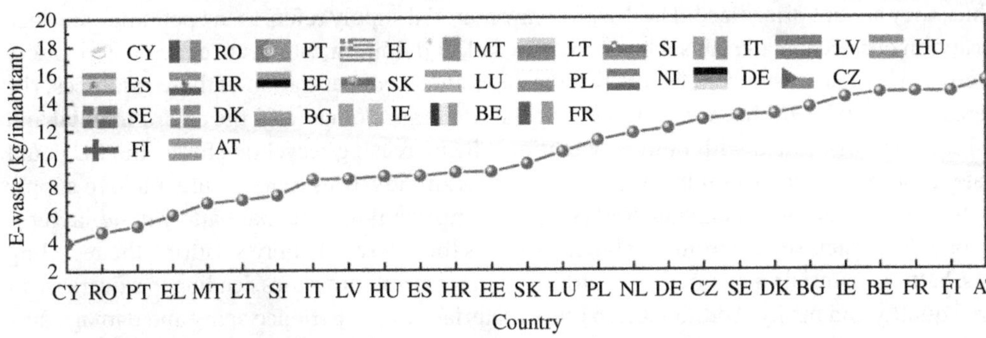

Figure 16.2 The amount of e-waste collected per inhabitant in EU countries. Source: Adapted from [14].

Recycling practices vary among EU countries. As shown in Figure 16.2, in 2021, Austria led EU countries in electronic waste collection, averaging over 15 kg per inhabitant. In 2021, the EU overall averaged 11 kg per inhabitant in electrical and electronic equipment waste collection [5]. However, this indicates that a significant amount of electronic waste is still not properly recycled, imposing a huge burden on the environment.

Batteries are essential components of electronic devices, and they contain valuable material resources as well as toxic substances. Improper disposal of batteries leads to significant environmental and economic losses. The highest collection and recycling rates are achieved with automotive lead-acid batteries, reaching 99%. The recycling rate for lead is between 90% and 100%, with most member states reporting rates of 97% or higher. The average collection rate for portable batteries in the EU is much lower, with nearly 48% of portable batteries sold in the EU collected for recycling in 2018. Approximately 35,000 tons of portable batteries enter municipal waste annually, potentially leaching hazardous substances [15]. The remaining portion is either stored in consumers' homes, exported outside the EU for use in products, or eventually enters electronic waste recycling.

The recycling rate for lithium-ion batteries is low, and recycling technology for them is challenging and costly. At present, lithium recycling has not been widely implemented, but battery companies are starting to use recycled lithium. Lithium-recycling technology is continuously developing, and there is a positive trend indicating that it is expected to become a key development direction for the battery industry in the future.

The recycling of lithium-ion batteries primarily focuses on recovering cobalt, nickel, and copper, as these metals are considered to have higher economic value than other materials found in batteries, such as lithium, manganese, aluminum, and graphite. Depending on the specific recycling process, the recovery efficiency of cobalt and nickel is approximately 95%, while copper's is around 80% [15], but graphite is typically not recovered. Although closing material loops as much as possible helps reduce the risk of raw material supply, the amount of recycled metals used in battery manufacturing within the EU remains low. Only 12% of aluminum is recycled, with recycling rates of 22% for cobalt, 8% for manganese, and 16% for nickel [16].

Recycling of lithium-ion batteries primarily focuses on recovering cobalt, nickel, and copper, which are considered economically more valuable, because they play critical roles in battery performance, particularly in improving energy density, longevity, and efficiency, and they are in high demand across multiple industries. In contrast, materials like graphite, lithium, and manganese have lower market value or recycling efficiency, making their recovery less economically viable, and thus, they are not prioritized. The "risk" of raw material supply refers to the potential shortage or disruption of critical materials such as cobalt, nickel, and lithium, which are essential for battery production. This risk arises from factors like the geopolitical concentration of these resources, rising demand driven by electric vehicles and renewable energy storage, and the environmental and ethical concerns associated with mining operations. By increasing recycling efforts, the EU could mitigate these risks, but current low recycling rates mean industries remain vulnerable to supply chain disruptions and increasing material costs. It is important to note that battery material recycling not only impacts the environment but also affects the safety of batteries. During the recycling process, battery materials may undergo multiple treatments and processes, leading to a decrease in material quality and purity. Additionally, battery materials may experience aging and damage during the recycling process, particularly during extraction, where they may be subjected to physical or chemical damage. Such aging and damage may compromise the stability and safety of the batteries. Furthermore, recycled battery materials may exhibit unstable performance, such as uneven electrical conductivity or unstable material structure. This may also result in safety hazards during battery usage.

With issues arising from electronic waste management, such as increased European electronic waste and illegal exports to underdeveloped countries, EU policymakers are increasingly focused on environmental issues and regulations. Such regulations can help expand electronic waste recycling or reduce the losses caused by electronic waste. Based on considerations of sustainable economic development and the pursuit of carbon neutrality goals, the EU has been trying to reduce the generation and increase the recycling rate of electronic waste.

Answers to the electronic waste problem can be divided into upstream and downstream solutions. Upstream solutions emphasize more rational and environmentally friendly consumption habits to reduce the quantity of electronic waste and improve its recyclability. Downstream solutions include taking appropriate actions through international cooperation and coordination, better enforcement of legislation, and improving more affordable and environmentally friendly electronic waste recycling technologies to reduce illegal trade in electronic waste [17].

16.2 Key Points of the EU Right to Repair Regulations

The European Union has several initiatives for sustainable consumption. On the supply side, there is the Ecodesign for Sustainable Products Regulation (ESPR) [18], which sets requirements for the design of products. On the demand side, there is the Directive on Empowering Consumers for the Green Transition (ECGT) [19], which requires businesses to provide information on product durability and related aspects when selling products to help consumers make purchasing decisions that align with sustainability goals. According to a statement from the European Commission, these two proposals did not address the repair issue in the after-sales stage.

In April 2024, the European Commission adopted the Right to Repair Directive (R2RD), which includes new common rules to promote product repair, including a new set of right to repair rules. This directive includes adding electronic devices, such as smartphones, to the list of products with repairability considerations. It requires hardware manufacturers to provide repairs for up to 10 years [20]. This will save consumers expenses, promote more environmentally friendly consumption habits, reduce electronic waste generation, and support the goals of the European Green Deal.

In the past few decades, replacement has often been prioritized over repair when products are defective, and there has been insufficient incentive for consumers to repair goods when the warranty expires. The R2RD will make it easier and more cost-effective for consumers to repair goods rather than replace them. Moreover, it will drive more demand for the repair industry while encouraging manufacturers and retailers to develop more sustainable business models. These three initiatives complement each other, working together to achieve common goals of sustainable consumption through a comprehensive approach, generating synergistic effects, and collectively covering the entire life cycle of products. The key points stipulated in the R2RD are listed in Table 16.1.

Table 16.1 The key points stipulated in the EU repair rights regulations.

Number	Key points in the EU Right to Repair Directive (R2RD)
1	• The new rules expand the list of goods that must be built with repairability in mind. • Portable electronic devices such as smartphones are now included in this list.
2	• Companies selling consumer goods in the EU must offer free repairs (instead of just replacing damaged devices) within the legal guarantee period. • Exceptions apply if it is cheaper to replace the damaged item.
3	• Repair such items for five to 10 years after purchase if a customer demands it and if the repair is possible. • Inform consumers about which products they are liable to repair. • Provide a Repair Information Form to make pricing and fees transparent.
4	• An online platform will help EU consumers find repair services. • A European repair standard will be established to identify repairers committed to higher quality.
5	• The EU aims to reduce the environmental impact of products. • The new rules address greenwashing, making it illegal to make green claims about a product without evidence.

According to EU regulations, the repairability requirements cover various appliances, such as washing machines, dishwashers, refrigeration equipment, vacuum cleaners, and modern electronic devices like smartphones and tablets. The purpose is to ensure consumers have access to reliable repair services when purchasing electronic devices, thereby promoting sustainability and efficient resource utilization. These regulations not only provide consumers with longer product lifespans but also offer important safeguards for environmental protection and resource management.

For consumer goods sold in the EU, manufacturers are legally required to provide free repair services during the statutory warranty period. Under this law, which aims to protect consumer rights, companies must offer free repair services to consumers during the statutory warranty period, instead of merely replacing damaged equipment. However, if replacing the damaged item is lower than the repair cost, companies may make an exception and choose to replace the item instead of providing repair services.

In accordance with repairability requirements, companies producing compliant products must commit to repairing these goods for 5–10 years after purchase, provided that the specific customer request and repair are feasible. To ensure transparency and consumer awareness, these companies must provide consumers with repair information sheets, clearly stating which products may require repairs and providing transparent information on pricing and costs. This requirement aims to ensure that consumers are aware of repair needs and associated costs when purchasing these goods and have the right to repair services for a certain period after purchase.

The R2RD aims to establish an innovative online repair matching platform to connect consumers with repair businesses in their regions, providing convenient repair services. Through this platform, consumers can easily contact reliable repair businesses and obtain a "European Repair Information Form" to compare quotes and services transparently. Additionally, a European quality standard will be developed to help consumers identify high-quality, reliable, and consistent repair service providers. Through the implementation of this innovative online platform and the quality standard, EU consumers will more easily be able to find and choose repair services with guaranteed quality.

Through the new regulations, the EU is committed to reducing the negative environmental impact of products and addressing greenwashing. These new regulations explicitly state that making green claims about products without sufficient evidence is illegal. Companies must provide reliable environmental information in their advertising and marketing and ensure that their claimed environmental characteristics are substantiated. It aims to prevent companies from using false environmental claims to mislead consumers, while encouraging companies to take more proactive environmental actions to reduce the negative environmental impact of their products.

16.3 Controversies and Discussions Triggered by the Right to Repair Rules

The growing right to repair legislative movement worldwide aims to make it easier and more affordable for consumers to repair products by requiring manufacturers to share repair information, provide diagnostic tools, and supply service parts. Regarding the new Right to Repair Directive introduced by the European Union, the Repair Rights Alliance [21] welcomes the introduction of the new rules. However, it points out that it "misses the opportunity to popularize the right to repair" [20]. Although these rules aim to reduce waste by increasing the repairability of products, they fail to address issues of affordability of repair costs and anti-repair practices.

The Repair Rights Alliance believes that spare parts and repair costs may still be priced too high, making it difficult for ordinary consumers to afford. This viewpoint highlights the shortcomings of

Table 16.2 The corresponding controversies and discussions arising from the formulation of legislation on maintenance rights in various sectors of society.

Sector of society	Corresponding controversies and discussions
Manufacturers	• Manufacturers argue that right to repair legislation could break their monopoly on repair services. They believe that controlling repairs ensures quality and safety. • Manufacturers claim that sharing repair information could infringe on their intellectual property rights. • Some products are intricate, and allowing consumers to repair them might lead to misuse or accidents, potentially holding manufacturers liable.
Market environment	• Manufacturers might flood the market with cheap, low-quality goods to meet repairability requirements, leading to more waste. • Alternatively, manufacturers might raise prices to cover repair costs, affecting consumers.
Consumer experience and safety	• Manufacturers claim that self-repair could compromise the user experience (e.g., poorly executed repairs). • They argue that authorized repair centers ensure quality and safety.
Insurance industry	• Insurers sometimes support right to repair clauses to curb contractor abuses. • Insurers may use these clauses to perform cheaper repairs, potentially leading to substandard work.
Legal ambiguities	• EU member countries have right to repair statutes, but there are ambiguities regarding their exclusivity and impact on common law claims. • Legislatures are attempting to address these ambiguities through amendments.
Economic considerations	• Some believe that imposing a general right to repair at the federal level could alter manufacturers' operations, affecting costs and user experience. There is no guarantee that consumers will always benefit. • Consumer Benefit Uncertainty: Universal consumer advantages from the right to repair remain uncertain. • Increased Production Costs: Manufacturers may face higher initial costs due to the need for higher-quality components to improve repairability. • Supply Chain Challenges: Extended spare parts inventory requirements may escalate logistical and administrative costs. • Profit Margin Reduction: Demand for new products could decline, negatively impacting profits from after-sales services. • Consumer Cost Implications: Initial price increases may occur; however, enhanced repairability could reduce long-term ownership expenses. • Market Competition Dynamics: The rise of independent repair shops and third-party suppliers will likely intensify market competition.

the regulations, as even if products are easier to repair, widespread adoption of the right to repair cannot be achieved if repair costs are too high. It is undeniable that right to repair legislation will be controversial in various sectors of society [22–25]; the main controversies and discussion points are summarized in Table 16.2.

The Repair Rights Alliance is a coalition of consumer advocacy groups, environmental organizations, and other stakeholders dedicated to advancing the right to repair movement. Their goal is to ensure that consumers have the tools, information, and access needed to repair their devices, thereby promoting sustainability and reducing electronic waste. The Alliance's expertise spans consumer rights, environmental sustainability, and technological policy, making their analysis of the Right to Repair Directive (R2RD) highly relevant. Their insights are particularly valuable because they represent the interests of consumers and the environment, shedding light on both the practical and policy-related gaps in the EU's new regulations.

By considering their perspective, the reader gains a better understanding of how the R2RD, while progressive, may still face challenges in achieving widespread, affordable repair access for consumers. The Repair Rights Alliance's critique highlights the importance of balancing repairability with affordability, ensuring that the legislation truly benefits society at large.

16.3.1 Manufacturers' Concerns

The impact of the right to repair legislation proposed by the European Union on manufacturers is multifaceted, covering aspects such as market competition, intellectual property protection, liability, and brand image. Manufacturers must find a suitable balance between meeting consumer demands, maintaining brand image, and ensuring intellectual property protection to adapt to regulatory changes and market competition challenges.

Manufacturers may have enjoyed a monopoly on repair services, but the implementation of right to repair legislation will lead to increased market competition. This could attract more repair service providers into the market, providing consumers with more choices and prompting manufacturers to offer more competitive services and products.

Manufacturers may be concerned that right to repair legislation could infringe upon their intellectual property. They possess proprietary information about product design and technical details, and they fear that sharing this information will lead to a loss of competitive advantage in the market. Therefore, they may seek more legal protection to ensure that intellectual property rights are not compromised while sharing repair information within compliance boundaries.

With the continuous advancement of product technology, many electronic devices are becoming increasingly complex, with more internal structure and functionality. Devices are typically composed of multiple precision components involving complex electronic, mechanical, and software systems. Therefore, maintenance work requires a variety of professional skills and equipment. Manufacturers may be concerned that opening up repair licenses could lead to consumers mishandling or improperly repairing their devices, resulting in product damage or accidents.

Taking smartphones as an example, an ordinary consumer may encounter minor issues such as abnormal screen displays or shortened battery life. In the past, these issues might have simply required sending the phone to an after-sales repair center, where repair personnel would be responsible for fixing it. However, with the enactment of legislation, manufacturers have begun offering open repair licenses, allowing consumers to choose to repair their smartphones themselves, purchase parts, and follow online repair tutorials. Although this openness may bring convenience to some, most consumers lack sufficient professional knowledge and skills. They may be unable to properly identify the root cause of the problem or operate repair tools correctly. For example, they may attempt to disassemble the phone but accidentally damage components in the process, leading to more serious issues. Even if consumers have some technical ability, repairing without professional equipment and environment may lead to accidents. Improper repair operations at home could damage the battery or other critical components, resulting in overheating, short circuits, or explosions, causing fires or serious personal injury.

In addition to the risks to personal safety, open repair licenses may also cause damage to the products themselves. Because consumers are not under the control of manufacturers, they may use inappropriate parts or nonstandard repair methods, which could lead to decreased product performance, shortened lifespan, or complete failure. For example, if a consumer replaces an original part with a low-quality battery, it could cause the phone battery to overheat, damaging other hardware components or causing fires.

Faced with these potential risks, manufacturers may pay more attention to product design and safety. They may strengthen the structural design of products to make maintenance easier and safer, such as adopting modular designs, which allow users to replace components more easily without the need for extensive disassembly of devices. They may also introduce automated systems to monitor device status and provide alerts. Furthermore, manufacturers may introduce stricter warranty liability clauses, limiting or denying warranty for losses caused by consumer self-repair, thereby mitigating their losses caused by repair errors.

The quality of manufacturers' repair services directly affects their brand image and market recognition. Opening up repair permissions may provide consumers with more choices, but manufacturers need to ensure that the quality of their repair services meets consumer needs and maintains their brand image. To do this, they may increase investment in and training for repair services to enhance service quality and strengthen market competitiveness.

16.3.2 Environmental Impact

Right to repair legislation may lead manufacturers to introduce inexpensive, low-quality goods to meet repairability requirements. These products may fail or break down shortly after purchase, being perceived as "disposable" items that are no longer economically viable to repair. In such cases, consumers may be more inclined to discard the product rather than repair it, leading to a waste of resources and an increased environmental burden.

The proliferation of cheap goods in the market shortens the lifespan of products, accelerating their obsolescence and disposal. This increases the demand for raw materials and energy, and it puts pressure on waste disposal and recycling, thereby increasing the product's carbon footprint and negative environmental impact.

To offset increased repair costs, manufacturers may pass the costs onto consumers, resulting in higher product prices. This may affect consumers' purchasing decisions, leading more people to opt for inexpensive, low-quality goods, exacerbating the environmental issues.

Although right to repair legislation aims to increase product repairability and reduce waste generation, improper regulation and guidance may result in effects contrary to environmental protection. Manufacturers and government agencies need to work together to promote the improvement of product quality and sustainability. Such cooperation will guide consumers to adopt more environmentally friendly consumption habits, in order to achieve environmentally friendly product repair and consumption patterns.

16.3.3 Consumer Experience and Safety

The implementation of right to repair legislation may change the relationship between consumers and products to some extent. While this legislation aims to provide consumers with more choices and flexibility, its implementation may also give rise to a series of issues related to product repair, use, and safety. Specifically, manufacturer concerns about self-repair and their endorsement of authorized repair centers will influence consumer decisions and experiences in repairing and using electronic devices. The impact of the right to repair on consumer experience and safety is multifaceted.

Manufacturers are concerned that self-repair may compromise user experience. Improper repair actions by consumers may lead to product damage or malfunction, affecting the intended normal usage experience. Incorrect repairs may result in device malfunctions, decreased performance, or even safety risks, causing inconvenience and frustration to users and reducing their perception of product quality.

Self-repair poses safety risks because consumers may lack the professional knowledge and experience to properly handle complex repair operations. Improper repairs may lead to electronic device failures, short circuits, or even safety hazards such as fires. Manufacturers worry that such situations may threaten consumers' personal and property safety, thereby affecting trust and usage of the product.

Manufacturers argue that specialized repair centers can ensure repair quality and safety. These centers are typically authorized by manufacturers and provide professional training, with qualified technicians and appropriate equipment for repairs. Compared to self-repair, authorized repair centers are more likely to offer reliable and safe services, ensuring the performance and safety of the product after repair.

The right to repair may have complex implications for consumer experience and safety. While self-repair offers more choices and flexibility, there may be quality and safety risks. In contrast, authorized repair centers may better guarantee repair quality and safety, but they may also limit consumer choice. Therefore, manufacturers must balance between self-repair and authorized repair centers, considering consumer experience, safety, and product quality, to meet consumer needs and ensure their trust and loyalty.

16.3.4 Insurance Industry Perspective

The purpose of right to repair legislation is to protect consumer rights and promote market competition, but insurance companies also play a crucial role. Insurance companies [26] may support certain right to repair provisions to curb contractor abuse, but they also face challenges as a result. The impact of right to repair legislation on the insurance industry is complex, influencing repair quality and costs.

Insurance companies often provide coverage to consumers, manufacturers, or even the products themselves, depending on the specific policy. When they support right to repair provisions, it is often because these provisions help them curb contractor abuse in repair processes. For example, by clearly defining the right to repair, insurance companies that insure consumers or products can better control repair costs and prevent contractors from overcharging or using unnecessary repairs. This benefits both consumers, who receive lower premiums or coverage, and the insurance companies, who reduce fraudulent claims. However, these companies may also face challenges since they must balance their need to seek less expensive repair solutions with their obligation to ensure quality.

Insurance companies insuring consumers, products, or manufacturers need to consider risk management and liability issues. They must ensure that services provided through the right to repair do not increase risks, such as unqualified repair work that may lead to safety issues or recurring faults. These risks could lead to increased claims, negatively affecting the insurer. As a result, insurance companies may take measures such as accrediting repair service providers to ensure repairs meet quality and safety standards.

In addition, insurance companies must ensure that repair services meet customer needs and improve satisfaction. If the right to repair is used to cut costs by selecting cheaper but lower-quality repairs, this could harm customer trust and loyalty. For companies insuring consumers, this might lead to fewer renewals of policies or a damaged reputation. In contrast, for those insuring products or manufacturers, poor repair quality could lead to product failure and higher claim rates.

The impact of the right to repair on the insurance industry touches on various factors such as cost control, risk management, customer satisfaction, and trust. Insurance companies must balance reducing costs and providing high-quality repair services to manage risks effectively. Achieving this balance will enhance customer satisfaction and maintain the company's reputation and steady business development within the framework of the right to repair.

16.3.5 Legal Ambiguities

In response to legal ambiguities and uncertainties surrounding right to repair rules, legislative bodies are actively taking action to address these challenges through amendments. Particularly in EU member countries where there is authority to modify regulations, there is ambiguity regarding the specific scope, exclusivity, and relationship with common law claims. This ambiguity may lead to confusion and uncertainty in legal enforcement, causing unnecessary trouble for manufacturers, consumers, and other stakeholders.

Legislative bodies are engaged in drafting amendments to clarify the specific provisions and scope of right to repair legislation, as well as its relationship with other legal provisions. The amendments aim to provide a clearer and more operational legal framework, ensuring effective protection of the right to repair and promoting a fair balance of interests between manufacturers and consumers.

16.3.6 Economic Considerations

Some argue that the EU mandating a general right to repair could trigger a series of changes at the economic level, involving manufacturers' operations, costs, and user experience. However, such changes also raise the question of whether the economic benefits also ensure benefits for consumers.

With the mandatory implementation of a general right to repair, manufacturers may need to increase resource allocation to meet repair requirements. This could entail providing more repair manuals and information, training staff, and establishing a more comprehensive after-sales service system. These additional costs will impact manufacturers' operational expenses, potentially leading to price increases and profit margin decreases.

Implementing right to repair legislation may affect manufacturer competitiveness. Some manufacturers may struggle to adapt to the new repair requirements, resulting in decreased product quality and performance, thus losing market share. On the other hand, some large manufacturers may find it easier to absorb these additional costs and enhance their competitive advantage by providing better repair services.

Manufacturers may need to compromise on product design and manufacturing to meet repair requirements, which could lead to a reduction in product performance and functionality, thereby affecting user experience. Moreover, the right to repair does not necessarily guarantee benefits for consumers. Some consumers may be dissatisfied due to price increases or product performance decreases, while others may not be willing or able to repair products. This means that the implementation of right to repair legislation will not necessarily benefit all consumers positively.

16.4 Measures Taken by the EU to Improve Consumer Ability to Replace Batteries in Portable Devices

The European Council has passed a new regulation called the New Battery Act, requiring original equipment manufacturers to equip smartphones with removable and replaceable batteries. This means that companies producing smartphones will need to reintroduce the classic removable backplate design, allowing users to replace batteries themselves [27]. This will have a significant

Table 16.3 Requirements for the removability and replaceability of batteries in portable electronic products.

Number	Key points of the regulation
1	• The European Council has adopted a new rule that requires smartphones to have removable and replaceable batteries by 2027. • Companies manufacturing smartphones will need to reintroduce the classic removable back panel design to allow users to replace batteries themselves.
2	• The EU's new right to repair rules expand the list of goods that must be built with repairability in mind. • Portable electronic devices such as smartphones are now included in this list. • Companies selling consumer goods in the EU must offer free repairs within the legal warranty period and ensure repairability beyond that period.
3	• The new regulation ensures that batteries become available as spare parts for five years after placing the last unit of a specific model on the market.
4	• These regulations aim to reduce electronic waste, promote sustainability, and empower consumers. • By making batteries replaceable, the EU intends to break monopolistic practices and enhance consumer rights.

impact on portable electronic products sold in Europe. Given the regulatory influence of the EU and the European market, this impact will likely extend globally. Table 16.3 summarizes the main points of the regulations to enhance consumers' ability to replace batteries in portable devices.

The EU has stated that this new regulation will take effect by 2027, giving OEMs a few years to prepare new designs that comply with the law. The purpose of the regulation is to reduce the environmental impact of batteries on the European continent. Smartphones must be "removable and replaceable by the end user [28]", meaning that battery replacement must be easy even without any technical expertise.

The EU has also set a requirement that, by 2027, manufacturers must collect about 63% of portable battery waste, reaching 73% by 2030. The collection of lithium portable battery waste must reach 50% of the target by 2027 and 80% by 2031. The new regulation mandating the use of removable and replaceable batteries in devices will fundamentally change the current trend in smartphone design. Reintroducing replaceable batteries for consumers will be the second major shift in EU smartphone design. The first shift occurred when the EU voted to adopt USB-C as the standard charging interface for Android phones, iPhones, and laptops. If OEM devices do not comply with the charging interface standard, manufacturers will need to make modifications and update them three years later to comply with the requirements for removable and replaceable batteries.

As mentioned earlier, the EU right to repair rules expand the list of goods that must consider repairability. Smartphones and tablets are now included in this list. Companies selling consumer goods in the EU must provide free repairs during the legal warranty period and ensure repairability, including replacement of batteries by the users, beyond that period [16]. The new regulations ensure that batteries are available as spare parts for five years after the last model of a specific type is produced [29].

The European Council has enacted this regulation with the aim of reducing waste and breaking the monopoly of technology companies, such as Apple, Samsung, and Google, as it is difficult to replace components without purchasing a brand-new phone. This is a major victory for the right to repair movement. Although this regulation applies only to the EU, its impact is global. Producing two different types of smartphones to meet the requirements of Europe and other regions is unrealistic. Therefore, this regulatory measure may affect the global market, drive changes in smartphone design, promote sustainability, and empower consumers further [30].

16.5 Arguments Against Allowing Consumers to Replace Smartphone Batteries

The new regulation passed by the European Council has sparked a series of controversies regarding the introduction of user-replaceable batteries in smartphone design. Although this change contributes to reducing electronic waste, manufacturers and consumers have raised various concerns. These concerns primarily revolve around aesthetic design, durability, water resistance, engineering challenges, user experience and safety, planned obsolescence, profit margins, quality control, and environmental impact [31–34], as shown in Table 16.4. Understanding this, manufacturers and regulatory bodies need to carefully balance various factors and seek appropriate solutions to reconcile repairability with other design requirements.

Table 16.4 Requirements for the removability and replaceability of batteries in portable electronic products.

Device aspect	Arguments against allowing consumers to replace smartphone batteries
Slimmer and sleeker designs	• Manufacturers prioritize slim, lightweight designs for smartphones. • Nonremovable batteries allow for sleeker profiles and better aesthetics. • Removable batteries would require additional space and impact the overall design.
Durability and water resistance	• Sealed designs enhance durability and water resistance. • Removable battery compartments create potential entry points for water and dust. • Manufacturers argue that nonremovable batteries improve overall device longevity.
Engineering challenges	• Designing a smartphone with a user-replaceable battery is more complex. • It requires additional components like latches, connectors, and protective covers. • Manufacturers must balance repairability with other design considerations.
User experience and safety	• DIY battery replacements can lead to user errors. • Incorrectly installed batteries may cause malfunctions, overheating, or even fires. • Manufacturers prioritize safety by controlling the repair process.
Planned obsolescence and profit margins	• Some argue [35] that nonreplaceable batteries align with planned obsolescence. • Manufacturers benefit from selling new devices rather than supporting older ones. • Repairability could impact profit margins.
Quality control and authorized repair centers	• Manufacturers claim that authorized repair centers ensure quality. • DIY repairs might lead to substandard work or voiding warranties. • Centralized repair services maintain consistent standards.
Environmental impact	• While replaceable batteries reduce e-waste, they may not always lead to better environmental outcomes. • Manufacturers might flood the market with inexpensive, low-quality goods to meet repairability requirements. • Raising prices to cover repair costs could affect consumers.

With the intense competition in today's smartphone market, slim design has become a mainstream trend, and nonremovable batteries can better achieve a streamlined appearance and lighter body. Introducing removable batteries may require additional space, affecting the overall design and potentially resulting in dissatisfaction with the smartphone's appearance.

Many smartphones today adopt sealed designs to enhance device durability and water resistance. However, removable batteries introduce additional compartments, which may serve as potential entry points for water and dust, thereby reducing the device's waterproof performance. Some manufacturers [36] argue that nonremovable batteries increase the overall lifespan of devices, as there is no risk of user mishandling during battery replacement.

Additional components, such as latches, connectors, and protective covers, are needed to enable users to replace batteries, increasing the complexity of the internal structure of the phone. Manufacturers must balance other design considerations while ensuring the repairability of the phone, which may increase production costs and engineering difficulty.

Self-replacement of batteries could lead to user errors, such as improper installation, which may cause malfunctions, overheating, or even fire hazards. If a person causes a fire during a repair, there could be various legal and liability implications depending on the circumstances. If the fire is directly caused by a mistake made during the repair process, the individual may be responsible for the damages, including property damage and potential injuries. However, there could be scenarios where blame is shifted to the manufacturer if it can be proven that the fire was caused by a defect in the product, inadequate safety warnings, or faulty instructions provided by the manufacturer. In such cases, the manufacturer may be held liable for the damages, and the affected party may seek compensation through legal channels. Ultimately, the outcome would depend on the specific details of the incident, the applicable laws, and the evidence presented to determine where the responsibility lies. Manufacturers ensure safety by controlling the repair process, but self-repair may compromise repair quality and even void warranties.

Some argue [35] that nonreplaceable batteries align with planned obsolescence, prompting consumers to purchase new devices rather than repairing and extending the lifespan of old devices. This planned obsolescence model may benefit manufacturers by promoting the sale of new devices rather than supporting repairability and environmental sustainability. However, while replaceable batteries reduce electronic waste, manufacturers may flood the market with inexpensive, low-quality batteries to meet repairability requirements, thereby lowering the overall quality of products.

In the event of fires caused by faulty repairs or substandard batteries, determining liability becomes complex. Manufacturers will likely incur significant operational costs due to providing repair services and bearing compensation for repair liabilities during the repair process. To offset the costs and liabilities associated with repairs, manufacturers may increase product prices, thereby impacting consumer purchase decisions and potentially affecting profit margins.

16.6 Summary

The EU is at the forefront of promoting a circular economy, outlining various measures in its action plans and legislation. Initiatives include the European Green Deal, Circular Economy Action Plan, and Right to Repair Directive. The EU aims to tackle challenges in electronic waste management through stricter regulations, improved recycling technologies, and the promotion of sustainable consumption habits.

The EU has enacted regulations to promote product repair, including extending warranty periods and ensuring access to repair services. Manufacturers are required to provide repair services for up to 10 years, and consumers have the right to access repair information and high-quality repair services. These provisions aim to reduce waste and promote sustainability, which aligns with the objectives of the European Green Deal.

Despite positive intentions, right to repair regulations have sparked debates, particularly regarding economic affordability, anti-repair practices, and the overall legislative impact. Manufacturers express concerns about market competition, intellectual property protection, liability, and product safety. Environmental impact is also a concern, as cheap goods may lead to increased waste and resource consumption. Consumer experience and safety may be affected due to varying repair quality and safety risks. The insurance industry faces challenges in managing repair costs and ensuring quality repairs under the new regulations. Legal ambiguities and uncertainties need to be addressed to clarify the provisions and scope.

The EU mandated the use of removable and replaceable batteries in smartphones by 2027. Manufacturers must collect a certain proportion of portable battery waste to promote sustainability and reduce environmental impact. These regulations aim to enhance consumer empowerment, promote sustainability, and reduce electronic waste globally.

Concerns include aesthetic design, durability, engineering challenges, user experience and safety, planned obsolescence, and environmental impact. Nonreplaceable batteries are favored for their streamlined design and durability, while replaceable batteries may compromise waterproofing and overall device lifespan. Engineering challenges and potential safety hazards associated with user replacement are major worries. Some argue [35] that nonreplaceable batteries promote planned obsolescence, while replaceable batteries could flood the market with inexpensive, low-quality replacements.

References

1 European Parliament (2023). *Circular economy: definition, importance and benefits*. European Parliament https://www.europarl.europa.eu/news/en/headlines/economy/20151201STO05603/circular-economy-definition-importance-and-benefits.

2 DS Smith (2024). Ce este economia circulară? DS Smith. https://www.dssmith.com/ro/sustenabilitate/leadership-in-economia-circular/ce-este-economia-circular.

3 European Commission (2015). First circular economy action plan, European Commission. https://environment.ec.europa.eu/topics/circular-economy/first-circular-economy-action-plan_en.

4 European Commission (2020). The European Green Deal, ISBN 978-92-76-17190-4, European Commission, April 20, 2020. https://data.europa.eu/doi/10.2775/19001.

5 European Parliament (2024). E-waste in the EU: facts and figures (infographic), European Parliament. https://www.europarl.europa.eu/news/en/headlines/society/20201208STO93325/e-waste-in-the-eu-facts-and-figures-infographic.

6 European Parliament (2011). Directive 2011/65/EU of the European Parliament and of the Council on the restriction of the use of certain hazardous substances in electrical and electronic equipment, EUR-Lex Document 32011L0065, European Parliament, August 6, 2011. http://data.europa.eu/eli/dir/2011/65/oj.

7 European Parliament (2023). Circular economy: definition, importance and benefits, European Parliament. https://www.europarl.europa.eu/news/en/headlines/economy/20151201STO05603/circular-economy-definition-importance-and-benefits.

8 Worldloop, About e-waste, Worldloop. https://worldloop.org/e-waste/.

9 Toner Buzz (2023). Staggering E-Waste Facts & Statistics 2024, Toner Buzz. https://www.tonerbuzz.com/blog/e-waste-facts-statistics/.

10 Anonymous (2023, October 19). Staggering E-Waste Facts & Statistics 2024 [Online]. https://www.tonerbuzz.com/blog/e-waste-facts-statistics/.

11 Anonymous (2023, April 25). Millions of tons of electronic waste flow from Europe to Asia and Africa every year [Online]. https://baijiahao.baidu.com/s?id=1764125388629055225&wfr=spider&for=pc.

12 Forti, V., Baldé, C.P., Kuehr, R., and Bel, G. (2020). The Global E-waste Monitor 2020: Quantities, flows and the circular economy potential, United Nations University/United Nations Institute for Training and Research, International Telecommunication Union, and International Solid Waste Association, Bonn, Geneva and Rotterdam.

13 Chancerel, P., Meskers, C.E.M., Hagelüken, C., and Rotter, V.S. (2009). Assessment of precious metal flows during preprocessing of waste electrical and electronic equipment. *Journal of Industrial Ecology* 13: 791–810.

14 E. Parliament (2024, March 21). E-waste in the EU: facts and figures [Online]. https://www.europarl.europa.eu/topics/en/article/20201208STO93325/e-waste-in-the-eu-facts-and-figures-infographic.

15 Halleux, V. (2023). New EU regulatory framework for batteries, in EU Legislation in Progress, European Parliament.

16 European Parliamentary Research Service, New EU regulatory framework for batteries: Setting sustainability requirements, EU Legislation in Progress, European Parliamentary Research Service. https://www.europarl.europa.eu/RegData/etudes/BRIE/2021/689337/EPRS_BRI(2021)689337_EN.pdf.

17 Bakhiyi, B., Gravel, S., Ceballos, D. et al. (2017). Has the question of e-waste opened a Pandora's box? An overview of unpredictable issues and challenges. *Environment International* 110: 173–192.

18 European Commission (2022). Ecodesign for Sustainable Products Regulation, European Commission. https://commission.europa.eu/energy-climate-change-environment/standards-tools-and-labels/products-labelling-rules-and-requirements/sustainable-products/ecodesign-sustainable-products-regulation_en.

19 João Antonucci Rezende (2024). New directive to empower consumers – what does it mean for repair? João Antonucci Rezende. https://repair.eu/news/new-directive-to-empower-consumers-what-does-it-mean-for-repair/.

20 Vigliarolo, B. (2023). Europe's right-to-repair law asks hardware makers for fixes for up to 10 years, The Register. https://www.theregister.com/2023/03/22/new_eu_right_to_repair/.

21 Vigliarolo, B. (2023, Wed 22). Europe's right-to-repair law asks hardware makers for fixes for up to 10 years [Online]. https://www.theregister.com/2023/03/22/new_eu_right_to_repair/.

22 Yang, L., Jin, C., and Zhu, C. (2023). Research: The Unintended Consequences of Right-to-Repair Laws, Harvard Business Review. https://hbr.org/2023/01/research-the-unintended-consequences-of-right-to-repair-laws.

23 Insurance Litigation Group (2018). An Insurance Company's 'Right to Repair,' Insurance Litigation Group. https://restorationlaw.com/an-insurance-companys-right-to-repair/.

24 Mr. Glucksman (2015). Right to Repair Statutes: the Fix to Construction Defect Litigation," Amwins. https://www.amwins.com/resources-insights/article/right-to-repair-statutes-the-fix-to-construction-defect-litigation_11-15.

25 Londoño, J. (2021). The Debate on Right to Repair: A Primer, American Action Forum. https://www.americanactionforum.org/insight/the-debate-on-right-to-repair-a-primer/.

26 Luyi Yang, C.J. and Zhu, C. (2023, January 19). The Unintended Consequences of Right-to-Repair Laws [Online]. https://hbr.org/2023/01/research-the-unintended-consequences-of-right-to-repair-laws.

27 Diaz, N. (2023). The EU mandates user-replaceable smartphone batteries by 2027, Android Central. https://www.androidcentral.com/phones/eu-mandates-replaceable-batteries-2027.

28 Diaz, N. (2023, July 15). The EU mandates user-replaceable smartphone batteries by 2027 [Online]. https://www.androidcentral.com/phones/eu-mandates-replaceable-batteries-2027.

29 Ganapini, C. (2022). The EU battery directive will make it easier to replace batteries, Repair.eu. https://repair.eu/news/the-eu-battery-directive-will-make-it-easier-to-replace-batteries/.

30 Mauran, C. (2023). All smartphones, including iPhones, must have replaceable batteries by 2027 in the EU, Mashable SEA. https://sea.mashable.com/tech/25136/all-smartphones-including-iphones-must-have-replaceable-batteries-by-2027-in-the-eu.

31 Collins, K. (2023). Phones With Replaceable Batteries: How They Might Make a Comeback, CNET. https://www.cnet.com/tech/mobile/phones-with-replaceable-batteries-how-they-might-make-a-comeback/.

32 Phillips, G. (2023). Why the EU Is Forcing Replaceable Batteries and How It Will Affect Smartphone Design, Make Use Of. https://www.makeuseof.com/why-eu-forcing-replaceable-batteries-how-affect-smartphone-design/.

33 Knutsson, K. (2024). Should you replace your phone battery or buy a new phone? Find out what is the best option for you, CyberGuy. https://cyberguy.com/ask-kurt/should-you-replace-your-phone-battery-or-buy-a-new-phone/.

34 Agarwal, S. (2021). To save the planet, phone makers need to bring back removable batteries, Digital Trends. https://www.digitaltrends.com/mobile/smartphones-removable-batteries-climate-impact-comeback-analysis/.

35 Agarwal, S. (2022, May 9). To save the planet, phone makers need to bring back removable batteries [Online]. https://www.digitaltrends.com/mobile/smartphones-removable-batteries-climate-impact-comeback-analysis/.

36 Collins, K. (2023, July 30). How They Might Make a Comeback [Online]. https://www.cnet.com/tech/mobile/phones-with-replaceable-batteries-how-they-might-make-a-comeback/.

17

Battery Reuse and Repurposing: Balancing Sustainability with Risk

Shalini Dwivedi, Aparna Akula, and Michael G. Pecht

The rising demand for lithium-ion batteries (LIBs) comes with limitations in resource availability, environmental impact, and safety risks, among other challenges [1, 2]. Lithium-ion batteries rely on lithium, cobalt, and many other materials that face constraints due to geopolitical factors, limited reserves, extraction difficulties, supply chain disruptions, and price fluctuations. These constraints often result in habitat destruction, water pollution, and carbon emissions [3], while improper disposal of LIBs poses safety, soil, and water contamination risks [4]. Recycling of spent batteries offers economic and environmental benefits [3, 5–9]. However, it is still in its early stages due to gaps in science and technology, economic challenges, a lack of clear regulation, and difficulties in collecting, moving, and sorting batteries [9, 10].

Considering the escalating demand for sustainable energy storage solutions and challenges in efficiently recycling retired batteries, companies have investigated battery repurposing from portable electronics, electric vehicles (EVs), and renewable energy systems as an alternate end-of-life management technique. Repurposing extends the useful life of batteries beyond their initial applications, maximizing economic and environmental value by reducing waste generation. It bridges the gap between battery remanufacturing and recycling, offering better opportunities for sustainable resource utilization and circular economy practices.

Hu et al. [11] discussed the repurposing of EV lithium-ion batteries (LIBs) in stationary energy storage applications. They highlighted the challenges of battery disposal and the benefits of repurposing for environmental protection and resource recovery. Stevan and Converti [12] emphasized the importance of disassembly procedures, maintaining optimal performance in the second life, and implementing safety protocols. They highlighted the need for safe and cost-effective methods to maximize the value of batteries beyond their primary use. Other studies [3, 11–15] examined the technical feasibility and economic viability of repurposing batteries, considering factors such as battery condition, performance degradation, and safety.

With the rising demand for high-energy-density batteries, ensuring safety throughout battery manufacturing, transportation, and use is critical to preventing accidents and safeguarding public health [5, 16]. To optimize the usefulness of retired batteries and reduce their bad environmental impacts, the risks linked with end-of-life management techniques need to be addressed.

The Safety Challenges and Strategies of Using Lithium-Ion Batteries, First Edition. Edited by Michael G. Pecht.

17.1 Discarding of Batteries

Lithium-ion batteries are essential for powering the modern world, but discarding them conventionally after a single use poses an environmental threat due to their toxic materials. Responsible and sustainable end-of-life management is crucial. This necessitates exploring options such as reusing, repurposing, and recycling retired batteries to capture their remaining potential and minimize environmental impacts.

The disposal of LIBs at various stages of their life cycle presents significant environmental challenges. During the manufacturing stage, scrap poses a serious threat to the environment. At the consumer level, batteries reach their end of life after a certain number of charge cycles or due to device replacement. At the end of a product's life cycle, such as electric vehicles, a large volume of batteries may be retired simultaneously [17], and their disposal significantly impacts the environment.

Conventional LIB disposal in landfills or incinerators presents a critical challenge due to environmental pollution [4, 18]. Improper disposal of used batteries can pollute soil, water, and air. The toxic materials in these batteries can also harm living things and human health [10]. The growing demand for LIBs has brought concerns about the long-term viability and self-sufficiency of the raw materials needed to produce them, especially for lithium and cobalt. Conventional disposal aggravates the scarcity of these valuable metals, which are essential for battery production and other applications. This concern is magnified by the growing number of electric vehicles and the consequent generation of battery waste [6, 19].

To address challenges posed by the anticipated surge in retired batteries from EVs or consumer electronics and their improper disposal, researchers and policymakers suggest sustainable alternatives such as recycling and repurposing the spent batteries. This promotes a circular economy approach that recovers valuable materials and extends battery life, thus minimizing environmental impact [6, 20].

Recycling provides a solution to the responsible end-of-life management of spent LIBs. Their incorporation into the circular economy involves a complex interplay of resource efficiency, economic viability, and technological advancement [5, 15]. Recycling LIBs reduces the environmental impact related to new battery production by conserving resources and reducing ecological footprint [15, 21]. Recycled LIBs offer a secondary source of crucial materials like lithium, cobalt, and nickel, which are vital for making new batteries. This is especially important due to resource limitations and increasing material demand.

The commercial advantages of including LIBs in mainstream e-waste recycling are acknowledged worldwide, with potential economic gains from recovered materials [4, 22]. There are three main methods commonly used to recycle retired LIBs: pyrometallurgical, hydrometallurgical, and direct recycling processes. The first two methods concentrate on recovering materials and salts, while the direct recycling process reuses and refurbishes electrode materials for making new batteries on a small scale in laboratories. Before any of these processes can be applied, the battery cell must be discharged, which requires pretreatment at various stages depending on the recycling technology employed [10, 15].

In conjunction with the previously described procedures, alternative processes exist wherein the recovery includes high-value-added metals or critical raw materials from the batteries and yields battery components in the form of salts as by-products. This type of recycling can be found in mechanical processing, wherein the electrolyte is extracted independently, and disintegration of the battery cells occurs to concentrate the metals, thereby mitigating the necessity for thermal separation steps [23].

Lithium-ion battery recycling primarily focuses on high-value electrode materials containing nonferrous metals like cobalt, lithium, and nickel. However, recycling of the electrolyte, due to its volatility, toxicity, and flammability, has received less attention. Mao et al. [24] have examined the state of battery recycling and emphasized that electrolytes contain lithium salts and organic solvents at specific concentrations, which can be economically valuable and suitable for recovery and reuse.

When considering battery recycling, various factors, such as economic viability, technological feasibility, and regulatory compliance, must be carefully evaluated to ensure effective and sustainable practices. Different studies have investigated various costs related to battery recycling. Liu et al. [25] highlighted purchase, transportation, and dismantling costs. They concluded that the cost of disassembling for recycling is lower when first used for echelon utilization compared to direct recycling of spent batteries without echelon utilization. However, they overlooked factors like labor and depreciation costs in their model, which may lead to an underestimation of overall cost. Lima et al. [7] provided a more detailed breakdown of costs, including depreciation, new battery, waste treatment, labor, taxes, electricity, and equipment. They concluded that encouraging battery recycling by providing tax exemptions or reductions and establishing joint ventures can enhance its economic viability and promote the circular economy. This would need to encompass logistics, transportation, and battery disposal. Jorges et al. [26] compared the equivalent annual cost (EAC) for recycling and second-life use. They concluded that a recycled battery may demonstrate a lower EAC than a second-life battery, even considering technological advancements. The literature [7, 25] has discussed the cost-based feasibility of battery recycling; however, it is crucial to incorporate additional factors, such as logistics, preprocessing, refining, external factors (such as insurance and permits), and material losses, for a complete recycling cost-estimation model.

Beyond cost analysis, it is crucial to assess the efficacy of reintroducing recycled materials into the battery supply chain. This involves optimizing collection via reverse logistics and integrating these materials throughout the product life cycle, including procurement, production, storage, distribution, and recovery [27].

Despite the economic and environmental benefits, LIB recycling is still in its early stages due to various obstacles. These include gaps in science and technology, economic challenges, difficulties in collecting, moving, and sorting batteries, and a lack of clear regulations. Fluctuating raw material prices and availability, along with decreasing battery production costs, play a significant role in the economics of recycling spent batteries [9, 10] (Figure 17.1).

17.2 Repurposing of Lithium-Ion Batteries

Reusing retired batteries for the same application or repurposing them for different applications presents alternative methods for managing batteries at the end of their lifespan. While reusing batteries for the same application may seem like a cost-effective solution, there are several drawbacks to consider. They may require frequent replacement due to wear and tear experienced during their initial use. Additionally, their decreased performance may not meet the power requirements of the same application [15].

Retired LIBs may no longer meet the performance requirements for their original applications, but they often retain significant capacity. Repurposing these batteries for different secondary applications contributes to conserving resources, reducing environmental impact, and potentially decreasing reliance on new battery materials. This approach bridges the gap between battery remanufacturing and recycling [28].

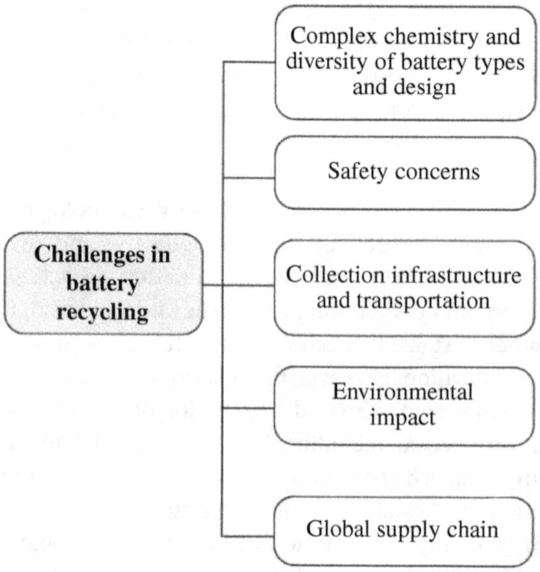

Figure 17.1 Factors impeding the sustainability and efficiency of battery recycling.

Realizing the significance of repurposing retired batteries, it is essential to examine various factors, such as economic feasibility, environmental implications, technological constraints, and regulatory frameworks, to ensure their efficacy. The technical characteristics, health, and usage history of retired LIBs are essential for optimizing their potential for second-life use. This knowledge supports safe, efficient, and sustainable redeployment in new applications.

Repurposing batteries retired from EVs has been explored in existing literature [14, 29–34]. Tang et al. [35] have conducted economic analyses to assess the viability of repurposed EV batteries in distributed photovoltaic (PV) systems under sharing business models. They concluded that sharing business models, which include battery storage and shared PV generation among various users, can lower community electricity costs. DeRousseau et al. [14] explored potential second-life applications for retired EV batteries, considering their varying state-of-health (SOH) levels and matching them with applications of different power requirements: high-SOH packs (70–80%) for grid services, medium-SOH (50–70%) for e-mobility such as e-scooters, rickshaws, and low-SOH (<50%) for home appliances. Mixed-SOH configurations offer further possibilities for repurposing.

The majority of the literature has concentrated on repurposing batteries retired from electric vehicles, overlooking other potential sources such as laptops and smartphones [9, 14, 35]. Repurposing these batteries for small-scale energy storage applications provides a valuable opportunity to utilize their remaining capacity. By being repurposed, LIBs can maximize their environmental and economic value while reducing waste, signifying a sustainable future for LIBs. Potential applications for batteries retired from various sources are described in Table 17.1.

17.3 Responsible Battery Repurposing: Navigating Resilience and Safety Concerns

Emphasizing the requisite factors for responsible battery repurposing, access to the histories of retired batteries is paramount. It enables SOH estimation, supports the development of aging models, informs the selection of applications, and aids the financial justification for repurposing batteries.

Figure 17.2 illustrates the factors that play a key role in responsible battery repurposing. It is observed that the health of retired batteries plays an important role in assessing the suitability of

Table 17.1 Possible reutilization of batteries retired from varying primary applications.

Sources of retired batteries	Possible second-life applications
Electric vehicles	Stationary energy storage for homes, grid support backup power system
Consumer electronics (laptops, smartphones)	Power bank (2 to 5 Ahr) portable chargers, remote controls, wireless mouse and keyboard, Bluetooth speakers, backup power for Raspberry Pi or Arduino
Renewable energy storage	Stationary energy storage, grid stabilization, backup power
Uninterruptible power supplies	Home energy storage, backup power for critical systems
Power tools	Portable lighting, small electronic devices, DIY projects
Medical devices	Remote monitoring devices, low-power medical equipment
Electric bikes	Portable power sources, smart appliances, backup power
Data centers	Backup power for remote facilities, off-grid power

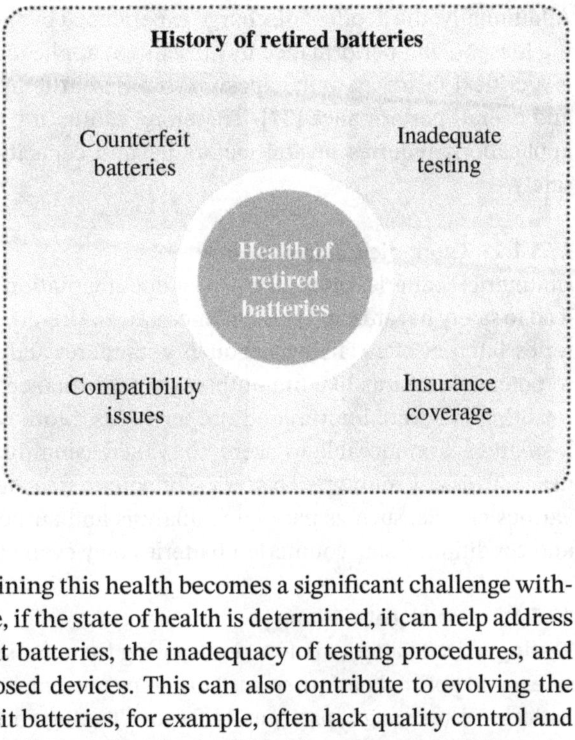

Figure 17.2 Determinants of battery repurposing.

second-life applications. Accurately determining this health becomes a significant challenge without access to a detailed history. Furthermore, if the state of health is determined, it can help address issues such as the prevalence of counterfeit batteries, the inadequacy of testing procedures, and potential compatibility issues with repurposed devices. This can also contribute to evolving the landscape of insurance coverage. Counterfeit batteries, for example, often lack quality control and pose safety risks, especially when repurposed. Knowing the health of batteries can mitigate the prevalence of counterfeiting by providing a basis for comparison and assessment with genuine batteries.

When the health of a battery is known, it allows for more targeted and effective testing procedures. Inadequate testing of retired batteries may fail to identify potential hazards, increasing the likelihood of compatibility issues like voltage mismatches. Inadequate testing can lead to equipment damage or malfunctions, raising concerns for insurance coverage. Insurers may question liability when accidents occur due to repurposed batteries' poor health, potentially affecting coverage terms and premiums. Since ensuring retired batteries' health through adequate testing is essential to mitigate risks associated with counterfeit products, compatibility issues, and insurance implications, it is imperative to have a detailed discussion of each factor.

17.3.1 Health of Retired Batteries

Before repurposing retired batteries, it is essential to know their history. This helps assess their state of health and suitability for second-life applications. In the literature [25, 31], the evaluation of battery health has mainly focused on capacity as a determining criterion. However, other factors also affect the health of batteries, such as temperature and accelerated operating conditions. Notably, not all of these factors directly impact battery capacity, necessitating assessments of each parameter, such as internal resistance and rate of charge-discharge cycles during their first-life applications.

Internal resistance, which increases with use and impacts efficiency and heat generation during operation, is crucial for performance assessment. As a battery's health degrades, internal resistance increases [36]. Charging characteristics, such as voltage and current profiles, influence the battery's performance in its second life and are important for developing health predictive models. Additionally, the depth of discharge experienced by the battery in its first use can affect its remaining lifespan and performance in subsequent applications. The temperature sensitivity of batteries is a critical factor as well; exposure to different temperature ranges can influence battery aging and overall performance [37]. Therefore, estimating the viability of used batteries in second-life applications requires several factors beyond capacity to assess their health and ensure battery safety.

17.3.1.1 Counterfeit Batteries

Counterfeit batteries often lack proper documentation of their history and health status, which can lead to safety hazards and performance issues. Unauthorized replicas closely resemble genuine batteries, but they often fail to meet quality standards and safety regulations. This discrepancy can lead to potential hazards like fire outbreaks, performance shortcomings, and environmental damage, resulting in severe injuries and property loss. Moreover, counterfeit batteries may lack adequate insulation, unnoticeable to users, thus increasing the risk of electric shocks during handling or usage. To assess counterfeit batteries for potential second-life applications, it is important to employ various checks, such as packaging analysis and authentication codes, to verify their authenticity and condition. Using counterfeit batteries may even void warranties on the devices they power.

17.3.1.2 Inadequate Testing

Testing retired batteries before repurposing them for second-life applications is essential to ensure safety and optimal performance. This helps prevent unexpected failures caused by overestimating batteries' remaining capacity and lifespan. Testing and sorting batteries during the transition from first to second life help ensure their suitability for applications like backup power, residential storage, EV charging, and utility-scale storage [32, 33]. Inadequate testing may not reveal the actual extent of performance degradation, resulting in suboptimal powering for second-life applications. This could result in an overestimation of economic benefits, leading to investments in applications that may not be cost-effective in the long term. For instance, batteries repurposed for stationary energy storage must handle specific load profiles and maintain sufficient capacity over time. Failure to maintain sufficient capacity would render the investment wasteful.

Various testing methods exist for assessing battery performance, including internal resistance measurement, electrochemical dynamic response testing, and frequency domain testing. However, relying solely on one testing technique may not fully reveal the actual condition of retired batteries. Each test complements the others, resulting in better performance estimates for second-life deployment. For instance, internal resistance measurement, while common, may not be sufficient for LIBs due to additives keeping resistance low and differing readings between AC and DC methods. It can be used along with electrochemical dynamic response testing, which measures ion flow mobility and response time during load pulses [38]. Proper testing ensures that retired batteries deliver maximum economic and environmental value while minimizing unexpected risks.

17.3.1.3 Compatibility Issues

When repurposing retired batteries for second-life applications, ensuring their compatibility with the intended application is crucial. Battery chemistries exhibit varying properties, including voltage, capacity, and discharge rate. This implies that a battery designed for one application may not be compatible with another due to their different characteristics. For instance, an LIB from an EV may not suit a stationary energy storage system due to differing voltage and discharge rate requirements. Retired batteries should match the voltage requirements of the intended application to prevent inefficiencies or safety hazards. Connecting cells with mismatched voltage ratings in series can lead to cell overloading and damage, while parallel connections risk unbalanced currents and safety hazards. The size and form factor of the battery pack must align with the available space in the second-life application to ensure efficient utilization. Additionally, selecting batteries with similar capacities is vital for optimal performance, as mismatched capacities can lead to imbalanced energy distribution and reduced system efficiency. Second-life LIBs can be effectively integrated into stationary energy storage systems by addressing cell compatibility issues, which maximizes their residual value and ensures safe operation [33].

17.3.1.4 Insurance Coverage

Battery safety is a critical concern, especially when dealing with retired batteries. To address potential risks associated with high-value batteries and devices, particularly those used in EVs or renewable energy systems, insurance companies provide specialized coverage. Insurance coverage is crucial, providing financial protection against potential risks such as malfunction, fire, or other hazards. This safeguards against unexpected costs from damage or accidents during battery operation [39].

However, determining liability for harm caused to battery users remains a concern that needs addressing. The degradation of battery components over time and usage can have adverse effects, which may not be visible during the initial stage of repurposing but might endanger the user's life. Therefore, factoring in insurance coverage for the user is essential when conducting economic analyses. This promotes transparency and accountability, leading to a safety-conscious approach and facilitating the wider adoption of sustainable second-life applications for LIBs.

17.3.2 Beyond "Can We?": Delving into the Imperatives and Challenges of Battery Repurposing

Retired LIBs, especially those from EVs, present opportunities and challenges, resulting in limited potential for second-life deployment. Figure 17.2 details the challenges and considerations for battery repurposing, which are also summarized here. The lack of traceability mechanisms in battery collection makes managing retired batteries difficult. The lack of awareness among customers regarding collection points for used batteries also impedes efforts in battery collection and repurposing [1, 40]. Clear regulations and guidelines help streamline the process of

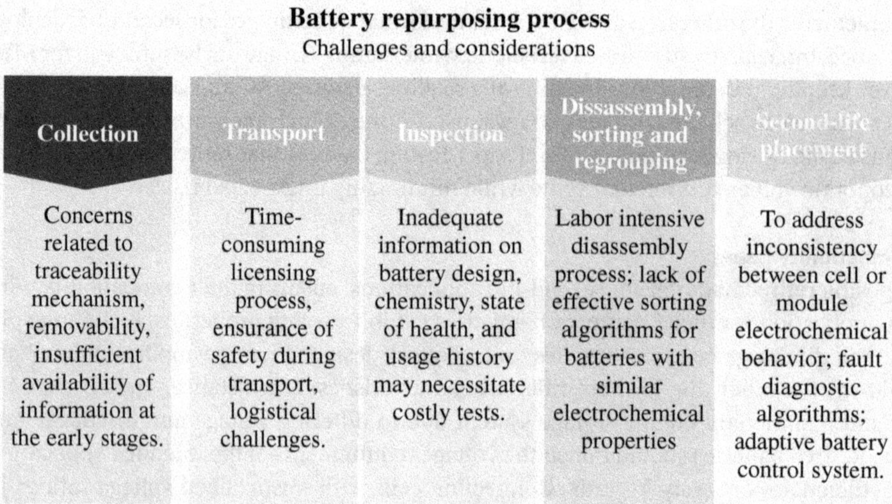

Figure 17.3 Challenges, risks, and design considerations for battery repurposing.

battery repurposing or second-life applications. However, regulatory ambiguity complicates this effort, as unclear guidelines lead to logistical challenges. Additionally, inadequate information during collection, including ownership and battery condition, makes redirecting batteries challenging. Transporting batteries also faces hurdles such as licensing processes, safety costs, and logistical complexities due to long distances to repurposing centers. Disassembly, likewise, presents challenges, with modules and cells not designed for easy disassembly alongside diverse battery designs, making the process labor-intensive. Importantly, assistance is needed to sort and regroup batteries of similar electrochemical properties to identify indicators and thresholds for optimal second-life battery deployment. Finally, second-life battery placement involves adapting control strategies, developing equalization strategies, and adjusting fault-diagnosis algorithms to address issues such as power stabilization, overheating, and discharging [41] (Figure 17.3).

17.4 Summary

Lithium-ion battery technology advancements and end-of-life management raise concerns over costs, raw material scarcity, environmental and social impacts, and disposal. Battery repurposing and recycling provide a solution by reducing the need for new materials. Despite the economic and environmental benefits, the recycling of LIBs is still in its early stages due to gaps in science and technology, economic challenges, a lack of clear regulations, and difficulties in collecting, moving, and sorting batteries. Several factors, such as fluctuations in raw material prices and availability and decreasing battery production costs, play significant roles in the economics of recycling spent batteries.

Before implementing large-scale battery repurposing, careful consideration should be given to the projected volumes, safety, and potential logistical, technological, and socioeconomic challenges. This understanding will guide policy and infrastructure development, facilitating a seamless transition toward a closed-loop battery ecosystem. Responsible retired battery repurposing necessitates thoroughly evaluating the battery's health conditions and compatibility with intended applications. Factors such as internal resistance, charging characteristics, depth of discharge, and temperature sensitivity play crucial roles in determining the viability of batteries for second-life applications. Safety risk analysis and mitigation methods, warranties, and insurance coverage will also need to be considered. In addition, poor quality and counterfeit batteries must not enter the repurposing process.

The potential for repurposing batteries reaching their end of life can come from EVs and consumer electronics like computers, laptops, and mobile phones. While initially, EVs will be more profitable, shifting focus toward smaller-format batteries presents an opportunity to utilize their residual value by identifying suitable second-life applications in small-scale energy storage (e.g., portable electronics, DIY projects, and off-grid solutions).

References

1 ACCC (2023). Lithium-ion batteries and consumer product safety. https://www.accc.gov.au/system/files/Lithium-ion%20Batteries%20report_3_0.pdf (accessed 3 April 2024).

2 Boyden, A., Soo, V.K., and Doolan, M. (2016). The environmental impacts of recycling portable lithium-ion batteries. *Procedia CIRP* 48: 188–193.

3 Harper, G.D.J., Sommerville, R., Kendrick, E. et al. (2023). Roadmap for a sustainable circular economy in lithium-ion and future battery technologies. *Journal of Physics: Energy* 5: 2.

4 Li, L., Fan, E., Wu, F., and Chen, R. (2022). Sustainable recycling of spent lithium ion batteries. *ECS Meeting Abstracts*, MA2022-01, 5, 614.

5 Sheth, R.P., Ranawat, N.S., Chakraborty, A. et al. (2023). The lithium-ion battery recycling process from a circular economy perspective—a review and future directions. *Energies* 16: 7.

6 Yang, T., Luo, D., Yu, A., and Chen, Z. (2023). Enabling future closed-loop recycling of spent lithium-ion batteries: direct cathode regeneration. *Advanced Materials* 35: 36.

7 Lima, M.C.C., Pontes, L.P., Vasconcelos, A.S.M. et al. (2022). Economic aspects for recycling of used lithium-ion batteries from electric vehicles. *Energies* 15: 6.

8 Mrozik, W., Rajaeifar, M.A., Heidrich, O., and Christensen, P. (2021). Environmental impacts, pollution sources and pathways of spent lithium-ion batteries. *Energy & Environmental Science* 14: 6099–6121.

9 Tao, Y., Rahn, C.D., Archer, L.A., and You, F. (2021). Second life and recycling: energy and environmental sustainability perspectives for high-performance lithium-ion batteries. *Science Advances* 7: 45.

10 Marchese, D., Parise, M., Curcio, V. et al. (2024). An overview of the sustainable recycling processes used for lithium-ion batteries. *Batteries* 10: 1.

11 Hu, X., Zhang, K., Liu, X. et al. (2022). A review of second-life lithium-ion batteries for stationary energy storage applications. *Proceedings of the IEEE* 110 (6): 735–753.

12 Stevan, S.L. and Converti, A. (2023). Second life of lithium-ion batteries of electric vehicles: a short review and perspectives. *Energies* 16 (2): 953.

13 Rangarajan, S.S., Sundaresan, S., Thirumalai, T., and Mohan, R. (2022). Lithium-ion batteries—the crux of electric vehicles with opportunities and challenges. *Clean Technologies* 4 (4): 908–930.

14 DeRousseau, M., Gully, B., Taylor, C. et al. (2017). Repurposing used electric car batteries: a review of options. *JOM* 69 (9): 1575–1582.

15 Kotak, Y., Kotak, K., Miranda, P., and Blanco, R. (2021). End of electric vehicle batteries: reuse vs. recycle. *Energies* 14: 8.

16 Montes, T., Etxandi-Santolaya, M., Eichman, J. et al. (2022). Procedure for assessing the suitability of battery second life applications after EV first life. *Batteries* 8: 9.

17 Huang, B., Pan, Z., Su, X., and An, L. (2018). Recycling of lithium-ion batteries: recent advances and perspectives. *Journal of Power Sources* 399: 274–286.

18 Karmazínová, K., Kazda, T., and Jandova, K. (2017). Li-ion batteries recycling—processes & technologies. *ECS Transactions* 81 (1): 255–260.

19 Celikler, D., Aksan, Z., and Yilmaz, A. (2023). Raising awareness in sustainable environment education: waste batteries. *Pedagogical Research* 8 (2): em0157.

20 Vu, F., Rahic, M., and Chirumalla, K. (2020). Exploring second life applications for electric vehicle batteries. *Advances in Transdisciplinary Engineering* 13: 273–284.

21 Hossain, R., Sarkar, M., and Sahajwalla, V. Technological options and design evolution for recycling spent lithium-ion batteries: impact, challenges, and opportunities. *Wiley Interdisciplinary Reviews: Energy and Environment* 12 (5): e481.

22 Ciez, R.E. and Whitacre, J.F. (2019). Examining different recycling processes for lithium-ion batteries. *Nature Sustainability* 2 (2): 148–156.

23 Sonoc, A., Jeswiet, J., and Soo, V.K. (2015). Opportunities to improve recycling of automotive lithium ion batteries. *Procedia CIRP* 29: 752–757.

24 Mao, Z., Song, Y., Gang, A., and Sun, W. (2024). Recycling of electrolyte from spent lithium-ion batteries. *Next Sustainability* 3: 100015.

25 Liu, Z., Liu, X., Hao, H. et al. (2020). Research on the critical issues for power battery reusing of new energy vehicles in China. *Energies* 13: 8.

26 Jorges, E.E.M., Quintino, A.M.N., and Santos, D.M.F. (2023). Economic analysis of lithium-ion battery recycling. *AIMS Energy* 11 (5): 960–973.

27 Syah, R., Surya, I., Hakim, M. et al. (2022). Design a multi period closed-loop supply chain program to supply recycled products. *Foundations of Computing and Decision Sciences* 47 (2): 163–176.

28 Cui, J., Tan, Q., Liu, L., and Li, J. (2023). Environmental benefit assessment of second-life use of electric vehicle lithium-ion batteries in multiple scenarios considering performance degradation and economic value. *Environmental Science & Technology* 57 (23): 8559–8567.

29 Casals, L.C., Amante García, B., and Canal, C. (2019). Second life batteries lifespan: rest of useful life and environmental analysis. *Journal of Environmental Management* 232: 354–363.

30 Jian, L., Zechun, H., Banister, D. et al. (2018). The future of energy storage shaped by electric vehicles: a perspective from China. *Energy* 154: 249–257.

31 Chen, X., Tang, J., Li, W., and Xie, L. (2020). Operational reliability and economy evaluation of reusing retired batteries in composite power systems. *International Journal of Energy Research* 44 (5): 3657–3673.

32 Nováková, K., Pražanová, A., Stroe, D.I., and Knap, V. (2023). Second-life of lithium-ion batteries from electric vehicles: concept, aging, testing, and applications. *Energies* 16: 5.

33 Hassan, A., Osigwe, E., Martinez, R., and Ogunbiyi, S. (2023). Second-life batteries: a review on power grid applications, degradation mechanisms, and power electronics interface architectures. *Batteries* 9: 12.

34 Liu, Z., Zhao, J., Wang, H., and Yang, C. (2020). A new lithium-ion battery SOH estimation method based on an indirect enhanced health indicator and support vector regression in PHMs. 13 (4): 830.

35 Tang, Y., Zhang, Q., Li, H. et al. (2019). Economic analysis on repurposed EV batteries in a distributed PV system under sharing business models. *Energy Procedia* 158: 4304–4310.

36 Huang, Y., Qin, L., and Wu, G. (2021). State of health estimation based on constant current charging profiles. *2021 40th Chinese Control Conference (CCC)*, 5899–5904.

37 Xiao, Z., Tang, Y., Yang, L. et al. (2023). Review—online monitoring of internal temperature in lithium-ion batteries. *Journal of the Electrochemical Society* 170 (5): 057517.

38 Battery University (2021). Testing lithium based batteries. https://batteryuniversity.com/article/bu-907-testing-lithium-based-batteries (accessed 3 April 2024).

39 Maiti, M. (2023). EV Motor Insurance: Here Are The Benefits Of Having A 'Battery Secure Add-On' Cover For Your EV. https://business.outlookindia.com/personal-finance/insurance/ev-motor-insurance-here-are-the-benefits-of-having-a-battery-secure-add-on-cover-for-your-ev (accessed 3 April 2024).

40 Dunn, J., Ritter, K., Velázquez, J.M., and Kendall, A. (2023). Should high-cobalt EV batteries be repurposed? Using LCA to assess the impact of technological innovation on the waste hierarchy. *Journal of Industrial Ecology* 27 (5): 1277–1290.

41 Tankou, A., Bieker, G., and Hall, D. (2023). Scaling up reuse and recycling of electric vehicle batteries: assessing challenges and policy approaches. *International Council on Clean* Transportation. www.theicct.orgcommunications@theicct.org (accessed 3 April 2024).

18

Risks Associated with Recycling and Disposal

Simin Peng, Jinkang Chen, Jie Wu, and Michael G. Pecht

Lithium-ion batteries contain valuable metals and materials, including cobalt, nickel, and lithium, among others; however, there is limited mining capacity for these raw materials [1]. Many of these materials require complex mining operations, which are generally harmful to the environment and human health. Hence, if batteries can be reused or recycled, their adverse impacts on the environment and human health will be effectively reduced [2]. Recycling is meant to reclaim resources, thus alleviating pressure on natural resource extraction. Nevertheless, disposal must be considered when recycling is not an option.

18.1 Retired Batteries

A "retired battery" typically refers to a battery that, for any reason, is no longer used in its original purpose, has reached or exceeded its design lifespan, is no longer able to function reliably, or cannot provide its expected performance due to factors such as capacity degradation, chemical reaction failure, or physical damage. The reasons for the retirement of lithium-ion batteries are diverse and influenced by various interconnected factors. Depletion of battery capacity is a primary factor, and people do not generally appreciate products that need to be recharged too often. The problem is that during use, changes occur in the electrolyte, electrodes, and other constituent materials, and these changes will reduce the battery's capacity. Overcharging, overdischarging, and generally improper use of chargers also accelerate battery aging [3, 4]. Overcharging may also lead to the formation of metallic lithium inside the battery, posing safety hazards; meanwhile, overdischarging can damage the battery's chemical structure. Elevated temperatures also expedite the pace of capacity loss. Elevated temperatures expedite the rate of undesirable chemical side reactions within the battery, resulting in the thickening of the solid electrolyte interphase (SEI) layer and the irreversible loss of lithium ions [5]. These alterations culminate in a decline in battery capacity. Furthermore, under high-temperature conditions, cyclic aging is proportional to the cumulative charge throughput, and the stress encountered by the battery during charge and discharge cycles also further accelerates capacity degradation [6].

Physical damage is a common cause of battery retirement, and damaged batteries often catch fire or explode; approximately 25–30% of battery failures are attributable to physical damage, which can precipitate thermal runaway and potentially result in fire or explosion [7, 8]. As a result, any product subject to physical impacts, compression, or other forms of damage should be discarded properly. Additionally, abused batteries can lead to electrolyte leakage or internal battery short-circuiting.

The Safety Challenges and Strategies of Using Lithium-Ion Batteries, First Edition. Edited by Michael G. Pecht.

Table 18.1 Sources, types, and characteristics of different kinds of retired lithium-ion batteries [9–13].

Type of battery	Source	Characteristic
Lithium cobalt oxide	Portable electronic devices (e.g., smartphones, laptops, tablets, portable audio devices)	• High energy density • Stable charge and discharge performance • High risk of thermal runaway
Lithium nickel manganese cobalt oxide	Electric vehicles (EVs), hybrid electric vehicles (HEVs), and plug-in hybrid electric vehicles (PHEVs)	• High balance performance • Long cycle life • Widely commercialized applications
Lithium iron phosphate	Electric bicycles, electric motorcycles, energy storage systems, solar energy storage systems	• High level of security • Long cycle life • Low environmental pollution
Lithium manganese oxide	Low-tier application scenarios (e.g., portable lighting equipment, handheld tools, and similar devices)	• Low-cost • High level of security

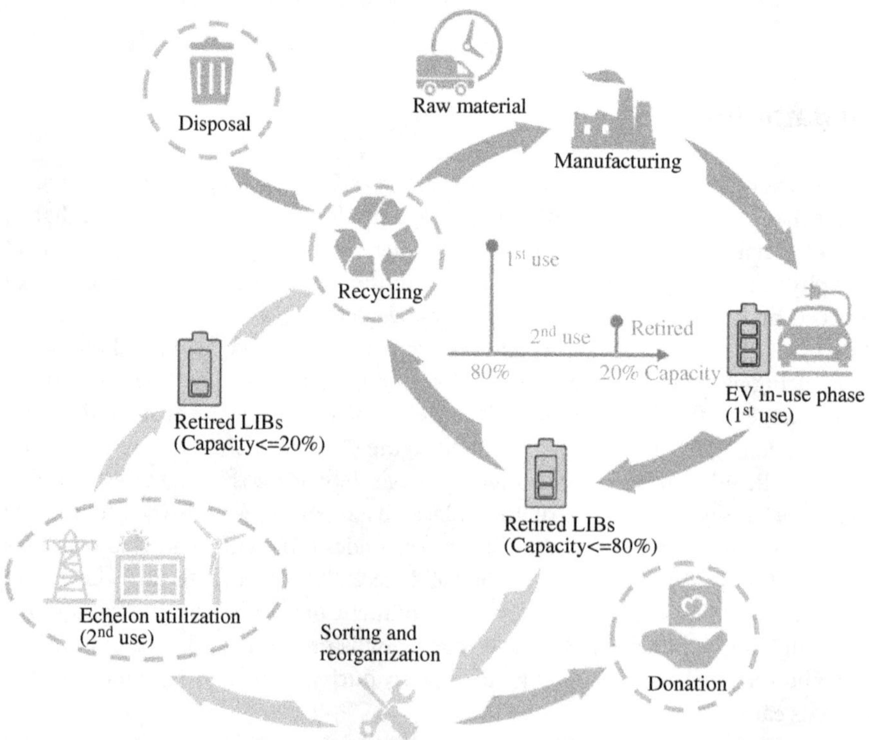

Figure 18.1 An example treatment of retired lithium batteries. Source: Adapted from Wang et al. [11].

Table 18.2 Comparison of different retired battery treatment methods [14, 15].

Method	Processing Objective	Procedure
Reuse	Prolong battery lifespan and minimize resource wastage	• Perform battery testing and maintenance • Repair possible damage • Put back into service
Recycling	Diminish environmental footprint, reclaim usable materials, and mitigate the necessity for fresh raw materials	• Perform battery disassembly • Extract recyclable materials • Send to the recycling facility
Disposal	Minimize potential environmental and health hazards	• Identify the battery type and classify it according to regulatory guidelines • Adhere to safe operating protocols • Choose the suitable disposal method for destruction

Discarded or disposed-of batteries may come from consumer electronics, industrial equipment, vehicles, and other devices. Consequently, retired lithium-ion batteries from different sources will require different reuse and recycling treatments to ensure the safety of the environment and human health, while maximizing the recovery and utilization of useful resources within the battery. Table 18.1 presents the sources, types, and characteristics of different kinds of retired lithium-ion batteries.

An example of the treatment of retired lithium-ion batteries used in electric vehicles is illustrated in Figure 18.1. In this case, the batteries are considered unusable when their capacity falls below 80%, but this value can be any number, depending on the reuse application requirements. Batteries with capacity drops of over 20% are often sent to be recycled [11].

Depending on the state of the retired battery, recyclers need to carefully select the appropriate method to mitigate potential hazards to human health and the environment, while enhancing resource recovery rates. Table 18.2 lists a comparison of different treatment methods for retired batteries.

18.2 Recycling

Battery recycling involves processing retired batteries to extract valuable materials, which are then utilized in manufacturing new batteries or other products. The objectives of battery recycling are to minimize the demand for new resources, mitigate environmental impact, and fully utilize the renewable materials found in batteries.

There are three main ways to recycle batteries: (a) pyrometallurgy, (b) hydrometallurgy, and (c) direct recycling [16]. Castelvecchi summarizes battery recycling thus:

> In a typical recycling plant, batteries are first shredded, which turns cells into a powdered mixture of all the materials used. That mix is then broken down into its elemental constituents, either by liquefying it in a smelter (pyrometallurgy [17]) or by dissolving it in acid (hydrometallurgy [18]). Finally, metals are precipitated from the solution as salts [19, 20].

The primary steps in battery recycling encompass collection and sorting, dismantling, material extraction, refining and processing, and reuse. Initially, batteries are safely collected and sorted according to their type, considering variations in chemical compositions and materials. Subsequently, the batteries are disassembled into constituent parts, including casings, electrodes, and electrolytes, to facilitate the extraction of renewable materials such as metals, plastics, and other chemicals. These extracted materials undergo further refining and processing to eliminate impurities and ensure compliance with manufacturing standards. Ultimately, the refined materials are reused, serving as inputs for producing new batteries, electronics, or other manufactured products [21].

The recovery and reuse of retired batteries present a promising avenue for effective resource recycling. The annual output of retired lithium batteries totals around 67,500 metric tons, comprising 22,950 metric tons of lithium cobaltate and lithium nickel cobaltate/lithium nickel cobalt manganese oxide, 6075 metric tons of copper, and 4050 metric tons of aluminum [22]. The different materials recovered from lithium-based batteries are illustrated in Figure 18.2.

Recycling lithium cobaltate can yield approximately 240 metric tons per year, valued at over five and a half million dollars, yielding cobalt, copper, aluminum, iron, and other metals. For context, cobalt constitutes about 15% of this composition. However, the complexity of retired lithium batteries, with varying specifications, materials from different manufacturers, and diverse formulations, makes their recycling and reuse challenging [22].

Considering mobile phones, over 100 million mobile phone batteries are retired worldwide annually, each weighing approximately 20 g. These batteries contain around 3 g of cobalt, 2.8 g of copper, 1 g of aluminum, and 5 g of iron. From every metric ton of retired mobile phone batteries, approximately 0.15 metric tons of cobalt, 0.14 metric tons of copper, and 0.05 metric tons of aluminum can be recovered. With cobalt priced at $55,400 per metric ton, copper at $9700 per metric ton, and aluminum at $2770 per metric ton, the prospect of recycling retired batteries appears highly lucrative [23].

Battery recycling contributes to reducing reliance on new resources, fostering the principles of a circular economy, and alleviating environmental strain. Nevertheless, it is imperative to exercise caution when recycling batteries, in order to ensure safe and eco-friendly disposal of waste and mitigate adverse impacts on the environment and human health.

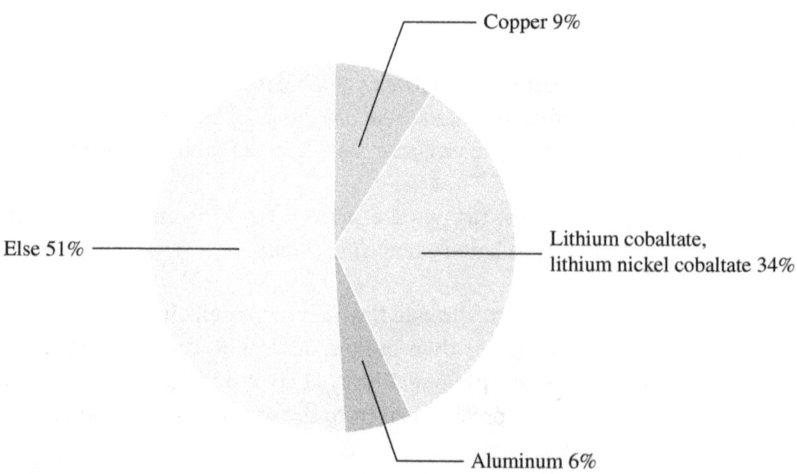

Figure 18.2 Composition of retired lithium batteries. Source: Adapted from Guo et al. [22].

18.3 Disposal

The disposal of retired batteries entails secure destruction of batteries that are no longer viable for reuse or recycling. This mitigates environmental and health risks, aiming to prevent the release of hazardous substances and minimize potential harm to the environment and human health.

The methods for disposal of retired batteries vary depending on the type of battery, local regulations, and environmental standards. However, the primary goal remains consistent: disposing of them in a manner that poses no harm to humans or the environment.

Mechanical shredding is a prevalent physical method in the disposal process, wherein retired batteries undergo compression and shredding, facilitating subsequent treatment and disposal [24]. Additionally, chemical methods for destroying retired batteries primarily encompass chemical dissolution and neutralization treatments. Chemical dissolution entails breaking hazardous components within batteries through appropriate chemical treatment, reducing their adverse environmental impact. Neutralization involves using suitable chemicals to stabilize acidic and alkaline substances within the battery, thereby minimizing potential hazards. Thermal treatment, another method, typically employs high temperatures for battery disposal, such as smelting, proving effective for addressing specific types of batteries, especially lithium batteries [25]. Furthermore, biological treatment offers a more environmentally sustainable approach, leveraging microbiological or biological methods to facilitate the degradation of hazardous substances [26].

Throughout the disposal process, it is imperative to adhere to regulations and environmental standards. This involves maintaining transparency, ensuring regulatory compliance, and minimizing adverse impacts on both the environment and human health. Importantly, meticulous consideration and enforcement of compliance are essential to guarantee that retired batteries are disposed of in a manner that not only meets regulatory mandates but also optimizes environmental preservation and public safety.

18.4 Safety Risk Assessment and Suggestions for Different Treatments

Life cycle management of spent lithium batteries, which includes reuse, recycling, and disposal. Improper handling of retired batteries poses significant safety risks due to the presence of hazardous materials and the potential for thermal runaway reactions. There needs to be a comprehensive analysis of the risks associated with each step in the treatment process, along with scientifically grounded recommendations to mitigate these dangers. By understanding and addressing these risks, the industry and policymakers can work together to develop sustainable, safe, and efficient management strategies for spent lithium batteries. Thus, they will protect human health and the environment, ensuring a secure future for advanced battery technologies.

Retired batteries may present a variety of safety concerns, such as leakage, overheating, and fire hazards. These issues can pose significant threats to both the environment and human health, and they underscore the need to evaluate safety risks associated with retired batteries. In conducting such assessments, several factors merit consideration, including battery type, physical condition, chemical composition, and more.

Evaluating retired batteries typically includes identifying their type, inspecting for damage, and testing voltage and temperature. Upon collecting retired batteries, the initial task is to determine their respective types, including alkaline, lithium-ion, nickel-cadmium, and others. This process can assess both chemical composition and potential hazards [27]. Subsequently, an appearance inspection must be conducted by observing the battery shell for obvious damage, deformation, or cracks, which will determine its overall state and potential internal problems. Following that, a voltage test should be performed to detect any abnormal voltage levels, which could indicate battery failure. Subsequently, temperature measurements of the battery casing must be performed to check for overheating, as it may signify internal battery issues. Then, an odor check should be performed to detect any unusual smells, indicating leakage or overheating.

Additionally, checking the battery's weight for abnormalities could suggest internal changes like liquid leakage. Conducting a discharge test to detect the remaining charge and assess its performance status can determine if the batteries are viable for potential reuse. Finally, referencing the manufacturing date will allow for an effective assessment of the battery's age and potential risk of failure.

During the safety analysis process, it is imperative to ensure that retired batteries are disposed of per pertinent regulations and environmental standards. This approach safeguards the environment while prioritizing human health and public safety.

18.5 Recycling of Batteries and Chemical Pollution Risks

During the recycling of lithium batteries, the accumulation of a large number of batteries presents inherent risks during storage and transport. While transporting recycled retired batteries, the mutual compression and collisions between batteries may result in the rupture of battery shells. This rupture allows the electrolytes to seep into the surrounding environment, posing a direct health threat to exposed human populations.

Furthermore, during transportation, compression and collision incidents can lead to short-circuiting the positive and negative connections in retired batteries, posing risks such as explosions and fires. Therefore, it is essential to establish comprehensive storage and transport protocols, as well as safety procedures for handling batteries, with a focus on effective process management.

The electrolyte is one of the major contributors to chemical pollution from retired batteries, and it is also a crucial component of power batteries. It ranks as the third-largest material in lithium batteries and typically constitutes approximately 15–25% of a battery's mass, though this ratio can vary among manufacturers. Lithium hexafluorophosphate is a commonly used electrolyte, appearing as a white crystalline or powder with a relative density of 1.50. It is soluble in water and dissolves in low concentrations of methanol, ethanol, acetone, carbonate, and other organic solvents. However, when exposed to air or heated, lithium hexafluorophosphate undergoes decomposition. This decomposition occurs rapidly in the presence of air and water vapor, leading to the release of phosphorus pentafluoride (PF5) and the formation of white fumes [28].

Two frequently employed organic solvents carry irritant and harmful properties when in contact with exposed skin, eyes, and the respiratory tract. Vinyl carbonate is a transparent, colorless liquid that solidifies into a crystalline form at room temperature. By comparison, propylene carbonate is a colorless and odorless flammable liquid. It exhibits miscibility with ether, acetone, benzene, chloroform, and vinyl acetate, among other substances. Propylene carbonate is soluble in water and carbon tetrachloride, and it possesses a notable ability to absorb carbon dioxide while maintaining stability.

When retired batteries containing metals such as mercury, cadmium, arsenic, and zinc undergo incineration at high temperatures, these metals volatilize and are carried away by the flue gas. As the flue gas cools, it can condense into particles smaller than 1 μm. Subsequently, this process results in the enrichment of metal content in the fly ash, which may lead to significant air pollution.

During the incineration process for battery recovery, high-temperature calcination will generate toxic gases such as carbon monoxide (CO), hydrogen chloride (HCl), hydrogen fluoride (HF), hydrogen cyanide (HCN), benzene, toluene, and others. Alternatively, in the wet recovery process, strong inorganic acid leaching of valuable metals produces toxic gases such as chlorine (Cl_2), sulfur oxides (SOx), and nitrogen oxides (NOx). These emissions contribute to environmental pollution and harm human health [29]. Therefore, it is necessary to strengthen the monitoring process, improve the sensor's sensitivity, solve the real-time news transmission, and effectively monitor and prevent leakage of harmful gases.

Efforts should be made to promote the efficient recycling of both consumer lithium batteries and automotive lithium power batteries. This includes advocating for the adoption of safety standards and raising awareness among the public about the importance of battery recycling. By implementing these measures, the risks associated with battery transportation can be mitigated, and the overall safety of battery recycling processes can be enhanced.

18.6 Disposal of Batteries and Environmental Pollution Risks

The disposal of retired batteries mainly concerns primary batteries, with typical treatment methods including incineration and landfills [30].

During incineration, partial metal compositions of retired lithium-ion batteries become volatile at high temperatures and are carried by the flue gases. As the flue gas cools and condenses into particles, these metals become significantly enriched in the resulting fly ash, posing a significant risk of atmospheric pollution. The distribution and form of heavy metals within the incineration system are primarily influenced by the volatility of the metals. The rate at which heavy metals are emitted with flue gas depends on factors such as the metals' volatility, the quantity of heavy metals present in the incinerated waste, and the efficiency of the flue gas treatment system. Additionally, the feed rate of heavy metals is determined by the amount of waste incinerated and the concentration of heavy metals within the waste.

When retired batteries are landfilled, heavy metals in retired batteries will dissolve into leachate and permeate into the soil and aquifers, leading to the direct contamination of water sources and soils. This contamination will pose health risks for nearby residents who consume the contaminated groundwater. Moreover, the slow migration of heavy metals in clay complicates their release into the environment from landfills equipped with natural clay liners.

Studies indicate that when disposable retired batteries are left in landfills alongside municipal solid waste, the health risks associated with various heavy metals remain within acceptable levels [30]. Conversely, treatment methods such as pure incineration or a combination of landfill and incineration may elevate certain risks to unacceptable levels, with incineration posing a risk of waste ash exceeding regulatory standards. Given the diverse compositions of batteries and the potential contamination of waste acids and alkalis, it is advisable to collect and treat retired batteries separately.

18.7 Examples of Companies That Deal with the Retired Batteries

With the growing demand for portable electronics and electric vehicles, the disposal of retired batteries has become a pressing environmental and economic issue. This section showcases internationally recognized companies that recycle and dispose of retired batteries, highlighting their strategies, capabilities, and contributions to the circular economy.

Umicore Valéas is a global materials technology group with four core businesses: catalysis, energy materials, high-performance materials, and recycling. Umicore Valéas has been operating a 7000-tonne battery recycling plant since 2011. From 2018 onward, Umicore entered into cooperation with BMW, Audi, Swedish battery company Northvolt, and other companies in battery recycling [31]. These automobile and battery manufacturers are responsible for supplying retired power batteries and cooperating in developing recycling processes. At the same time, Umicore recovers valuable metals from the batteries and reprocesses the recovered materials into cathode materials and precursors.

Founded in 1896, Accurec of Germany uses a hydrometallurgical process to extract lithium from retired batteries, including electric vehicles, electric bicycles, electronic equipment, and conventional household batteries. Now, the company is realigning its "total process technology chain" to focus on lithium recovery, making the Krefeld-based company the first in Europe to recover lithium on an industrial scale.

Table 18.3 Recycling methods used by major recycling companies [31, 33, 36, 37].

Company	Capacity	Methodology	Process
Umicore	Recycling business of 640 million euros in 2021 and total disposal of 70 million metric tons of retired batteries in 2023	Pyrometallurgy, Hydrometallurgy	The pyrometallurgical method is utilized to directly restore the battery at high temperatures, with the battery shell, cathode material, plastic diaphragm, and other parts providing the reducing agent and energy, respectively, and the final metal is recovered in the form of an alloy.
Accurec	The annual processing capacity is up to 4000 metric tons of retired batteries	Pyrometallurgy, Hydrometallurgy	Retired batteries are first sorted, then heat-treated and mechanically processed to separate iron, aluminum, and copper. Then, nickel and cobalt are extracted from the remaining material, and lithium is washed out using a hydrometallurgical process.
Recupyl	The annual processing scale can be up to 110 metric tons	Hydrometallurgy	Mechanical processes are used to reduce and separate materials into different parts. Copper, aluminum, and plastics are removed by physical processes, while lithium and cobalt are recovered by hydrometallurgical methods.
Retriev Technologies	Annual production capacity can reach 4500 metric tons	Freezing, Hydrometallurgy	Recovery of valuable metals in lithium-ion batteries, such as copper, aluminum, iron, cobalt, and nickel, occurs using mechanical and hydrometallurgical processes.
GEM	Recycled and dismantled power batteries reached over 27,000 metric tons in 2023; full-year lithium carbonate recycling capacity expansion to 10,000 metric tons/year	Pyrometallurgy, Hydrometallurgy	After dismantling and sorting retired batteries, metal salts are obtained by extraction after acid leaching, and metal monomers such as copper, aluminum, iron, cobalt, and nickel are recovered through high-temperature reduction.

SungEel HiTech, the largest battery recycler in South Korea, can currently process 24,000 metric tons of lithium-ion battery materials per year. Moreover, the company plans to increase its annual production to 56,000 metric tons through its South Korean hydrometallurgical plant by recycling metals such as nickel, cobalt, manganese, and lithium [32]. At the end of 2020, the Australian graphite materials company EcoGraf announced that it had signed a memorandum of understanding (MoU) with South Korean lithium battery recycler SungEel HiTech to enter into a partnership in the field of recycling retired power batteries. Under the agreement, SungEel HiTech can utilize EcoGraf's proprietary purification process to recover high-purity graphite materials from retired batteries.

Recupyl is an international company specializing in recycling retired batteries. The Recupyl process, with a nominal capacity of 110 metric tons per year, is developed with a low-temperature lithium battery recycling technology, which directly addresses the gas emissions resulting from pyrometallurgy [33].

Retriev Technologies is a pioneer and market leader in the battery recycling industry in North America, with a capacity of 4500 metric tons per year [33]. They recover valuable metals from lithium-ion batteries, such as copper, aluminum, iron, cobalt, nickel, and others, attained by mechanical and hydrometallurgical processes [34].

Founded in 2001 in China, GEM Co., Ltd. has a planned annual dismantling capacity of 215,000 metric tons for power battery packs, an annual production capacity of 4.1 GWh for laddering products, and an annual processing capacity of 200,000 metric tons for recycling black powder. The black powder is an electronic waste consisting of crushed retired batteries, and it represents a highly valuable metal mixture from batteries [35]. In the treatment of retired batteries, there are some key technologies developed for the recycling of ultrafine cobalt powder and nickel powder from retired batteries and cobalt-containing waste. This involves utilizing cobalt-containing waste, such as retired batteries, to produce spherical cobalt powder directly.

Table 18.3 summarizes the recycling methods used by the major recycling companies described here.

18.8 Standards for Retired Battery Treatment

The European Union (EU), the United States, Japan, and other developed regions have established mature regulations and standard systems for the recycling of retired batteries.

The EU boasts the world's most stringent regulations regarding retired battery recycling. It has implemented a series of directives and regulations, including the EU Battery Directive 2006/66/EC, the Waste Electrical and Electronic Equipment Directive 2012/19/EU, the Battery Revision Directive 2013/56/EU, and the Battery Regulation 2023/1542 [38]. These regulations set forth specific requirements concerning the content of toxic and hazardous substances, battery labels, and the recycling of retired batteries and materials. The primary goals are to regulate batteries in the EU market, reduce the usage of hazardous substances in batteries, and enhance the collection, treatment, recycling, and disposal of retired batteries within member states.

The United States has relatively comprehensive legislation in retired battery recycling. It has enacted laws such as the Mercury-Containing and Rechargeable Battery Management Act, the Resource Conservation and Recycling Act, and the Universal Waste Rule, in addition to various state and local laws. These regulations provide detailed guidelines for the entire battery recycling process, encompassing collection, transportation, and storage. The laws aim to bolster the management of hazardous waste from retired batteries and enhance the efficiency and safety of recycling.

Furthermore, some industry organizations and institutions in the United States have developed voluntary standards for retired battery recycling, such as the Responsible Business Alliance (RBA) Code of Conduct.

Japan established its battery recycling system early; it initiated a battery recycling program in 1994 and developed a "battery production-sales-recycling" framework. To date, Japan has developed a recycling channel primarily led by battery manufacturers. This channel allows battery producers to collect retired batteries from consumers free of charge through networks of retailers, automobile dealers, and gas stations, which are then processed by specialized battery recycling companies [39]. Automakers such as Toyota, Nissan, and Mitsubishi actively invest in research and development of battery recycling to align with the Japanese government's principle that "new energy vehicle manufacturers are obligated to recycle used batteries." These laws promote retired battery recycling and support the development of related industries, fostering the industrialization of spent battery recycling.

The International Organization for Standardization (ISO) and the International Electrotechnical Commission (IEC) have issued numerous international standards for retired battery recycling. These standards provide valuable guidelines and references for the proper recycling of retired batteries on a global scale.

The EU Battery Regulation 2023/1542 concerning batteries and waste batteries, amending directive 2008/98/EC and regulation 2019/1020, and repealing directive 2006/66/EC (hereinafter referred to as the "new EU regulation"), entered into force in August 2023 [38]. The new EU regulation applies to all types of batteries (except for special applications such as military, aerospace, and nuclear energy) within the EU, including retired portable batteries, electric vehicle batteries, and industrial batteries. Batteries are classified into various types, each tailored to specific usage scenarios.

Portable batteries, typically weighing less than 5 kg, are designed for convenient and on-the-go power solutions [38]. These compact and sealed units provide energy for a wide range of handheld devices and small electronics, ensuring mobility and flexibility in powering everyday gadgets. Moving up in weight, light means of transport (LMT) batteries, weighing less than 25 kg, serve as energy sources for electric bicycles, scooters, and other similar modes of transportation [38].

In the automotive realm, starting, lighting, and ignition batteries play a vital role in vehicle operation. Whether powering ignition systems, lighting, or facilitating smooth starts, these batteries ensure reliable performance in various weather conditions and driving situations. Industrial batteries, exceeding 5 kg, are engineered for robustness and endurance, catering to the power needs of industrial machinery and equipment [38].

At the forefront of sustainable transportation, EV batteries represent a significant technological advancement. Weighing over 25 kg, these batteries supply hybrid and pure electric vehicles, enabling efficient and emissions-free traction for extended journeys [38]. EV batteries are integral to the transition toward cleaner and greener transportation solutions, reducing reliance on fossil fuels and mitigating environmental impact.

The unassembled battery cells/modules that are used in the market are also included in the scope of control of the bill. Notably, the new EU regulation also mentions that from the year 2027 onward, all portable batteries should have a removable battery design that allows consumers to open a unit and remove or replace a battery cell on their own without the use of special tools (e.g., solvents or heat) or training. All LMT batteries, on the other hand, should be able to be easily removed and replaced by professionals [38].

As a point of consideration, the new EU regulation needs to set specific recycling targets. The law focuses on the risk of supplying raw materials and guarantees the security of some raw material supplies by setting recycling targets. Importantly, the new law stipulates that recycling efficiency, material recovery, and recycling targets will be gradually introduced from 2025 onward. All collected retired batteries should be recycled, and a high level of recovery should be achieved, especially for key raw materials such as cobalt, lithium, and nickel. In adherence with the material recovery targets outlined in the new EU regulation, manufacturers will face increasingly stringent requirements. For retired portable batteries, the law stipulates recycling targets of 45% by 2023, 63% by the end of 2027, and 73% by the end of 2030. Similarly, for the batteries used in light vehicles, the recycling targets are set at 51% by the end of 2028 and 61% by the end of 2031 [38]. Figure 18.3 illustrates the increasing recycling targets of different batteries over time.

In terms of recycling efficiency targets, the new EU regulation will establish a recycling efficiency target of 80% for nickel-cadmium batteries and 50% for other batteries by the end of 2025. By December 31, 2027, the recycling targets include recovering at least 90% of cobalt, copper, and

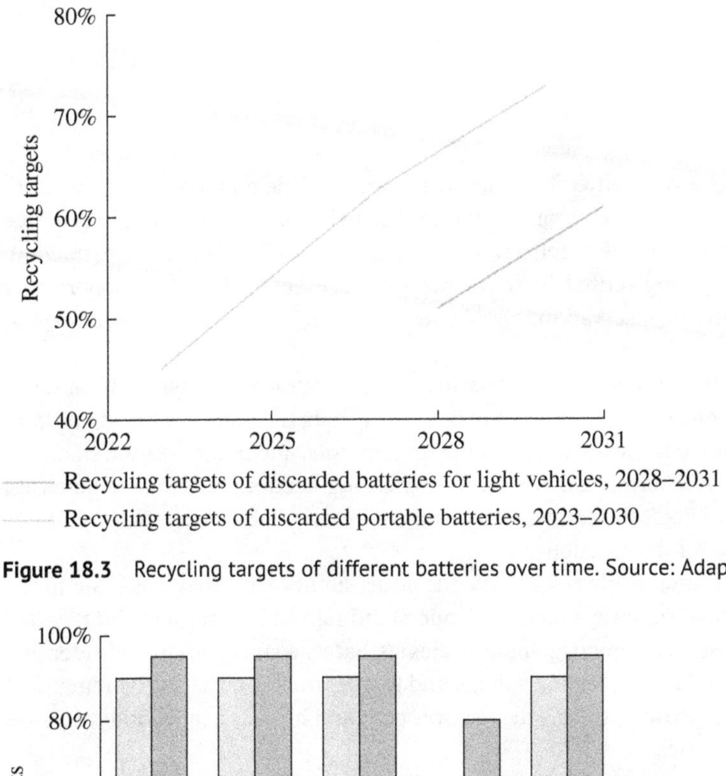

Figure 18.3 Recycling targets of different batteries over time. Source: Adapted from [38].

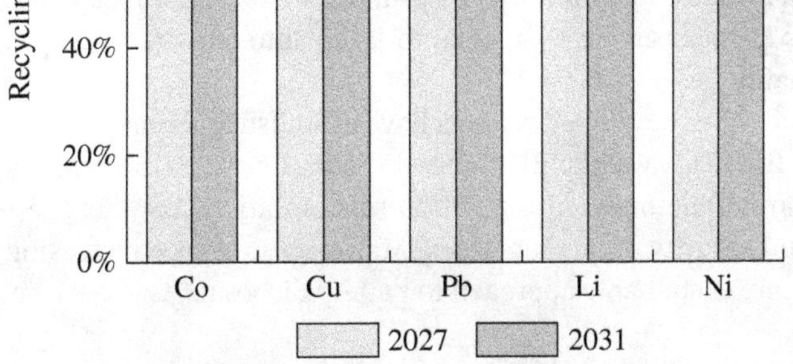

Figure 18.4 Recycling targets of recirculation materials, 2027–2031. Source: Adapted from [38].

lead content, 50% of lithium, and 90% of nickel content. Further, by December 31, 2031, the targets will escalate to 95% recovery for cobalt, copper, and lead, 80% for lithium, and 95% for nickel [38]. Figure 18.4 compares the recycling targets of these recirculation materials.

Evaluating the regulation, the European Council said that, first and foremost, it is a fair rule for all market participants. Officially, the Council stated that "the new regulations aim to improve the internal functioning of the battery market and ensure fairer competition through safety, sustainability and labeling requirements." This will be achieved through regulations on battery performance, durability and safety standards, strict limits on hazardous substances such as mercury, cadmium, and lead, and mandatory information on the carbon footprint of batteries.

The new EU regulation aims to reduce the environmental and social impact of batteries throughout their life cycle. Accordingly, the law establishes strict due diligence rules for operators other than small and medium-sized enterprises, who need to verify the origin of raw materials used in batteries on the market.

18.9 Summary

Battery recycling involves processing retired batteries to extract valuable materials for reuse, aiming to minimize resource demand and environmental impact and maximize renewable resource utilization. The process mainly includes collection, sorting, dismantling, material extraction, refining, and processing. Recycling retired batteries not only presents a lucrative opportunity but also contributes to resource conservation, circular economy principles, and environmental sustainability.

The disposal of retired batteries involves various methods tailored to the types of batteries and local regulations. Mechanical shredding compresses and shreds the batteries for subsequent treatment, chemical dissolution breaks down hazardous components, and neutralization stabilizes acidic and alkaline substances. Thermal treatment, like smelting, is effective for some batteries, notably lithium types, while biological treatment offers an environmentally sustainable option by leveraging biological methods for degradation.

Life cycle management of retired lithium-ion batteries poses significant safety risks due to the hazardous materials and thermal runaway potential. Understanding and addressing these risks are crucial for developing safe, sustainable management strategies. Safety evaluations include identifying battery type, inspecting for damage, testing voltage and temperature, and assessing remaining charge. This comprehensive approach protects the environment and human health, and it ensures the safe disposal of retired batteries.

The disposal of retired batteries, particularly primary batteries, presents environmental challenges mainly addressed through incineration and landfills. During incineration, heavy metals within lithium-ion batteries can become volatile and enrich in fly ash, posing risks of atmospheric pollution. Landfilling retired batteries also causes heavy metals to leach into soil and aquifers, endangering water sources and soil quality.

The EU, the United States, Japan, and other developed regions have established stringent regulations on retired battery recycling. In the EU, laws like EU Battery Regulation 2023/1542 ensure proper collection and recycling standards. The new EU regulation sets ambitious recycling targets, emphasizing material recovery and security of raw materials. Manufacturers face increasing recycling requirements, aiming for a more sustainable approach to battery disposal.

References

1 European Commission (2024). RMIS—Raw Materials Information System: Battery Supply Chain Challenges [Online]. https://rmis.jrc.ec.europa.eu/analysis-of-supply-chain-challenges-49b749

2 Wang, M., Tan, Q., Yu, J. et al. (2023). Pollution-free recycling of lead and sulfur from spent lead-acid batteries via a facile vacuum roasting route. *Green Energy and Resources* 1 (1): 100002.

3 Ouyang, D., Weng, J., Chen, M. et al. (2022). Sensitivities of lithium-ion batteries with different capacities to overcharge/over-discharge. *Journal of Energy Storage* 52: 104997.

4 Feng, L., Jiang, L., Liu, J. et al. (2021). Dynamic overcharge investigations of lithium ion batteries with different state of health. *Journal of Power Sources* 507: 230262.

5 Luo, G., Zhang, Y., and Tang, A. (2023). Capacity degradation and aging mechanisms evolution of lithium-ion batteries under different operation conditions. *Energies* 16: 4232.

6 Alipour, M., Ziebert, C. et al. (2020). A review on temperature-dependent electrochemical properties, aging, and performance of lithium-ion cells. *Batteries* 35: 6.

7 Shahid, S. and Agelin-Chaab, M. (2022). A review of thermal runaway prevention and mitigation strategies for lithium-ion batteries. *Energy Conversion and Management: X* 16: 100310.

8 Feng, X.N., Ouyang, M.G. et al. (2018). Thermal runaway mechanism of lithium ion battery for electric vehicles: a review. *Energy Storage Materials* 10: 246–267.

9 Wang, K., Wan, J., Xiang, Y. et al. (2020). Recent advances and historical developments of high voltage lithium cobalt oxide materials for rechargeable Li-ion batteries. *Journal of Power Sources* 460: 228062.

10 Zhang, R., Xia, B., Li, B. et al. (2018). Study on the characteristics of a high capacity nickel manganese cobalt oxide (NMC) lithium-ion battery—an experimental investigation. *Energies* 11 (9): 2275.

11 Wang, M., Liu, K., Dutta, S. et al. (2022). Recycling of lithium iron phosphate batteries: status, technologies, challenges, and prospects. *Renewable and Sustainable Energy Reviews* 163: 112515.

12 Feng, T., Guo, W., Li, Q. et al. (2022). Life cycle assessment of lithium nickel cobalt manganese oxide batteries and lithium iron phosphate batteries for electric vehicles in China. *Journal of Energy Storage* 52: 104767.

13 Zhu, Z., Wang, M., Meng, Y. et al. (2020). A high-rate lithium manganese oxide-hydrogen battery. *Nano Letters* 20 (5): 3278–3283.

14 Wen, D.Y., Fu, L.B. et al. (2015). Research on the management and disposal of discarded batteries. *Environmental Science and Management* 40 (12): 37–41.

15 Wang, Y., Wang, Z. et al. (2019). Closed-loop supply chain models with product recovery and donation. *Journal of Cleaner Production* 227: 861–876.

16 Arat, H.T. (2023). Recycling and reusing batteries: a significant way for effective sustainability of FCEVs and EVs. *International Journal of Hydrogen Energy* 48 (60): 23272–23278.

17 Cornelio, A., Zanoletti, A., Bontempi, E. et al. (2024). Recent progress in pyrometallurgy for the recovery of spent lithium-ion batteries: a review of state-of-the-art developments. *Current Opinion in Green and Sustainable Chemistry* 46: 2452–2236.

18 Yao, Y., Zhu, M., Zhao, Z. et al. (2018). Hydrometallurgical processes for recycling spent lithium-ion batteries: a critical review. *ACS Sustainable Chemistry & Engineering* 6 (11): 13611–13627.

19 Castelvecchi, D. (2021). Electric cars: the battery challenge. *Nature* 596: 336–339.

20 Harper, G., Sommerville, R., Kendrick, E. et al. (2019). Recycling lithium-ion batteries from electric vehicles. *Nature* 575 (7781): 75–86.

21 Jo, C.H. and Myung, S.T. (2019). Efficient recycling of valuable resources from discarded lithium-ion batteries. *Journal of Power Sources* 426: 259–265.

22 Guo, H.J., Shi, Z.W., and Tian, G. (2015). Brief description of the current situation of discarded battery recycling. *World Nonferrous Metals* 09: 18–20.

23 Schneider, E., Oliveira, C., Brito, R. et al. (2014). Classification of discarded NiMH and Li-Ion batteries and reuse of the cells still in operational conditions in prototypes. *Journal of Power Sources* 262: 1–9.

24 Zhang, T., He, Y., Ge, L. et al. (2013). Characteristics of wet and dry crushing methods in the recycling process of spent lithium-ion batteries. *Power Sources* 240: 766–771.

25 Lei, L., Wei, L. et al. (2020). Retired lithium-ion power battery recycling technology and safety risk analysis. *Scientific and Technological Innovation* 25: 189–190.

26 Mokarian, P., Bakhshayeshi, I., Taghikhah, F. et al. (2022). The advanced design of bioleaching process for metal recovery: a machine learning approach. *Separation and Purification Technology* 291: 120919.

27 Nigl, T., Schwarz, T.E., Walch, C. et al. (2020). Characterisation and material flow analysis of end-of-life portable batteries and lithium-based batteries in different waste streams in Austria. *Waste Management & Research* 38: 649–659.

28 Hanisch, C., Loellhoeffel, T., Diekmann, J. et al. (2015). Recycling of lithium-ion batteries: a novel method to separate coating and foil of electrodes. *Journal of Cleaner Production* 108: 301–311.

29 Vest, M. (2016). Weiterentwicklung des Pyrometallurgischen IME Recyclingverfahrens für Li-Ionen Batterien von Elektrofahrzeugen, RWTH Aachen University: Shaker Verlag.

30 Xiang, W. and Chao, X. (2019). Environmental hazards and recycling of waste batteries. *China Collective Economy* 19: 188–189.

31 Lin, W. (2023). Environmentally friendly battery recycling technology research and development is emerging worldwide, China Energy News, Beijing, CNKI Rep. 001962, 2023.

32 CNGR Advanced Material Co. (2021). Announcement on the signing of the Memorandum of Strategic Cooperation with SungEel HiTech [Online]. https://static.cninfo.com.cn/finalpage/2021-11-11/1211552121.PDF.

33 Velázquez-Martínez, O., Valio, J., Santasalo-Aarnio, A. et al. (2019). A critical review of lithium-ion battery recycling processes from a circular economy perspective. *Batteries* 5 (4): 68.

34 Smith, W. N. and Swoffer, S. (2013). Recovery of Lithium-Ion Batteries, U.S. Patent 861475 B1, June 18, 2013.

35 GEM, Power battery life cycle value chain business [Online]. https://www.gem.com.cn/UsedBatteryRecycling/index.html.

36 Randall, C., Accurec announces new process to recycle lithium from old batteries [Online]. https://www.electrive.com/2022/12/09/accurec-announces-new-proces-to-recycle-lithium-from-old-batteries/.

37 Securities Times (2024). Ltd's core product shipments to grow more than 10% by 2023 [Online]. https://finance.sina.com.cn/jjxw/2024-01-05/doc-inaakyaq3431011.shtml.

38 European Union (2023). Regulation (EU) 2023/1542 of the European Parliament and of the Council of 12 July 2023 Concerning Batteries and Waste Batteries, Amending Directive 2008/98/EC and Regulation (EU) 2019/1020 and Repealing Directive 2006/66/EC. Brussels, Belgium: European Union.

39 Japanese cabinet (1991). Act on the Promotion of Effective Utilization of Resources (Act No. 48) [Online]. https://www.meti.go.jp/policy/it_policy/kaden/index03.html.

Epilog: An Executive Summary

Over the past decade, there have been significant advancements in lithium-ion batteries in terms of volumetric energy density—the amount of energy stored within a given volume. Higher volumetric energy density allows for smaller battery packs while maintaining the same energy content. This is critical to optimize space in products like smartphones, laptops, and electric vehicles. In addition, there has also been an improvement in gravimetric energy density—the amount of energy stored for a given weight, which is critical for airplanes, drones, and space applications.

In 2010, the volumetric energy density was around 90 Wh/L, and the gravimetric energy density was around 250 Wh/kg. By 2020, the volumetric energy density was around 450 Wh/L, and the gravimetric energy density was around 300 Wh/kg. From these numbers, one can see that the industry has made great strides in putting more energy into a smaller package, as well as modest gains in increasing the gravimetric energy density. These numbers will increase and likely at a more exponential rate. But as we put more energy into a smaller space, we are also likely to see more fires due to the higher energy density if the manufacturing processes are not properly controlled, and if the proper safety mechanisms and precautions are not implemented.

It is relatively inexpensive to buy the basic equipment and materials to make batteries. As a result, there are now thousands of companies making high energy-density batteries, and these companies can likely make at least a few batteries that pass the current testing standards. Nevertheless, some of these companies have very questionable quality processes. We have already seen recalls due to battery fires in cars, ranging from the Chevy Bolt, the Hyundai Iconic, and the Stellantis electric vehicles, as well as fires in smartphones, laptops, and the Boeing Dreamliner, and in the increasing number of fires due to batteries in e-bikes, e-scooters, and vaping devices. While any company can have problems in manufacturing that can result in a battery that subsequently causes a fire or explosion, companies with poor quality control are more at risk.

Selecting the right battery cells and packs is crucial for any application. The process involves a multistep evaluation, considering the intended usage environment, performance targets, desired life span, supplier capabilities, battery reliability and safety, and various business decisions. It is recognized that there will be trade-offs between these factors, and a balancing act may be needed to reach the optimal decision, although safety is always a major priority.

The first step is to define the battery requirements, which lays the groundwork for the entire battery source selection process. The intended use of the battery significantly impacts the selection process, whether it is for a portable electronic device, a medical instrument, an electric vehicle, an energy storage system, or something else entirely. Understanding the application will guide the size, weight, allowable application conditions, and durability requirements.

Life cycle application conditions are a key part of the requirements and include the operational and non-operational temperature ranges, temperature extremes, humidity levels, and potential for

The Safety Challenges and Strategies of Using Lithium-Ion Batteries, First Edition. Edited by Michael G. Pecht.

shock, vibration, and abuse that the battery may encounter. Performance objectives require determination of the capacity, voltage range, discharge current, charging current and charging time, and power output needed from the battery pack. These will influence the selection of the cell chemistry, capacity, and the number and configuration of cells required. Target service life issues include the desired operational lifespan for the battery, calendar ageing, and charge–discharge profiles. Safety considerations include potential hazards associated with the application and selecting cells and pack designs that meet relevant safety standards.

Once the battery requirements are established, the next step is the selection of a battery that aligns with the specific performance demands of the intended application and associated electronics or systems. This includes assessing information from the datasheets, including voltage, current, capacity, energy density, power density, operating temperature range, charging/discharging rate, storage, and cycle life. In addition, the environmental impact of the battery products should be considered. This includes prioritizing batteries with a high percentage of recycled materials to minimize their environmental footprint and to support a circular economy (some EU regulations mandate or incentivize using recycled materials in batteries); ensuring the selected batteries have responsible end-of-life management plans to prevent environmental hazards from improper disposal; and favoring suppliers committed to environmentally friendly practices throughout the battery's life cycle, encompassing responsible material sourcing, manufacturing processes, and recycling initiatives. In the context of negotiating bulk or long-term contracts, the geopolitical conditions and trade regulations of the manufacturer's originating country may also need to be assessed.

Datasheets furnished by battery manufacturers are required to assess and choose batteries tailored to the defined requirements. The datasheets must incorporate information on the battery's functionalities, operational boundaries, performance metrics, dependability, and safety protocols [1], including the cell's nominal capacity, charging/discharging voltage, energy density, cycle life, operating temperature range, and safety features [2, 3]. Additionally, they often provide insights into performance parameters like charge and discharge rates, internal resistance, and capacity retention [2, 3]. Despite their critical role, examining datasheets from leading battery manufacturers uncovers numerous issues. Challenges, including ambiguous terminology, inconsistent data presentation, and incomplete information, collectively undermine the process of battery selection and management [1]. For instance, Panasonic (NCR18650B [4]) and Sony (US18650VC7 [5]) only offer basic information, such as storage or operating temperature, while datasheets of Samsung (INR21700-30T [2] and INR18650-25S [3]) include data on storage duration or capacity retention under different temperature conditions. In addition, manufacturers use a variety of terms to describe similar parameters or employ the same term with different meanings, leading to user confusion. For example, the term "capacity" is variously referred to as rated capacity, nominal capacity, and minimum capacity without consistent definitions for these measurements [6, 7].

Temperature parameters for battery storage and operation present another area of inconsistency. The user must understand whether the temperature reference point relates to the cell surface or the ambient/environmental temperature. For example, Samsung [8] and CATL [9] refer to the cell surface temperature in their datasheets, whereas Panasonic [10] and Sony [11] refer to the ambient or environment temperature. Some manufacturers like BYD [12] and Huawei [13] do not specify whether the temperature data provided pertains to the cell surface or the ambient temperature. This lack of standardization or absence of temperature reference data complicates system designers' efforts to optimize thermal management, thereby affecting battery performance, lifespan, and safety.

When selecting a battery, one must beware that it is not uncommon to encounter contradictory information on physical dimensions, electrical performance, and safety characteristics across datasheets for the same battery model [1]. In addition, many manufacturers fail to provide information on critical parameters, such as the operating/storage temperature range and operating current limits under various conditions [4, 5]. These variations could stem from different testing methodologies, unreported changes in electrode and electrolyte compositions, or manufacturing inconsistencies [1]. Such discrepancies not only question the reliability of the datasheets but also challenge users to integrate these batteries into their systems safely and efficiently.

Accessing datasheets, especially for low-volume purchasers, often proves difficult, and often leads users to encounter multiple versions online with varying degrees of accuracy and detail. This situation is exacerbated by the absence of a standardized format for these documents, hindering users' ability to extract necessary information efficiently [1]. This is why, as will be discussed later, it is necessary to audit the cell manufacturer before using any cells from them.

At the same time, candidate batteries are selected, there should be an assessment of the cell manufacturer and the pack manufacturer. In some cases, these may be the same company. Cell companies manufacture cells and may market them to pack manufacturers or directly to final product manufacturers. Pack houses assemble battery packs using pre-manufactured cells sourced from other manufacturers (upstream suppliers). They focus on pack design, configuration, and integration based on customer specifications. When working with a pack house, the product company may have the option to specify the desired cell source or allow the pack house to recommend cells based on their expertise and supplier relationships. This decision will depend on the product company's access to cell information, and specific needs. The choice between these options depends on the product company's specific needs and priorities. Pack houses often offer greater flexibility in pack design and customization, while cell and pack manufacturers may provide greater control over cell quality and consistency.

Regardless of whether pack houses or cell and pack manufacturers are chosen, an evaluation of the manufacturers and their supply chain is crucial. This includes an evaluation of their track record, depth of experience in the field, and feedback from other customers to gauge their reliability and commitment to quality. An evaluation of the reliability capability of the battery manufacturer can be conducted, for example, according to the IEEE 1642 Standard (IEEE Standard for Organizational Reliability Capability). Additionally, assess the financial stability and reputation of the supplier, recognizing that a financially sound supplier is more likely to provide consistent support and fulfill warranty obligations. Audits should also be conducted to assess the management of quality, health, safety, and environmental systems, crucial for risk management and sustainability, as well as the alignment of manufacturing processes with global quality standards and identifying areas for improvement. Company audits by well-trained inspectors are needed, to assess the battery manufacturer's quality assurance processes to assess the potential for defects to enter the batteries. This is essential because, at the time of this writing, there are no cost-effective screens to ensure that there are no defects in a battery when it comes off the manufacturing line.

A review of the warranty terms provided by the manufacturer is needed, recognizing that the warranty coverage indicates the manufacturer's confidence in the product's reliability. Pay close attention to the duration of coverage, terms for repair or replacement, and any limitations or exclusions outlined in the warranty policy. Fourth, review the manufacturer's warranty terms, as comprehensive warranty coverage can signal confidence in the product's reliability. Additionally, inquire about the availability of technical support and customer service. Understand the expected life cycle of the battery products, including cycle life, maintenance requirements, and the availability of replacement parts.

When received, the cells must be authenticated, especially if the cells or battery pack comes from a distributor, warehouse, or retailer. The purpose is to determine whether the received cells are genuine and unused. This is especially of concern when purchasing battery packs incorporating batteries. This can be conducted through the document provided by a trustworthy organization or test verification but may require a structural comparative analysis against a known authentic cell using CT-scanning or deconstruction methods.

As part of the authentication process, documentation and safety standards compliance need to be checked. Check whether the supplier provides a datasheet with all the required information and other necessary documents for your application. For instance, details on charging and discharging current limits at various temperatures are required in low-temperature applications. Prioritize safety and compliance by confirming that the battery products meet pertinent industry standards and regulations. This encompasses obtaining necessary safety certifications, adhering to environmental directives, and complying with industry-specific standards applicable to your intended application. Verify that the battery products have undergone rigorous testing and hold certifications from organizations such as UL (Underwriters Laboratories), IEC (International Electrotechnical Commission), CE (Conformité Européenne), RoHS (Restriction of Hazardous Substances), and others required for certain regions. Assess the safety features integrated into the battery products, including thermal management systems, overcharge, and over-discharge protection, short-circuit prevention mechanisms, and safety cutoff features.

In general, it is wise to conduct your own performance testing before embedding them into their final products, even if the battery has some form of certification. This is especially important if the battery manufacturer is new to your business or if the batteries received were transported and in storage for a long time after the cells were made. Typically, the manufacturer's testing protocols are outlined in the manufacturer's specifications, but they must include for performance under temperature, C-rate, and voltage ranges. If the original equipment manufacturer (OEM) directly sources the battery cells and packs from the original manufacturers, the OEM may define their own performance and qualification tests beyond the battery manufacturer's specifications.

Safety testing should be conducted to validate the design of the battery products considering the robustness of the casing, incorporation of fail-safe mechanisms (e.g., positive temperature coefficient device (PTC), current interrupt device (CID), mandrels, and vents), the effectiveness of thermal management systems, provisions for overcharge and over-discharge protection, as well as mechanisms for preventing short-circuits. Not all battery cells incorporate all of these safety mechanisms. Furthermore, if there is a battery management system (BMS), its functionality and reliability should also be evaluated.

Finally, the supplier of cells and battery packs must provide a product change notice (PCN) if there are any changes in form, fit, function, or reliability. This should include changes in manufacturing location, changes in materials and chemistries, as well as changes in test procedures and datasheets. While it's tempting to choose cheaper cells and packs, balancing cost with the quality and safety features provided is crucial. Investing a bit more can often lead to better life cycle performance, better warranty, and enhanced safety. It's important to assess not only the upfront cost of the battery but also total life cycle cost, which includes maintenance, charging expenses, and replacement costs. The reader is recommended to read the Prolog to see the numerous safety issues and associated costs that have occurred due to lithium-ion batteries that caused fires and explosions.

There is a tradeoff between the implementation of higher energy-density batteries and safety. To solve these problems, engineers are going to need to think about ways to ensure that the manufacturing processes don't introduce any defects, and this will be one of the greatest challenges to the industry. The next generation of battery engineers will likely need to look at the processes used by the semiconductor industry to get more ideas about the role of cleanliness, environmental controls, and automation in high-quality manufacturing. If measures are not taken care of, communities are likely to take more draconian measures such as banning electric vehicles (cars, scooters, e-bikes) in underground parking garages or near apartments.

Finally, as a brief summary of things to remember for the safe use of batteries:

- Only purchase batteries from a reputable manufacturer or supplier. Cheap, re-wrapped, and counterfeit batteries may not undergo proper quality control processes and will have a higher likelihood of failing.
- Read all documentation supplied with your battery, and never reverse polarity.
- When not in use, store loose batteries in a container. Do not put batteries in contact with conductive materials, water, seawater, strong oxidizers, or strong acids.
- Avoid excessively hot and humid conditions, especially when batteries are fully charged. Do not place batteries in hot locations (e.g., in a car on a hot day).
- Inspect batteries for any signs of damage before use. Never use and promptly dispose of damaged batteries. This includes any wear in cylindrical wrappers, tears in tabs, and signs of corrosion.
- Adhere to applicable electrical protection standards (terminal protection, shielding, and personal protective equipment (PPE)) is required to avoid exposure to electrical hazards, especially in lithium-ion batteries with voltages greater than 60 V.
- Do not mix different types of batteries or mix new and old ones together (e.g., in a power pack).
- Do not open or repair battery modules unless you have proper training.
- Do not use any products that incorporate batteries without a BMS.
- Immediately disconnect products/batteries from chargers if, during operation or charging, they emit an unusual smell, get hot, start to swell, or behave abnormally.
- Exercise caution with low-cost products (e-bikes, e-scooters, hoverboards) that may encourage cheap knock-offs built without adherence to safety standards.

References

1 Diao, W., Kulkarni, C., and Pecht, M. (2021). Development of an informative lithium-ion battery datasheet. *Energies* 14 (17): 5434.

2 Samsung. INR21700-30T. https://www.imrbatteries.com/samsung-30t-21700-3000mah-35a-battery/ (accessed 13 March 2023).

3 Samsung. INR18650-25S. https://www.18650batterystore.com/products/samsung-25s (accessed 13 March 2023).

4 Panasonic. NCR18650B Standard data. https://enerpower.de/wp/wp-content/uploads/2016/03/Specification_NCR18650B.pdf (accessed 27 March 2023).

5 Sony. Lithium ion rechargeable battery (Model Number US18650VC7). https://www.imrbatteries.com/content/sony_us18650vc7.pdf (accessed 1 April 2023).

6 Samsung. INR18650-30Q. https://www.18650batterystore.com/products/samsung-30q (accessed 2 March 2024).

7 Samsung. INR18650-20S. https://www.imrbatteries.com/content/samsung_20S.pdf (accessed 24 March 2024).

8 Samsung. INR21700-50G. http://www.batteryexpert.org/wp-content/uploads/2021/03/INR21700-50G.pdf (accessed 5 July 2023).

9 CATL. LEP71H3L7-01. https://www.microcharge.de/anleitungen/CATL_LEP71H3L7-01.pdf (accessed 25 October 2023).

10 Panasonic. NCR18650B. https://www.imrbatteries.com/content/panasonic_ncr18650b-2.pdf (accessed 25 October 2023).

11 Sony. US18650VTC5A. https://www.datasheet-pdf.com/PDF/US18650VTC5A-Datasheet-Sony-1097066 (accessed 25 October 2023).

12 BYD. BATTERY-BOX PREMIUM HVS/HVM. https://www.bydbatterybox.com/uploads/downloads/230530_BYD_Battery-Box_Premium_HVS&HVM_Datasheet_V1.7_EN-647eedf90f9c3.pdf (accessed 5 July 2023).

13 Huawei. LUNA2000. https://solar.huawei.com/en-GB/download?p=%2F-%2Fmedia%2FSolar%2Fattachment%2Fpdf%2Feu%2Fdatasheet%2FLUNA2000-5-15-S0.pdf (accessed 5 July 2023).

Index

The Safety Challenges and Strategies of Using Lithium-Ion Batteries, First Edition. Edited by Michael G. Pecht.

d

www.ingramcontent.com/pod-product-compliance
Lightning Source LLC
Chambersburg PA
CBHW081950221025
34337CB00005B/53
9781394342907